THE 68000 MICROPROCESSOR

Hardware and Software Principles and Applications

Fourth Edition

JAMES L. ANTONAKOS
Broome Community College

Prentice Hall

Upper Saddle River, New Jersey   Columbus, Ohio

Library of Congress Cataloging-in-Publication Data

Antonakos, James L.
    The 68000 microprocessor: hardware and software principles and
applications / James L. Antonakos.—4th ed.
      p.   cm.
    Includes index.
    ISBN 0-13-668120-4
    1. Motorola 68000 (Microprocessor)   I. Title.
QA76.8.M6895A57   1999
004.165—dc21   97-46805
                                                    CIP

Cover art/photo: © Phil Matt
Editor: Charles E. Stewart, Jr.
Production Coordination: Custom Editorial Productions, Inc.
Design Coordinator: Karrie M. Converse
Cover Designer: Brian Deep
Production Manager: Deidra M. Schwartz
Marketing Manager: Ben Leonard

This book was set in Times Roman by Custom Editorial Productions, Inc., and was printed and bound by Quebecor/
Book Press. The cover was printed by Phoenix Color Corp.

 © 1999, 1996 by Prentice-Hall, Inc.
Simon & Schuster/A Viacom Company
Upper Saddle River, New Jersey 07458

Earlier editions © 1993 by Macmillan Publishing Company and ©1990 by Merrill Publishing Company.

Printed in the United States of America

10 9 8 7 6 5 4 3 2 1

ISBN   0-13-668120-4

Prentice-Hall International (UK) Limited, *London*
Prentice-Hall of Australia Pty. Limited, *Sydney*
Prentice-Hall Canada Inc., *Toronto*
Prentice-Hall Hispanoamericana, S. A., *Mexico*
Prentice-Hall of India Private Limited, *New Delhi*
Prentice-Hall of Japan, Inc., *Tokyo*
Simon & Schuster Asia Pte. Ltd., *Singapore*
Editora Prentice-Hall do Brasil, Ltda., *Rio de Janeiro*

To Alan C. Dixon, my mentor and friend,
who has taught me so much

PREFACE

INTRODUCTION

The rapid spread of microprocessors in society has both simplified and complicated our lives. Whether we rely on a computer at work or come in contact with one for other reasons, most of us have used a computer at one point or another. Most people know that a microprocessor is lurking somewhere inside the machinery, but what a microprocessor is and what it does remain a mystery.

This book is intended to help remove the mystery concerning the 68000 microprocessor through detailed coverage of its hardware and software and by means of examples of many different applications. Some of the more elaborate applications include the Apple Macintosh computer, commercial video games, and network communications controllers. Industry has also adopted the 68000 for use in digital flight control computers and other high-level applications.

The book is intended for two- or four-year electrical engineering, engineering technology, and computer science students. Professionals such as engineers and technicians would also find it a handy reference. The material is intended for a one-semester course in microprocessors.

Prior knowledge of digital electronics, including combinational and sequential logic, decoders, memories, Boolean algebra, and operations on binary numbers, is required. This presumes knowledge of standard computer-related terms, such as RAM, EPROM, TTL, and so on.

CHAPTER TOPICS

For those individuals who have no previous knowledge of microprocessors, Chapter 1, "Microprocessor-Based Systems," is a good introduction to the microprocessor, how it functions internally, and how it is used in a small system. Chapter 1 is a study of the overall operation of a microprocessor-based system, and introduces the ASM68K and EMU68K programs. A programming example is included to familiarize you with the 68000.

Chapter 2, "An Introduction to the 68000 Microprocessor," highlights the main features of the 68000. Data types and instructions are surveyed. Also, the 68000 is compared with another processor, the 8086. The calculator project is started in this chapter.

Chapter 3, "Software Details of the 68000," provides the foundation for all programming in the remaining chapters. The 68000's instruction set and addressing modes are covered in detail, with more than 75 examples provided to help the student grasp the material.

Chapter 4 covers exception processing. The basic sequence of an exception is covered, as are multiple exceptions, prioritized exceptions, and exception handlers.

Chapter 5 provides an introduction to the data structures commonly used when programming and how they are represented in assembly language. Both static and dynamic storage allocation are covered, with examples showing how to use each structure.

The first real programming efforts are found in Chapter 6, "An Introduction to Programming the 68000." Numerous programming examples are included to show how the 68000 performs routine functions involving binary and BCD mathematics, string operations, data table manipulation, and control applications. Instruction timing is also covered. Each program is written in such a way that its operation may be grasped quickly. Most examples, however, leave much room for improvement. The improvements are deliberately left for the student. The end-of-chapter study questions require modifications or additions to existing routines and the creation of new ones. Chapter 6 was written in this way to challenge the student to write his or her own code.

The hardware operation of the 68000 is covered in Chapter 7, "Hardware Details of the 68000." All CPU pins are discussed, as are timing diagrams.

Chapter 8, "Memory System Design," and Chapter 9, "I/O System Design," extend the information presented in Chapter 7 and utilize it in the design of custom memory and I/O circuitry. Memory system topics include bus buffering, full- and partial-address decoding, RAM and EPROM interfacing, and dynamic RAM. I/O system topics include memory-mapped I/O, parallel data transfer, serial data transfer, and memory-mapped video.

Advanced programming using 68000 peripherals is the subject of Chapter 10. In this chapter, a number of peripherals designed for the 68000 are examined. These peripherals implement serial and parallel I/O, memory management, DMA, and floating-point operations. Programming and interfacing are discussed for each peripheral. Interfacing a non-Motorola peripheral is also covered.

Other textbooks rarely pay equal attention to both hardware and software. This book was written to give equal treatment to both, culminating in a practical exercise: building and programming your own single-board computer! Chapter 11, "Building a Working 68000 System," is included to give students a chance to design, build, and program their own 68000-based computer. The system contains 2K words of EPROM, 2K words of RAM, and a serial I/O device. The hardware is designed first, followed by design of the software monitor program. Some books choose to explain the operation of a commercial system, such as the MC68000 Education Computer. This approach is certainly worthwhile, but it does not give the student the added advantage of knowing *why* certain designs were used. The hardware and software designs in Chapter 11 are sprinkled with many questions, which are used to guide the design toward its final goal.

The single-board computer presented in Chapter 11 has been built by a number of senior students at Broome Community College, in an average of six weeks. It is reasonable to say that most students can build a working system in one semester.

Finally, Chapter 12, "An Introduction to the Advanced 680x0 Series Microprocessors," provides a brief overview of the features and improvements in the 32-bit successors to the 68000.

Because of the information presented, some chapters are much longer than others. Even so, it is possible to cover certain sections of selected chapters out of sequence, or to pick and choose sections from various chapters. Chapter 3 could be covered in this way, with emphasis placed on additional addressing modes, or groups of instructions, at a rate deemed appropriate by the instructor. Also, some sections in Chapter 10 may be skipped, depending on the instructor's choice of peripheral. Some instructors may wish to cover hardware before programming. There is no reason this cannot be done.

The appendixes present a full list of 68000 instructions, their allowed addressing modes, flag usage, and instruction times. In addition, they contain data sheets for three 68000-based peripherals, the 68681 Dual UART, the 68230 PI/T, and 68881 floating-point coprocessor. This avoids the need for secondary references.

In summary, more than 200 illustrations and 45 different applications are used to give the student sufficient exposure to the 68000. The added benefit of Chapter 11, where a working system is developed, makes this book an ideal choice for a student wishing to learn about microprocessors. The old saying that 8-bit machines are easier to learn on is outdated now. The instruction sets of the newer 16-bit machines, although more complex, are easier to learn and code with. Furthermore, even though this book deals only with the 68000 family, the serious microprocessor student should be exposed to other CPUs as well. But to try to cover two or more different microprocessors in one text does not do justice to either. For this reason, attention is focused on the 68000 family and not on other CPUs.

CHANGES AND ADDITIONS FOR THE FOURTH EDITION

A number of changes and additions were made to the text in this edition:

- A Troubleshooting Techniques section has been added to every chapter. These are often in the form of tips and suggestions on how to work with various hardware and software topics.
- New examples of stack-based parameter passing, recursion, source-code assembly, and stack frames have been added to Chapters 3, 4, and 6.
- Material on the 68060 and the 683xx microcontrollers has been added to Chapter 12.
- Many new figures, examples, and study questions have been added.
- The ASM68K and EMU68K programs included on the companion disk are the newest versions, with many improvements. EMU68K now allows the DOS interrupt mechanism to be used inside a 68000 program. This opens up more of the PC's power and allows more sophisticated programming assignments.
- New appendix material on the ASCII table, binary number systems, DOS function calls, and the ASM68K and EMU68K programs has also been added.

THE COMPANION DISK

The disk included with the book contains *all* of the source files presented in the book. The files are stored in separate directories related to their specific chapters. In addition, the various binary and executable files related to the single-board computer from Chapter 11 are also included. Also, two programs, ASM68K and EMU68K, are provided on the disk to allow students to assemble and execute (emulate) 68000 programs on an MS-DOS personal computer.

The disk contains an executable file called README.COM that explains the disk contents in detail.

ACKNOWLEDGMENTS

I would like to thank my editor, Charles E. Stewart, Jr., and his assistants, Kate Linsner and Kim Yehle, for all their help while I was putting this book together. I would also like to thank the many students and instructors who used the third edition, and who e-mailed me with comments and suggestions. I also deeply thank Professor Sol Rosenstark of the New Jersey Institute of Technology for his challenging suggestions and tireless testing of the emulator. Tony Plutino, from Motorola, was very helpful with photographic samples, and I appreciate his effort. In addition, I would like to thank Jim Reidel, who managed the book through production, and my copyeditor, Cindy Lanning, for doing an excellent job.

The following individuals provided many useful comments during the revision, and I am grateful for their advice: Venkata Anandu, Southwest Texas State University, and Craig A. Bergman, Binghamton University.

James L. Antonakos
antonakos_j@sunybroome.edu
http://www.sunybroome.edu/~antonakos_j

BRIEF CONTENTS

CONTENTS

SOLUTIONS AND ANSWERS TO SELECTED ODD-NUMBERED STUDY QUESTIONS

INDEX

PART 1

Introduction

The Atlas Communication Engine's ACE360/QM embedded router.

CHAPTER 1

Microprocessor-Based Systems

OBJECTIVES

In this chapter you will learn about:

- The block diagram of a microprocessor-based system and the function of each section
- The processing cycle of a microprocessor
- The way software is used to initialize hardware and peripherals
- The history of the microprocessor and of the different generations of computers
- How 68000 code is created and used in a 68000-based system
- The calculator project
- Some typical errors encountered during program development

1.1   INTRODUCTION

The invention of the microprocessor has had a profound impact on many aspects of our lives, since today even the most mundane chores are being accomplished under its supervision—something that allows us more time for other productive endeavors. Even a short list of the devices using the microprocessor shows how dependent we have become on it:

- Pocket calculators
- Digital watches (some with calculators built in)
- Automatic tellers (at banks and food stores)
- Smart telephones
- Compact disk players
- Home security and control devices
- Realistic video games

- Talking dolls and other toys
- VCRs
- Personal computers

The purpose of this chapter is to show how a microprocessor is used in a small system and to introduce you to the operation of the personal computer. We will see what types of hardware may be connected to the microprocessor, and why each type is needed. We will also see how software is used to control the hardware, and how that software can be developed.

Section 1.2 shows how the microprocessor has evolved over time, from the initial 4-bit machines to today's 32-bit processors. Section 1.3 covers the block diagram of a typical microprocessor-based system and explains each functional unit. Section 1.4 explains the basic operation of a microprocessor. Section 1.5 discusses the hardware and software requirements of a small microprocessor control system. Section 1.6 brings the material of the first five sections together in a technical description of the first Macintosh personal computer. Section 1.7 shows how software is developed for, and utilized by, a 68000-based computer system. An introduction to the calculator project is given in Section 1.8. Finally, Section 1.9 introduces the first in a series of troubleshooting techniques.

1.2   EVOLUTION OF MICROPROCESSORS

We have come a long way since the early days of computers, when ENIAC (for Electronic Numerical Integrator and Computer) was state of the art and occupied thousands of square feet of floor space. Constructed largely of vacuum tubes, it was slow, prone to breakdown, and performed a limited number of instructions. Even so, ENIAC ushered in what was known as the **first generation** of computers.

Today, thanks to advances in technology, we have complete computers that fit on a piece of silicon no larger than your fingernail that far outperform ENIAC.

When the transistor was invented, computers shrank in size and increased in power, leading to the **second generation** of computers. **Third-generation** computers came about with the invention of the integrated circuit, which allowed hundreds of transistors to be packed on a small piece of silicon. The transistors were connected to form logic elements, the basic building blocks of digital computers. With third-generation computers we again saw a decrease in size and increase in computing power. Machines like the 4004 and 8008 by Intel found some application in simple calculators, but they were limited in power and addressing capability. When improvements in integrated circuit technology enabled us to place *thousands* of transistors on the same piece of silicon, computers really began to increase in power. This new technology, called large-scale integration (LSI), is even faster than the previous medium- and small-scale integration (MSI and SSI) technologies, which dealt with only tens or hundreds of transistors on a chip. LSI technology has created the **fourth generation** of computers that we use today. An advanced form of LSI technology, VLSI, meaning very-large-scale integration, is now being used to increase processing power.

The first microprocessors that became available with third-generation computers had limited instruction sets and thus restricted computing abilities. Although they were suitable for use in electronic calculators, they simply did not have the power needed to operate more complex systems, such as guidance systems or scientific applications. Even some of the early fourth-generation microprocessors had limited capabilities because of the lack of addressing modes and instruction types. Eight-bit machines like the 8080, Z80, and 6800 were indeed more advanced than previous microprocessors, but they still did not possess multiply and divide instructions. How frustrating and time consuming to have to write a program to do these operations when needed!

Within the last decade microprocessor technology has improved tremendously, however. Thirty-two-bit processors can now multiply and divide, operate on many different data types (4-, 8-, 16-, and 32-bit numbers), and address *billions* of bytes of information. Processors of the 1970s were limited to 64KB, a small amount of memory by today's standards.

Each new microprocessor to hit the market boasts a fancier instruction set and faster clock speed, and indeed our needs for faster and better processors keep growing. A new technology called RISC (for reduced instruction set computer) has recently gained acceptance. This technology is based on the fact that most microprocessors utilize only a small portion of their entire instruction set. By designing a machine that uses only the more common types of instructions, processing speed can be increased without the need for a significant advance in integrated circuit technology. The 68060 microprocessor, manufactured by Motorola, uses many of the architectural techniques employed by RISC machines.

Why the need for superfast machines? Consider a microprocessor dedicated to displaying three-dimensional color images on a video screen. Rotating the three-dimensional image around an imaginary axis in real time (in only a few seconds or less) may require millions or even billions of calculations. A slow microprocessor would not be able to do the job.

Eventually we will see fifth-generation computers. The whole artificial intelligence movement is pushing toward that goal, with the desired outcome being the production of a machine that can think. Until then we will have to make the best use of the technology we have available.

1.3   SYSTEM BLOCK DIAGRAM

Any microprocessor-based system must of necessity have some standard elements such as memory, timing, and input/output (I/O). Depending on the application, other exotic circuitry may be necessary as well. Analog-to-digital (A/D) converters and their counterpart, digital-to-analog (D/A) converters, interval timers, math coprocessors, complex interrupt circuitry, speech synthesizers, and video display controllers are just a few of the special sections that may also be required. Figure 1.1 depicts a block diagram of a system containing some of the more standard circuitry and functions normally used.

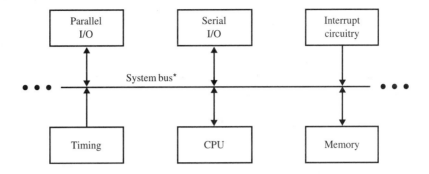

* Data, address, and control signals

FIGURE 1.1   Standard block diagram of a microprocessor-based system

As the figure shows, all components communicate via the **system bus.** The system bus is composed of the processor address, data, and control signals. The **central processing unit (CPU)** is the heart of the system, the master controller of all operations that can be performed. The CPU executes instructions that are stored in the **memory** section. For the sake of future expansion, the system bus is commonly made available to the outside world (through a special connector). Devices may then be added easily as the need arises. Commercial systems have predefined buses that accomplish this. All devices on the system bus must communicate with the processor, usually within a tightly controlled period of time. The **timing** section governs all system timing and thus is really responsible for the proper operation of all system hardware. The timing section usually consists of a crystal oscillator and timing circuitry (counters designed to produce the desired frequencies) set up to operate the processor at its specified clock rate. Using a high-frequency crystal oscillator and dividing it down to a lower frequency provides for greater stability.

The CPU section consists of a microprocessor and the associated logic circuitry required to enable the CPU to communicate with the system bus. These logic elements may consist of data and address bus drivers, a bus controller to generate the correct control signals, and possibly a math coprocessor. **Coprocessors** are actually microprocessors themselves; their instruction set consists mainly of simple instructions for transferring data and complex instructions for performing a large variety of mathematical operations. Coprocessors perform these operations at very high clock speeds with a great deal of precision (80-bit results are common). In addition to the basic add/subtract/multiply/divide operations, coprocessors are capable of finding square roots, logarithms, a host of trigonometric functions, and more.

The actual microprocessor used depends on the complexity of the task that will be controlled or performed by the system. Simple tasks require nothing more complicated than an 8-bit CPU. A computerized cash register would be a good example of this kind of system. Nothing more complicated than binary coded decimal (BCD) addition and subtraction—and possibly some record keeping—is needed. But for something as complex as a flight control computer for an aircraft or a digital guidance system for a missile, a more powerful 16- or 32-bit microprocessor must be used.

The **memory** section usually has two components: **read-only memory (ROM)** and **random access memory (RAM).** Some systems may be able to work properly without RAM, but all require at least a small amount of ROM. The ROM is included to provide the system with its intelligence, which is ordinarily needed at start-up (power-on) to configure or initialize the peripherals, and sometimes to help recover from a catastrophic system failure (such as an unexpected power failure). Some systems use the ROM program to download the main program into RAM from a larger, external system, such as a personal computer (PC) or a mainframe computer. In any event, provisions are usually made for adding additional ROM as the need arises.

There are three types of RAM. For small systems that do not process a great deal of data, the choice is static RAM. Static RAM is fast and easy to interface with, but comes in small sizes (as little as 16 bytes per chip). Larger memory requirements are usually met by using dynamic RAM, a different type of memory that has high density (256K bits per chip or more) but that unfortunately requires numerous refreshing cycles to retain the stored data. Even so, dynamic RAM is the choice when large amounts of data must be stored, as in a system gathering seismic data at a volcano or in one receiving digitized video images from a satellite.

Both static and dynamic RAM lose their information when power is turned off, which may cause a problem in certain situations. Previous solutions involved adding battery backup circuitry to the system to keep the RAMs supplied with power during an outage. But batteries can fail, so a better method was needed. Thus came the invention of **nonvolatile memory (NVM),** which is memory that retains its information even when power is turned off. NVM comes in small sizes and therefore is used to store only the most important system variables in the event of a power outage.

Another type of storage media is the floppy or hard disk. Both provide the system with large amounts of storage (for programs and data), although the data is accessed at a much slower rate than that of RAM or ROM. Floppy and hard disks also require complex hardware and software to operate, and are not needed in many control applications.

When a microprocessor is used in a control application, sometimes the system must respond to special external circumstances. For example, a power failure on a computer-controlled assembly line requires immediate attention by the system, which must contain software designed to handle the unexpected event. The event actually *interrupts* the processor from its normal program execution in order to service the unexpected event. The system software is designed to handle the power-fail interrupt in a certain way and then return to the main program. An interrupt thus is a useful way to grab the processor's attention, get it to perform a special task, and then resume execution from where it left off.

Not all types of interrupts are unexpected. Many are used to provide the system with useful features such as real-time clocks, multitasking capability, and fast input/output operations.

The interrupt circuitry needed from system to system will vary depending on the application. A system used for keeping time has to use only a single interrupt line connected to a timing source. A more complex system, such as an assembly line controller, which may need to monitor multiple sensors, switches, and other items, may require many different prioritized interrupts and would therefore need more complex interrupt circuitry.

Some systems may require serial I/O for communication with an operator's console or with a host computer. In Figure 1.2 we see how a small system might communicate with other devices or systems via serial communication. While this type of communication is slow, it has the advantage of simplicity: Only two wires (for receive and transmit) plus a ground are needed. Serial communication is easily adapted for use in fiber-optic cables. Parallel I/O, on the other hand, requires more lines (at least eight) but has the advantage of being very fast. A special parallel operation called **direct memory access (DMA)** is used to transfer data from a hard disk to a microcomputer's memory. Other uses for parallel I/O involve reading switch information, controlling indicator lights, and transferring data to A/D and D/A converters and other types of parallel devices.

All of these sections have their uses in a microprocessor-based system. Whether or not they are actually utilized depends on the designer and the application.

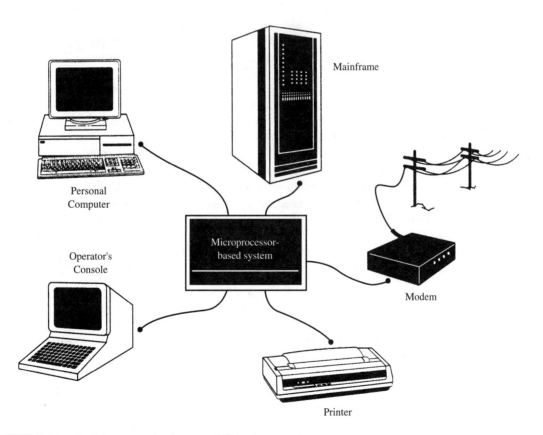

FIGURE 1.2   Serial communication possibilities in a small system

1.4   MICROPROCESSOR OPERATION

No matter how complex microprocessors become, they will still follow the same pattern of operations during program execution: endless fetch, decode, and execute cycles. During the fetch cycle the processor loads an instruction from memory into its internal instruction register. Some advanced microprocessors load more than one instruction into a special buffer to decrease program execution time. The idea is that while the microprocessor is decoding the current instruction, other instructions can be read from memory into the instruction **cache,** a special type of internal high-speed memory. In this fashion the microprocessor performs two jobs at once, thus saving time.

During the decode cycle the microprocessor determines what type of instruction has been fetched. Information from this cycle is then passed to the execute cycle. To complete the instruction, the execute cycle may need to read more data from memory or write results to memory.

While these cycles are proceeding, the microprocessor is also paying attention to other details. If an interrupt signal arrives during execution of an instruction, the processor will usually latch onto the request, holding off on interrupt processing until the current instruction finishes execution. The processor also monitors other signals such as WAIT, HOLD, or READY inputs. These are usually included in the architecture of the microprocessor so that slow devices, such as memories, can communicate with the faster processor without loss of data.

Most microprocessors will also include a set of control signals that allow external circuitry to take over the system bus. In a system where multiple processors share the same memory and devices, these types of control signals are necessary to resolve **bus contention** (two or more processors needing the system bus at the same time). Multiple-processor systems are becoming more popular now as we continue to strive toward faster execution of our programs. **Parallel processing** is a term often used to describe multiple-processor systems and their associated software.

Special devices called **microcontrollers** are often used in simple control systems because of their many features. Microcontrollers are actually souped-up microprocessors with built-in features such as RAM, ROM, interval timers, parallel I/O ports, and even A/D converters. Microcontrollers are not used for really big systems, however, because of their small instruction sets. Unfortunately, we have yet to get everything we want on a single chip!

1.5   HARDWARE/SOFTWARE REQUIREMENTS

We saw earlier that it is necessary to have at least some ROM in our system to take care of peripheral initialization. What type of initialization is required by the peripherals? The serial device must have its baud rate, parity, and number of data and stop bits programmed. Parallel devices must be configured because most allow the direction (input or output) of their I/O lines to be programmed in many different ways. It is then necessary to set the direction of these I/O lines when power is first applied. For a system containing a digital-to-analog converter it may be important to output an initial value required by the external hardware. Since we can never assume that correct conditions exist at power-on, the microprocessor is responsible for establishing them.

Suppose a certain system contains a video display controller. Start-up software must select the proper screen format and initialize the video memory so that an intelligent picture (possibly a menu) is generated on the screen of the display. If the system uses light-emitting diode (LED) displays or alphanumeric displays for output, they must be properly set as well. High-reliability systems may require that memory be tested at power-on. While this adds to the complexity of the start-up software and the time required for initialization, it is a good practice to follow. Bad memory devices will certainly cause a great deal of trouble if they are not identified.

Other systems may employ a special circuit called a **watchdog monitor.** This circuit operates like this: During normal program execution the watchdog monitor is disabled by special instructions placed at the appropriate locations within the executing program. Should the program veer from its proper course, the special instructions will no longer disable the watchdog monitor, causing it to automatically reset the system. A simple way to make a watchdog monitor is to use a binary counter, clocked by a known frequency. If the counter is allowed to increment up to a certain value, the processor is automatically reset. The software's job, if it is working correctly, is to make sure the counter never reaches this count. A few simple logic gates can be used to clear the counter under microprocessor control, possibly whenever the CPU examines a certain memory location.

For flexibility, the system may have been designed to download its main program from a host computer. If this is the case, the system software will be responsible for knowing how to communicate with the host and place the new program into the proper memory locations. To guarantee that the correct program is loaded, the software should also perform a running test on the incoming data, requesting the host to retransmit portions of the data whenever it detects an error.

Sometimes preparing for a power-down is as important as doing the start-up initialization. A power supply will quite often supply voltage in the correct operating range for a few milliseconds after the loss of AC. It is during these few milliseconds that the processor must execute the shutdown code, saving important system data in nonvolatile RAM or doing whatever is necessary for a proper shutdown. If the system data can be preserved, it may be possible to continue normal execution when power is restored.

For systems that will be expanded in the future, the system bus must be made available to the outside world. To protect the internal system hardware, all signals must be properly buffered. This involves using tri-state buffers or similar devices to isolate the internal system bus from the bus available to the external devices. Sometimes optoisolators are used to completely separate the internal system signals from the external ones. The only connection in optoisolators is a beam of light, which makes them ideal when electrical isolation is required.

Figure 1.3 sums up all of these concepts with an expanded block diagram of a microprocessor-based control system. Notice once again that all devices in the system communicate with the CPU via the system bus.

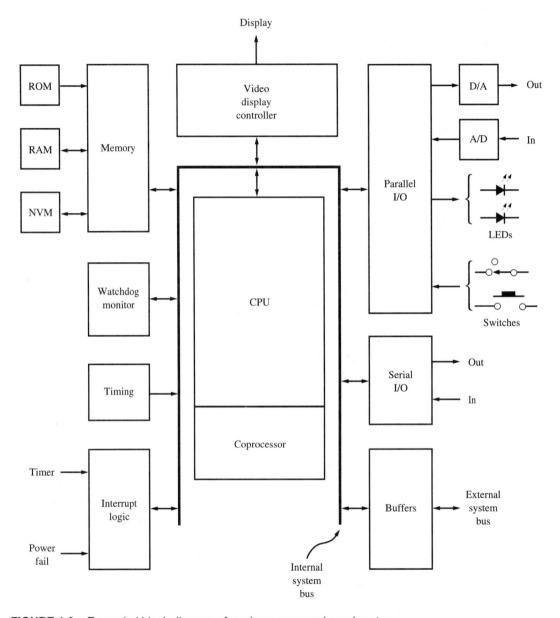

FIGURE 1.3   Expanded block diagram of a microprocessor-based system

1.6   THE MACINTOSH PERSONAL COMPUTER

All of the material in this chapter has, up to this point, dealt with general microprocessor-based systems. In this section we will see how a specific microprocessor-based system, the Macintosh personal computer, uses many of the hardware features already described. Although the

Macintosh has been around now for many years, and has evolved into a powerful machine containing very advanced 680x0 technology, it began as a much simpler machine constructed around the 16/32-bit Motorola 68000 microprocessor. The 68000 came out in the late 1970s and offered a higher level of computing power than the 8-bit processors of the time. When Apple chose the 68000 for use in the Macintosh personal computer, it upgraded the computing power of its earlier machines significantly (the Apple II used the 8-bit 6502 processor).

Let us take a historical look at how the 68000 was used. The initial Macintosh contained a keyboard for entering commands and data, a monochrome video display for viewing text and graphics, one or two floppy disk drives for storing information and running programs, and a memory large enough for many useful applications. It also came equipped with a software program called an **operating system,** which made it possible to access files on the disk drives and run programs with the use of a mouse device.

Most of the electronics within the Macintosh were contained on a single printed circuit board called the **motherboard.** Memory chips, timing circuitry, interrupt logic, the 68000 microprocessor, and other hardware all resided on the motherboard.

Let us now take a detailed look at the inside of the Macintosh. Figure 1.4 is the block diagram for a Macintosh 512KB computer motherboard. As shown, all communication is through the system bus. The heart of the machine is the 68000 microprocessor, running at

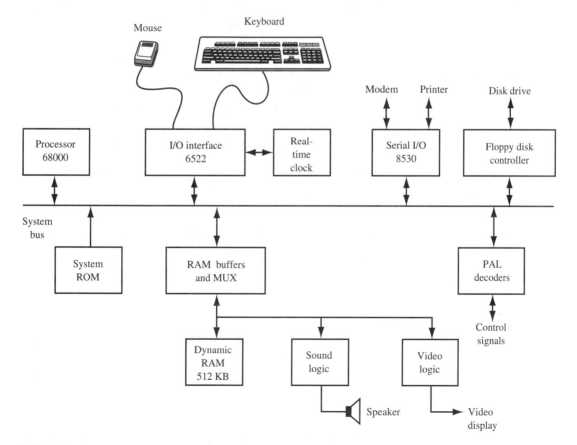

FIGURE 1.4 Block diagram of 512KB Macintosh computer motherboard

8 MHz. The operating system is partially contained in a ROM referred to as the **System ROM,** which is used to control the Macintosh when it is turned on. The System ROM is responsible for checking and initializing all peripherals and devices on the motherboard, and for starting up the disk drive to load the rest of the operating system.

512KB of dynamic RAM is included to allow execution of custom programs, although not all of it is available for the programmer's use. Some of the RAM is reserved for display memory and operating system functions. Notice that the sound and video circuits are directly connected to the RAM, since both rely heavily on RAM data for their operation.

Mouse, keyboard, and real-time clock functions are controlled by the 6522, an I/O adapter chip. Other adapters control the modem, printer (the 8530), and floppy disks. Special chips called PALs (for Programmable Array Logic) are used to decode address and control information.

It is not necessary to go into great detail about the hardware operation of each motherboard component, since many are controlled by a few bytes of data output to the correct memory locations. In later chapters we will see how I/O devices can be controlled through software.

1.7   DEVELOPING SOFTWARE FOR THE 68000

In order to get the most use out of a 68000-based computer, it is necessary to understand the software capabilities of the processor. We will explore the software architecture of the 68000 in great detail in the following chapters, first by examining the 68000's instruction set and then by looking at programming examples. What we will see is that the 68000 speaks a different language than we do.

Machine Language vs. Assembly Language

Our language is one of words and phrases. The 68000's language is a string of 1s and 0s. For example, the instruction

```
ADD.B   D2,D3
```

contains a word, ADD, that means something to us. Apparently we are adding D2 and D3 together, whatever they are. So, even though we might be unfamiliar with the 68000's instruction set, the instruction ADD.B   D2,D3 has meaning.

If we were instead given the binary string

```
1101 0110 0000 0010
```

or the hexadecimal equivalent

```
D6 02
```

and asked its meaning, we might be hard-pressed to come up with anything. We associate more meaning with ADD.B   D2,D3 than we do with D6 02, which is the way the instruction is actually represented by the processor. All programs for the 68000 will simply be long strings of binary numbers.

Because of the 68000's internal decoders, different binary patterns represent different instructions. Here are a few examples to illustrate this point:

```
D6 02    ADD.B    D2,D3          ;add D2 to D3, result in D3
96 42    SUB.W    D2,D3          ;subtract D2 from D3, result in D3
C6 82    AND.L    D2,D3          ;D3 equals D2 AND D3
42 02    CLR.B    D2             ;clear lower byte of D2
52 03    ADDI.B   #1,D3          ;add 1 to D3
36 02    MOVE.W   D2,D3          ;put copy of D2 into D3
```

The .B, .W, and .L terms stand for *byte, word,* and *longword.* These are the three main data sizes the 68000 is capable of working with. A word represents 2 bytes and a longword represents 4 bytes of information.

Can you guess the meaning of each instruction just by reading it? Do the hexadecimal codes for each instruction mean anything to you? What we see here is the difference between **machine language** and **assembly language.** The machine language for each instruction is represented by the hexadecimal codes. This is the binary language of the machine. The assembly language is represented by the wordlike terms that mean something to us. Putting groups of these wordlike instructions together is how a program is constructed. In the next few sections we will see how an assembly language program is written, converted into machine language, and executed.

HELLO: Our First Machine Language Program

When the single-board computer of Chapter 11 is first turned on, instructions in the start-up software output a message that indicates the machine is up and running. The message is stored in memory as a sequential group of bytes, with each byte representing a different message character. The message is displayed with the help of a character output routine that has been included in the single-board computer's software. The output routine is executed when the 68000 encounters the instruction TRAP   #3. Let's see how a 68000 program can be developed to use TRAP   #3 to output the message "Hello!"

Creating Machine Code with ASM68K

The companion disk contains two IBM-PC executable programs: ASM68K (a 68000 assembler) and EMU68K (a 68000 emulator). These two programs allow you to assemble and execute 68000 programs on any IBM-compatible personal computer. These programs are not needed if your system is 68000-based (such as a Macintosh or Amiga) and you have access to a 68000 assembler for your machine. The remainder of this discussion assumes that you have access to an IBM-compatible PC.

Using a PC-based word processor or text editor, enter the following text file exactly as you see it. Save the file as a plain ASCII text file under the name HELLO.ASM.

```
        ORG      $8000          ;starting address of data
HMSG    DC.B     'Hello!'       ;message characters
        DC.B     0              ;end-of-message marker
        ORG      $8100          ;starting address of program
START   MOVEA.L  #HMSG,A3       ;load A3 with message address
        TRAP     #3             ;output message
        TRAP     #9             ;return to command processor
        END      START          ;end of source file
```

The eight statements in HELLO.ASM constitute a **source file,** the starting point of any 68000 program.

To convert HELLO.ASM into a group of hexadecimal bytes that represent the corresponding machine language, use the ASM68K program found on the companion disk. ASM68K is an **assembler,** a program that takes a source file as input and determines the machine language for each source statement. ASM68K creates two additional files. These are the **list, object,** and **hex** files. The list file contains all of the text from the source file, plus additional information, as we will soon see. The object file contains only the machine language. The hex file is a printable version of the object file.

To assemble HELLO.ASM, enter the following command at the DOS prompt:

```
ASM68K HELLO
```

This instructs ASM68K to assemble HELLO.ASM and create HELLO.LST (the list file), HELLO.OBJ (the object file), and HELLO.HEX. The list file created by ASM68K looks like this:

```
008000                              ORG       $8000        ;starting address of data
008000   4865 6C6C 6F21   HMSG      DC.B      'Hello!'     ;message characters
008006   00                         DC.B      0            ;end-of-message marker
008100                              ORG       $8100        ;starting address of program
008100   267C 0000 8000   START     MOVEA.L   #HMSG,A3     ;load A3 with message address
008106   4E43                       TRAP      #3           ;output message
008108   4E49                       TRAP      #9           ;return to command processor
00810A                              END       START        ;end of source file
```

ASM68K has determined the machine language for each source statement. The first column is the set of memory locations where the instructions are stored. The second column is the group of machine language bytes or words that represent the actual 68000 instructions.

Running Machine Language Programs with EMU68K

To execute the HELLO program we run the EMU68K **emulator** program. EMU68K examines the information found in the HELLO.HEX file and simulates the operation of the 68000 microprocessor.

The DOS command to do this is:

```
EMU68K HELLO
```

The emulator will display the following after you press Return:

```
68000 Emulator V3.00
55736 (0xD9B8) bytes allocated for emulator memory.
HELLO.HEX loaded into emulator memory.
-
```

The – sign is EMU68K's command prompt. To execute HELLO, use the **g** (go) command, as follows:

```
68000 Emulator V3.00
55736 (0xD9B8) bytes allocated for emulator memory.
HELLO.HEX loaded into emulator memory.
-g
Hello!

Program exit at address 00008108
-
```

The emulator outputs the Hello! message and terminates normally. The address associated with the program exit (8108) is the location of HELLO's TRAP   #9 instruction (a predefined exit mechanism). This is indicated in the list file. This is the preferred way of ending a program that will be executed with EMU68K.

Programming Exercise 1.1:   Remove the DC.B   0 statement from HELLO.ASM and reassemble the program. What happens when the new program is executed by EMU68K?

Programming Exercise 1.2:   Write a program that displays your first and last name, using two TRAP   #3 calls.

1.8   THE CALCULATOR PROJECT

Although the following chapters contain numerous hardware and software design examples and applications, a book-long example called the *calculator project* will be utilized to show how an entire 68000-based system is developed. The common features and operations of a calculator are known to all, and this familiarity is used to bring in new concepts and ideas as they are presented in each chapter. For example, the display and keypad sections of the calculator are designed in the chapter on peripherals. The calculator's mathematical routines are developed in the chapters on programming and exceptions. Throughout the project, numerous questions are asked and answered to guide the design toward a final product. Room for improvement is deliberately included, to encourage original thought.

A simplified block diagram of the calculator is shown in Figure 1.5. Specific interfaces for the RAM, EPROM, and input/output sections will be designed, as will the soft-

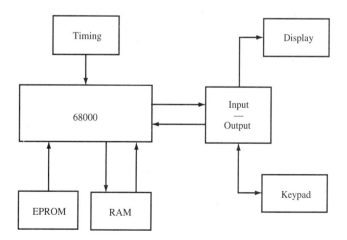

FIGURE 1.5   Calculator block diagram

ware to run the entire system. Since the design is broken up into specific sections, every effort is made to ensure that out-of-sequence treatment does not affect the reader's understanding of any single section.

1.9   TROUBLESHOOTING TECHNIQUES

You may think it premature to begin discussing troubleshooting techniques, when we have been exposed to so little of the 68000 architecture. Even so, we have already seen a number of places where errors can occur, and it would be worthwhile to discuss them. For example, the HELLO.ASM source file could have contained one or more *typographical* errors, such as a misspelled instruction (MOVA versus MOVEA), or a missing comma, or a comma where a semicolon was expected. Generally, when errors such as these are present in a source file, the assembler will report them with a brief error message.

Even if the source file does not have any typographical errors, we could still run into trouble. We could enter the command to invoke EMU68K incorrectly, or not use the correct options.

When the source file correctly assembles, there is still the possibility of a *run-time error* in the program. Run-time errors are typically caused by incorrect sequences of instructions and incomplete or faulty logical thinking.

To avoid a loss of time and effort, it is good to keep these common stumbling blocks in mind. Paying attention to the details will really pay off, as you learn to create a working program with a minimum of time and effort.

SUMMARY

In this chapter we have examined the operation of microprocessor-based systems. We saw that the complexity of the hardware, and thus of the software, is a function of the type of application. Through the use of many different types of peripherals, such as parallel and serial devices, analog-to-digital converters, a system can be tailored to perform almost any job. We also reviewed the basic fetch, decode, and execute cycle of a microprocessor, and examined the other duties the CPU performs, one of which was interrupt handling.

We also covered the initialization requirements of peripherals used in a microprocessor-based system, and why it is necessary to perform initialization in the first place. Other types of hardware and software requirements were also examined, such as the use of a watchdog monitor and a nonvolatile memory.

Four different generations of computers were presented and their differences highlighted. Current computing trends dealing with parallel processing and artificial intelligence were also introduced.

This was followed by an introduction to the motherboard hardware of a 512KB Macintosh personal computer. Since software is needed to control the hardware, we finished with a quick look at the technique for creating and executing machine language programs. This involved the use of the ASM68K and EMU68K programs, which are included on the companion disk.

The chapter finished with a description of the calculator project, a hardware/software design concept that will reappear in the following chapters, and a short set of troubleshooting tips.

STUDY QUESTIONS

1. Make a list of 10 additional products containing microprocessors that we use everyday.
2. Why would an oscillator circuit utilizing a resistor-capacitor network to control its frequency be unstable and unsuitable for use in a microprocessor-based system?
3. Speculate on the uses for timing signals in the serial I/O, memory, and interrupt sections.
4. Why do coprocessors enhance the capabilities of an ordinary CPU?
5. Draw a block diagram for a computerized cash register. The hardware should include a numerical display, a keyboard, and a compact printer.
6. What kind of initialization software would be required for the cash register of Question 5?
7. What would be the difference in system RAM requirements for two different cash registers, one without record keeping and one with?
8. What type of information should be stored in NVM during a power failure in a system designed to control navigation in an aircraft?
9. What types of interrupts may be required in a control system designed to monitor all doors, windows, and elevators in an office complex?
10. Name some advantages of downloading the main program into a microprocessor-based system. Are there any disadvantages?
11. Suppose that a number of robots making up a portion of an automobile assembly line are connected to a master factory computer. What kinds of information might be passed between the factory computer and the microprocessors controlling each robot?
12. A certain hard disk transfers data at the rate of 8 million bits per second. Explain why the CPU would not be able to perform the transfer itself, thus requiring the use of a DMA controller.
13. What kinds of problems arise if two devices attempt to use the system bus at the same time?
14. Explain how two microprocessors might be connected so that they share the same memory and peripherals.
15. Suppose that three microprocessors are used in the design of a new video game containing color graphics and complex sounds. How might each microprocessor function?
16. Why did processing speed increase with each new generation of computers?
17. List five different applications that might need the fast computing power of a RISC-based machine.
18. One reason 16-bit processors are faster than 8-bit machines is that they operate on twice as many data bits at the same time. Why doesn't everyone using an 8-bit machine just switch over to a 16-bit processor?

19. A backward-compatible microprocessor is one that can execute instructions from earlier models. How would a designer of the new CPU implement upward compatibility?
20. Which of the statements in the HELLO.ASM source file actually generate code?
21. It takes a certain amount of time to execute each instruction in the HELLO program. Which instruction do you think takes the longest?
22. What do the .B, .W, and .L extensions stand for?
23. List all the HELLO files created in Section 1.7.
24. What do you notice about the machine code for the MOVEA.L instruction in HELLO.ASM? Does the first ORG statement have anything to do with it?
25. What advantages does a microcontroller have over a microprocessor? What disadvantages?

CHAPTER 2

An Introduction to the 68000 Microprocessor

OBJECTIVES

In this chapter you will learn about:

- The register set of the 68000
- The addressing capabilities and data types that may be used
- The different instruction types available
- The signals generated/used by the 68000
- Some of the hardware and software advantages of the 68000

2.1 INTRODUCTION

The introduction of the 68000 into the arena of microprocessors came at a time when we were reaching the limits of what an 8-bit machine could do. With their restricted instruction sets and addressing capabilities, it was obvious that something more powerful was needed. The 68000 contains instructions previously unheard of in 8-bit machines, a very large address space, many different addressing modes, and an architecture that easily lends itself to multiprocessing or multitasking (running many programs simultaneously).

In this chapter we will examine the features of the 68000 microprocessor. Only basic material will be covered, leaving the hardware and software details for upcoming chapters. From reading this chapter you should become aware that the 68000 is a machine with many possibilities.

Section 2.2 covers the software model of the 68000. Section 2.3 provides a brief functional description of the processor. Section 2.4 explains the various data types that may be used (8-, 16-, and 32-bit data sizes for many instructions). Section 2.5 lists the numerous instruction types that comprise the 68000's instruction set. Section 2.6 compares the 68000 with another 16-bit machine—the 8086—and states its advantages. Section 2.7 begins the calculator project, and Section 2.8 presents some troubleshooting techniques.

| 2.2 | THE SOFTWARE MODEL OF THE 68000 |

The 68000 microprocessor contains eight data registers, referred to as D0 through D7, eight address registers, A0 through A7; a program counter (PC); and a status register (SR). All registers except for the status register are 32 bits in length.

The data registers are used to store information *within* the 68000 processor itself, much like the memory found on a calculator. Address registers are used to store the location where data can be found *outside* the processor, in external memory chips. These registers are all indicated in the 68000's software model, which is shown in Figure 2.1. Notice that the 32-bit data registers are subdivided into two smaller portions of 16 and 8 bits each. Also notice that, although the program counter is 32 bits wide, only the lower 24 bits are used by the 68000. More advanced 680x0 architectures use the full 32 bits.

Technically speaking, there are two A7 address registers. One makes up the **user** stack pointer (USP) and the other the **supervisor** stack pointer (SSP). Only one of these address registers is ever in effect, depending on the processor's processing state. These two processing states, user and supervisor, make the 68000 an ideal processor for applications involving multiprogramming, multitasking, and the creation of operating systems. Programs running in the user state are denied access to a few special instructions, two of them being STOP and RESET. Trying to execute these privileged instructions in the user state causes an error condition. The processor responds to this error condition by entering the supervisor state and taking appropriate action, determined by software, in a routine called an **exception handler.**

The processing state is determined by a special bit in the status register. Figure 2.2 portrays the details of the 16-bit status register. The lower 8 bits form the user byte. These bits contain the five status bits that may be directly tested by the programmer. These 5 bits are commonly called **condition codes,** or **flags.** X, N, Z, V, and C represent the processor's extended, negative, zero, overflow, and carry conditions, as determined by previous instruction execution. The upper 8 bits make up the system byte. These bits are unavailable to the programmer unless the processor is in the supervisor state. The processing state is controlled by the S bit (0 for user, 1 for supervisor), and there are special instructions available for manipulating this bit. A nice addition to the 68000, and something not available on previous 8-bit machines, is the T bit, which enables trace mode in the processor. Trace mode makes it possible to single-step through a program, instruction by instruction. This is very useful when debugging a new piece of code. A trace is actually an **exception** that allows the user to debug (or monitor execution of) an executing program.

I_0 through I_2 make up the processor's **interrupt mask.** They are used to determine what levels of external interrupts the 68000 will respond to.

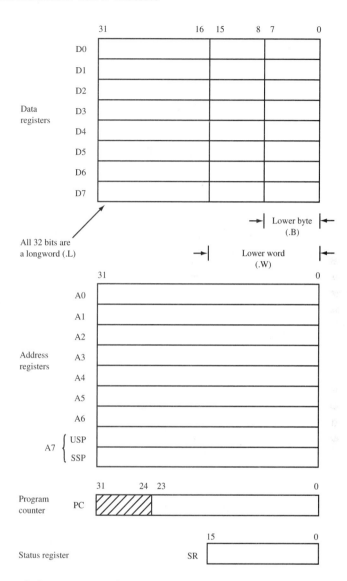

FIGURE 2.1 Software model of the 68000

2.3 A FUNCTIONAL DESCRIPTION OF THE 68000

The 68000 is most commonly referred to as a 16-bit machine, even though it can perform a wide variety of operations on 32-bit data words. As shown in Figure 2.3, all external data enters the CPU on 16 bidirectional data lines (D_0 through D_{15}). The 68000's 23 address lines (A_1 through A_{23}), together with two other signals—$\overline{UDS}$ and $\overline{LDS}$ (for upper and lower data strobe)—give the processor the ability to address over 16 *million* bytes of mem-

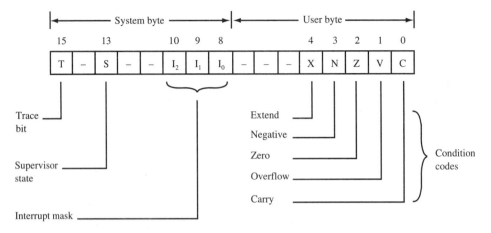

FIGURE 2.2   68000 status register

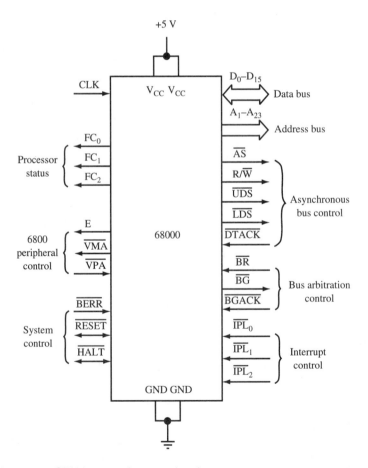

FIGURE 2.3   68000 CPU input and output signals

ory. This large addressing space makes the handling of large databases convenient, and also supports the development of multiuser systems. Suppose that a customer needs a word processing system capable of supporting 16 users simultaneously. The 68000 makes it possible to give each user almost 1 million bytes of memory, enough storage space to hold approximately 218 pages of text (with each page containing 60 lines of 80 characters each). Some memory must, of necessity, be reserved for the operating system functions.

Other signals on the 68000 provide for two different types of data transfers: synchronous and asynchronous, with asynchronous being faster. The synchronous transfers enable the 68000 to communicate with peripherals designed for the earlier 6800 microprocessor.

Furthermore, the 68000 will respond to seven levels of external hardware interrupts (via $\overline{IPL_2}$–$\overline{IPL_0}$) and has bus arbitration logic that supports its use in multiprocessor systems.

All of these features come in a 64-pin package that runs on a single 5-volt power supply. Figure 2.4 shows the pinout of the 68000's 64-pin package. In addition, clock speeds for current versions of the 68000 run from 4 MHz up to 12.5 MHz. The high clock frequency, together with the ability to load 16 bits of data at once (twice that of an 8-bit machine), greatly increases the processing speed beyond the barriers encountered in the 8-bit machines.

FIGURE 2.4   68000 pinout

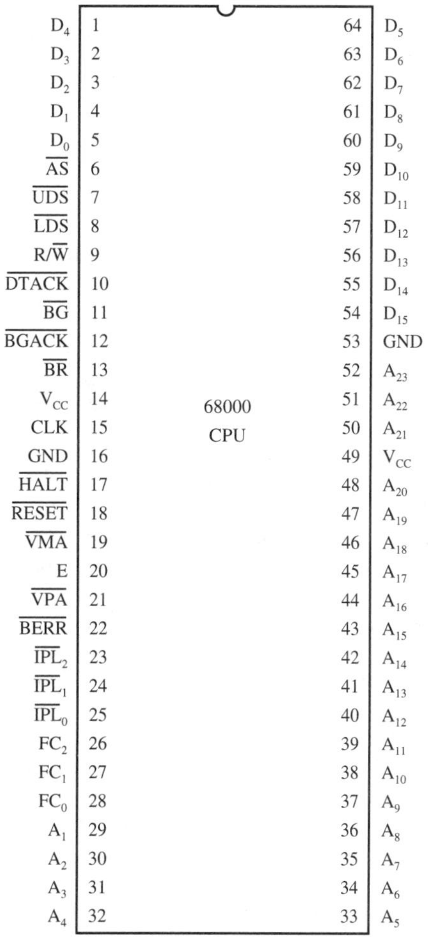

Left signal	Pin		Pin	Right signal
D_4	1		64	D_5
D_3	2		63	D_6
D_2	3		62	D_7
D_1	4		61	D_8
D_0	5		60	D_9
$\overline{AS}$	6		59	D_{10}
$\overline{UDS}$	7		58	D_{11}
$\overline{LDS}$	8		57	D_{12}
$R/\overline{W}$	9		56	D_{13}
$\overline{DTACK}$	10		55	D_{14}
$\overline{BG}$	11		54	D_{15}
$\overline{BGACK}$	12		53	GND
$\overline{BR}$	13		52	A_{23}
V_{CC}	14		51	A_{22}
CLK	15	68000	50	A_{21}
GND	16	CPU	49	V_{CC}
$\overline{HALT}$	17		48	A_{20}
$\overline{RESET}$	18		47	A_{19}
$\overline{VMA}$	19		46	A_{18}
E	20		45	A_{17}
$\overline{VPA}$	21		44	A_{16}
$\overline{BERR}$	22		43	A_{15}
$\overline{IPL_2}$	23		42	A_{14}
$\overline{IPL_1}$	24		41	A_{13}
$\overline{IPL_0}$	25		40	A_{12}
FC_2	26		39	A_{11}
FC_1	27		38	A_{10}
FC_0	28		37	A_9
A_1	29		36	A_8
A_2	30		35	A_7
A_3	31		34	A_6
A_4	32		33	A_5

2.4   68000 DATA ORGANIZATION

Even though the 68000's data and address registers are 32 bits wide, it is possible to work with smaller bit quantities. Many of the processor's instructions can be directed to operate on 16, 8, or even 1 bit at a time! We generally refer to 8-, 16-, and 32-bit data lengths as bytes, words, and longwords. Telling an assembler what data type to use in an instruction is done by following the instruction mnemonic with a .B, .W, or .L extension. For example, MOVE.B means to move a byte of data, and MOVE.L means to move a longword of data. When the 68000 is directed to use only 8 bits of a register, it will use the lower 8 bits (bits 0 through 7). Words occupy the lower 16 bits of the register. Operations on these lower bits do not affect the higher bits in the register. By convention, we refer to bit 0 as the **least significant bit,** and bit 7, 15, or 31 (for byte, word, or longword operations) as the **most significant bit.**

Figure 2.5 shows how the 68000 stores 16-bit data words in memory. The upper 8 bits of the word occupy a memory location that has an even address. The lower 8 bits are stored in an odd memory location one higher than that of the upper byte. The figure indicates that the processor has stored the word value B52D in memory locations 0 and 1, with the next available word boundaries starting at addresses 2, 4, and 6. The hardware signals $\overline{UDS}$ and $\overline{LDS}$ are used to indicate which portion of the data bus contains information: $\overline{UDS}$ for bits 15–8 and $\overline{LDS}$ for bits 7–0. Note that the 68000 is a *big endian* processor, due to the fact that words are stored with their lower 8 bits in the higher of the two storage locations. *Little endian* processors, like the Intel 80x86 family, store the lower byte in the first memory location.

The 68000 is capable of detecting operations that may try to violate this storage technique (for example, trying to write a word to an odd location), resulting in another type of exception, called an **address error.** Longwords are stored with the upper word occupying the first two memory locations and the lower word, the next two locations. Bytes may be stored too, in any location, regardless of the even/odd requirement.

Using a diagram such as that shown in Figure 2.5 is very useful for indicating the contents of a few memory locations. For example, Figure 2.6 shows where a data table of the first six prime numbers are stored in memory. The first prime number is stored as a byte in location $1A00, the last in location $1A05. Notice the use of the $ preceding each address in Figure 2.6. This is the preferred way of representing hexadecimal numbers for Motorola products. When no $ is used, we assume the number is a decimal number. Thus, the last prime number in the data table is 11 decimal. If the value were $11 instead, the decimal equivalent would be 17. Unless it is obvious that we are dealing with decimal or hexadecimal numbers, the $ will be used from here on to indicate a hexadecimal number.

FIGURE 2.5   Storing words and longwords in memory

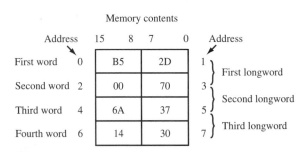

FIGURE 2.6   Data table of
first six prime numbers

$1A00	1
$1A01	2
$1A02	3
$1A03	5
$1A04	7
$1A05	11

A diagram such as Figure 2.6 is fine for representing small amounts of data. When hundreds, thousands, or even millions of bytes of data must be represented, it is difficult to show each and every location. Instead, *groups* of locations are shown as a single block, in a diagram called a **memory map.** Figure 2.7 shows an example of a memory map for the 512KB Macintosh. Hexadecimal addresses are shown on the left side of the memory map. These addresses indicate the *beginning* of each new block in the memory map. For example, the RAM block begins at address $000000 (all address lines are low). The first unused block begins at address $080000. This means that the last address of the RAM block is one less than $080000, or $07FFFF. This range of addresses for the RAM block, $000000 to $07FFFF, represents $080000 locations, which corresponds to 512KB of memory. By similar reasoning, the ROM block has a range of addresses from $400000 to $40FFFF, which is a 64KB range.

The only address in the memory map that does not represent the beginning of a new block is $FFFFFF, which is the *last* memory location in the 68000's addressing space (all address lines are high). This makes the range of the last unused block of memory begin at $F00000 and end at $FFFFFF, a total of $100000 locations (an entire megabyte worth of addresses).

FIGURE 2.7   Memory map
of the 512KB Macintosh

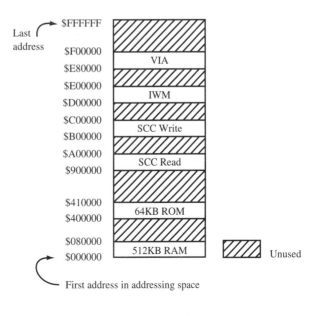

2.5   68000 INSTRUCTION TYPES

The 68000 contains several groups of instructions designed to make the task of writing source code less tedious for the programmer. Older machines often contained many different forms of instructions that all did the same thing. For example, 8- and 16-bit additions were handled by different instructions, requiring the use of different mnemonics in the source code. The 68000 eliminates the need for the programmer to keep track of these differences. A single ADD instruction mnemonic can be easily coded to perform 8-, 16-, and 32-bit additions on either data or address registers. The same is true for data transfer instructions as well. Instead of using different mnemonics (and therefore instructions) for data transfers, register to register, register to memory, and immediate to register/memory use only *one* mnemonic to handle everything. Figure 2.8 shows the three ADD instructions used to add 8-, 16-, or 32-bit portions of D4 to D5.

The 68000 has an impressive instruction set, with over 70 instructions provided for the programmer's use. Table 2.1 gives a brief description of each instruction, with the entire set broken up into eight different groups. In practice, many programmers use a small subset of processor instructions when writing programs. So, if Table 2.1 looks too complicated at this point, just remember that only a few instructions from each group are needed to begin writing useful 68000 programs. For example, a routine that converts temperature in degrees Fahrenheit into degrees Celsius might look like this:

```
FTOC SUBI.W    #32,D0     ;subtract 32 from D0
     MULS      #5,D0      ;multiply D0 by 5
     DIVS      #9,D0      ;divide D0 by 9
     RTS                  ;return from subroutine
```

FIGURE 2.8   Adding different data sizes with ADD

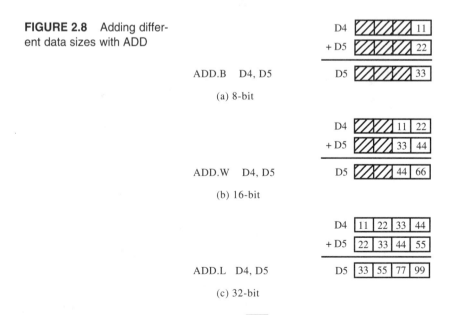

ADD.B   D4, D5

(a) 8-bit

ADD.W   D4, D5

(b) 16-bit

ADD.L   D4, D5

(c) 32-bit

⧄ = No change in contents

TABLE 2.1 68000 instruction set

Data transfer group		Shift and rotate group	
EXG	Exchange registers	ASL	Arithmetic shift left
LEA	Load effective address	ASR	Arithmetic shift right
LINK	Link and allocate	LSL	Logical shift left
MOVE	Move data	LSR	Logical shift right
MOVEA	Move address	ROL	Rotate left
MOVEM	Move multiple registers	ROR	Rotate right
MOVEP	Move peripheral data	ROXL	Rotate left with extend
MOVEQ	Move quick	ROXR	Rotate right with extend
PEA	Push effective address	*Bit manipulation group*	
SWAP	Swap register halves		
UNLK	Unlink	BCHG	Bit change
Arithmetic group		BCLR	Bit clear
		BSET	Bit set
ADD	Add binary	BTST	Bit test
ADDA	Add address	*Binary coded decimal group*	
ADDI	Add immediate		
ADDQ	Add quick	ABCD	Add BCD
CLR	Clear operand	NBCD	Negate BCD
CMP	Compare	SBCD	Subtract BCD
CMPA	Compare address	*Program control group*	
CMPI	Compare immediate		
CMPM	Compare memory	B*cc**	Conditional branch
DIVS	Divide signed numbers	DB*cc**	Decrement and branch
DIVU	Divide unsigned numbers	S*cc**	Conditional set
EXT	Extend sign	BRA	Branch always
MULS	Multiply signed numbers	BSR	Branch to subroutine
MULU	Multiply unsigned numbers	JMP	Jump
NEG	Negate	JSR	Jump to subroutine
NEGX	Negate with extend	RTR	Return and restore
SUB	Subtract binary	RTS	Return from subroutine
SUBA	Subtract address	*System control group*	
SUBI	Subtract immediate		
SUBQ	Subtract quick	ANDI SR	AND immediate to SR
SUBX	Subtract with extend	EORI SR	EOR immediate to SR
TAS	Test and set	MOVE SR	Move to/from SR
TST	Test	MOVE USP	Move to/from USP
Logical group		ORI SR	OR immediate to SR
		RESET	Reset processor
AND	Logical AND	RTE	Return from exception
ANDI	AND immediate	STOP	Stop processor
OR	Logical OR	CHK	Check register
ORI	OR immediate	ILLEGAL	Illegal instruction
EOR	Exclusive OR	TRAP	Trap call
EORI	Exclusive OR immediate	TRAPV	Trap on overflow
NOT	Logical complement	ANDI CCR	AND immediate to CCR
		ORI CCR	OR immediate to CCR
		EORI CCR	EOR immediate to CCR
		MOVE CCR	Move to/from CCR
		NOP	No operation

*Note: *cc* stands for condition code

Recall that C = (5/9)*(F – 32).

Notice that all work is done with a single data register (D0) and that we only needed instructions from two groups (arithmetic and program control) to do the job. Also, the *signed* multiply and divide instructions were used to allow for negative temperatures.

2.6   ADVANTAGES OF THE 68000

The great number of microprocessors currently available forces a designer or programmer to choose a particular one based on its advantages over the others. Let us briefly examine some of the 68000's advantages over another popular 16-bit machine, the 8086.

The 8086 is Intel's offering in the world of 16-bit machines. It connects to the outside world via a 16-bit data bus that is, unfortunately, multiplexed with 16 of the 8086's address lines. Even though the remaining four address lines are directly available, processing speed is somewhat reduced by the need to demultiplex the address and data buses. With 20 address lines, the 8086 is capable of accessing over 1 million bytes of memory, only a sixteenth of the 68000's address space.

Furthermore, the registers of the 8086 are only 16 bits in length, and there are fewer of them. Table 2.2 compares the entire register set of both machines. The four 8086 segment registers CS (code segment), DS (data segment), ES (extra segment), and SS (stack segment) are used to point to 64KB blocks of memory. Thus, even though the 8086 can address over 1 million bytes of storage, it only does so in 64KB chunks. This requires careful

TABLE 2.2   Register sets of the 8086 and 68000

8086 (16-bit registers)	68000 (32-bit registers)	
General-purpose registers		
AX	D0	A0
BX	D1	A1
CX	D2	A2
DX	D3	A3
BP	D4	A4
SI	D5	A5
DI	D6	A6
	D7	
Segment registers		
CS		
DS		
ES		
SS		
System control		
SP (stack pointer)	A7 (USP, SSP)	
IP (instruction/program counter)	PC	

FIGURE 2.9 Generating a 20-bit address in the 8086

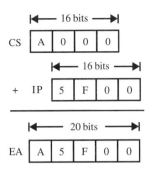

manipulation of the segment registers, via software, and is a confusing procedure for the beginning programmer. Figure 2.9 shows how a full 20-bit address is formed in the 8086. Notice how the segment register is shifted 4 bits to the left. This causes all memory references to begin on 16-byte boundaries called **paragraphs.** Although this technique makes program relocation much simpler, it is still easier to directly refer to *any* memory location, a capability we enjoy on the 68000.

The 8086 supports 256 levels of interrupts, which operate similar to the 255 exceptions available with the 68000. In the case of the 8086, a 1KB block of memory is dedicated to storing the interrupt vector locations. Externally, only two interrupts may be generated, one of them nonmaskable. This is much less than the seven levels of external hardware interrupts available on the 68000.

The 8086 runs at a slower clock speed than the 68000, and requires an external clock generator chip to control its internal timing.

In summary, the 68000's 32-bit arithmetic, 32-bit registers, large addressing space, nonmultiplexed address and data buses, external interrupt features, and faster clock speed make it a better choice than the 8086.

2.7 BEGINNING THE CALCULATOR PROJECT

We have only been briefly introduced to the capabilities of the 68000 in this chapter, but there is enough information at our disposal to begin asking questions about the calculator project. For example, how much memory will the calculator need? How much of the calculator's memory will be EPROM, and how much will be RAM? Even though the 68000 can access over 16 million bytes of memory, it would be impractical and prohibitively expensive to design a calculator with this much memory.

To get a clearer idea of our memory requirements, we need to know what kind of operations will be available on the calculator. A calculator capable of performing only the standard mathematical functions +, −, *, and / will require less EPROM than a calculator that can compute sine and cosine values, square roots, and logarithms.

Furthermore, will the calculator perform its math using binary numbers or BCD numbers? The 68000 can perform the standard four operations on binary numbers, but can only add and subtract BCD numbers. BCD multiplication and division require software subroutines (stored in EPROM on the calculator) for implementation. We must also decide on how a number will be stored in the calculator. Storing a BCD number in a register results in a range of 0 to 99,999,999 (since each BCD digit requires 4 bits and a 68000 register is 32 bits wide). Storing a binary number in a register gives an unsigned range of 0 to 4,294,967,206 (2^{32}). Storing numbers in scientific notation almost certainly necessitates the use of multiple memory locations (say 8 bytes per number) and even more software for coding and manipulation.

Will the calculator have its own internal memory where the user can store numbers for use in later calculations? If so, where will these memories reside, in registers or actual RAM locations? How many memories should there be and how will they be accessed?

Some questions require a knowledge of the 68000's digital architecture. For example, the calculator could have an automatic power-down feature, a mechanism that shuts the calculator off if no keypad keys have been pushed for a number of minutes. This feature would require a timing circuit capable of signaling the processor when the allotted time is up, possibly through the use of a dedicated hardware interrupt.

Clearly, we have only touched the surface of what is involved in the calculator project. In the next chapter we will be introduced to the instruction set of the 68000. This will enable us to begin answering some of the questions raised here, and to begin asking new ones.

2.8   TROUBLESHOOTING TECHNIQUES

A great amount of material was presented in this chapter regarding the Motorola 68000 architecture. It would be helpful to commit some of the most basic material about the 68000 to memory. At a minimum, you should be able to do the following without much thought:

- Name all of the processor registers, their bit sizes, and the meaning of the .B, .W, and .L suffixes.
- Be familiar with several architectural features, such as the processor's addressing space (16MB), interrupt mechanism, and data bus operation.
- Explain why the 68000 is a big endian processor.
- List the names and meanings of the most common flags, such as zero, carry, and sign.
- State the difference between the user and supervisor states.
- Show how an instruction is composed of an operation, a set of operands, and a particular addressing mode.

Knowing these basics thoroughly will assist you in mastering the 68000 instruction set that we will examine in Chapter 3.

SUMMARY

In this chapter we have been introduced to a powerful new processor, the 68000. We saw that the 68000 contains eight data registers and eight address registers, all 32 bits in length.

Furthermore, the 68000 communicates via a 16-bit data bus and 24-bit address bus (A_1 through A_{23}, $\overline{UDS}$, and $\overline{LDS}$), capable of addressing over 16 million bytes of memory. The 68000 is equipped with seven levels of external hardware interrupts and 255 different types of exceptions. The processor's large instruction set gives the programmer flexibility when writing new code.

Two states of operation, user and supervisor, make the 68000 an ideal choice for multiuser and multitasking systems.

We were introduced to the concept of a memory map for representing the contents of memory, and saw an example of the memory map for the 512KB Macintosh computer. In addition, the data types and instructions available in the 68000 were surveyed, and an example program that performed temperature conversion was shown. A short list of troubleshooting tips on the 68000 architecture was also examined.

By now we have an idea of what working with the 68000 will involve. In the next chapter, we will jump into the details of the 68000's instruction set, our first step toward writing our own programs!

STUDY QUESTIONS

1. Why restrict certain instructions, such as STOP and RESET, from programs running in the user state? What kind of problems could occur in a multiuser system that did not have any restrictions of this kind?
2. How might the trace function help in debugging a new piece of code?
3. A 68000-based system contains 1024KB of RAM to be divided among 32 users. How much RAM does each user get? Should any of the RAM be dedicated to the operating system?
4. Name the two types of memory transfers available in the 68000.
5. What are the .B, .W, and .L extensions used for?
6. How many bytes of storage are available just using the 68000's data registers? How many words of storage are possible?
7. Table 2.3 shows the data in a few selected memory locations. What is the longword stored at location 30? What is the word stored at location 30?
8. Regarding Table 2.3, what is the byte stored at address 2F? What is the byte stored at address 33?
9. What does the $ signify in the number $100? Is the number 100 the same as $100?
10. When the word value $1234 is stored in memory beginning at address $9000, what byte value resides in location $9001?

TABLE 2.3 For Questions 2.7 and 2.8

Address	Data
2E	CB05
30	9AFC
32	3007

11. Classify the following values as byte, word, or longword. It may be possible to use more than one classification for a single value.
    a) $3C29A0
    b) 1554290
    c) $10600
    d) 10600
    e) 6
12. What happens when the 68000 attempts to perform a word access to an odd location?
13. What is contained in the user byte of the status register?
14. What is contained in the system byte of the status register?
15. The status register contains $238C. What is the state of each condition code?
16. Regarding Question 15, is the processor in the user or supervisor state?
17. Why can the 68000 directly access any memory location with one of its registers? Can the 8086 do the same?
18. Name three differences between the 68000 and the 8086.
19. A 68000-based telecommunication system has 128KB of ROM beginning at address $000000, 64KB of RAM beginning at address $200000, and a 2KB I/O block beginning at address $7C0000.
    a) What is the last address in each block?
    b) Draw a memory map of the system.
20. The first 10 perfect squares (1, 4, 9, 16, 25, etc.) are stored in memory beginning at address $207C. Draw a memory map containing each of the squares and label the address of each value.
21. What instructions are needed to convert from speed in miles/hour to speed in feet/second? Use the FTOC routine as a guide.
22. Make a list of functions you would like to see in the calculator project. Assume that each function (+, −, *, /, etc.) requires 256 bytes of machine code. If the control program requires an additional 2KB, how much EPROM is needed for your calculator?
23. What is the difference between big endian and little endian storage formats?
24. Why are there two A7 registers?
25. How are $\overline{\text{UDS}}$ and $\overline{\text{LDS}}$ associated with the 68000's 16-bit data bus?

PART 2

Software Architecture

Microsoft Windows 95 desktop showing a program under development.

Let us look at a sample source file, a subroutine designed to find the sum of 16 bytes stored in memory. It is not important at this time that we understand what each instruction does. We are simply trying to get a feel for what a source file might look like and what conventions we should follow when we write our own programs.

```
        ORG     $8000
TOTAL   CLR.W   D0              ;clear result
        MOVE.B  #16,D1          ;init loop counter
        MOVEA.L #DATA,A0        ;init data pointer
LOOP    ADD.B   (A0)+,D0        ;add data value to result
        SUBI.B  #1,D1           ;decrement loop counter
        BNE     LOOP
        MOVEA.L #SUM,A1         ;point to result storage
        MOVE.W  D0,(A1)         ;save sum
        RTS                     ;and return
SUM     DC.W    0               ;save room for result
DATA    DS.B    16              ;save room for 16 data bytes.
        END
```

The first line of source code contains a command that instructs the assembler to load its own program counter with $8000. The ORG (for origin) command is known as an assembler **pseudo-opcode,** a fancy name for a mnemonic that is understood by the assembler but not by the microprocessor. ORG does not generate any source code; it merely sets the value of the assembler's program counter. This is important when a section of code must be loaded at a particular place in memory. The ORG statement is a good way to generate instructions that will access the proper memory locations when the program is loaded into memory.

The second source line contains the major components normally used in a source statement. The label TOTAL is used to point to the address of the first instruction in the subroutine. Other labels in this example are LOOP, SUM, and DATA. Single-line assemblers do not allow the use of labels.

The opcode is represented by CLR.W and the operand field by D0. So far we have three fields: label, opcode, and operand. The fourth field, if it is used, usually contains a comment explaining what the instruction is doing. Comments are preceded by a semicolon (;) to separate them from the operand field. In writing source code, you should follow the four-column approach. This will result in a more understandable source file.

Two more pseudo-opcodes, DC.W and DS.B, appear at the end of the source file. DC.W (define constant word) is used to define a 2-byte value that will be placed in the object file. DC.B and DC.L may also be used.

DS.B (define storage byte) is used to reserve a section of memory. When we are writing our program, we may not know what data will be used, so we simply reserve room for it with the DS.B statement. DS.W and DS.L may also be used.

The final pseudo-opcode in most source files is END. The END statement informs the assembler that it has reached the end of the source file. This is important, since many assemblers usually perform two *passes* over the source file. The first pass is used to determine the lengths of all instructions and data areas, and to assign values to all symbols (labels) encountered. The second pass completes the assembly process by generating the machine code for all instructions, usually with the help of the symbol table created in the first pass. The second pass also creates and writes information to the list and object files. The list file for our example subroutine looks like this:

```
008000                          1              ORG      $8000
008000   4240                   2    TOTAL     CLR.W    D0
008002   123C 0010              3              MOVE.B   #16,D1
008006   207C 0000 801E         4              MOVEA.L  #DATA,A0
00800C   D018                   5    LOOP      ADD.B    (A0)+,D0
00800E   5301                   6              SUBI.B   #1,D1
008010   66FA                   7              BNE      LOOP
008012   227C 0000 801C         8              MOVEA.L  #SUM,A1
008018   3280                   9              MOVE.W   D0,(A1)
00801A   4E75                  10              RTS
00801C   0000                  11    SUM       DC.W     0
00801E   ??                    12    DATA      DS.B     16
00802E                         13              END
```

Normally the comments would follow the instructions, but they have been removed for the purposes of this discussion.

The first column of numbers represents the memory address of each instruction or data area. Notice that the first address matches the one specified by the ORG statement.

The second column of numbers are the machine codes generated by the assembler. The machine codes are intermixed with data and address values. For example, 4240 on line 2 represents the instruction CLR.W D0. On line 3, the 4-byte string 123C 0010 is a mix of machine code and data. 123C means MOVE.B to D1, and 0010 is the hexadecimal representation of the decimal number 16. If you examine the rest of the list file carefully, you should be able to spot other interesting points. For instance, can you determine the relationship between line 4 and line 12, and between lines 8 and 11?

Following the code on each line is the original source line. Having all of this information available is very helpful during the debugging process.

A final point about the source file concerns the placement of reserved data areas. The storage for SUM and DATA is declared at the end of the subroutine. In the past it was impossible for some assemblers to correctly generate machine code when source files were written in this way. The 68000 assembler does not have this problem because the data size is indicated in the instruction. Even so, a generally accepted programming practice is to define the data areas *before* the instructions that will use them. Rewriting the example slightly results in this source file:

```
          ORG      $8000
SUM       DC.W     0               ;save room for result
DATA      DS.B     16              ;save room for 16 data bytes.


          ORG      $8100
TOTAL     CLR.W    D0              ;clear result
          MOVE.B   #16,D1          ;init loop counter
          MOVEA.L  #DATA,A0        ;init data pointer
LOOP      ADD.B    (A0)+,D0        ;add data value to result
          SUBI.B   #1,D1           ;decrement loop counter
          BNE      LOOP
          MOVEA.L  #SUM,A1         ;point to result storage
          MOVE.W   D0,(A1)         ;save sum
          RTS                      ;and return
          END      TOTAL
```

The ORG $8000 statement tells the assembler where to put the data areas. It is not necessary for this ORG value to be smaller than the ORG of the subroutine. It is all a function of

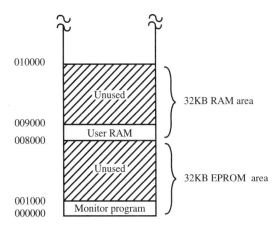

FIGURE 3.2 Partial memory map of the single-board computer

where RAM exists in your system. If you plan on building the single-board computer (SBC) in Chapter 11, now would be a good time to discuss the memory map of the SBC, so that programs you write will run correctly on it. Figure 3.2 shows a partial memory map of the SBC. The SBC's monitor program resides in 4KB of EPROM beginning at address 000000. Room is available for 28KB of additional EPROM. User (and monitor) RAM begins at address 008000, with 4KB available and 28KB unused. This information is important, since this is where your programs must load and execute. Also, the monitor portion of user RAM occupies the last 256 bytes of the first 4KB block, addresses 008F00 through 008FFF. This leaves addresses 008000 to 008EFF available for user programs. The EMU68K program supports this range of addresses also.

Knowing that user RAM begins at address 008000 explains why the ORG statement at the beginning of the source file is so important. It is necessary to match the addresses generated during assembly with the addresses the program code will occupy in physical memory.

The addition of the TOTAL label in the END statement informs the assembler that TOTAL, not SUM, is the starting execution address. This information is also included in the object file. When a large program must be written by a team of people, each will be assigned a few subroutines to write. They must all assemble and test their individual sections to ensure the code executes correctly. When all portions of the program (called **modules,** after a technique called modular programming) are assembled and tested, their object files are combined into one large object file via a program called a **linker.** Figure 3.3 represents this process. The linker examines each object file, determining its length in bytes, its proper place in the final object file, and what modifications should be made to it.

In addition, a special collection of object files is sometimes available in a **library** file. The library may contain often-used subroutines, or patches of code. Instead of continuously reproducing these code segments in a source file, a special pseudocode is used to instruct the assembler that the code must be brought in at a later time (by the linker). This helps keep the size of the source file down, and promotes quicker writing of programs.

When the linker is through, the final code is written to a file called the **load** module. Another program called a **loader** takes care of loading the program into memory. Usually the linker and loader are combined into a single program called a **link-loader.**

FIGURE 3.3 Linking multiple object files together

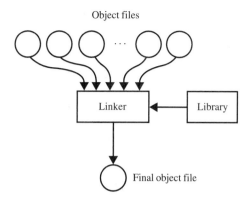

So, writing the source file is actually only the first step in a long process. But even before a source file can be written, the programmer must obtain an understanding of the instructions that will be used in the source file. The remaining sections will cover this important topic.

3.3   68000 INSTRUCTION TYPES

The instruction set of the 68000 microprocessor is composed of eight different groups:

- Data transfer
- Arithmetic
- Logical
- Shift and rotate
- Bit manipulation
- BCD operations
- Program control
- System control

The data transfer group contains instructions that transfer data from memory to register, register to register, and register to memory. Data may be 8, 16, or 32 bits in length. Address and data registers may be used, as well as the system registers CCR (condition code register), USP (user stack pointer), and SR (status register). The data transfer instructions are:

EXG	Exchange registers
LEA	Load effective address
LINK	Link and allocate
MOVE	Move data
MOVEA	Move address

MOVEM	Move multiple registers
MOVEP	Move peripheral data
MOVEQ	Move quick
PEA	Push effective address
SWAP	Swap register halves
UNLK	Unlink

The arithmetic group provides addition and subtraction of 8-, 16-, and 32-bit values; signed and unsigned 16-bit by 16-bit multiplication; signed and unsigned 32-bit by 16-bit division; 8-, 16-, and 32-bit clears, compares, tests; and negation (a 2's complement operation). An additional operation called sign extension is also available. These operations may be performed on all address and data registers and memory locations. The arithmetic instructions are:

ADD	Add binary
ADDA	Add address
ADDI	Add immediate
ADDQ	Add quick
CLR	Clear operand
CMP	Compare
CMPA	Compare address
CMPI	Compare immediate
CMPM	Compare memory
DIVS	Divide signed numbers
DIVU	Divide unsigned numbers
EXT	Extend sign
MULS	Multiply signed numbers
MULU	Multiply unsigned numbers
NEG	Negate
NEGX	Negate with extend
SUB	Subtract binary
SUBA	Subtract address
SUBI	Subtract immediate
SUBQ	Subtract quick
SUBX	Subtract with extend
TAS	Test and set
TST	Test

The logical group is used to perform AND, OR, EOR (exclusive OR), and NOT (one's complement) operations on 8-, 16-, and 32-bit data contained in data registers, memory, and the CCR and SR system registers. The logical instructions are:

AND	Logical AND
ANDI	AND immediate
OR	Logical OR
ORI	OR immediate
EOR	Exclusive OR
EORI	Exclusive OR immediate
NOT	Logical complement

Shift and rotate operations may be performed on 8-, 16-, and 32-bit data values contained in data registers or memory. The shift and rotate instructions are:

ASL	Arithmetic shift left
ASR	Arithmetic shift right
LSL	Logical shift left
LSR	Logical shift right
ROL	Rotate left
ROR	Rotate right
ROXL	Rotate left with extend
ROXR	Rotate right with extend

The bit manipulation group works only with 8- and 32-bit data types. The data to be manipulated may be in a data register or memory. Individual bits may be tested, set, cleared, or complemented. The bit manipulation instructions are:

BCHG	Bit change
BCLR	Bit clear
BSET	Bit set
BTST	Bit test

BCD operations are implemented with instructions that add, subtract, and negate (10's complement) 8-bit data values (two BCD digits). Data registers or memory may be used to hold the data. The BCD instructions are:

ABCD	Add BCD
NBCD	Negate BCD
SBCD	Subtract BCD

The program control group makes use of the processor condition codes in its branch and set-byte instructions; it also contains unconditional branches and jumps. Subroutine call and return instructions are also included in this group. The program control instructions are:

Bcc	Conditional branch
DBcc	Decrement and branch
Scc	Conditional set
BRA	Branch always

Example 3.1:   If register D3 contains 100030FF and register D4 contains 8E552900, what is the result of MOVE.W D3,D4?

Solution: Since a word operation is specified, the upper half of D4 will not be changed. The lower half will be loaded with the lower half of D3. Thus, D3 will remain unchanged and D4 will contain 8E5530FF.

Address Register Direct

This addressing mode uses an address register as the operand.

Example 3.2:   What will A2 contain after execution of MOVEA.L   A5,A2?

Solution: A2 will contain the same value as register A5, since a 32-bit transfer was specified by the .L suffix on the instruction.

Whenever an address register is used as a destination, the size of the operation must be .W or .L. When .W is used, the address register will be loaded with a **sign extended** value. Sign extension is performed as follows: If the most-significant bit of the word value is low (word values 0000 through 7FFF), the upper 16 bits of the address register will be cleared. Thus, if A0 contains the address 00006800, the instruction MOVEA.W   A0,A1 will put 00006800 into A1. However, if the most-significant bit is high (word values 8000 through FFFF), the upper 16 bits of the address register will be set. To illustrate, if A0 contains 0000C580, the instruction MOVEA.W   A0,A1 will place FFFFC580 into A1. This convention is known as **short addressing,** and allows two 32KB blocks of memory to be accessed easily. The first block is from 000000 to 007FFF. The second block is from FF8000 to FFFFFF. Notice that these two ranges are the only ones available when sign extension is used. This is a useful addressing mode when the system is small and requires only a small amount of RAM and EPROM, such as an electronic gas pump, an automated vending machine, or an assembly-line controller.

Memory Addressing Modes

The next five addressing modes are used to access data that resides in memory, not one of the processor's internal registers. These addressing modes will all be a variation of *indirect* addressing, where we use an address register to point to a particular memory location. This is important when we must access data stored in memory in particular locations. For example, consider the small database of telephone numbers shown in Figure 3.4. The database is composed of four **records,** with each record specifying a person's first name (up to eight letters), a seven-digit phone number, and an end-of-database flag. Thus, each record consists of three **fields,** and each field has a fixed length.

Notice that the label TOP has been associated with the address of the first byte in the first record. Since each record has a fixed size of 16 bytes, the starting address of any

Address   Memory contents (each box is a byte)

TOP	'A'	'L'	'A'	'N'					5	5	5	1	2	1	2	0
TOP+16	'C'	'H'	'A'	'R'	'L'	'I'	'E'		5	5	5	2	6	6	3	0
TOP+32	'M'	'I'	'C'	'H'	'E'	'L'	'E'		5	5	5	8	9	7	7	0
TOP+48	'J'	'A'	'M'	'E'	'S'				5	5	5	0	1	4	3	1

|◄─────────── Name field ───────────►|◄─────── Number field ───────►| | |

End of database flag ─┘

FIGURE 3.4   Small database of telephone numbers

record in the database will always equal the address of TOP plus some multiple of 16. By a similar fashion, the starting address of each field within a record can also be determined. For example, ALAN's phone number begins at address TOP + 8 and ALAN's end-of-database flag is stored at address TOP + 15. To access data within the database, use any of the five addressing modes covered next.

Address Register Indirect

In this addressing mode an address register holds the address of the memory location that contains the operand data. The assembler will recognize this addressing mode whenever the address register is surrounded by parentheses, as in (A0), (A3), and (A7).

Example 3.3:   If A0 contains 00007F00, what happens when MOVE.B   (A0),D7 is executed?

Solution: The contents of memory location 007F00 are copied into the lower byte of D7. The upper 24 bits of D7 remain unchanged. Figure 3.5 illustrates this principle, assuming that D7 initially contains 1234FEDC. Note that even though address registers are 32 bits long, only the lower 24 are used to address memory in the 68000. There are no external address lines for the upper eight address bits!

FIGURE 3.5   For Example
3.3, address register indirect

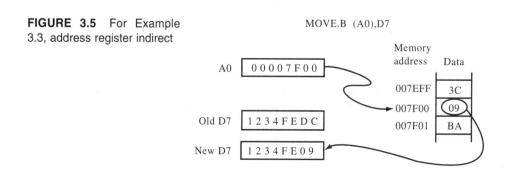

Regarding Figure 3.4, if the address of TOP is 00008600, we can use these two instructions to read the first letter of ALAN's name into data register D0:

```
MOVEA.L    #$8600,A0
MOVE.B     (A0),D0
```

Similarly, if we use:

```
MOVEA.L    #$8608,A0
MOVE.B     (A0),D0
```

we will read the first digit of ALAN's phone number into D0.

Address Register Indirect with Postincrement

This addressing mode works in much the same way as address register indirect, except that the address register is incremented after the data transfer is accomplished. The 68000 will automatically increment the address register by 1, 2, or 4, depending on the data type specified. The assembler will use this addressing mode if the indirect register notation is followed by a plus sign, as in (A0)+, (A4)+, and (A6)+.

Example 3.4:   If A5 contains the value 00007F00 and D2 contains 4E4F2000, what will each register contain after MOVE.W   (A5)+,D2 executes? See Figure 3.6 for more details.

FIGURE 3.6 For Example 3.4, address register indirect with postincrement

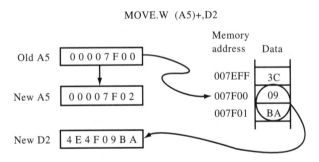

Solution: The upper word of D2 will remain unchanged and the lower word will be replaced by 09BA. Since this is a word operation, A5 will be incremented by 2, making its value after the instruction 0007F02.

Regarding Figure 3.4, suppose we want to read the letters of ALAN's name one at a time. If address register A0 is initialized to point to the first letter in ALAN's name (address 00008600), the instruction:

```
MOVE.B     (A0)+,D0
```

will get the first letter, and then advance A0 to the address of the second letter. When the MOVE.B   (A0)+,D0 instruction is executed a second time, we get the second letter, and A0 is advanced to the address of the third letter. If MOVE.B   (A0)+,D0 is executed eight times, we will have completely read ALAN's name and advanced A0 to the next field in the database (the address of the first digit of ALAN's phone number).

Address Register Indirect with Predecrement

The operand in this addressing mode is found by first decrementing the specified address register by 1, 2, or 4, depending on the data type involved. The address register is then used to point to the memory location that contains the operand data. Precede indirect register notation with a minus sign to specify this addressing mode, as in –(A2) and –(A7).

Example 3.5:   A2 and D4 initially contain 00007F00 and F3052BC9. What will they contain after execution of MOVE.B   –(A2),D4? See Figure 3.7 for details.

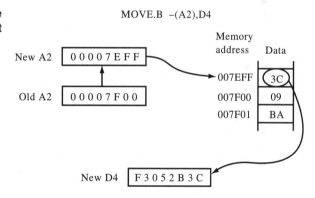

FIGURE 3.7   For Example 3.5, address register indirect with predecrement

Solution: Since a byte transfer is called for, A2 is decremented by 1, making its new value 00007EFF. Data is then copied from memory location 007EFF into the lower byte of D4.

These two addressing modes, address register indirect with postincrement and predecrement, are useful for maintaining stacks or queues in memory. When an address register is used as a stack pointer, it will be incremented or decremented by 2 even if byte addressing is specified, so that the processor can maintain an even address within the register.

Address Register Indirect with Displacement

In this addressing mode, the operand address is found by adding a 16-bit signed displacement value to the address register, and then using this result as the operand address. The 16-bit signed displacement may range from –32768 to 32767. The contents of the address register remain unchanged. In general, the assembler recognizes $d_{16}(An)$ as the form specifying this mode, although other assemblers may require (d_{16},An) instead.

Example 3.6:   Registers A0 and D0 initially contain 00007F00 and 02040608. What will D0 contain after MOVE.W $100(A0),D0 executes? Use Figure 3.8 for the operand information you will need.

FIGURE 3.8   For Example 3.6, address register indirect with displacement

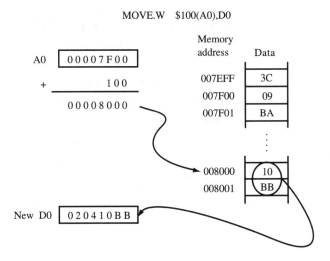

Solution: The operand address is found by adding 100 to A0, giving 00008000. This becomes the address of the operand in memory. Locations 008000 and 008001 contain the data that will replace the lower word of D0.

We can use this addressing mode to quickly access the first byte of information in each field of our sample database. Refer to Figure 3.4 again. The following instructions can be used to read the first name character, the first phone number digit, and the end-of-database flag from any record, assuming that A0 already contains the address of the first byte in the record:

```
MOVE.B    0(A0),D0        ;name byte
MOVE.B    8(A0),D1        ;phone number digit
MOVE.B    15(A0),D2       ;end flag
```

Address Register Indirect with Index

This addressing mode combines many of the ones we have just seen. The operand address is specified by the sum of an address register, a signed index register (which may be an address *or* a data register), and a signed 8-bit displacement. This particular addressing mode is very useful for implementing two-dimensional arrays in memory. A two-dimensional array can be thought of as a matrix, containing a number of rows and columns. A technique is used to combine the row and column numbers into a unique index into the data area. The required assembler syntax is either $d_8(An,Xn)$ or (d_8,An,Xn) where Xn is an address or data register.

Example 3.7: If address register A0 contains 00007F00 and data register D4 contains 00000100, what will D3 contain after MOVE.L   2(A0,D4.W),D3 executes? See Figure 3.9 for more details.

FIGURE 3.9 For Example 3.7, address register indirect with index

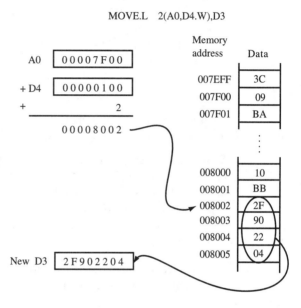

Solution: The operand address is found by adding 007F00 + 100 + 2, giving 008002 as the final location. Data is then read from locations 008002 through 008005 to complete the 32-bit transfer.

Let us take one last look at Figure 3.4. We are already familiar with the ability to use a displacement value to access any individual component (byte) of a record. For example, to access the first byte of the phone number field, we use a displacement of 8, as in the instruction MOVE.B   8(A0),D0. Remember that A0 contains the address of the first byte in the record.

If we now add an index register to the instruction, we get something like:

```
MOVE.B   8(A0,D1.W),D0
```

which allows the integer value in the lower word of D1 to specify which phone number digit will be read. If D1 equals 0, we get the first digit. If D1 equals 1, we get the second digit, and so on. We are now treating the phone number field as an **array,** a sequential group of storage locations. Can you imagine how this addressing mode might be used for other types of records, such as an array of longwords?

Absolute Short Address

This addressing mode is used to directly access data in two specific ranges of memory. Both memory spaces are 32KB in length. The first memory space occupies addresses 000000 through 007FFF. The second memory space occupies locations FF8000 through FFFFFF. Addresses in either space may be specified by a single 16-bit quantity. The 68000 will automatically *sign extend* the 16-bit value into 24 bits, the size required for an address. If the address supplied in the instruction is between 0000 and 7FFF, the MSB (most-significant bit) is zero. The 68000 extends this zero into the upper 8 address bits. If the address is between 8000 and FFFF, the processor will fill the upper 8 bits with 1s (because the MSB is now high). A direct benefit of this addressing mode is that less machine code is required to fetch the operand address, thus saving valuable storage space and decreasing execution time. Recall that address register direct also uses this type of sign extension.

Example 3.8:   What memory locations are accessed when these two instructions are used: MOVE.B $3C00,D1 and MOVE.W $9AE0,D2?

Solution: The sign extended address for the first instruction is 003C00. The byte stored at this location is copied into the lower 8 bits of register D1. The sign extended address for the second instruction is FF9AE0. The bytes at this location and the one at FF9AE1 are copied into the lower 16 bits of D2.

Notice that the physical memory address is directly specified in the instruction, and *not* contained in an address register.

Absolute Long Address

The difference between this addressing mode and absolute short address is that no sign extension takes place. The full 24-bit address is included in the machine code (by using two additional words after the opcode). Any address within the range 000000 to FFFFFF may be used with this instruction. For example, MOVE.B $2E000,D0 would cause the byte stored at location 02E000 to be copied into the lower 8 bits of register D0. This addressing mode increases execution time, because of its 6-byte machine code length (2 bytes for the opcode and 4 bytes for the address), so it should be used sparingly.

Program Counter with Displacement

This addressing mode uses the 68000's program counter, together with a signed 16-bit displacement, to form the operand address. The signed 16-bit displacement allows accessing of memory 32,768 locations behind and 32,767 locations forward of the program counter. While the 68000 uses this addressing mode to implement branch instructions, it is also useful for memory references as well. It should be possible to place the program's data area within the displacement range, thus avoiding the need for other types of addressing modes to access the data. The required assembler syntax is either $d_{16}(PC)$ or (d_{16},PC).

Example 3.9: Figure 3.10 shows how this addressing mode might be used to read data from a memory location. In this example, a data byte pointed to by the label DATA is referenced in the MOVE instruction. The actual machine code for MOVE.B DATA(PC),D4 is 183A   001C, where 183A is the opcode that instructs the processor to MOVE.B into D4 using program counter with displacement addressing mode, and 001C is the displacement. D4 initially contains 00000000.

   During assembly, the value of DATA and the program counter will be subtracted. The result becomes the displacement used in the machine code. During program execution, the displacement value is added to the current program counter (000902) to determine the operand address. It is important to note that the program counter is advanced by 2 after each instruction fetch. This accounts for the difference between the instruction address (000900) and the current program counter (000902).

FIGURE 3.10 For Example 3.9, program counter with displacement

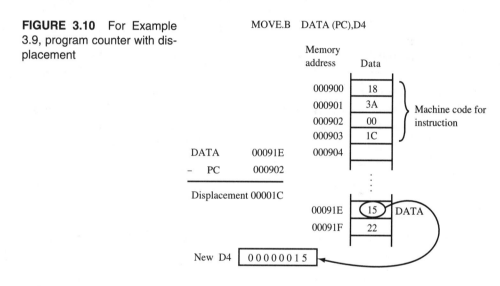

Using this addressing mode allows the program to be placed anywhere in memory and still access the data correctly. This is because the displacement value specifies how far away the data is from the current address. So, as in Example 3.9, the data will always be located 1C locations away from the MOVE.B   DATA(PC),D4 instruction that reads it. This type of programming leads to programs that are **relocatable.** The program has the ability to load and execute correctly at any address, no matter what ORG value was used during assembly.

Program Counter with Index

There are two differences between this addressing mode and the previous one. First, the signed displacement is now only 8 bits wide. Second, any address or data register may be used as an index register. The operand address is found by adding the current program counter, the signed displacement, and the signed index register.

Example 3.10: What is the address of the operand for MOVE.W   DATA(PC,A0.L),D1? The instruction is located at address 002044, register A0 contains 00000002, and the displacement is 10.

Solution: Adding the displacement and the current program counter gives us 002056. Remember that the program counter is always advanced by 2 after the instruction fetch: Adding the index register (A0) value to 002056 gives 002058. This is the operand address.

The signed displacement gives a −128/+127 location range and the signed index register gives a range that exceeds 8 million bytes in either direction. Another example will show how data areas located *before* the instruction may be accessed.

Example 3.11: What is the operand address for MOVE.B   SINE(PC,D6.L),D0? The instruction is located at address 001058, D6 contains 00000004, and the displacement is F2.

Solution: Adding the displacement to the current program counter gives 00104C. Note that this address is smaller than the address of the instruction. The signed displacement of F2 has caused this. Adding the index register (D6 this time) to 00104C gives 001050 as the operand address. Figure 3.11 summarizes this address calculation.

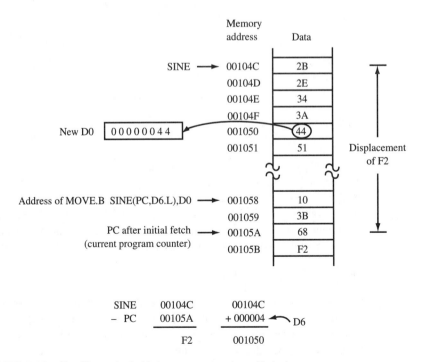

FIGURE 3.11   For Example 3.11, program counter with index

If all this appears a bit tricky, remember that it is done automatically by the 68000. You need only write the source code, making sure that the data area, SINE in this case, is not located outside the –128/+127 location range.

Immediate Data

It is possible to include the operand value within an instruction. This form of addressing is useful for loading constant values into registers. Immediate data is preceded by a # sign for recognition by the assembler. Without it, the assembler would not be able to tell the difference between immediate data and absolute short or long addressing.

Example 3.12: Data register D5 initially contains 12345678. What are its contents after the following instructions are executed?

a) `MOVE.B    #$3A,D5`
b) `MOVE.W    #$9E00,D5`
c) `MOVE.L    #1,D5`

Solution: The byte transfer in (a) only affects the lower 8 bits of D5. Thus, D5 contains 1234563A after execution. The word transfer in (b) causes D5 to become 12349E00. The longword transfer in (c) loads D5 with 00000001.

Quick Immediate Data

This addressing mode is similar to the previous one, except that only byte values may be specified in the instruction. These byte values are sign extended to 32 bits before use. The advantage of this addressing mode is that only 2 bytes are needed for the entire instruction! An ordinary MOVE.L immediate instruction requires 6 bytes instead.

Example 3.13: What will register D3 contain after execution of

a) `MOVEQ   #$2C,D3`
b) `MOVEQ   #$8F,D3`

Solution: Since the MSB of the immediate data in (a) is zero, D3 is loaded with 0000002C. In (b) the MSB is high, indicating a signed number. D3 is thus loaded with FFFFFF8F.

Implied Addressing

The last of the 14 addressing modes is implied addressing. In this mode the instruction makes a reference to a processor register within its mnemonic. For example, ANDI #$27,SR (AND immediate data to the status register), MOVE   CCR,CODES (move a copy of the condition code register to memory), and TRAPV (TRAP if the overflow bit is true) all make reference to processor registers or flags within their mnemonics.

In the next section we will see how these addressing modes are used with the 68000's many instructions.

3.5   THE 68000'S INSTRUCTION SET

As with any microprocessor, a detailed presentation of available instructions is important. Unless you have a firm grasp of what can be accomplished with the instructions you may use, your programming will not be efficient. Indeed, you may even create problems for yourself.

Still, there is no better teacher than experience. If you have a 68000-based computer, you should experiment with these instructions and examine their results. Compare what you see with the manufacturer's data. A difficult concept often becomes clear in practice.

Each of the following sections deals with a separate group of instructions. Information about the instruction, how it works, how it is used, what its mnemonic looks like, how it affects the condition codes, and more will be presented for each instruction. Even so, it is strongly suggested that you constantly refer to Appendix B as you read about each new instruction. Most of the material in this appendix, such as allowable addressing modes and condition code effects, is not reproduced here.

In some cases the machine code for the instruction will be included. This is for no purpose other than to compare instruction lengths and explain new features about the 68000.

Examples will also be given for each instruction.

The Condition Codes

Most 68000 instructions affect the state of the five flags that make up the condition code register. The flags and their meanings are as follows:

> *N: The Negative Flag.*   The **negative** flag is set if the MSB of the result is set, and cleared otherwise. Note that the actual MSB used depends on the size of the operand involved. Byte operations use bit 7. Word operations use bit 15, and longword operations use bit 31.

Example 3.14:   Consider the following two pairs of instructions:

```
MOVE.B  #$3F,D0        MOVE.B  #$7F,D0
ADDI.B  #1,D0          ADDI.B  #1,D0
```

In each case, the final value in D0 will be interpreted as a *signed* binary number.

In the first pair of instructions, the value 3F is incremented to 40, which is 01000000 in binary. Notice that the MSB (bit 7) is zero, indicating that 40 is a *positive* number. The negative flag will be cleared in this case.

In the second pair of instructions, the value 7F is incremented to 80, which is 10000000 in binary. The MSB is now 1, indicating that 80 is a *negative* number. The negative flag will be set by this result.

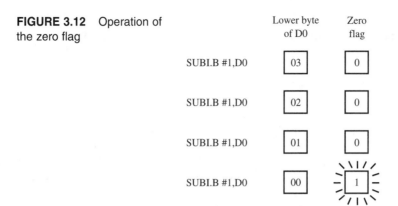

FIGURE 3.12   Operation of
the zero flag

Z: The Zero Flag.   The **zero** flag is set if the result of an operation equals zero, and
cleared otherwise. As shown in Figure 3.12, the lower byte of data register D0 begins
at the value 3, and is decremented until it becomes zero. Only then does the zero flag
become set.

Example 3.15:   From a hardware perspective, it is not difficult to determine if a group
of bits are all zero. As Figure 3.13 shows, a NOR gate is used to generate an output signal
called ZERO. Recall that a NOR gate outputs a one when all of its inputs are zero. Think
of the eight NOR inputs as if they were connected to the individual lower 8 bits of D0.
When these 8 bits become 00000000, ZERO will go high. Any other 8-bit pattern in the
lower byte of D0 will cause ZERO to be low (indicating a *not-zero* condition). Can you
imagine what is needed to check for zero in a 32-bit register?

FIGURE 3.13   Hardware generation of zero condition

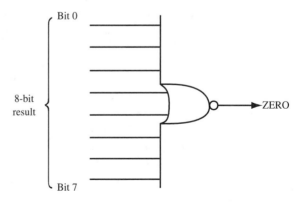

V: The Overflow Flag.   The **overflow** flag is set whenever a result represents the
possibility of a sign change. This is illustrated in the following example.

Example 3.16:   Consider the following sequence of instructions:

```
MOVE.B  #$77,D0      ;load lower byte of D0 with 77.
ADDI.B  #3,D0        ;D0 = 7A now. V is cleared.
ADDI.B  #9,D0        ;D0 = 83 now. V is set by sign change.
SUBI.B  #1,D0        ;D0 = 82 now. V is cleared.
SUBI.B  #4,D0        ;D0 = 7E now. V is set by sign change.
```

Recall that the MSB of an operand may be interpreted as a sign bit, where 0 means positive and 1 means negative. When 3 is added to D0, the result 7A still has its MSB low. The result is still positive. When 9 is added next, the result 83 has its MSB set. This indicates a negative result (if we want to interpret it that way) and thus a **sign change** has just occurred. The overflow flag is set to indicate this.

When 1 is subtracted from D0, the result 82 still has its MSB set, so no sign change occurred. But when 4 is subtracted next, the final result 7E has gone back to being positive (the MSB is zero). So we have another sign change if we are interpreting the lower byte of D0 as a signed integer.

C: The Carry Flag.   The **carry** flag is the carry out of the MSB of the result during an addition.

Example 3.17:   The carry flag is also used to indicate a *borrow* as a result of a subtraction. Consider these two pairs of instructions:

```
MOVE.B   #6,D0                MOVE.B    #6,D0
SUBI.B   #1,D0                SUBI.B    #9,D0
```

Remember that subtraction in binary is found by adding the 2's complement of one number to another.

In the first pair of instructions, subtracting 1 from 6 leaves 5. The carry flag is cleared in this case, since we subtracted a smaller number from a larger one.

In the second pair of instructions, subtracting 9 from 6 gives FD (or 11111101 in binary, the 2's complement representation of –3). Since we subtracted a larger number from a smaller one, the carry flag will be set, indicating a borrow.

X: The Extend Flag.   The **extend** flag is not directly set by the results of an operation. Many instructions have no effect on the extend flag, and those that do usually set it the same as the carry flag.

The condition codes contain valuable information concerning the operation of a program, on an instruction-by-instruction basis. Thus, you should make good use of the flags when writing programs. Employ the conditional branch instructions where possible, and pay attention to how the flags are affected by all instructions in your program. Sometimes a well-written program that appears completely logical in its method will still yield incorrect results because a flag condition was overlooked.

Keep the condition codes in mind as you study the remaining sections.

Using Motorola's Instruction Descriptions

In Appendix B you will find descriptions of all instructions supported by the 68000. These descriptions are useful for many different reasons. Let us examine one of them in detail so that you will know what to look for as you learn about the instruction set.

Figure 3.14 shows a copy of the ADDI instruction description found in Appendix B. Its various components are described as follows:

1. The instruction mnemonic ADDI is given, with a brief description (Add Immediate) of what it stands for.

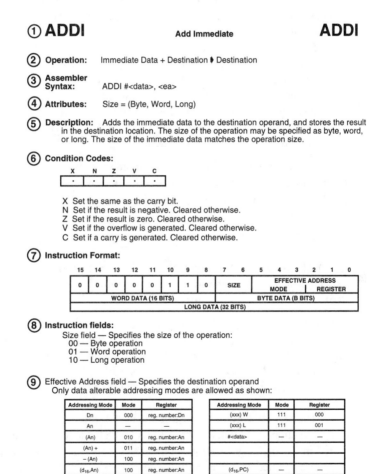

FIGURE 3.14   ADDI instruction description

2. The operation performed by the instruction is shown here.
3. The required assembler syntax is given as ADDI #<data>,<ea>, where the # sign indicates immediate data, <data> is the actual data value (the source operand), and <ea> is the **effective address,** a general term for the memory location specified by the destination addressing mode.
4. The allowed data sizes are given here. ADDI supports .B, .W, and .L data sizes.
5. A more detailed description of what the processor does when executing the ADDI instruction is given here.
6. The effect of the ADDI instruction on the five condition codes is given here. All condition codes are updated when ADDI executes.
7. The format of the ADDI instruction is shown here, indicating what specific bit values are used to represent the instruction in binary. Notice that information from items 8 and 9 is required to fill in the missing bit values in the lower byte of the instruction word. The instruction word is followed by one or two more words of data, which represent the immediate data given in the instruction.
8. The two bits specifying the ADDI instruction's data size are defined here.
9. The table shown here indicates what destination addressing modes are allowed. For example, data register direct (D*n*) is allowed, but address register direct (A*n*) is not. This table is very useful for debugging program code when an illegal addressing mode has been used with a particular instruction.

Many instructions allow multiple source addressing modes as well. A similar table is given for allowed source addressing modes in those instructions (look at MOVE as an example). ADDI does not supply this table, since only one source addressing mode is allowed: immediate.

Example 3.18: Let us combine all of the information just presented into a practical example. Suppose that the following two words are fetched by the 68000 for decoding: 065F 1234. What does the processor do?

Solution: The 06 portion of the first word indicates that the ADDI instruction is being used. Close examination of the 5F portion of the first word gives the following bit patterns: 01—011—111 (simply 5F in hexadecimal). The first two bits (01) specify .W as the data size. The next three bits (011) specify (A*n*)+ as the addressing mode (address register indirect with postincrement), and the last three bits (111) indicate that A*n* is actually A7.

The second word of the instruction (1234) represents the immediate data. So, the entire instruction is ADDI.W   #$1234,(A7)+. The processor adds 1234 to the data word stored in the memory locations pointed to by A7, and then adds 2 to A7. The condition codes are updated as indicated in the description.

The Data Transfer Group

The first instruction in this group is EXG (exchange registers). Any of the 16 general-purpose data or address registers may be exchanged. The exchange is always a 32-bit transfer. None of the condition codes are affected.

Example 3.19: How can registers D3 and D5 be swapped?

Solution: Either EXG D3,D5 or EXG D5,D3 may be used to swap these registers. Swapping address registers A3 and A5 could be accomplished by EXG A3,A5.

The second instruction in the data transfer group is LEA (load effective address). It is used to load a 24-bit address into an address register. Whenever an addressing mode is used to compute the address of an operand, the result is called the **effective address.** Usually this address is used internally by the processor and then forgotten. LEA gives us a way to obtain the effective address used in an instruction. No condition codes are affected.

Example 3.20: What are the effective addresses for each of these instructions?

a) LEA $8500,A1
b) LEA $10(PC),A1

Solution: In (a), the address is sign extended into 32 bits, giving A1 the value FFFF8500. In (b), A1 is loaded with a value equal to $10 plus the current program counter.

The third instruction in the data transfer group is LINK (link and allocate). This instruction is used to allocate stack space and implement linked-lists. An example of this instruction might be LINK A3,#$20. The operation of LINK is as follows: The specified address register is pushed onto the stack. Next, the stack pointer is copied into the specified address register. Then the signed 16-bit displacement is added to the stack pointer. No condition codes are affected.

Example 3.21 Address registers A0 and A7 contain 00006200 and 0000FFC4, respectively. What will their values be after execution of LINK A0,#$FFF0?

Solution: In any stack push, the stack pointer is first decremented. Since an address register value is being pushed, the stack pointer is decremented by 4, making A7 equal to 0000FFC0. A0 is then written into memory locations 00FFC0 through 00FFC3. Next, A0 is loaded with 0000FFC0, the current stack pointer. Then the signed displacement (FFF0) is added to the stack pointer, giving 0000FFB0 as the final stack pointer (the value in A7). Note: The processor uses A7 by default in stack operations.

The next instruction in the data transfer group is MOVE (move data). This instruction is used to move byte, word, and longword data between data registers, address registers, and memory.

Examples of the MOVE instruction are:

```
MOVE.B    #$29,D3
MOVE.W    D3,D6
MOVE.L    (A0)+,D0
```

Many more MOVEs are possible because this instruction supports all addressing modes in the source field. Since this instruction was used in many different ways in Section 3.4 to illustrate addressing modes, no additional examples will be presented here. Instead, we will examine the effect on the condition codes during a MOVE instruction.

Example 3.22:   What are the states of the five condition codes after execution of MOVE.B   #$86,D2?

Solution: Since the MSB of the immediate data ($86) is high, the data is interpreted as negative, and N becomes 1. Z is cleared because the data is not zero. Both V and C are also cleared, and X remains unchanged.

The next instruction in the data transfer group is MOVEA (move address). The destination operand in this instruction is always an address register. Word and longword sizes may be used. When the data size is word, the source operand is sign extended into 32 bits before being placed into the address register. For example, MOVEA.L   #$9F00,A2 causes 00009F00 to be placed into A2, and MOVEA.W   #$9F00,A2 causes FFFF9F00 to be stored. Recall that a negative number, represented via 2's complement, has its MSB set. No condition codes are affected by MOVEA.

The next instruction in the data transfer group is MOVEM (move multiple registers). This instruction is used to transfer data and address registers to and from memory. Only word and longword operations are allowed. When a word operation is specified during a read from memory, the word data is sign extended before being loaded into the register. For example, if 3500 is read from memory, the register is loaded with 00003500. Likewise, if AF10 is read, the register is loaded with FFFFAF10.

Two addressing modes commonly used with MOVEM are postincrement and predecrement. Postincrement addressing is used for transferring data from memory to specified registers. Predecrement addressing may only be used for register-to-memory transfers. These two addressing modes may be used to implement a stack with MOVEM. Since the order of reading and writing are reversed in a stack operation (last in becomes first out), predecrement addressing may only be used when writing register data to memory, and postincrement for reading.

No matter what addressing mode is used, the user must specify a list of registers to be transferred or loaded. Transfer of individual registers is accomplished by separating the registers by a /. For example, registers D0, D2, D3, D5, A4, and A6 may be specified by D0/D2/D3/D5/A4/A6 in the source statement. Using A6/D5/D0/A4/D3/D2 instead will do the same job, regardless of the different order. The 68000 writes from A7 to A0, and then from D7 to D0, and reads in reverse order (D0 through D7, then A0 through A7). D0 occupies the lowest memory location, no matter how the order is specified by the programmer. A7 is stored in the highest memory location.

Specifying a sequence of registers is done differently. Suppose that D0 through D4 and A2 through A5 are to be transferred. All nine registers may be easily listed by using D0–D4/A2–A5 in the source statement. Once again the order is unimportant. No condition codes are affected by this instruction.

Example 3.23: Suppose that data registers D0 through D3 are loaded as follows: D0: 55556666, D1: 77778888, D2: 9999AAAA, D3: BBBBCCCC, and that TAB1 refers to a data area in memory that is located at address 0030B8. What data is written into memory if MOVEM.W   D0–D3,TAB1 is executed?

Solution: The indicated registers are written into memory starting at address 0030B8. Since the data size is word, the processor uses two locations for each register value. Figure 3.15 shows the contents of memory after execution. Notice that only the lower half of each register has been transferred.

FIGURE 3.15 For Example 3.23

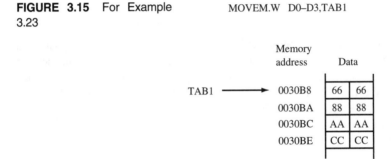

Example 3.24: This is a repetition of Example 3.23 except that the instruction executed is now MOVEM.L   D0–D3,TAB1. What are the contents of memory this time?

Solution: Figure 3.16 shows the new results. Since the data size is longword, twice as many memory locations are used.

FIGURE 3.16 For Example 3.24

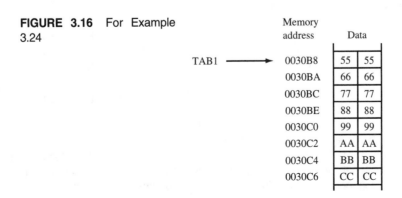

Exactly the same results may be accomplished by the following sequence of instructions:

```
MOVEA.L  #TAB1,A0
MOVEM.L  D0-D3,(A0)
```

Here the 68000 will automatically increment A0 by 4 during execution of MOVEM. On completion, A0 is restored to its original value (that of TAB1).

Another method that will yield the same results is to use predecrement addressing. Since memory locations will be used in a decreasing direction, the address register must initially point to the *end* of the reserved data area. The following two instructions will do the same job as the previous ones:

```
MOVEA.L  #$30C8,A1
MOVEM.L  D0-D3,-(A1)
```

Even though D0 is specified first in the instruction mnemonic, the 68000 automatically writes D3 into memory first, having already decremented A1 by 4. When done, A1 will contain 000030B8.

Example 3.25:   The previous two examples involved writing data into memory. This example illustrates how MOVEM reads data *from* memory. Consider a data area, called DATA, set up as shown in Figure 3.17. Data registers D0 and D1 may be loaded via MOVEM.L   DATA,D0/D1. In this case, D0 will contain 11112222, and D1 will contain 88889999. These two registers may be loaded the same way by using this sequence of instructions:

```
MOVEA.L  #DATA,A0
MOVEM.L  (A0),D0/D1
```

or by using

```
MOVEA.L  #DATA,A0
MOVEM.L  (A0)+,D0/D1
```

The use of postincrement addressing in the last two instructions causes A0 to be incremented by 4, twice.

FIGURE 3.17 For Example 3.25

Four data registers may be loaded from the same set of data by using a word size instruction: MOVEM.W   DATA,D0–D3. Remember that each 16-bit value now read from memory will be sign extended into 32 bits. The data registers will now be loaded as follows: D0: 00001111, D1: 00002222, D2: FFFF8888, and D3: FFFF9999.

D0 through D3 can also be loaded by either of these two pairs of instructions:

```
MOVEA.L   #DATA,A0
MOVEM.W   (A0),D0-D3
```

or

```
MOVEA.L   #DATA,A0
MOVEM.W   (A0)+,D0-D3
```

with both pairs causing A0 to be incremented by 2 a total of four times.

The next instruction in the data transfer group is MOVEP (move peripheral data). When a system's hardware has been designed so that all peripherals communicate with only one-half of the system data bus, this instruction becomes useful. Systems such as these may very well have been leftovers from the 8-bit era that were upgraded to 16-bit architecture.

MOVEP accesses memory in groups of two locations at a time. For example, if an even address is specified in the instruction, only even locations will be accessed. Only word and longword data sizes may be used with MOVEP, and no condition codes are affected. Furthermore, only register indirect with displacement addressing is allowed.

Example 3.26:   What memory locations are affected by MOVEP.W   D2,0(A1)? The contents of D2 are 12345678, and A1 contains 00045000.

Solution: Two memory locations will be accessed since word size is indicated in the instruction. The two locations begin at the address pointed to by the contents of address register A1 plus the displacement. The zero displacement causes the first location to be 045000. The data byte written to this location is 56, obtained from bits 8–15 of D2. The second location accessed is 045002 (the next even address). Bits 0–7 of D2 (78) are written into this location.

If the instruction had been coded as MOVEP.W   0(A1),D2, data from locations 045000 and 045002 would have been loaded into the lower word of D2, with location 045000's data going into bits 8–15, and 045002's into bits 0–7.

The next instruction in the data transfer group is MOVEQ (move quick). This instruction is only used to move 8 bits of data into a data register. Before the transfer is made, the immediate data is sign extended into a full 32-bit value. All condition codes except X are affected. The advantage of MOVEQ over MOVE is that MOVEQ requires fewer bytes of machine code. Consider the following example: MOVE.L   #36,D0 assembles into 6 bytes of machine code: 20 3C 00 00 00 36. The 20 3C bytes represent the instruction and the other 4, the data. MOVEQ   #$36,D0 is much more efficient, assembling into only 2 bytes of machine code: 70 36. The MOVEQ into D0 is represented by 70, and the data by 36. Good assemblers will examine the source instruction and use the more efficient form (MOVE or MOVEQ) where possible. Although most assemblers are two-pass assemblers, there are also multipass optimizing assemblers that are capable of performing other types of improvements (loop unrolling, addressing mode substitutions) that may optimize for speed or code size.

Example 3.27: What does register D4 contain after execution of MOVEQ   #$B7,D4?

Solution: Since B7 has its MSB set, the 68000 assumes it is a negative number and loads D4 with FFFFFFB7.

The next instruction in the data transfer group is PEA (push effective address). The effective address of the operand is computed and translated into a 32-bit value before being pushed onto the processor's stack. No condition codes are affected.

Example 3.28: What is the effective address pushed onto the stack during execution of PEA $40(A5)? Address register A5 contains 00003060.

Solution: Adding 40 to 003060 gives 0030A0. This is the effective address computed by the 68000. Translation into 32 bits yields 000030A0, the 4 bytes pushed onto the stack.

The next instruction in the data transfer group is SWAP (swap register halves). The upper and lower words of a data register are swapped. No condition codes are affected.

Example 3.29: If D5 contains 3CFF9100, what are its contents after execution of SWAP D5?

Solution: Swapping the halves of D5 results in 91003CFF.

The last instruction in the data transfer group is UNLK (unlink). It is used to complement the operation LINK. The value contained in the address register specified in the instruction is copied into the stack pointer. The longword on top of the new stack is then popped and placed in the specified address register. No condition codes are affected.

Example 3.30: Address register A2 contains 0009FFB4. What occurs when UNLK A2 executes?

Solution: The processor stack pointer is loaded with 0009FFB4. The longword stored at locations 0009FFB4 through 0009FFB7 is then copied into A2. The final stack pointer is 0009FFB8.

The Arithmetic Group

The first instruction in this group is ADD (add binary). It is used to add 8-, 16-, and 32-bit values. A data register must be specified as the source or destination. All condition codes are affected.

Example 3.31: If D2 and D3 contain 12345678 and 5F02C332, respectively, what are the results of ADD.B   D2,D3? How are the condition codes affected?

Solution: The instruction specifies that D2 should be added to D3 and the result stored in D3. Since the data size is byte, only the lower 8 bits of D3 will be affected. D2 will remain unchanged, and D3 will have a final value of 5F02C3AA. The condition codes after execution will be N: 1, Z: 0, V: 1, C: 0 and X: 0. The N flag is set because the MSB of the lower 8 bits of D3 is high.

The second instruction in the arithmetic group is ADDA (add address). This instruction is used to add data to an address register. All addressing modes may be used, but only word and longword data sizes are possible. The condition codes are not affected.

Example 3.32: What are the results of ADDA.W   A0,A3, if A0 contains CE001A2B and A3 contains 00140300?

Solution: The lower 16 bits of both address registers are added, and the result replaces the lower word of A3. A0 remains unchanged, and A3 has a final value of 00141D2B.

The third instruction in the arithmetic group is ADDI (add immediate). Byte, word, or longword values may be added to the destination operand. PC relative addressing is not allowed and the destination may not be an address register (hence the need for ADDA). All condition codes are affected.

Example 3.33: What is the difference between ADDI.B   #$10,D2 and ADDI.W #$10,D2, when D2 contains 250C30F7?

Solution: The first instruction causes only the lower 8 bits of D2 to be affected. Adding 10 to F7 gives 07 (with the C flag set). This causes D2 to become 250C3007. The second instruction uses the lower 16 bits in the addition, resulting in 250C3107 as D2's final value. The C flag is cleared in this case.

The fourth instruction in the arithmetic group is ADDQ (add quick). This instruction is identical to ADDI, except that the immediate data must be in the range 1 to 8. Adding immediate data to an address register with this instruction affects all 32 bits of the address register, no matter what data size is specified.

The fifth instruction in the arithmetic group is ADDX (add extended). When using this instruction, only two forms of addressing are allowed: data register to data register, and memory to memory using address register indirect with predecrement. The contents of the X flag are included in the addition operation. All three data sizes may be used. All condition codes are affected.

Example 3.34:   What occurs during execution of each of these instructions: ADDX.B D2,D3 and ADDX.W   –(A0),–(A1)?

Solution: In the first instruction, the lower byte of each data register, plus the X bit, are added together, and the result stored in D3. In the second instruction, both address registers are first decremented by 2. Then the data at the locations pointed to by A0 and A1 is fetched and added, including the X bit. The result is stored in the location pointed to by A1.

The next instruction in the arithmetic group is CLR (clear an operand). This instruction writes zeros into the location specified. All three data sizes may be used. All condition codes except X are affected.

Example 3.35:   What are the results of these three instructions?

a) `CLR.B   D0`
b) `CLR.W   A4`
c) `CLR.L   ARRAY`

Solution: The first instruction clears the lower 8 bits of D0. The second instruction clears the lower 16 bits of A4. The third instruction writes 4 bytes of zeros into memory starting at location ARRAY.

The next instruction in the arithmetic group is CMP (compare). This instruction is used to compare data with a data register and set the condition codes accordingly. The compare is accomplished by subtracting the source operand from the destination operand, without affecting either operand. All three data sizes may be used. All condition codes except X are affected.

Example 3.36: What is the state of the zero flag after execution of CMP.W #$29AF,D6, if D6 contains 485C29AF?

Solution: Since the immediate data and the lower word of D6 are identical, the subtraction produces zero, which causes the Z flag to be set.

The next instruction in the arithmetic group is CMPA (compare address). This instruction operates similarly to CMP, except that data is compared with an address register only. Both word and longword data sizes may be used. When a word operand is specified, it is sign extended into 32 bits before the comparison is made. All condition codes except X are affected.

Example 3.37:   What does this instruction do: CMPA.L   A2,A3?

Solution: The contents of both address registers are compared. The Z flag will be set if both registers contain identical data.

The next instruction in the arithmetic group is CMPI (compare immediate). This instruction is used to compare immediate data with the destination operand and set the condition codes accordingly. PC relative addressing may not be used to specify the destination. In addition, the destination operand may not be an address register (CMPA should be used in that case). All three data sizes may be used, and all condition codes are affected.

Example 3.38:   If A3 contains 00015030, what occurs when CMPI.W   #5,(A3) executes?

Solution: The word stored in locations 015030 and 015031 is compared with 5, and the condition codes set accordingly.

The next instruction in the arithmetic group is CMPM (compare memory). This instruction operates similarly to the other compare instructions we have seen, except that the source and destination operands must be specified using address register indirect with postincrement. All three data sizes may be used, and all condition codes except X are affected.

Example 3.39:   Two data areas of 10 bytes each reside in memory. Address registers A1 and A2 point to the first byte in each data table. How can we tell if the data tables are identical?

Solution: Since we are comparing two pieces of memory data, we should use CMPM.B (A1)+,(A2)+ to compare a byte from each data table. If we can execute the CMPM instruction 10 times and never see a zero in the Z flag (which becomes 1 when the data bytes are the same), the tables are identical.

The next instruction is DIVS (signed divide). This instruction allows a signed 32-bit number to be divided by a signed 16-bit number. The destination operand, which must be a data register, is divided by the source operand. Only the word data size may be used. After execution, the lower 16 bits of the destination contain the quotient, and the upper 16 bits the remainder. The sign of the remainder is always the same as the sign of the dividend, unless the remainder is equal to 0. Division by 0 causes a special exception to be generated (see Chapter 4). All condition codes except X are affected.

Example 3.40:   Four data registers are loaded in the following way:

D2:	FFFFFC18	(–1000 decimal)
D3:	000186A0	(100000 decimal)
D4:	000001F4	(500 decimal)
D5:	000009C4	(2500 decimal)

What are the results of DIVS   D2,D3 and DIV   D4,D5?

Solution: The first instruction divides D3 by D2 and stores the result in D3, whose final value is 0000FF9C. The upper word is zero because the two registers divide evenly. The lower word (FF9C) represents –100 decimal in 2's complement notation.

The second instruction divides D5 by D4, storing the result in D5. After execution, D5 contains 00000005. Again the upper word is zero because the two registers divide evenly. Since both registers contained positive numbers, the result (5) is also positive.

The next instruction in the arithmetic group is DIVU (unsigned divide). This instruction is almost identical to DIVS. The difference is that the operands are treated as unsigned binary numbers. The condition codes are affected in the same way as they are for DIVS.

Example 3.41:   What is the result of DIVU   D4,D5 if D4 and D5 initially contain 0000019A and 0007A120, respectively?

Solution: The value 7A120 in D5 equals 500000 decimal. The value 19A in D4 equals 410 decimal. Thus, the DIVU instruction is dividing 500000 by 410. Since the registers do not divide evenly, the remainder will be placed in the upper word of D5. After execution, D5 contains 00D204C3. The 04C3 word equals 1219 decimal, which is correct. The 00D2 word equals 210 decimal, the remainder left over as a result of the non-even division. To check this for yourself, subtract 410*1219 from 500000. You will get 210.

The next instruction in the arithmetic group is EXT (sign extend). This instruction is used to extend the sign bit of a data register into the remaining upper bits of the register. Word and longword operations are allowed. When extending the sign of a byte, bits 8 through 15 will match the state of bit 7. When extending the sign of a word, bits 16 through 31 will match the state of bit 15. All condition codes except X are affected.

Example 3.42:   What is the result of EXT.W   D3, if D3 initially contains 000000C6? What if EXT.L   D3 were used instead?

Solution: The first instruction extends the sign into bits 8 through 15 only, resulting in D3 containing 0000FFC6. The second instruction causes D3 to be 000000C6.

The next instruction in the arithmetic group is MULS (signed multiply). This instruction performs a 16-bit by 16-bit signed multiplication between a data register (used as the destination) and the source operand. The lower word of the data register is used during the multiply. All condition codes except X are affected.

Example 3.43:   What is the result of MULS   D4,D5, if D4 contains 0000FFF0 and D5 contains 0000FFF6?

Solution: FFF0 represents −16 decimal, and FFF6 represents −10. The product of these two numbers is positive 160. The result placed in D5 is 000000A0 (A0 equals 160 decimal).

The next instruction in the arithmetic group is MULU (unsigned multiply). This instruction is identical to MULS, except that now the operands are treated as unsigned binary numbers. Larger products may be produced this way. The condition codes are affected the same way they are for MULS.

Example 3.44:   What will D5 contain after MULU   D4,D5 executes, if D4 and D5 initially contain 0000000A and 00000064?

Solution: D4 and D5 represent 10 and 100, respectively. The product of 1000 is saved in D5 as 000003E8.

The next instruction in the arithmetic group is NEG (negate). This instruction is used to generate the 2's complement of the destination by subtracting the destination from zero. All three data sizes may be used. All condition codes are affected.

Example 3.45: What is the result of NEG.B   D2, when D2 contains 052055C6?

Solution: Negating the lower byte of D2 (C6) gives 3A. The final result is 0520553A.

The next instruction in the arithmetic group is NEGX (negate with extend). In this instruction, the destination operand and the X bit are subtracted from zero. Since the X bit may have been affected by a previous NEGX, we see that this instruction may be used to implement multiprecision operations on large numbers. All three data sizes may be used. All condition codes are affected.

The next instruction in the arithmetic group is SUB (subtract binary). The source operand is subtracted from the destination operand. One of the operands must be a data register. All three data sizes may be used. All condition codes are affected.

Example 3.46: What will register D1 contain after SUB.W   D0,D1 executes, if D0 contains 0000E384 and D1 contains CC3EF385?

Solution: Subtracting the lower word of D0 from the lower word of D1 gives 1001. The upper word of D1 remains unaffected. The final result in D1 is CC3E1001.

The next instruction in the arithmetic group is SUBA (subtract address). In this instruction, the source operand is subtracted from the destination operand, which must be an address register. Only word and longword sizes may be used. When the data operand size is word, the operand is sign extended to 32 bits before the subtraction is performed. The condition codes are not affected.

Example 3.47: A unique data table has entries that consist of 7-byte groups of data. Suppose that A3 points to any entry in this data table. How can A3 be made to point to the previous entry?

Solution: To point to the previous entry, A3 must be decreased by 7. This may be accomplished by using SUBA.W   #7,A3.

The next instruction in the arithmetic group is SUBI (subtract immediate). In this instruction the immediate data are subtracted from the destination operand. All three data sizes may be used. All condition codes are affected.

Example 3.48:   What will D2 contain after execution of SUBI.B   #$2C,D2, if D2 initially contains 03059A2E?

Solution: Subtracting 2C from the lower byte of D2 results in 02. The result placed in D2 is 03059A02.

The next instruction in the arithmetic group is SUBQ (subtract quick). This instruction should be used whenever the immediate data is between 1 and 8. All three data sizes may be used. All condition codes are affected.

The next instruction in the arithmetic group is SUBX (subtract with extend). In this instruction the source operand and the X bit are subtracted from the destination operand. Only two choices of operands are available. The operands must be data registers, or memory locations pointed to by address register indirect with predecrement addressing mode. All three data sizes may be used. All condition codes are affected.

Example 3.49:   What does SUBX.W   D2,D2 do? What does SUBX.L   –(A0,)–(A1) do when A0 and A1 contain the same address? Before execution, X equals 0.

Solution: The first instruction clears the lower word of D2. The second instruction first decrements both address registers by 4, and then clears four successive memory locations.

The next instruction in the arithmetic group is TAS (test and set an operand). The lower 8 bits of the operand are tested and the condition codes modified accordingly. Bit 7 of the destination operand is then set. Only byte-sized operands are tested. All condition codes except X are affected.

Example 3.50:   What are the condition codes, and the contents of D5, after TAS D5 executes? D5 initially contains 2CC3E500.

Solution: Since the lower byte of D5 is 00, the zero flag will be set and the negative flag cleared. Then bit 7 of D5 will be set, resulting in a final register value of 2CC3E580.

The last instruction in the arithmetic group is TST (test an operand). The destination operand is compared with zero and the condition codes affected accordingly. The dest nation operand is not changed. All three data sizes may be used. All condition codes except X are affected.

The Logical Group

The first instruction in the logical group is AND (AND logical). A logical AND is performed on the source and destination operands and the result placed in the destination. All three data sizes may be used. One of the operands must be a data register. Neither operand may be an address register. All condition codes except X are affected.

Example 3.51: What is the result of AND.W   D0,D1, if D0 and D1 contain 3795AC5F and B6D34B9D, respectively?

Solution: Only the lower words of each register will be used. The logical AND of AC5F and 4B9D is 081D. The final value of D1 is B6D3081D.

The next instruction in the logical group is ANDI (AND immediate). This instruction operates similarly to AND, except that immediate data is supplied for the source operand. All three data sizes may be used. The condition codes are affected the same way they are for AND.

Example 3.52:   What will ANDI.B   #$F0,D7 do to the lower byte of D7?

Solution: ANDing the lower byte of D7 with F0 will clear bits 0 through 3 and leave bits 4 through 7 unaffected.

The next instruction in the logical group is OR (inclusive OR logical). This instruction is used to OR the source and destination operands together. One of the operands must be a data register. Neither operand may be an address register. All three data sizes may be used. All condition codes except X are affected.

Example 3.53:   What is the result of OR.L   D3,D4, when D3 contains 55555555 and D4 contains AAAAAAAA?

Solution: The logical OR of these two registers is FFFFFFFF. This is the result that will be placed in D4.

The next instruction in the logical group is ORI (inclusive OR immediate). This instruction is similar to OR, except the source operand is now supplied as immediate data. All three data sizes may be used.

The condition codes are affected the same way as they are for OR.

Example 3.54:   How can the MSB of the lower byte of D2 be set using ORI?

Solution: To set bit 7 of D2 without affecting any other bits, use ORI.B   #$80,D2.

The next instruction in the logical group is EOR (exclusive OR logical). In this instruction the source operand must be a data register. The source and destination operands are exclusive-ORed and the result placed in the destination location. All three data sizes may be used. All condition codes except X are affected.

Example 3.55:   What is the result of EOR.W   D2,D3, if D2 contains 02040608 and D3 contains 10121416?

Solution: The exclusive OR of 0608 and 1416 is 121E. The final contents of D3 are 1012121E.

The next instruction in the logical group is EORI (exclusive OR immediate). As in EOR, the source and destination operands are exclusive-ORed, except the source operand must be immediate data. All three sizes may be used. The condition codes are affected the same way they are for EOR.

Example 3.56:   How can bits 0, 3, 4, 8, 12, and 15 be complemented in register D4, with all other bits remaining unchanged?

Solution: Exclusive-ORing a data bit with 0 preserves its value. Exclusive-ORing it with 1 complements its value. The immediate data must then contain 1s in bit positions 0, 3, 4, 8, 12, and 15; and 0s elsewhere. This results in the following instruction: EORI.W #$9119,D4.

The last instruction in the logical group is NOT (logical complement). Each bit in the destination is complemented. All three data sizes may be used. All condition codes except X are affected.

Example 3.57:   What is the result of NOT.B   D1, when D1 contains 000000FF?

Solution: The lower byte of D1 is complemented, resulting in D1 containing 00000000.

The Shift and Rotate Group

All of the instructions in this group may use byte, word, and longword data sizes. Shifts and rotates are possible in both directions. Only data registers and memory locations may be used as operands. When a data register is to be shifted or rotated, a count must be supplied that specifies the number of bits to shift or rotate. When a memory location is the destination, only 1 bit may be shifted at a time, and the data size is restricted to word operands only.

The shift count may be specified in two ways. When shifting or rotating between 1 and 8 bits, use the immediate form. All of the following instructions illustrate this form:

ASL.B	#4,D2	Arithmetic shift left of 4 bits
LSR.W	#6,D1	Logical shift right of 6 bits
ROL.L	#3,D5	Rotate left 3 bits
ROXR.B	#5,D4	Rotate right with extend 5 bits

When the shift/rotate count is greater than 8, it must be placed in a data register. ASL.L   D2,D3 is an example of this form, with D2 containing the number of bits that D3 should be shifted.

When the operand is a memory location, only 1 bit may be shifted or rotated at a time, through a word operand. ROR.W   (A0) will rotate the word pointed to by A0 1 bit to the right.

The N and C flags are affected by all eight instructions in the same way. If the MSB of the result is high, the N flag will be set. It will be cleared otherwise. If the result is 0, the Z flag will be set. It will be cleared otherwise. The V flag will always be cleared, except when ASL or ASR is used. In this case, it will be set if a sign change occurs during shifting (if the MSB changes value at any time).

Figure 3.18 shows how bits are moved through an operand for each of the eight instructions in the shift and rotate group. Notice that all eight instructions affect the C bit, and that six of them also change the X bit.

Figure 3.18(a) shows the effect of an ASL (arithmetic shift left). In this instruction, a zero is shifted into the LSB, while all bits move to the left. The bit shifted out of the most-significant bit position replaces the X and C flags. Note that the MSB is bit 7 for byte operations, bit 15 for word, and bit 31 for longword.

Figure 3.18(b) shows the effect of an ASR (arithmetic shift right). All bits are shifted to the right. The bit that leaves the LSB position replaces the X and C flags. The MSB is shifted back into itself. This is very important for preserving the sign of the original binary number. If a register initially contained a negative value (MSB high), and a zero was shifted in during an ASR, the sign of the number would change.

Figures 3.18(c) and 3.18(d) show how the LSL (logical shift left) and LSR (logical shift right) instructions, respectively, operate. These instructions are used to move 0s into either end of a register or memory location, much as a shift register would do in a digital circuit.

Figures 3.18(e) and 3.18(f) show the effect of the ROL (rotate left) and ROR (rotate right) instructions, respectively. Unlike the shift instructions, no new data enters the register or memory location. Instead, the data is circulated through the register or memory location.

FIGURE 3.18 Shift and rotate instructions: (a) ASL; (b) ASR; (c) LSL; (d) LSR; (e) ROL; (f) ROR; (g) ROXL; (h) ROXR

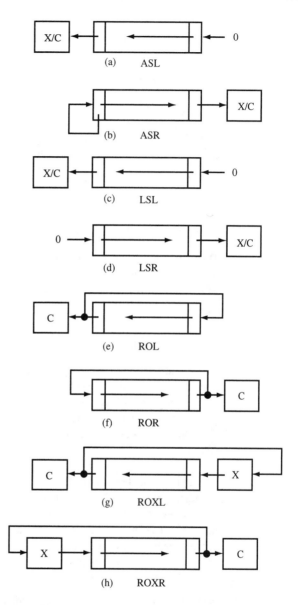

After enough data rotates, the original binary data appears again. For example, rotating a byte eight times gets all the bits back into the same place again. These two instructions are the only ones that do not also affect the X bit.

Figures 3.18(g) and 3.18(h) show the last two instructions in the shift and rotate group, ROXL (rotate with extend left) and ROXR (rotate with extend right), respectively. The operand data in these two instructions rotates through the X bit, giving the programmer the chance to insert data during rotation by playing with the value of the X bit. Thus, the final contents of the register or memory location may be very different from what they were initially.

Example 3.58:   What is the result of ASL.B   #3,D2, if D2 contains 2375A2B4?

Solution: The lower 8 bits will be shifted three positions to the left. As they shift, zeros will enter from the right. Thus, the final contents of D2 are 2375A2A0.

Example 3.59:   What does D0 contain after execution of ROR.W   D4,D0? The count in D4 is 9, and D0 contains 3F2E5983 to start.

Solution: Rotating the lower word right 9 bits results in 3F2EC1AC.

Example 3.60:   What is caused by execution of LSR.L   #4,D2, if D2 contains 31415926?

Solution: The LSR instruction causes all bits in D2 to move four positions to the right, with zeros entering from the left. This has the effect of moving all 4-bit nybbles to the right. The final D2 contents are 03141592.

The Bit Manipulation Group

The first instruction in the bit manipulation group is BCHG (test a bit and change). A specified bit in the destination is examined and the Z flag is adjusted accordingly. Actually, the bit is tested to see if it is 0. If so, the Z flag is set. If the tested bit is 1, the Z flag is cleared. It is easy to see that the Z flag is loaded with the complement of the bit that is tested.

Once the specified bit is tested and the Z flag adjusted, it is replaced by its complement. For instance, if bit 6 is 0 when it is tested, it will be a 1 when the instruction finishes execution.

Only byte and longword data sizes may be used. When the bit to be tested is in the lower byte of the destination (bit positions 0 through 7), the bit number may be specified as immediate data within the instruction. For example, BCHG.B   #3,D1 causes bit 3 in register D1 to be tested. When the bit position is from 8 to 31, it must be placed into a data register and that register used in the instruction as follows: BCHG.L   D0,D1. In this example, D0 contains the number of the bit to be tested in D1. Only the Z flag is affected by BCHG.

Example 3.61:   Data register D5 contains 2C3459A7. What is the state of the Z flag, and what are the contents of D5, after BCHG   #6,D5 executes?

Solution: The binary representation of the lower byte of D5 is 10100111. Bit 6, 1<0>100111, is a 0. This will cause the Z flag to be set and the final state of bit 6 to be a 1. This will result in D5 containing 2C3459E7 once bit 6 is complemented.

The second instruction in the bit manipulation group is BCLR (test a bit and clear). This instruction is similar to BCHG, except the specified bit is examined and then cleared. Only the Z flag is affected, just as it is in BCHG.

Example 3.62: Data registers D6 and D7 contain 0000000C and 75793290, respectively. What is the result of BCLR   D6,D7?

Solution: Register D6 specifies that the 12th bit position should be tested and cleared. The lower 16 bits of D7, with the 12th bit position indicated, are 001<1>001010010000. Since this bit is a 1, the Z flag will be cleared. Then bit 12 of D7 will be cleared, resulting in a final value of 75792290.

The third instruction in the bit manipulation group is BSET (test a bit and set). This instruction is identical to BCLR, except the specified bit is tested and then set. Once again only the Z flag is affected.

Example 3.63: What is accomplished by BSET   #2,(A3)?

Solution: The bit in position 2 of the memory location pointed to by A3 is tested and then set.

The last instruction in the bit manipulation group is BTST (test a bit). In this instruction, the specified bit in the destination is tested and the Z flag adjusted depending on its state. No other flags are affected, and the tested bit retains its value.

Example 3.64: Register D4 contains 0429595A. What is the state of the Z flag after BTST   #5,D4 executes?

Solution: The lower 8 bits of D4 with bit 5 highlighted are 01<0>11010. Since this bit is zero, the Z flag will be set. The contents of D4 are unchanged.

The BCD (Binary Coded Decimal) Group

The first instruction in the BCD group is ABCD (add decimal with extend). This instruction is used to add BCD numbers together. Using the X flag in the operation allows multiprecision additions to be performed. Only bytes may be added together. No other data sizes may be used. Also, only two addressing modes may be used: register direct (using data registers) or address register indirect with predecrement. All condition codes are affected.

Example 3.65: Registers D0 and D1 contain 00000034 and 00000068. What are the contents of D1 and the condition codes after ABCD   D0,D1 executes? Assume that X is cleared to begin with.

Solution: A straight binary addition of D0 and D1 would result in 9C. This value, however, is not a BCD number. The result of ABCD using decimal addition instead results in 00000002 replacing the contents of D1. Since 34 plus 68 is 102, we see that D1 contains only a partial answer, because the entire result cannot fit into the lower byte of D1. This is reflected in the state of the condition codes. Since the result is non-zero, the Z flag is cleared. The C flag (and thus X also) is set because the result did not fall into the range 00 to 99.

The second instruction in the BCD group is NBCD (negate decimal with extend). When we wish to represent a negative number in binary, we use 2's complement notation. This is accomplished by the NEG instruction. When working with BCD, we use 9's or 10's complement notation to represent a negative BCD number. The number we wish to find the 10's complement of is subtracted from 0. Since the X bit is also included in the subtraction, we get the 10's complement when $X = 0$ and 9's complement when $X = 1$. NBCD operates only on byte operands, and all condition codes are affected, with N and V undefined.

Example 3.66:   What is the 10's complement of the lower byte in D3, if D3 contains 00000034?

Solution: To get the 10's complement, we must make sure the X flag is cleared before executing NBCD   D3. The result is 00000066. Notice that adding 34 to 66 (the 10's complement of 34) gives 00. This is what we hope to get when we find the 10's complement of a BCD number. The 9's complement, which we would get if $X = 1$ prior to executing NBCD, would be 00000065.

The last instruction in the BCD group is SBCD (subtract decimal with extend). Like ABCD, only two addressing modes are possible. In this case, the source operand and the X flag are subtracted from the destination. All condition codes are affected, with N and V undefined.

Example 3.67:   Registers D2 and D4 contain 00000034 and 00000068. What is the result of SBCD   D2,D4 if the X flag is set prior to execution?

Solution: Subtracting D2 and the X flag from D4 gives 00000033. This result is placed in D4.

The Program Control Group

The first three instructions in the program control group make use of a set of predefined conditions whose individual states are determined by the current value of the processor's condition codes. These 16 conditions are shown in Table 3.2. The instructions that make

TABLE 3.2 Condition code tests

Condition	Meaning	Flag(s) tested	Encoding pattern
T	True	None	0000
F	False	None	0001
HI	High	$C + Z = 0$	0010
LS	Lower or same	$C + Z = 1$	0011
CC	Carry clear	$C = 0$	0100
CS	Carry set	$C = 1$	0101
NE	Not equal	$Z = 0$	0110
EQ	Equal	$Z = 1$	0111
VC	Overflow clear	$V = 0$	1000
VS	Overflow set	$V = 1$	1001
PL	Plus	$N = 0$	1010
MI	Minus	$N = 1$	1011
GE	Greater than or equal	$N \oplus V = 0$	1100
LT	Less than	$N \oplus V = 1$	1101
GT	Greater than	$Z + (N \oplus V) = 0$	1110
LE	Less than or equal	$Z + (N \oplus V) = 1$	1111

use of these conditions are called **conditional instructions,** because they may or may not perform a desired function, depending on the state of the condition.

The first conditional instruction is B*cc*. In this mnemonic the B stands for branch and *cc* the condition. Thus, BCC means branch if carry clear, BNE means branch if not equal, and so on as shown in Table 3.3. All of the conditions may be used in this instruction except for T and F. The purpose of the conditional branch is to transfer control to a new location in the program, depending on a certain condition. The range of the condi-

TABLE 3.3 Conditional branch instructions

Conditional instruction	Meaning
BHI	Branch if high
BLS	Branch if lower or same
BCC	Branch if carry clear
BCS	Branch if carry set
BNE	Branch if not equal
BEQ	Branch if equal
BVC	Branch if overflow clear
BVS	Branch if overflow set
BPL	Branch if plus
BMI	Branch if minus
BGE	Branch if greater than or equal
BLT	Branch if less than
BGT	Branch if greater than
BLE	Branch if less than or equal

FIGURE 3.19 Use of conditional branch in a loop

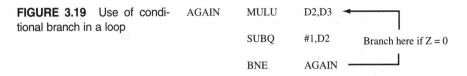

tional branch is represented by a signed 8- or 16-bit displacement. The 8-bit displacement permits transfer to locations up to 127 forward and –128 behind the current program counter. The 16-bit displacement allows a 32767, –32768 location range.

The condition codes are not affected by any of the conditional branches.

Conditional branches are very useful when implementing loops. Figure 3.19 shows how a BNE instruction is used to perform a loop operation as long as the Z flag is cleared. In this example, control is passed back to the address of AGAIN (the MULU instruction) as long as D2 is not 0. When the condition fails to be met, the conditional branch is ignored, and program execution continues at the address following the branch instruction.

As a point of interest, if data register D3 is loaded with 1 prior to entering the loop of Figure 3.19, the result is that D3 contains the factorial of the number in D2 when the loop completes. Thus, if D2 started at 5, then D3 will contain 120 (1 * 2 * 3 * 4 * 5) at the end of the loop.

The conditional branches are also useful for determining what to do with the result of a comparison. For example, suppose that data registers D0 and D1 are each loaded with a word of data. The instruction

```
CMP.W    D1,D0
```

compares the lower word of D1 to the lower word of D0 by using an internal subtract operation (D0 – D1). The subtract operation does not affect either register, but it does update the condition codes. It is easy to see that if D0 and D1 are equal (both contain the same word value), the zero flag will be set. In this case, following CMP.W   D1,D0 with a BEQ instruction will cause the branch to take place.

When the numbers in both registers are different, we often must resort to using combinations of condition codes to determine a branch condition. This is due to the nature of the register data. The data values in each register may be signed or unsigned. For this reason, the conditional branches are designed to interpret the results of both signed and unsigned comparisons. Furthermore, both numbers being compared must each be signed or unsigned. It is invalid to compare mixed types of data because there is no way to accurately interpret the condition codes.

When both values are signed, we may use any of the following branches after CMP.W   D1,D0:

Signed Comparison	Branch That Will Be Taken
D0 > D1	BGT (Branch if Greater Than)
D0 ≥ D1	BGE (Branch if Greater or Equal)
D0 = D1	BEQ (Branch if Equal)
D0 ≠ D1	BNE (Branch if Not Equal)
D0 ≤ D1	BLE (Branch if Lower or Equal)
D0 < D1	BLT (Branch if Lower Than)

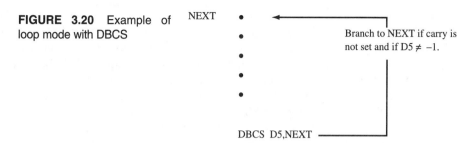

FIGURE 3.20 Example of loop mode with DBCS

Branch to NEXT if carry is not set and if D5 ≠ −1.

DBCS D5,NEXT

If both numbers are unsigned, we may use these branch instructions:

Unsigned Comparison	Branch That Will Be Taken
D0 > D1	BHI (Branch if Higher)
D0 ≥ D1	BCC (Branch if Carry Clear)
D0 = D1	BEQ (Branch if Equal)
D0 ≠ D1	BNE (Branch if Not Equal)
D0 ≤ D1	BLS (Branch if Lower or Same)
D0 < D1	BCS (Branch if Carry Set)

Keep these groupings in mind when you use a conditional branch after a compare operation.

The second conditional instruction in the program control group is DB*cc* (test condition, decrement, and branch). In general DB*cc* repeats a loop until the condition (*cc*) becomes true or the register used by DB*cc* becomes −1. Table 3.4 lists the various DB instructions. This instruction provides a looping function that may be terminated in

TABLE 3.4 Decrement-and-branch instructions

Conditional instruction	Meaning
DBT	Decrement and branch if true
DBF	Decrement and branch if false
DBHI	Decrement and branch if high
DBLS	Decrement and branch if lower or same
DBCC	Decrement and branch if carry clear
DBCS	Decrement and branch if carry set
DBNE	Decrement and branch if not equal
DBEQ	Decrement and branch if equal
DBVC	Decrement and branch if overflow clear
DBVS	Decrement and branch if overflow set
DBPL	Decrement and branch if plus
DBMI	Decrement and branch if minus
DBGE	Decrement and branch if greater than or equal
DBLT	Decrement and branch if less than
DBGT	Decrement and branch if greater than
DBLE	Decrement and branch if less than or equal

two ways. Figure 3.20 shows the use of this instruction in a loop. Here the DBCS instruction is used to perform the looping function. If the carry flag is set during a pass through the loop, the DBCS instruction will automatically cause the loop to terminate, and execution will continue with the first instruction after DBCS. If the carry flag is not set, the DBCS instruction will then decrement the specified data register (D5 in this case). If the lower word of D5 does not equal −1, program execution will resume at the beginning of the loop (NEXT in this example). If D5 did equal −1, the loop will terminate. The use of a counter in this instruction limits the maximum number of passes through the loop to 32768. We get an extra pass because the counter is tested for −1 instead of 0. No condition codes are affected by this instruction.

Example 3.68:   Assuming that the tested condition is always false, how many passes through the loop will be performed in Figure 3.20, if D5 initially contains 00000064?

Solution: D5 contains a count of 100. Since it must go negative for the loop to terminate, 101 passes will be performed.

Note that many assemblers accept another mnemonic for DBF (decrement and branch if false), which is DBRA (decrement and branch always). Both DBF and DBRA allow the number of passes through the loop to be controlled strictly by the loop register.

The last conditional instruction in the program control group is S*cc* (set according to condition). These instructions are shown in Table 3.5. S*cc* first tests the specified condition. If true, the destination byte is loaded with 1s. If false, the destination byte is loaded with 0s. No condition codes are affected by this instruction.

TABLE 3.5 Set-according-to-condition instructions

Conditional instruction	Meaning
ST	Set if true
SF	Set if false
SHI	Set if high
SLS	Set if lower or same
SCC	Set if carry clear
SCS	Set if carry set
SNE	Set if not equal
SEQ	Set if equal
SVC	Set if overflow clear
SVS	Set if overflow set
SPL	Set if plus
SMI	Set if minus
SGE	Set if greater than or equal
SLT	Set if less than
SGT	Set if greater than
SLE	Set if less than or equal

Example 3.69:   What happens to D3 if SEQ   D3 executes, and the Z flag was set prior to execution?

Solution: The lower byte of D3 is loaded with 1s.

The next instruction in the program control group is BRA (branch always). This type of instruction is unconditional; it will always execute. We also refer to this as an unconditional branch. BRA directly affects the processor's program counter with execution continuing at the address specified in the instruction. Since BRA is a relative instruction, the range of locations that we may transfer control to is limited. If an 8-bit signed displacement is used to specify the new execution address, the range of locations becomes 127 forward and –128 backward. If the address to branch to is outside this range, a signed 16-bit displacement will be used, resulting in a range of 32767 to –32768 locations. None of the condition codes are affected.

Example 3.70:   Consider these short sections of code:

```
        ORG   $1000            ORG   $2000
        BRA   HERE     THERE   NOP
         .                      .
         .                      .
         .                      .
        ORG   $1030            ORG   $7000
HERE    NOP                     BRA   THERE
```

What type of signed displacement is needed for the BRA   HERE instruction? How is the BRA   THERE instruction different?

Solution: The machine code for the BRA   HERE instruction is 60 2E. The first byte (60) is the opcode for BRA. The second byte (2E) is the signed 8-bit displacement. The assembler computes the displacement by subtracting $1002 (2 plus the address of the BRA instruction) from the address of the HERE label ($1030). Subtracting $1002 from $1030 gives $002E. This value represents a positive signed displacement, which causes the processor to branch forward.

The machine code for the BRA   THERE instruction is 60 00 AF FE. Subtracting $7002 from $2000 gives $AFFE. Since the MSB of the displacement is a 1, the processor branches backward. Notice that the lower byte of the instruction opcode is 00. This indicates that the next word in memory is the 16-bit displacement.

The next instruction in the program control group is BSR (branch to subroutine). This instruction is similar to BRA, except that the address immediately following the BSR instructions is pushed onto the stack (as the return address for the subroutine). Again the displacement is stored in 8- or 16-bit signed notation.

Example 3.71: A BSR instruction at address $80006 calls a subroutine located at address $80500. What address is pushed onto the stack? Address register A7 contains $6004.

Solution: The processor uses A7 as the default stack pointer. When the BSR instruction is executed, A7 will be decremented by 4 and its address used to point to the memory locations where the return address will be stored. The return address for the BSR will be $8000A, since the BSR codes into a 4-byte instruction. Once the return address is written into stack memory, the program counter will be loaded with $80500, the address of the subroutine. Figure 3.21 shows the resulting stack activity during execution of the BSR instruction.

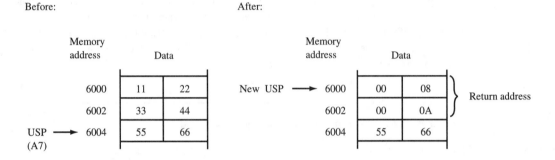

FIGURE 3.21   Stack activity during execution of BSR

The next instruction in the program control group JMP is (jump). This unconditional instruction causes execution to continue at the location specified. Any location in the entire address space of the processor may be reached with this instruction. Also, a limited number of addressing modes may be used with JMP. Address register indirect (and with displacement), absolute, indexed, and PC relative may all be used. None of the condition codes are affected.

Example 3.72: If A5 contains 0001F400, at what location does the processor resume execution if JMP (A5) is encountered?

Solution: This form of JMP instructs the processor to resume execution at the location specified in A5. Thus, 01F400 becomes the new address in the program counter.

The next instruction in the program control group is JSR (jump to subroutine). This is similar to JMP, except that the subroutine's return address is pushed onto the stack. The return address is the address following the JSR instruction. None of the condition codes are affected.

The next instruction in the program control group is RTR (return and restore condition codes). During execution, the condition codes are popped off the stack. The condition codes are replaced by the stack information. Then the new program counter value is popped off the stack (a return address previously pushed).

Example 3.73: The stack pointer contains $75800 when an RTR instruction is encountered. The contents of stack memory are shown in Figure 3.22. What is loaded into the condition codes? What is the return address?

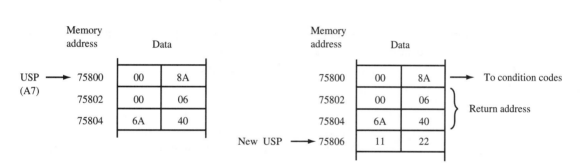

FIGURE 3.22 Execution of RTR

Solution: Figure 3.22 shows the word on top of the stack to be $008A. This value is popped first and written into the user byte of the status register, changing the condition codes. The next two words are popped off the stack and placed into the program counter, giving a return address of $66A40. The final value of the stack pointer is $75806.

The last instruction in the program control group is RTS (return from subroutine). The new program counter is popped off the stack. None of the condition codes are affected.

Example 3.74: Memory locations 7F80 through 7F83 contain the following bytes: 00, 04, 3E, and 2C. The stack pointer contains 007F80. Where will program execution resume when RTS executes?

Solution: The address popped off the stack is 043E2C. Execution resumes at that address. The stack pointer is incremented by 4.

The System Control Group

The system control group is the last group of instructions in the 68000's instruction set. These instructions are used to perform privileged operations on the system byte of the status register, reset or stop the processor, implement some useful expectations, and alter the condition codes.

The first privileged instruction is ANDI   SR (AND immediate to the status register). This instruction, like all privileged instructions, must be executed in the supervisor state to avoid the generation of a privilege violation exception. The purpose of ANDI   SR is to alter the contents of the status register by ANDing it with 16 bits of immediate data. Zeros in the immediate data will clear bits in the status register; ones will preserve the same bits. How the condition codes are affected depends on the immediate data used.

Example 3.75:   What does ANDI   #$FFFB,SR do?

Solution: The only zero present in the immediate area is located in bit 2. This corresponds to the position of the Z flag in the status register, which is cleared.

The next privileged instruction in the system control group is EORI   SR (exclusive OR immediate to the status register). This instruction has an effect on the status register similar to ANDI   SR, except for the change in logical operation.

Example 3.76:   What immediate data is needed to complement the state of the X bit in the status register?

Solution: The X bit is in position 4 of the status register. The instruction needed to complement this bit without affecting the other 15 bits is EORI   #$10,SR.

Another privileged instruction in the system control group lets you load the entire status register with new data. MOVE to SR (move to the status register) transfers 16 bits of data into the status register from the source operand.

Example 3.77:   During the execution of a subroutine, the condition codes are stored in memory at a location called STWORD. How might the status register be reloaded at a later time using this information?

Solution: MOVE   STWORD,SR will transfer data from memory into the status register.

The next privileged instruction in the system control group is MOVE from SR (move from the status register). This instruction copies the contents of the status register into memory or the specified data register. The condition codes are not affected.

Example 3.78:   What memory locations are affected by MOVE   SR,–(A2), when A2 contains 000C9008?

Solution: Since predecrement addressing is used, A2 is first decremented by 2 (because this is a word operation). The status register contents are then copied into memory locations 0C9006 and 0C9007.

The next privileged instruction in the system control group is MOVE   USP (move user stack pointer). Setting up the user stack pointer is important when subroutines compose parts of the user program. Using MOVE   USP,A3 will transfer a copy of the user stack pointer into A3. Setting the USP is accomplished by MOVE   A3,USP. All addressing modes are possible. The condition codes are not affected.

The next privileged instruction in the system control group is ORI   SR (inclusive OR immediate to the status register). Like ANDI   SR and EORI   SR, this instruction is used to modify bits in the status register. Sixteen bits of immediate data are ORed with the status register to generate its new contents. The condition codes are affected by the immediate data used.

Example 3.79:   What instruction is needed to set the interrupt priority level to 6?

Solution: The interrupt priority level bits in the status register are bits 8, 9, and 10. They need to be set to 110. This can be done using ORI   #$600,SR.

The next privileged instruction in the system control group is RESET (reset external devices). This instruction causes the $\overline{\text{RESET}}$ line on the processor to come active. Since the $\overline{\text{RESET}}$ line is normally connected to external system devices (such as peripherals), this instruction provides you with a way to reset external circuitry through software.

The next privileged instruction in the system control group is RTE (return from exception). This instruction is used as the last instruction in an exception handling routine. It restores the status register and program counter to their contents at the time the exception occurred. The data to restore the registers is popped off the stack when RTE is encountered. The condition codes are affected by the data popped off the stack.

The last privileged instruction in the system control group is STOP (load status register and stop). The immediate data included in the instruction is transferred into the status register, and the processor is halted. There are three ways to resume execution. If the processor is in the trace state when STOP is encountered, a trace exception is initiated. If an external $\overline{RESET}$ is requested, the processor begins a reset exception, leaving the halted state. Also, if an external interrupt arrives while the processor is stopped, it will be ignored unless its priority is higher than the current priority. If the interrupt is of sufficient priority, the processor will exit the halted state and initiate exception processing to handle the interrupt.

Example 3.80:   How may the processor be stopped with all condition codes cleared?

Solution: Use STOP   #0 to stop the processor and clear all condition codes.

Three instructions in the system control group are used specifically to generate TRAPs (exceptions). Unlike the privileged instructions, which cause an exception when they are encountered while the processor is in the user state, CHK, TRAP, and TRAPV can cause exceptions in either state.

CHK (check register against bounds) is used to compare a data register with a range of values that go from 0 to an upper bound specified in the source operand. Exception processing is initiated if the data register contains a value less than 0 or greater than the upper bound. The upper bound is a 2's complement integer. Only word operands may be used.

Example 3.81:   Is an exception generated by the following instruction: CHK   D4,D5? Registers D4 and D5 contain 3E552000 and 400C15A9.

Solution: Comparing the lower words of both registers shows that 15A9 does not exceed the upper bound of 2000 specified by D4. 15A9 is also larger than 0, so no exception processing occurs.

Using the ILLEGAL (illegal instruction) instruction will always cause an exception to occur. The designers of the 68000 chose a unique bit pattern, 4AFC, to implement this instruction, while at the same time reserving other patterns for future expansion.

Another way to generate an exception is to use the TRAP (trap) instruction. A vector number from 0 to 15 is supplied, which causes the processor to look up the address for the specified TRAP exception routine in an exception vector address table. This address table is located in memory from locations 000000 to 0003FF. The addresses for each of the 16 TRAP vectors are located in locations 000080 through 0000BF. For example, if TRAP #0 is encountered, the processor will look up the address stored in locations 000080 through 000083 and continue execution at that address.

A conditional TRAP instruction is available. TRAPV (trap on overflow) will initiate exception processing only if the overflow flag is set.

We will cover exception processing in detail in the next chapter.

Three instructions are included that operate only on the condition code byte of the status register: ANDI CCR, EORI CCR, and ORI CCR. All require a byte of immediate data to perform the logical operation with the condition codes. All three instructions will affect the condition codes according to the immediate data that is used.

Example 3.82: What are the five condition codes after execution of these instructions? The initial states of each bit are not important.

a) ANDI #$C,CCR
b) ORI   #1,CCR
c) EORI #$10,CCR

Solution: The ANDI instruction clears the X, V, and C flags. It does not affect N or Z. The ORI instruction sets the C flag. Finally, the EORI causes X to change its state to a 1. So, when done, we have X: 1, N: unchanged, Z: unchanged, V: 0, and C: 1.

One last way to affect the condition codes is to load them with new data. MOVE to CCR (move to condition codes) is used to do this. The data contained in the source operand is moved into the condition codes.

Example 3.83: What does MOVE (A2),CCR do?

Solution: The data byte stored in the memory location pointed to by A2 is copied into the condition code register.

Saving the condition codes for future use is done with MOVE from CCR (move from the condition code register). The condition codes are not affected. An example would be MOVE CCR,D2. The lower byte of D2 will contain a copy of the current condition codes.

The last instruction in the entire set is NOP (no operation). This instruction does nothing at all but provide a delay in execution time. It does not affect any condition codes or processor registers. Nonetheless, NOP is an ideal instruction to use in a timing loop.

3.6 HOW AN ASSEMBLER GENERATES MACHINE CODE

At the beginning of this chapter we looked at a sample source file and its assembled list file. During assembly, source statements are read one line at a time and examined. If they contain legal 68000 instructions, the assembler can determine the required machine code by filling in missing bits in a basic opcode format. For example, the instructions MOVE.B and MOVE.W both have the same basic opcode format: 00<*size*><*dst*><*src*>, where the leading pairs of zeros indicate MOVE. The remaining 14 bits (remember that the processor always fetches a word) are determined by the rest of the information contained in the operand fields. *Size* is a variable that takes on 01 for a byte operation, 11 for a word operation, and 10 for a longword operation. Thus, if the assembler reads in a source statement that contains MOVE.B, once it recognizes the MOVE part of the instruction, it will then look for a .B, .W, or .L to determine the correct bit pattern for <*size*>. It is easy to see that the machine code for each type of MOVE will be different. If the assembler does not see a .B, .W, or .L after MOVE, it will generate an error.

The <*dst*> and <*src*> fields each contain 6-bit positions and represent the destination and source operands. Each field is divided into a 3-bit register number and a 3-bit mode number. These fields are described in detail in Appendix B and explained initially in Figure 3.14. Example 3.84 indicates how these bits are assigned.

Example 3.84:   What is the machine code produced when MOVE.B   D3,(A5)+ is assembled?

Solution: Refer to the MOVE instruction description in Appendix B and to Figure 3.23. The .B extension indicates that the size field should be loaded with 01. The source operand is D3. The source effective address table calls for a mode of 000 when a data register is used, and the register number is 011 (D3). The destination effective address table calls for a mode of 011, which indicates postincrement addressing. The register number is 101 (for A5). Combining these individual bit patterns gives the final opcode pattern 00 01 101 011 000 011, which is 1AC3.

FIGURE 3.23 Creation of MOVE.B   D3,(A5)+ opcode

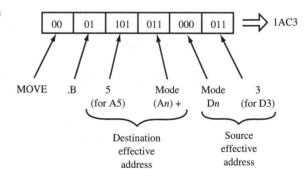

Example 3.85:   Consider the following source file:

```
        ORG     $2000
        CLR.W   D4
        BSR     $2070
NEXT    BSR     $5000
```

```
ADDI.W   #1000,D4
MULU     D2,D4
CMP.B    $20(A6),D2
BNE      NEXT
END
```

Let us assemble the entire source file, line by line, just as an assembler would.

When ORG $2000 is processed, the assembler sets its internal program counter to 2000.

When CLR.W   D4 is processed, the assembler loads the opcode format for CLR, which is 01000010*ssmmmrrr,* where *ss* are two size bits, *mmm* are three mode bits, and *rrr* are three register bits. The size .W sets *ss* to 01. The destination is a data register, which sets *mmm* to 000 (for data register direct addressing). Finally, *rrr* is set to 100 to indicate data register 4. This gives a 16-bit opcode of 0100001001000100, or 4244. Since the entire instruction is coded into a single word, the program counter is advanced to 2002.

When BSR $2070 is processed, the assembler subtracts the next physical program counter value (2002 + 2, or 2004) from the target address specified in the instruction to compute the displacement. The assembler automatically adds 2 to its current program counter (2002) to compute the next possible instruction address (2004). This is the address subtracted from the target specified in the BSR instruction. So, we have 2070 – 2004 = 6C as the 8-bit displacement. The opcode format for BSR is 01100001*dddddddd,* where *dddddddd* represents the displacement. So, the resulting opcode is 616C. The assembler's program counter is advanced by 2 to 2004.

The first thing encountered on the next line of the source file is the NEXT label. Since the assembler's program counter currently equals 2004, this is the value assigned to NEXT. Then BSR $5000 is processed. The next possible program counter value (2006) is subtracted from 5000, giving 2FFA as the displacement. Note that this is a 16-bit displacement. When the assembler sees this, it uses the second form of the BSR opcode, which uses an 8-bit displacement of 0 to indicate that a *second word* of information is required to specify the 16-bit displacement, which is 2FFA. Thus, we get 6100 2FFA as the resulting machine code. The program counter is now advanced to 2008.

When ADDI.W   #1000,D4 is processed, the assembler loads the opcode format for ADDI, which is 00000110*ssmmmrrr.* The size bits are set to 01 by the .W extension. Since D4 is specified as the destination, *mmm* and *rrr* are set to 000 and 100, respectively, giving a 16-bit opcode of 0000011001000100, or 0644. But the immediate data (#1000) must also be included as part of the instruction, so the assembler converts 1000 decimal into 03E8 and places this word after the instruction opcode. So, we end up with 0644 03E8 as the two-word instruction. The program counter is advanced to 200C.

When MULU   D2,D4 is processed, the assembler uses the following opcode format: 1100*nnn*011*mmmrrr,* where *nnn* refer to the number of the destination data register. These three bits are set to 100 to specify D4 as the destination. The *mmm* and *rrr* bits specify the source operand in this instruction. Since this operand is D2, *mmm* and *rrr* are set to 000 and 010, respectively. Notice that no size bits are required, since the size of MULU is always .W by definition. The resulting opcode is 1100100011000010, or C8C2. The program counter is advanced to 200E.

When CMP.B $20(A6),D2 is processed, the assembler loads the opcode format for CMP, which is 1011*nnnooommmrrr,* where the new bits *ooo* refer to the op-mode of the instruction. These bits are 000 for .B operations. Again, *nnn, mmm,* and *rrr* refer to the destination data register, the source addressing mode, and the source register, respectively.

The destination register D2 causes *nnn* to be set to 010. The source operand uses address register indirect addressing with displacement. The *mmm* and *rrr* bits are set to 101 and 110, respectively. This gives a 16-bit opcode of 1011010000101110, or B42E. This word is followed by 0020, the 16-bit displacement. So, the entire instruction becomes B42E 0020. The program counter is advanced to 2012.

When BNE   NEXT is encountered, the next possible program counter value (2014) is subtracted from the target address (NEXT, which equals 2004), giving FFF0. Note that the MSB is high, indicating a signed negative displacement. This is due to the fact that the NEXT label was encountered *before* the BNE   NEXT instruction.

To the assembler, F0 is as good as FFF0 for representing this particular displacement; thus, we can use the signed 8-bit displacement form of the instruction. The opcode format for the conditional branches is 0110*ccccdddddddd,* where *cccc* represent the 4-bit condition code. This is 0110 for the NE condition (see Table 3.2). All together we have 0110011011110000, or 66F0. The program counter is advanced to 2014.

The last line in the source file, END, terminates assembly. Putting all of the instructions and data generated by the assembler together produces the following list file:

```
002000                    1           ORG      $2000
002000   4244             2           CLR.W    D4
002002   616C             3           BSR      $2070
002004   6100 2FFA        4    NEXT   BSR      $5000
002008   0644 03E8        5           ADDI.W   #1000,D4
00200C   C8C2             6           MULU     D2,D4
00200E   B42E 0020        7           CMP.B    $20(A6),D2
002012   66F0             8           BNE      NEXT
002014                    9           END
```

After even a short example like this, we begin to appreciate the operation of an assembler.

The ASM68K assembler included on the companion disk performs the same types of operations while assembling 68000 code as we have just examined. The ASM68K statements used to assemble the MOVEA instruction are shown here for your interest. Look up the MOVEA instruction in Appendix B to help understand what the code is doing.

```
void MOVEA(int opsize)
{
        unsigned int mode;        //addressing mode
        unsigned int An;          //address register
        unsigned int reg;         //any register

        //check opsize for data size
        if ((opsize != WORD) && (opsize != LONG))
                warning_error(SIZE_NOT_SPECIFIED);

        //load base instruction pattern
        inst_word = 0x0040;

        //adjust size bits
        if (opsize == WORD)
                inst_word |= 0x3000;
        else
                inst_word |= 0x2000;
```

```
                    //determine source addressing mode
                    //and adjust instruction bits
                    get_address_mode(&mode,&reg);
                    inst_word |= (mode << 3) | reg;
                    eat(',');

                    //get destination address register
                    //and adjust instruction bits
                    getAn(&An);
                    inst_word |= (An << 9);

                    //clean up
                    write_list_file(WORD,inst_word);
                    finish_address_mode();
}
```

So we see that the process of assembling a program involves a great deal of time picking the right combination of bits to place into opcode patterns. Let us keep this in mind when we assemble our own programs later in the book. When the assembler indicates an error, it's not because it picked the wrong bits, it's because the information supplied *by the programmer* was incorrect, missing, or of the wrong type.

3.7 TROUBLESHOOTING TECHNIQUES

Before beginning our study of 68000 programming, it would be good to stop here and review some key points to remember when working with 68000 instructions.

- Examine the relationship between the instruction mnemonic and the resulting machine code. You will begin to see patterns. These patterns will help you discover errors such as leaving the $ off a hexadecimal number like $16 in the instruction MOVE.B #16,D0. The assembler will convert the 16 into $10, and generate the machine code 103C 0010 instead of 103C 0016. Knowing what to expect is a good place to start.

- Know your data sizes. It is frustrating to look at the instruction MOVE.B #500,D3 and wonder why it does not assemble. It looks OK, does it not? But remember that .B refers to the lower 8 bits of D3, and can only handle integers as large as 255. Knowing your data sizes will help you avoid problems like this.

- Do not make the assembler guess. For example, the instruction MOVE #5,D7 looks OK, but it actually generates an error because the assembler does not know if the '5' is an 8-bit 5, a 16-bit 5, or a 32-bit 5. This affects how register D7 is updated and could lead to trouble if the assembler does not use the correct size. Remember to use the .B, .W, and .L size extensions. Other instructions may generate errors because an illegal addressing mode was used. Check the allowed source and destination addressing mode tables in Appendix B when in doubt.

- Learn how to use several of the most basic addressing modes.

- Once again, always be aware of how an instruction uses, or affects, the processor flags. Many programs do not work simply because the flags are ignored by the programmer. This will be especially true in programs that must make decisions.

- Periodically review the entire instruction set. Programmers tend to fall in love with certain instructions and methods of writing code, and it is easy to forget that the processor already contains an instruction to fill your need.
- Understanding the binary number system and its associated operations is just as important as knowing the addressing modes and instruction types. It is difficult to perform any kind of interfacing, design, or programming without a good background knowledge of 1s and 0s. Take the time to review Appendix F if you have not already done so; it may save you some effort in the future.
- It may be a good idea to keep a logbook of the problems you encounter when working with assembly language. Until you have seen the same problem enough times, or learn how to avoid it, keeping track of the problem and its solution will save a lot of time and effort if you encounter it again.

These points should go a long way toward eliminating many of the common errors encountered when working with assembly language.

SUMMARY

In this chapter we examined the format of a 68000 assembly language source file. We saw that there are a number of predefined fields for information on any given source line. Label, opcode, operand, and comment fields must be properly maintained so that your source file may be easily examined.

We then looked at the operation of an assembler and the types of files generated, which were list (.LST) and object (.OBJ) files. Many object files are combined into one, and executed, by a link-loader.

Most of the chapter dealt with the addressing modes and instruction set of the 68000 microprocessor. The examples presented were intended to give you ideas about writing your own instructions and how the instructions and addressing modes work. The chapter finished with a number of examples that show how an assembler generates 68000 machine code.

STUDY QUESTIONS

1. Explain the use of the ORG, DC.W, DS.B, and END pseudo-opcodes.
2. What happens when a source file is assembled?
3. What two files are created by the assembler?
4. Name the opcode, data type, and operand(s) in this instruction:

   ```
   EOR    #$5C,D6
   ```

5. Why is it important to know what instructions are relocatable?
6. Which of these pseudo-opcodes produce code or data: ORG, DC.B, DC.W, DS.B, END?
7. What does a linker do?

8. List the eight basic instruction types.
9. Identify the destination addressing mode in each of these instructions:
   a) `EXG`   `D0,A2`
   b) `MOVE.B` `#5,D1`
   c) `MOVE.W` `D6,(A0)`
   d) `MOVE.W` `D6,(A0)+`
   e) `MOVE.W` `D6,-(A0)`
   f) `JMP`   `(PC)`
   g) `ADD.L`   `A0,$10(A1,D1.W)`
10. Suppose that the 68000 did not have predecrement and postincrement addressing modes available. Show how MOVE.L D0,–(A0) and MOVE. L (A0)+,D0 could be synthesized using other addressing modes and instructions.
11. Why are the condition codes so important in a control-type program?
12. What does this two-instruction sequence do?

    ```
    EXG   D0,A0
    EXG   D0,A1
    ```

13. What does this instruction accomplish?

    ```
    MOVE.B  -(A3),(A3)+
    ```

14. Memory locations 000490 through 000495 contain, respectively, 0A, 9C, B2, 78, 4F, and C3. What does D2 contain after each instruction? Assume that A0 contains 00000492 and that D2 contains 00000000 before each instruction executes.
    a) `MOVE.B (A0),D2`
    b) `MOVE.B -(A0),D2`
    c) `MOVE.W (A0),D2`
    d) `MOVE.W (A0)+,D2`
    e) `MOVE.W -(A0),D2`
15. What is contained in A0 after each instruction in Question 14?
16. What is accomplished by each of these instructions? Initially A0 contains 0028C504.
    a) `MOVEA.L (A0)+,A1`
    b) `MOVE.L A0,(A0)`
    c) `MOVE.L (A0),A0`
    d) `MOVEA.L -(A0),A1`
17. What are the sign extended addresses for these address values?
    a) `004000`
    b) `007C00`
    c) `008400`
    d) `00CB00`
    e) `140000`
    f) `EF0000`
18. What is the source operand address in each of these instructions? Assume that A0 contains 00003800 and D0 contains 00000200.
    a) `MOVE.B $10(A0),D2`
    b) `MOVE.B $1400(A0),D2`
    c) `MOVE.B $9F00(A0),D2`

   d) `MOVE.B $10(A0,D0.L),D2`

   e) `MOVE.B $84(A0,D0.L),D2`

19. Memory locations 004000 through 004007 contain, respectively, 11, 22, 33, 44, 99, AA, 55, and 66. What are the results of:

   a) `MOVEM.L $4000,D0/D1`

   b) `MOVEM.W $4000,D0-D3`

20. Suppose that these two instructions execute in sequence:

```
MOVEM.W   D3-D7,-(A2)
MOVEM.W   (A0)+,D3-D7
```

   What exactly has happened? Be very specific.

21. How does MOVEQ result in shorter, more efficient code?

22. What is the result of SWAP   D0, when D0 contains 042959FD?

23. What changes are made to D0 and D1 in the following series of instructions?

```
MOVE.W D1,D2
SWAP   D1
MOVE.W D0,D1
SWAP   D1
MOVE.W D2,D0
SWAP   D0
```

24. Write the instructions needed to add the lower word of registers D0, D2, and D6 together, with the result saved in D6. D0 and D2 must remain unchanged.

25. What is the difference between CLR.B   D0 and SUB.B   D0,D0?

26. Multiply the contents of D3 by 0.125. Since fractional multiplication is not available, you must think of an alternative way to solve this problem. Assume that D3 contains an unsigned binary value.

27. What does this sequence of instructions do?

```
MULS   D0,D1
DIVS   D1,D2
```

28. What are the contents of D2 after NEG.W   D2, if D2 contains 55555555?

29. Show the results of EXT.L   D6, when D6 contains 12345678 and 9ABCDEF0.

30. What are the condition codes after TAS   D2, if D2 contains 0000007F? What does D2 contain after execution?

31. What are the largest two decimal numbers that may be multiplied using MULU? Repeat for MULS.

32. Find the volume of a cube whose length on one side has been placed in register D3. The volume should be in D2 when finished.

33. Write the appropriate AND instruction to preserve bits 0, 3–9, and 13 of register D2, and clear all others.

34. What OR instruction is needed to always set bits 2, 3, and 5 of D1?

35. What is the result of this small sequence of instructions? Data register D5 initially contains 3B25AC89.

```
ANDI.B   #$2C,D5
ORI.W    #$C45,D5
EORI.L   #$789ABCDE,D5
NOT.W    D5
```

36. Show the instructions needed to find the exclusive OR of the upper byte of registers D4 and D5. Do not affect any of the lower 24 bits in either register. D5 is the destination register.

37. Data register D4 contains C9AE23A5. What are its contents after these two instructions execute?

```
ASR.L   #3,D4
ROL.W   #5,D4
```

38. Show how ROXL can be used to rotate a 64-bit register composed on D2 and D3, with D3 holding the upper 32 bits.

39. Why is it important for the ASR instruction to maintain the same value in the MSB position?

40. Data register D1 contains 00000019 (25 decimal). What is its value after LSL.W #6,D1?

41. Data registers D3 and D4 contain 56789ABC and 00000013. Determine the state of the Z flag and the contents of D3 for:
   a) BTST #4,D3
   b) BCLR D4,D3
   c) BSET #1,D3
   d) BCHG D4,D3

42. What instruction is needed to add registers D2 and D5 together using BCD arithmetic? The result should appear in D2.

43. What does D4 contain after NBCD   D4, if D4 initially contains 00000051? Assume that the X flag is cleared to begin with.

44. What does SBCD   –(A3),–(A5) do? Assume that X = 0 prior to execution.

45. What instruction will set the memory byte pointed to by A6 when the carry flag is clear?

46. What advantage does JSR have over BSR?

47. Why must a subroutine contain RTS as the final instruction?

48. What instruction (in the supervisor state) is capable of clearing the zero flag while leaving all other 15 bits in the status register unaffected?

49. What immediate data is needed to set the trace bit in the system byte of the status register?

50. Show the instruction needed to copy the condition codes into memory location OLDCODES.

51. Write the instructions needed to compute the area of a right triangle whose side lengths are stored in D2 and D3. Return the result in D1.

52. Write the instructions needed to swap nybbles in a byte operand. For example, bits 0–3 trade places with bits 4–7 in register D0.

53. Repeat Question 52, except now swap the 2 lower bytes of D0.

54. Use MULU and multiprecision addition to perform a 32-bit by 32-bit multiply. The numbers to be multiplied are in D0 and D1.

55. Assemble the following instructions:
   a) MOVE.B   #$10,(A3)
   b) EXT.L   D6
   c) LSL.W   #2,D6

    d) BTST.B   #7,D5

    e) DIVU $100(A4),D3

56. Assemble this BRA instruction: HERE BRA HERE. Explain why the address of HERE is not required to generate the machine code.

57. Make a list of instructions that might be useful for the calculator project.

58. Explain the problem with each of these instructions:

    a) MOVE.W   #$28E00,D5

    b) ANDI.B   D3,#7

    c) EOR.L   D1,A1

    d) MULU   #5,(A7)

59. Write a set of instructions that converts the single-digit Roman numeral stored in D6 into its equivalent decimal value (e.g., 'X' becomes 10).

60. Decode the following instructions. The type of instruction is indicated. Determine the operand size, and any source/destination operands.

Address	Machine Code			Instruction	
001000				ORG	$1000
a) 001000	303C	0064		MOVE	?
b) 001004	D803			ADD	?
c) 001006	0498	CFFF	FFFF	SUBI	?
d) 00100C	6000	FFF2		BRA	?

CHAPTER 4

Exception Processing

OBJECTIVES

In this chapter you will learn about:

- The three processing states: normal, halted, and exception
- The differences between user and supervisor states
- Methods used to change the privilege state
- Exception processing procedures
- The vector address table
- Multiple exceptions and exception priorities
- Special exceptions such as reset, bus and address error, and trace
- The general requirements of all exception handlers

4.1   INTRODUCTION

The 68000 microprocessor provides a very flexible method for recovering from what are known as **catastrophic** system faults. Through the same mechanism, external and internal interrupts may be handled and other events not normally associated with program execution may be taken care of. The method that does all of this for us is the 68000's **exception handler.** In this chapter we will see that there are many kinds of exceptions. Some of these deal with issues that have always plagued programmers (such as the divide-by-zero operation), while still others may be defined by the programmer. The emphasis in this chapter is on the definition of the numerous exceptions available. Actual programming examples designed to handle exceptions will be covered in the next chapter.

Section 4.2 discusses the different states of execution in the 68000. Section 4.3 explains the difference between the user and supervisor states. Section 4.4 shows how the 68000 may be switched from user to supervisor and back, and the events that might cause this to happen during normal program execution. Section 4.5 explains how exception processing works, and what is required by the programmer for correct handling of exceptions. Multiple exceptions are dealt with in Section 4.6. Section 4.7 deals with important, built-in exceptions such as reset, trace, and bus error, as well as interrupt handling exceptions, and stack-based parameter passing for exceptions and subroutines. Section 4.8 discusses exception handlers and methods of preserving registers during exception processing. Troubleshooting techniques are presented in Section 4.9.

4.2   EXECUTION STATES

The 68000 is always functioning in one of three states: normal, halted, or exception. When in the normal processing state, the 68000 is executing instructions that may be part of a user program. When in the halted state, the processor is not executing any instructions. The processor may have entered this state due to some kind of catastrophic system failure (such as a double bus fault or some kind of external hardware failure), or for reasons determined by the user (the $\overline{\text{HALT}}$ line may have been asserted via external hardware). The third state is the exception state, where the 68000 handles (processes) all exceptions. An exception may be loosely thought of as an interrupt, but we will see that this definition does not do justice to the entire range of exceptions and their uses.

It is possible for the 68000 to enter the exception processing state, execute some instructions, and then return to the normal execution state. It is also possible for the 68000 to enter the halted state from the exception state, if the conditions are correct. Figure 4.1 shows an example of a divide-by-zero exception encountered during normal program execution. In this example, the DIVU instruction in the user program caused the exception.

The exception state provides a method of saving the current processor **context** whenever exception handling is called for. The context of the processor, at any instant of time, is the state of all internal flags, together with the contents of all CPU registers, including the program counter. Knowing all of these values, and being able to copy them into memory for safe-

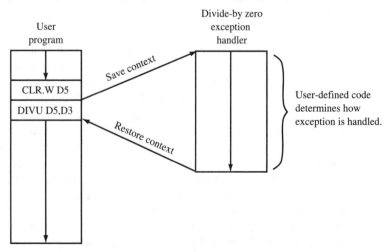

FIGURE 4.1   An exception occurring during normal program execution

keeping (usually via a software stack), provides us with a snapshot of exactly what the processor was doing when the exception occurred. Thus, reloading the processor's context at the end of exception handling enables us to continue processing right where we left off.

4.3 PRIVILEGE STATES

Associated with the three processing states are two privilege states: user and supervisor. These two states are provided so that designers of operating systems, or other complex programs that deal with user programs, may enjoy some sense of security. This is accomplished by making certain instructions **privileged.** A privileged instruction may execute only in the supervisor state. Thus, if user programs are forced to execute in the user state, their access to the processor's instruction set is limited and there is less chance that they might upset some important code in memory by writing over or changing it. Furthermore, we can design memory circuitry that distinguishes between user and supervisor accesses, and restrict certain memory accesses if we so desire. The 68000 uses a single bit in its status register to determine the privilege state.

In Figure 4.2 we see how the processor's status register is organized. Bit 13 is the S bit, which is used to determine the type of privilege: user or supervisor. If this bit is a 0, the processor executes instructions in the user state, the lower state of privilege. If the S bit is a 1, the 68000 executes instructions in the supervisor state, the higher state of privilege. It is very difficult for a program executing in the user state to change the S bit. The trace bit, bit 15, and the interrupt mask bits, I_0 through I_2 (bits 8, 9, and 10), also play an important role in certain exceptions. These bits will be covered later in this chapter. For now, we will concern ourselves with the privileged instructions of the supervisor state. The actual instructions that are restricted from execution when in the user state are the following:

STOP

RESET

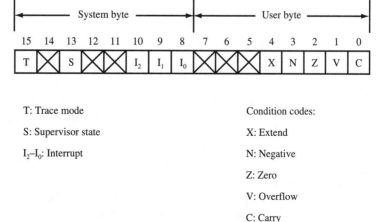

T: Trace mode

S: Supervisor state

I_2-I_0: Interrupt

Condition codes:

X: Extend

N: Negative

Z: Zero

V: Overflow

C: Carry

FIGURE 4.2 68000 status register

RTE

Move to SR

AND immediate to SR

EOR immediate to SR

OR immediate to SR

MOVE USP

The STOP and RESET instructions are excluded from execution in the user state because we cannot have user programs stopping or resetting the processor at will. This would certainly cause havoc, especially if the system is set up for multiple users. The RTE instruction is excluded because *all exception processing takes place in the supervisor state,* and RTE has no meaning in the user state. The rest of the instructions that are restricted all serve to manipulate bits in the status register, and thus would provide a way to set the S bit from the user state if they were allowed.

Another difference between user and supervisor states is that they both have their own stack pointer registers. These two registers are the USP (user stack pointer) and the SSP (supervisor stack pointer). The USP is used whenever the processor is in the user state. Even if the user program uses address register A7 for stack operations, the USP is still used. The 68000 really contains two A7 registers, one for each privilege state, accessed according to the S bit in the status register. The SSP is used when the 68000 is executing instructions in the supervisor state.

As mentioned before, it is possible to restrict portions of memory via some carefully written software and some simple circuitry in the memory section. Entire blocks of code, or data, can be excluded from access when a program is in the user state. A useful application for this technique would be in an operating system environment. Important data concerning users on the system can and should be hidden from the users, to provide system security and to prevent a user from wiping out the operating system. If we use the SSP for all protected memory references, we accomplish only half our goal. There is nothing to stop the user code from accessing protected memory, unless we provide the memory circuitry with some information about the current privilege state. Example 4.1 shows one simple way this may be accomplished.

Example 4.1:   In Figure 4.3, we see a simplified block diagram showing how access to user and supervisor memory sections is controlled via the processor's FC_2 status bit. FC_2 is low whenever the 68000 is in the user state and high when it is in the supervisor state. Since FC_2 is not used at all in the user memory section, this memory may be accessed when the processor is in either privilege state. This is what we want to achieve, for the supervisor state should not be denied access to *any* memory in the system. On the other hand, supervisor memory should be accessed only when the processor is in the supervisor state, thus the need for the inverter and the OR gate. The only way to enable the supervisor memory section is to output a low on the OR gate, which itself requires two low inputs, and FC_2 must also be high. Of course, both memory sections must contain the required address decoding circuitry; this material will be covered in Chapter 8.

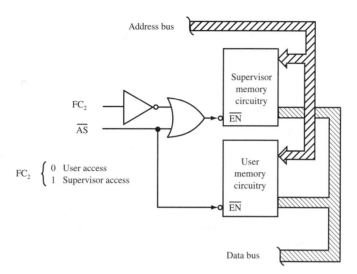

Address bus

FC_2

$\overline{AS}$

Supervisor
memory
circuitry

$\overline{EN}$

FC_2 {
0   User access
1   Supervisor access
}

User
memory
circuitry

$\overline{EN}$

Data bus

FIGURE 4.3   User/supervisor memory partitioning

Because we have two privilege states, we refer to memory references within the states as user state references and supervisor state references. Once again, these references may be distinguished from one another by the use of the processor's FC_2 status bit.

4.4   CHANGING THE PRIVILEGE STATE

We saw earlier that the instructions required to change the S bit in the status register are excluded from execution in the user state. So how does one go from the user state to the supervisor state? The only way this can be done is for an **exception** to occur (an interrupt, an illegal instruction, the occurrence of a bus error, etc.). The first step in any exception process that the 68000 performs is to save the current contents of the status register, and then to set the S bit high, so that the processor switches to the supervisor state. Remember, all exception processing takes place in the supervisor state. If the current state *is* the supervisor state and a switch to the user state is needed, there are many ways the processor can make the switch. First, the execution of an RTE instruction may take us back to the user state (if that is the state in which the exception occurred). Second, we could load the status register with a new set of status bits, clearing the S bit as we do so, by using the MOVE to SR instruction. The other ways involve the use of logical instructions AND and EOR to modify bits (specifically the S bit) in the status register. Note that we cannot use the OR instruction here, since there is no way to use the OR operation to make a 0. Example 4.2 shows how the AND instruction is used to clear the S bit.

Example 4.2:   What is the bit mask needed to clear the S bit in the status register while leaving all other bits unaffected?

Solution: When using the AND operation, a 0 input will always produce a 0 output, while a 1 input produces an output dependent on the other input. Thus, the binary mask needed is 1101 1111 1111 1111. The actual instruction would then be:

```
ANDI.W    #$DFFF,SR
```

Since ANDI is a privileged instruction, it may be executed only in the supervisor state.

4.5   EXCEPTION PROCESSING

So far we have been looking at the background information we need to be familiar with before we look into the full operation of exception processing on the 68000. We have seen that there are two important privilege states—user and supervisor—and have learned what the limitations of the user state are. We are also familiar with the status register and with how the S bit is used to control the privilege state. Let us now examine the sequence that is repeated every time an exception occurs.

Exception Processing Sequence

Processing an exception usually involves these four steps:

1. The contents of the status register are saved and the S bit is set so that the 68000 may enter the supervisor state.
2. The exception vector is obtained.
3. The program counter and status register are saved on stack.
4. Execution resumes at the address specified in the exception vector.

The next four sections will expand on these steps, which are all automatically performed by the processor.

Step 1: Adjust Status Register

A number of tasks are actually performed in this step. First, a copy of the status register is made (stored for the time being inside the processor). Then the S bit is set so that the processor may enter the supervisor state. In addition, the status register's T bit (bit 15: the trace bit) is cleared. This prevents tracing from occurring during processing of the current exception. If the exception type is a reset or an interrupt exception, the status register's interrupt mask bits (8–10) are also updated.

Step 2: Get Vector Number

Exceptions are generally referred to by their **vector number,** an 8-bit value either determined by the processor or supplied by an external device. For most exceptions, the vector number can be determined internally by the processor. The exception vector number ranges from 0 to 255 (00 to FF), and is used to point to a group of memory locations that contain the address of the routine that will handle the exception. The vector numbers are assigned as shown in Table 4.1. From Table 4.1 we see that the exception vector address-

TABLE 4.1 Exception vector asssignments

| Vector Numbers | Address | | Space[2] | Assignment |
	Dec	Hex		
0	0	000	SP	Reset: Initial SSP[3]
	4	004	SP	Reset: Initial PC[3]
2	8	008	SD	Bus error
3	12	00C	SD	Address error
4	16	010	SD	Illegal instruction
5	20	014	SD	Zero divide
6	24	018	SD	CHK instruction
7	28	01C	SD	TRAPV instruction
8	32	020	SD	Privilege violation
9	36	024	SD	Trace
10	40	028	SD	Line 1010 emulator
11	44	02C	SD	Line 1111 emulator
12[1]	48	030	SD	(Unassigned, reserved)
13[1]	52	034	SD	(Unassigned, reserved)
14	56	038	SD	Format error[4]
15	60	03C	SD	Uninitialized interrupt vector
16–23	64	040	SD	(Unassigned, reserved)
	92	05C		—
24	96	060	SD	Spurious interrupt[5]
25	100	064	SD	Level 1 interrupt autovector
26	104	068	SD	Level 2 interrupt autovector
27	108	06C	SD	Level 3 interrupt autovector
28	112	070	SD	Level 4 interrupt autovector
29	116	074	SD	Level 5 interrupt autovector
30	120	078	SD	Level 6 interrupt autovector
31	124	07C	SD	Level 7 interrupt autovector
32–47	128	080	SD	TRAP instruction vectors[6]
	188	0BC		—
48–63[1]	192	0C0	SD	(Unassigned, reserved)
	255	0FF		—
64–255	256	100	SD	User interrupt vectors
	1020	3FC		—

[1]Vector numbers 12, 13, 16 through 23, and 49 through 63 are reserved for future enhancements by Motorola. No user peripheral devices should be assigned these numbers.

[2]SP denotes supervisor program space, and SD denotes supervisor data space.

[3]Reset vector (0) requires four words, unlike the other vectors, which only require two words, and is located in the supervisor program space.

[4]MC68010 only. This vector is unassigned, reserved on the MC68000 and MC68008.

[5]The spurious interrupt vector is taken when there is a bus error indication during interrupt processing.

[6]TRAP #n uses vector number 32 + n.

es occupy the first kilobyte of storage in the 68000's address space (locations 000 to 3FF). Since this area of storage is reserved by the processor, we cannot use these locations for anything else. A common practice is to put EPROM memory in this space, so that the exception vector addresses are always present. When RAM is used instead, it is necessary to initialize the vector table at power-on. Special hardware must be used to jam the initial PC and SSP onto the data bus. All exception vector numbers except vector 0 point to a 4-byte block of memory that contains the address of the routine the processor will execute when handling the exception. Vector 0 is the reset exception and is always performed at power-on. The reset exception sequence is slightly different, in that the first 4 bytes contain the initial SSP data and the second 4 bytes contain the address of the start-up code. For this reason, we do not see vector 1 in the table.

Multiplying the vector number by 4 provides the address of the vector table entry.

Example 4.3:   Consider the user program of Figure 4.1. The DIVU instruction references a source operand, register D5, that contains zero at execution time. This causes a divide-by-zero exception, vector 5. The address the 68000 fetches the exception handler routine address from is vector 5, times 4. Thus, address 20 ($014) must contain the program counter for the start of the divide-by-zero exception handler. Note that this address must have been written into vector 5's area by the user program or some other initialization code. The processor does not automatically initialize any vectors by itself.

For exceptions resulting from external interrupts, the processor uses the interrupt acknowledge cycle to fetch the vector number from external circuitry, which must provide the 8-bit vector number on data lines D_0 through D_7.

Step 3: Save Processor Information

For all exceptions except reset, this step is used to save the current program counter and status register. Figure 4.4 shows how the values are saved in the stack area of memory pointed to by the SSP. The SSP is used because the processor has already entered the supervisor state. Remember that the SSP will decrement as we push items onto the stack. From Figure 4.4 we should be able to tell that the low word of the program counter is the first item pushed. In all exceptions, except for bus error and address error, the program counter points to the next instruction in memory that should execute when exception processing is completed. In the cases of the bus and address error exceptions, the value of the program counter is unpredictable, and even more information about the processor is saved on the stack. This information includes the first word of the aborted instruction and a record of the type of bus cycle that was attempted. This information is included to allow for some later diagnosis, via software, to determine the exact cause of the failure.

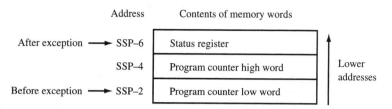

FIGURE 4.4 Exception stack contents

The high word of the program counter is pushed next, followed by the contents of the status register that existed *at the time the exception occurred.*

Example 4.4: Suppose that the SSP points to location 3FC0. If an exception occurs, the SSP is decremented and the low word of the program counter written into memory locations 3FBE and 3FBF. Then the SSP is decremented by 2, and the high word of the program counter is written to locations 3FBC and 3FBD. The SSP is decremented by 2 again and the status register is written to locations 3FBA and 3FBB. 3FBA is the final value of the SSP (the new top of the system stack).

Step 4: Fetch New Program Counter

All exceptions perform this step the same way: The new program counter is fetched from memory locations pointed to by the exception vector, and normal processing (in the supervisor state) resumes at the new address. It is assumed that the end of the exception processing code contains the RTE instruction, so that we may get back to the task we were running before the exception occurred. The addresses for the exception handlers are stored with the high word first, followed by the low word. Once again, the processor does not automatically initialize the vector addresses.

Example 4.5: A user writes code for a divide-by-zero exception handler and places the routine at 7A238. The exception vector table must be filled in the following way: Addresses 014 and 015 must contain 0007 (the high word of the starting address), and locations 016 and 017 must contain A238 (the low word of the starting address). Even though the 68000's address bus is only 24 bits wide, the address must occupy 32 bits in the vector table. This is accomplished by making the upper 8 bits 0.

The following instructions may be used to initialize the vector table entry for divide-by-zero:

```
MOVEA.L   #$14,A0      ;set up address of vector 5
MOVE.L    #$7A238,(A0) ;write handler address into vector table
```

4.6   MULTIPLE EXCEPTIONS

If you were standing on a street corner and noticed that, simultaneously, from one direction a runaway car was heading directly for you and from the other direction a large, snarling dog with big teeth was also running toward you, what would you do? This unfortunate situation is an example of what the processor must feel like when more than one exception occurs at the same time. For example, suppose that an interrupt request is received during execution of an instruction that is going to generate a divide-by-zero exception. Which exception gets executed first? Is one ignored? Let us see if we can answer these questions.

Table 4.2 shows the exception grouping and priorities that have been established by the designers of the 68000. Group 0 exceptions have higher priorities than those of the other two groups. Group 1 exceptions take priority over group 2 exceptions. This means that if a divide-by-zero exception were to occur at the same time as an interrupt, the interrupt exception would be processed first. Within a group, there are also levels of priorities. For example, if reset and address error exceptions occur at the same time, the reset exception is processed and the address error exception is ignored. There is a very simple reason why there are no priorities within group 2, but this is left for you to ponder as a homework problem. Before we take a look at some special exceptions in detail, let us see what happens during a multiple exception.

Example 4.6:   During execution of a program, a multiple exception condition occurs when the processor receives an interrupt request during execution of an instruction in trace mode (for example, the T bit in the status register is high). In this case, the trace exception will be executed first, because it has a higher priority than the interrupt exception. While the trace exception is being handled, the processor saves the interrupt request internally. We call this a **pending** request. When the trace exception finishes, the interrupt exception is finally allowed to take place. Normal program execution will

---

TABLE 4.2 Exception grouping and priority

Group	Exception	Priority
0	Reset	(Highest)
	Bus error	
	Address error	(Lowest)
1	Trace	(Highest)
	Interrupt	
	Illegal	
	Privilege	(Lowest)
2	TRAP	(All four have
	TRAPV	same priority)
	CHK	
	Zero divide	

resume when the interrupt handler finishes execution. Figure 4.5 illustrates this example. The interrupt occurs during execution of the EXG D0,D1 instruction.

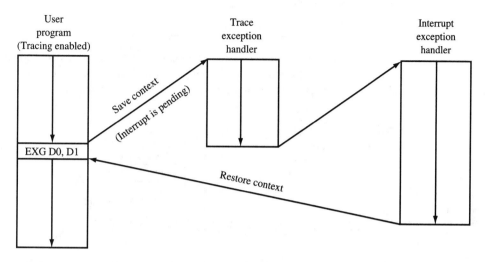

FIGURE 4.5 A multiple exception occurring during normal execution

4.7 SPECIAL EXCEPTIONS

In this section we will look at some of the more important exceptions. All of these exceptions may occur normally during execution, and it is the programmer's job to fully understand them and even plan for them, so that the executing program does not fail when one of them does occur. We will cover the special exceptions in order of their appearance in the priority table (Table 4.2).

Reset

This exception must be generated at power-on by the system hardware. Since the values of the SSP and program counter at power-on are undefined, the reset exception loads them for us. The SSP is loaded from locations 000 through 003, and the program counter from the next four locations (the locations referenced by vectors 0 and 1). Execution begins at the address contained in the program counter. In addition, the interrupt priority mask is set to level 7 to inhibit all lower interrupts from occurring during reset, and the status register is further modified so that the processor enters the supervisor state, with tracing disabled. Note that, if a reset exception occurs at a time when *any* kind of processing is taking place, that processing is terminated and *never* resumed.

The reset exception is not generated automatically by the 68000 at power-on. The external system hardware must pull $\overline{\text{RESET}}$ and $\overline{\text{HALT}}$ to ground for at least 100 ms to initiate a reset exception.

Example 4.7:   The contents of the first eight memory locations are shown as words in Figure 4.6. What is loaded into the initial program counter and what is loaded into the initial SSP when a reset exception occurs?

FIGURE 4.6 Reset exception vector contents

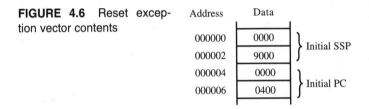

Address	Data	
000000	0000	} Initial SSP
000002	9000	
000004	0000	} Initial PC
000006	0400	

Solution: As Figure 4.6 indicates, the longword stored at address 000000 is 00009000. This is the address loaded into the SSP. The next longword (at address 000004) is 00000400. This value is loaded into the program counter. This means that the 68000 will fetch the very first instruction from address 00000400.

Bus Error

We have seen before that a bus error is one type of exception that pushes more than the standard information about the processor onto the supervisor stack. This extra information is required, since the processor was not able to completely finish its current bus cycle. It is the job of the external hardware to detect the bus error and to inform the processor about it by activating the $\overline{\text{BERR}}$ line (see Chapter 7). Since bus errors are random by nature (we never know when one might show up), there is no surefire way for the processor to determine where to resume execution when bus error exception processing is completed. Therefore, the programmer must include the necessary code for determining what to do after a bus error has occurred.

Furthermore, if a second bus error occurs while the 68000 is involved in exception processing for an initial bus error, we refer to this condition as a **double bus fault.** In this case, the processor stops all exception processing and enters the halt mode, where only a reset exception may be used to restart the processor. This is done to prevent runaway processors from destroying the contents of memory, possibly by writing bad data into it. It is not difficult to imagine the runaway processor getting stuck in a loop that constantly writes to memory.

Double bus faults are also caused by bus errors occurring during address error or reset exception processing. The information written to the stack during a bus error is shown in Figure 4.7.

FIGURE 4.7 Stack contents during bus error exception

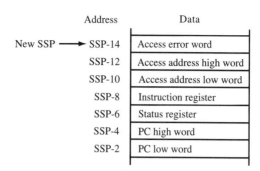

Address	Data
New SSP → SSP-14	Access error word
SSP-12	Access address high word
SSP-10	Access address low word
SSP-8	Instruction register
SSP-6	Status register
SSP-4	PC high word
SSP-2	PC low word

Address Error

Address errors occur whenever the processor tries to fetch data (whether program data or an instruction) longer than 1 byte in length from an odd addressed memory location. For example, an address error occurs if the 68000 tries to read a word (2 bytes) from an odd location. Once the address error is initiated, the current bus cycle is terminated and address error exception processing takes over. Once again, if any group 0 exception, except for reset, occurs when the address error exception is being processed, a double bus fault occurs and the processor halts.

Example 4.8: Consider the following instructions:

```
MOVEA.L    #$7000,A4
MOVE.B     (A4)+,D0
MOVE.W     (A4)+,D1
```

The second instruction uses postincrement addressing, which causes A4 to become 00007001 after the memory byte is read into D0. Since this is an odd address, the third instruction will cause an address error exception when it attempts to read a word from memory into D1.

Trace

The 68000 provides a very useful way for the programmer to debug new software. It is possible, by setting the trace bit in the status register, to generate a trace exception at the end of *every* instruction that is executed. A clever programmer will then write code that allows the trace exception to provide information about the processor address, data registers, and condition codes as each instruction executes.

Some instructions, such as illegal and privileged instructions, disable the trace facility. All group 0 exceptions disable the trace facility, too. Otherwise, the trace exception follows the priority scheme within group 1, meaning that if a trace exception is called for and an interrupt has been received, the trace exception is processed first, followed by the interrupt exception. We should remember here that tracing is disabled during *all* exception processing, to avoid cluttering up the supervisor stack.

Interrupt

The 68000 provides seven levels of external interrupts as shown in Table 4.3. The processor's three $\overline{IPL}$ inputs are used to select the interrupt level. Level 7 is referred to as a **nonmaskable** interrupt, because it cannot be masked out via the interrupt priority mask in the status register. Normally, the priority mask is set to a certain interrupt level, say level 4, to disable lower priority interrupts (levels 1–4) from occurring. Of course, the level is set by the programmer to satisfy program constraints. Level 0 is a noninterrupt condition, meaning that no interrupt is requested, and also has the lowest priority. Level 7 has the highest priority among the external interrupts.

When external logic indicates an interrupt request, the processor treats it as a pending interrupt, and will not act on the request until the end of the current instruction. Then the processor will compare the pending interrupt priority with that of the priority mask in the status register, and unless the pending interrupt level is of a higher priority than the priority mask, the interrupt is ignored.

Example 4.9:   The status register is loaded with information during execution of the following instruction (in the supervisor state):

```
MOVE    #$2300,SR
```

What are the contents of each flag, and what interrupts are allowed?

Solution: The immediate data $2300 is 0010001100000000 in binary. The upper 8 bits set the trace flag to 0, the supervisor flag to 1, and the three interrupt priority mask bits to 011. This priority mask will allow only interrupt levels 4 through 7 to be acknowledged.

The lower byte of the immediate data clears all condition code flags.

The circuit in Figure 4.8 shows one way 7 different devices may request interrupts. In this figure a 74LS148 priority encoder is used to generate the proper levels for the $\overline{IPL}$

TABLE 4.3 Hardware interrupt levels

Processor inputs			
$\overline{IPL_2}$	$\overline{IPL_1}$	$\overline{IPL_0}$	Interrupt level
0	0	0	7 (Highest)
0	0	1	6
0	1	0	5
0	1	1	4
1	0	0	3
1	0	1	2
1	1	0	1
1	1	1	0 (Lowest, none)

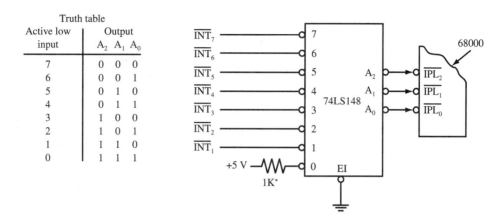

Truth table

Active low input	Output A_2 A_1 A_0
7	0 0 0
6	0 0 1
5	0 1 0
4	0 1 1
3	1 0 0
2	1 0 1
1	1 1 0
0	1 1 1

*Input 0 may also be connected to ground without changing the operation of the encoder.

FIGURE 4.8 Interrupt priority encoder

lines. The LS148 has eight inputs, all normally high, and three outputs, which are also normally high due to the pull-up resistor on input 0. As an example, imagine that $\overline{INT_7}$ is pulled low. This causes the 74LS148 to encode this into three low levels on outputs A_2 through A_0 (the 3-bit 1's complement of 7, the input that is now activated). This in turn causes the 68000 to register a level-7 interrupt request. When more than seven interrupt lines are needed, the 74LS148s may be daisy-chained together (via control inputs and outputs), but the designer is limited to a maximum number of 192 external interrupts. Remember from Table 4.1 that there are a number of reserved exceptions—63 in fact—that keep us from using all 255 available interrupt vectors externally.

When we use an external interrupt, the vector table address will be generated in one of two ways. Examine Figure 4.9 for a moment, and notice how the 68000 communicates with its external interrupt circuitry. The sequence of events that takes place when an external device requests an interrupt is lengthy. Initially, the interrupt logic responds to the interrupting device by generating a priority level for it and placing this 3-bit code on the 68000's $\overline{IPL}$ inputs. The processor will recognize the pending interrupt request but will not act on it until it completes execution of the current instruction. Then the processor will enter an interrupt acknowledge bus cycle. The external interrupt logic recognizes the interrupt acknowledge bus cycle by monitoring the function code outputs. At the beginning of the interrupt acknowledge cycle, the processor will output a zero on $\overline{AS}$ and $\overline{LDS}$, a high level on R/$\overline{W}$, and the requested interrupt level on address lines A_3 through A_1. At this time, the external interrupt logic may do one of two things. It may supply the vector number to the 68000, or it may request an **autovector.** To supply the vector number, the interrupt logic must place the 8-bit vector number on data lines D_7 through D_0 and pull $\overline{DTACK}$ low. The 68000 will monitor $\overline{DTACK}$ and load the supplied vector number accordingly. If the interrupt logic instead requires that the 68000 use an autovector, it will pull $\overline{VPA}$ low instead of $\overline{DTACK}$. When the 68000 sees $\overline{VPA}$ go low, it will generate its own vector, based on the interrupt level first supplied to the IPL inputs. The autovectors occupy locations 064 through 07F in the vector

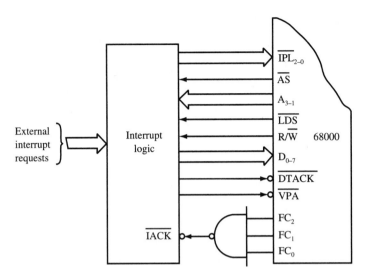

FIGURE 4.9   External interrupt circuitry block diagram

address table (see Table 4.1). Autovectoring should be used whenever the external hardware requires only a few interrupts (or a maximum of seven).

Example 4.10:   The original Macintosh computer had three sources of hardware interrupts: the VIA (Versatile Interface Adaptor), the SCC (Serial Communication Controller), and an external interrupt pushbutton. When one or more of these devices issued an interrupt request, motherboard hardware assigned an interrupt level chosen from those shown in Table 4.4. This allowed the processor to respond to special events, such as mouse movements, serial data transfers, and user debugging requests.

TABLE 4.4   Macintosh interrupt assignments

Interrupt level	Interrupt source
0	None
1	VIA
2	SCC
3	SCC + VIA
4	Pushbutton
5	Pushbutton + VIA
6	Pushbutton + SCC
7	Pushbutton + SCC + VIA

If the external logic does not respond with either $\overline{DTACK}$ or $\overline{VPA}$, but instead indicates a bus error, this indicates that a **spurious** interrupt has occurred, and the 68000 will use vector 24 as it begins exception processing.

If more than seven levels of external interrupts are needed, autovectoring cannot be used. Figure 4.10 gives an example of how 16 external interrupts might be generated. This circuit employs two 74LS148 priority encoders. Each encoder handles eight interrupt lines. You may wish to refer to the truth table for the LS148 before continuing. The $\overline{EI}$ (input enable) of the upper 74LS148 is grounded, so that $\overline{INT}_{15}$ through $\overline{INT}_8$ will always be recognized. $\overline{EO}$ (enable output) of the upper 74LS148 controls the lower priority encoder. If none of the upper interrupt lines are active, $\overline{EO}$ will be low, enabling the lower priority encoder to respond to $\overline{INT}_7$ through $\overline{INT}_0$. Thus $\overline{EI}$ and $\overline{EO}$ provide a method for prioritizing all 16 inputs.

If none of the interrupt lines are active (for example, they are all high), the $\overline{GS}$ outputs of both encoders will be high. This has the effect of placing a level 0 request on the $\overline{IPL}$ inputs of the processor. This represents a noninterrupt condition. If any of the $\overline{INT}$ lines go low, one of the $\overline{GS}$ outputs will go low also, which generates a level-7 interrupt request. When either $\overline{GS}$ output goes low, the A_2 through A_0 outputs of the priority encoder represent the encoded number of the activated input. These three ouputs, together with $\overline{EO}$, are combined to form a true 4-bit code that represents the number of the $\overline{INT}$ line that requested service. This code is latched by the 74LS374 octal D-type flip-flop. When the 68000 enters the interrupt acknowledge cycle, the $\overline{INTACK}$ signal will go low. This causes two things to happen. The 74LS374, previously tri-stated, now places the vector number on the data bus. In addition, $\overline{DTACK}$ is asserted. When the processor completes the interrupt acknowledge cycle, $\overline{AS}$ will go high, the function code outputs will change, and $\overline{INTACK}$ will return to a high state, tri-stating the LS374 and removing $\overline{DTACK}$.

This circuit can be further analyzed and improved, and these exercises are left for you to do as homework.

Illegal Instructions

The 68000 provides for detection of illegal instructions encountered during program execution. An illegal instruction is any code read from memory, during an instruction fetch, that does not match the bit patterns of any of the predefined opcodes in the 68000's instruction set. When an illegal instruction is encountered, exception processing begins at the address contained in vector 4, locations 010 through 013 (see Table 4.1). Illegal instructions may be encountered if the processor does an instruction fetch from a data area, or if the processor begins a fetch at an undefined address.

Three opcodes are reserved and will generate an illegal instruction exception if encountered. They are 4AFA, 4AFB, and 4AFC. Only 4AFC should be used by the programmer (via the ILLEGAL instruction), because the first two are reserved by Motorola.

Unimplemented Instructions

Unimplemented instructions are not illegal instructions. Instead, they are provided to give Motorola and the user a chance to enhance the instruction set of the 68000, by adding custom (or additional) instructions at a later time. All unimplemented instructions begin with

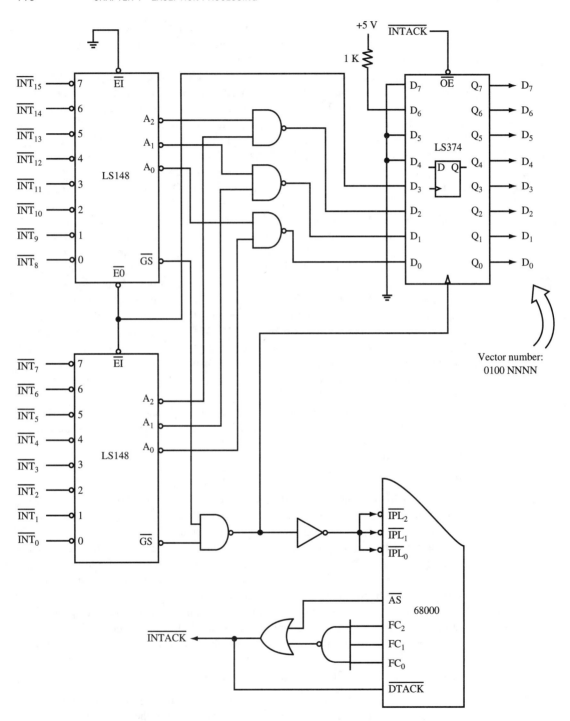

FIGURE 4.10   External interrupt circuitry providing 16 levels of interrupts

either A or F. F345, A200, FF5C, and AB00 are all examples of valid unimplemented instructions. They must contain A or F in their upper nybble.

When either of these types is encountered, exception processing will begin. If the unimplemented instruction begins with A, exception vector 10 is used. If it begins with F, exception vector 11 is used. Any instruction not beginning with A or F, and not contained in the instruction set, is an illegal instruction. See Table 4.1 for the vector addresses.

Example 4.11: The original Macintosh computer made good use of the A-line unimplemented instruction format. Different opcodes were assigned for various operating system functions (file management, timekeeping, memory management) and other important software tasks (screen graphics, arithmetic). These opcodes all began with 1010 in the upper 4 bits. Table 4.5 shows just a few of the assignments.

TABLE 4.5 Sample Macintosh A-line traps

Trap word	Function name
A00C	GetFileInfo
A00D	SetFileInfo
A10C	FreeMem
A11D	MaxMem
A03A	SetDateTime
A8F6	DrawPicture
A8F8	ScalePt
A8F9	MapPt
A8FA	MapRect
A8FC	MapPoly
A8FE	InitFonts
A992	DetachResource
A99E	CountTypes
A9A0	GetResource

Since opcodes are 16 bits long, there are 12 bits left over for generating unique patterns A000 through AFFF. This corresponds to 4,096 different opcodes, all of which begin with A. This means that we have the potential of performing 4,096 different tasks. What the designers of the Macintosh did was to use a subset of these 4,096 opcodes to represent different functions to be performed. Figure 4.11 shows how the designers of the Macintosh assigned bits within the 12-bit field. Bit 11 controls the interpretation of the remaining 11 bits. When bit 11 is low, the lower 8 bits of the opcode represent an **operating system trap.** This allows a total of 256 different operating system functions, with each function being identified by a unique pattern from 00 to FF.

When bit 11 is high, the lower 9 bits of the opcode represent a **Toolbox trap.** A total of 512 Toolbox functions are possible (functions 00 through 1FF). Whenever either type of trap is encountered (any instruction beginning with A on the Macintosh), the 68000 will

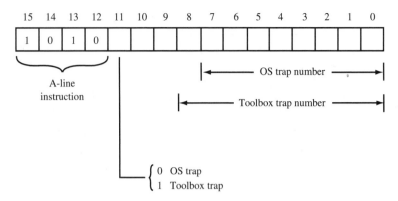

FIGURE 4.11   Macintosh trap word format

enter into exception processing. Recall that the processor automatically pushes the return address and flags onto the stack at the beginning of exception processing.

In order to process the exception, the Macintosh must be able to obtain a copy of the A-line instruction that caused the trap. Figure 4.12 shows one way this may be done. Figure 4.12(a) indicates the A-line instruction causing the trap (A10C). When this instruction is encountered, the 68000 begins exception processing. The return address of the next instruction (303C at address 007030) is pushed onto the stack along with the flags in Figure 4.12(b). Note that the SSP is being used for stack access now, since all exception processing takes place in the supervisor state. The processor then reads the A-line handler address out of the vector table and begins execution of the handler code. The first two instructions of the handler code are:

```
MOVEA.L    2(A7),A0
MOVE.W     -(A0),D0
```

In Figure 4.12(c) the MOVEA.L   2(A7),A0 instruction is used to read the return address (007030) out of stack memory and into A0. This address is equal to the address of the previous A10C instruction, plus 2. So, we must read a word from address 00702E to get the original A-line instruction back. This is accomplished in Figure 4.12(d) where the MOVE.W   –(A0),D0 instruction first subtracts 2 from A0 before reading the A-line instruction opcode out of memory and into D0.

Once we have a copy of the instruction, we can test bit 11 to determine if the instruction is an OS trap or a Toolbox trap. This is done as follows:

```
BTST.L    #11,D0        ;Zero flag is set if bit 11 equals 0
BEQ       OSTRAP        ;Bit 11 equals 0
BRA       TOOLBOXTRAP   ;Bit 11 equals 1
```

Then, to guarantee that only the trap number is left in D0, the upper bits of the register are cleared with an appropriate AND instruction, as in:

```
ANDI.W     #$00FF,D0     ;D0 equals OS trap number
```

or

```
ANDI.W     #$01FF,D0     ;D0 equals Toolbox trap number
```

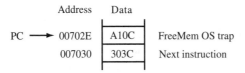

(a) Processor fetches A-line instruction.

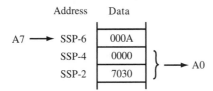

(b) Processor pushes return address and flags onto stack.

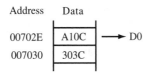

(c) MOVEA.L   2(A7),A0 reads return address into A0.

Address   Data

00702E   A10C ⟶ D0
007030   303C

(d) MOVE.W   –(A0),D0 reads A-line instruction into D0.

FIGURE 4.12   Implementation of A-line exception

The trap number in D0 can then be used to select and execute the associated routine.

A technique similar to that used in Example 4.11 may be applied to the F-line exception. However, since the 68881 floating-point coprocessor utilizes instruction words that always begin with F, the F-line exception is usually reserved for software emulation of the coprocessor when the physical chip is not included in the hardware of the system. This means that all of the floating-point instructions are emulated with appropriate software routines, which are in turn activated through the F-line exception handler provided by the user.

Privilege Violation

We saw in Section 4.3 that there are a number of instructions that the 68000 considers to be privileged. STOP, RESET, RTE, ANDI to SR, and others may only execute in the supervisor state. Any attempt to use these instructions from the user state results in a privilege violation exception. The 68000 will use vector 8 to process this error condition. This exception is very valuable for a programmer involved in the creation of an operating system, where strict lines must be drawn between users and **superusers,** special users that contain privileges that other users do not have. If a method did not exist for separating these two kinds of users, the operating system would have a very low level of security.

TRAP

The TRAP instruction may be used by the programmer to generate exceptions from *within* a program. A common term for these types of exceptions is the **software interrupt** or **supervisor call.** The TRAP instruction gives us a method to enter the supervisor state through a process under our control. Since TRAP is similar in operation to an interrupt, we apply the nonmaskable term to it, because there is no way to disallow a TRAP from occurring when it is encountered.

The form of the TRAP instruction is:

```
TRAP    #N
```

where 'N' is a value from 0 to 15. Thus, there are only 16 allowable TRAPs. Each TRAP causes the 68000 to generate a unique address within the vector address table (Table 4.1). TRAP #0 causes the CPU to vector to the address contained in memory locations 080 through 083. Likewise, TRAP   #15 causes a vector to the address contained in locations 0BC through 0BF. Figure 4.13 shows how these addresses may be computed from the TRAP vector number. The actual vector numbers for the TRAP instructions (per Table 4.1) range from 32 to 47.

A direct application of this technique is in the creation of assemblers, which must somehow generate unique codes from simple input data, such as the 0 to 15 vector number. Further examination of Figure 4.12 shows that the starting address of any TRAP vector is equal to $80 plus 4 times the vector number.

TRAPs work like any other kind of exception. The program counter and status register contents are saved on the supervisor stack, and the 68000 continues execution at the address supplied by the vector address table.

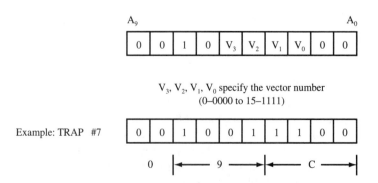

FIGURE 4.13   Calculating the vector table address from the TRAP vector number

Example 4.12:   The single-board computer designed in Chapter 11 uses the first 10 TRAP vectors for its own purposes to perform a wide variety of tasks. As shown in Table 4.6, these TRAPs can be used to perform character I/O, read and display hexadecimal numbers of various sizes, and restart the SBC monitor program. Since these functions may be useful to a user program, they can be executed by placing the appropriate TRAP instruction within the user program. To illustrate, examine the HELLO program from Chapter 1 again.

TABLE 4.6   Predefined SBC TRAPs

Trap #	Function
0	CHAR_IN
1	CHAR_OUT
2	CRLF
3	PRINT_MSG
4	PRINT_BYTE
5	PRINT_WORD
6	PRINT_LONG
7	GET_BYTE
8	GET_ADDR
9	GETCMD

```
        ORG     $8000           ;starting address of data
HMSG    DC.B    'Hello!'        ;message characters
        DC.B    0               ;end-of-message marker
        ORG     $8100           ;starting address of program
START   MOVEA.L #HMSG,A3        ;load A3 with message address
        TRAP    #3              ;output message
        TRAP    #9              ;return to command processor
        END     START           ;end of source file
```

Now we might have a better understanding of why TRAP   #3 and TRAP   #9 were used in the HELLO program. The address of the "Hello!" message is passed to TRAP   #3 through A3. TRAP   #3 requires that the last byte in the message be 00 (to know when to stop outputting characters). TRAP   #9 automatically restarts the monitor program, giving control back to the user in command mode.

A similar mechanism has been used in the EMU68K program to allow simple character I/O on the PC. As Table 4.7 shows, TRAPs 0, 1, 3, and 9 perform the same function in EMU68K as they do on the SBC. This allows you to write 68000 programs that are

TABLE 4.7   Predefined EMU68K TRAPs

Trap #	Function
0	CHAR_IN (from PC keyboard)
1	CHAR_OUT (to PC display)
3	PRINT_MSG (to PC display)
9	GETCMD (return to emulator)
10	OS CALL (operating system functions)

somewhat interactive. For example, assemble the following program with ASM68K and run it with EMU68K:

```
        ORG     $8000       ;starting address of program
START   TRAP    #0          ;read PC keyboard character
        CMPI.B  #'Q',D1     ;did user enter Q?
        BEQ     EXIT        ;branch if true
        TRAP    #1          ;else, output character to PC display
        BRA     START       ;repeat from beginning
EXIT    TRAP    #9          ;return to EMU68K command processor
        END     START       ;end of source file
```

The program should simply echo all keys pressed until a Q is entered.

Trap #10 on EMU68K allows the programmer to access the PC's operating system functions through DOS INT functions. This provides a wide assortment of additional features to EMU68K. More details of TRAP #10 are given in Chapter 6.

TRAPV

This instruction is similar to the TRAP instruction, except that TRAPV does not require an operand field and will generate an exception only if the overflow (V) flag is set. If an exception is generated, the 68000 will continue with execution at the address specified in locations 01C through 01F. If the V bit is not set, TRAPV will not initiate exception processing, and execution will continue at the next instruction in the program.

TRAPV is very useful in routines that crunch numbers. For example, following every ADD instruction with a TRAPV will ensure that any out-of-bound sums will be identified.

CHK

This instruction is used to check the value in a data register to see if it is within a certain range. Only the lower word of the data register is checked. The upper limit of the range is specified in the operand field of the instruction. The lower limit is always 0. CHK will initiate exception processing if the register contents do not fall within the specified range (because it is either negative or greater than the upper limit). If exception processing is called for, the 68000 will use vector 6; thus, the CHK service routine address should be stored at locations 018 through 01B. An example using CHK follows.

Example 4.13   During execution of a program, CHK #$300,D6 is encountered. Register D6 contains the value 4E3001F7. Since the lower word of D6 (01F7) does not exceed 0300, no exception processing is initiated.

Divide-by-Zero

Even the best programmers cannot foresee all occurrences of errors in their programs, or test a working version of the program with all possible input data conditions. The divide-by-zero exception is therefore a useful tool to help the programmer deal with future events that may supply bad data. DIVU and DIVS are the only two instructions that may cause a divide-by-zero exception. In both cases, the instructions will examine the contents of the source operand and initiate the exception if the source operand is zero. Locations 014 through 017 in the vector address table (vector 5) must contain the address of the exception handler.

A Note About Subroutines

The similarities between exceptions and subroutines are numerous. Both use the stack for return addresses and temporary storage, and both change the normal flow of instructions into the processor. When we use a planned exception, such as a TRAP, it is very similar to using a subroutine, with the difference that the processor flags are pushed, along with the return address.

It is useful to use the stack to save the parameters we want to send into the subroutine (or planned exception). One way to pass parameters via the stack is illustrated in the following code:

```
LINK     A4,#-6              ;set up stack frame
MOVE.W   #$1234,-2(A4)       ;store first parameter
MOVE.L   #$EEEEFFFF,-6(A4)   ;store second parameter
BSR      ASUB                ;call subroutine
UNLK     A4                  ;release stack frame
```

These instructions make use of a *stack frame,* a portion of stack memory reserved by the programmer for storage of local variables and parameters used by a subroutine. Figure 4.14 shows how the LINK   A4,#–6 instruction sets up a stack frame containing 6 bytes of reserved storage *within* stack memory. The current contents of register A4 ($00028000 for this example) are pushed onto the stack first. Then A4 is loaded with the current value of the stack pointer (register A7). The immediate operand from the LINK instruction (–6) is added to the stack pointer, moving it to the position shown.

The contents of the stack frame after the subroutine parameters have been stored are shown in Figure 4.15. Address register indirect with displacement addressing is used

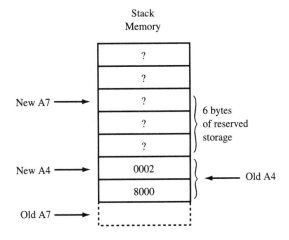

·**FIGURE 4.14** Execution of LINK   A4,#–6

to save the input values in the appropriate locations. The word $1234 requires a displacement of −2. The longword $EEEEFFFF requires an additional displacement of −4, for a total of −6. The beauty of this method is that the input values are accessed *independently* of the stack pointer. For instance, Figure 4.16 indicates that the return address ($0003E900) for ASUB's call has been pushed onto the stack. This in no way affects our ability to access the input values; they are still at the same displacements relative to A4 (−2 and −6). This is just as useful if the BSR instruction is replaced with a TRAP or some other exception. The additional information pushed onto the stack during exception processing will not change the relative displacements either. The same is not true when data is accessed relative to the stack pointer.

The stack frame is useful for passing parameters back from a subroutine (or planned exception) as well. Suppose ASUB modifies the word at displacement -2 (to send a result back to the caller). This value can be retrieved before the stack frame is discarded by adding an additional instruction to our short example sequence.

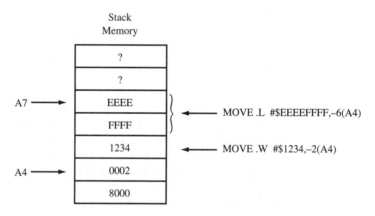

FIGURE 4.15 Saving parameters in stack memory

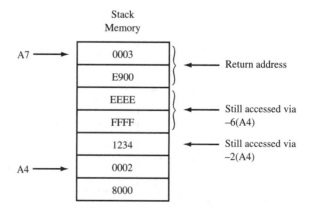

FIGURE 4.16 Calling a subroutine that uses stack parameters

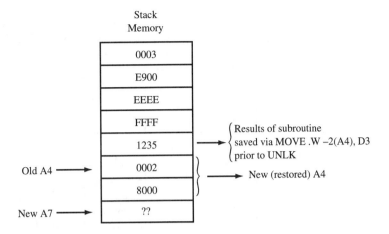

FIGURE 4.17 Execution of UNLK   A4

```
LINK     A4,#-6                  ;set up stack frame
MOVE.W   #$1234,-2(A4)           ;store first parameter
MOVE.L   #$EEEEFFFF,-6(A4)       ;store second parameter
BSR      ASUB                    ;call subroutine
MOVE.W   -2(A4),D3               ;save result returned by ASUB
UNLK     A4                      ;release stack frame
```

Figure 4.17 shows why it is important to save the results before using UNLK. When UNLK executes, the stack pointer (A7) is loaded with the value of A4. This positions the stack pointer at the beginning of the stack frame. Then a longword is popped off the stack and stored in A4 (the original contents of A4). The displacements have no meaning after the value of A4 is changed.

In general, take care when using the stack to pass parameters. It is very easy to misjudge a displacement and damage valuable stack information. Further details of this technique, and how it is used to support C functions, are provided in Chapter 6.

4.8   EXCEPTION HANDLERS

The exception handler is the actual section of code that takes care of processing a specific exception when it is encountered. The exception handler for a divide-by-zero exception would then be drastically different from a handler written to process a privilege violation or another that handles a level-5 interrupt.

Even though these handlers are written to accomplish different goals, there are portions of each that, for the sake of good programming, look and operate the same. Recall that any time an exception occurs, the 68000 enters the supervisor state, and pushes the program counter and status register onto the supervisor stack, before vectoring to the address of the exception handler. Clearly, we must see that the exception handler will change the contents of various data or address registers while it is processing the exception. Since

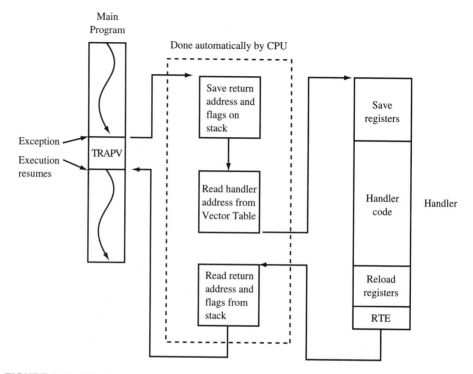

FIGURE 4.18   Typical exception handler operation

we desire to return to the same point in our program where we left off *before* the exception occurred and resume processing, we insist that all prior conditions exist upon return. This means that we must return from the exception with the state of all data and address registers preserved. It is now the responsibility of the exception handler to preserve the state of any registers that it alters. Figure 4.18 shows how this is done. In this example, TRAPV causes an exception. The first thing the exception handler does is save the data and address registers on the stack. These registers may be saved individually, using an instruction like

```
MOVE.L    D3,-(A7)
```

or all data and address registers except for A7 may be saved at the same time, with the instruction

```
MOVEM.L    D0-D7/A0-A6,-(A7)
```

which performs 15 pushes onto the supervisor stack. We do not bother to save A7 because it is designed to be used as a stack pointer. Using it as a general purpose register automatically lowers the performance of your software.

When the body of the exception handler finishes, it is necessary to reload the registers that were saved at the beginning of the routine. Again, this may be done singly, with

```
MOVE.L    (A7)+,D3
```

or all at once, with

```
MOVEM.L    (A7)+,D0-D7/A0-A6
```

as you so desire. Saving all registers is preferable, and will save you much heartache in the future, when you find that saving one or two registers was insufficient as the needs of the routine became more complex.

Example 4.14   A 60-Hz clock is connected to circuitry that generates a level-2 interrupt with each pulse. An exception handler for this interrupt must be written to call the subroutine ONESEC whenever 60 interrupts have been received. The handler shown here is one possible solution:

```
LEVEL2     MOVEM.L    D0-D7/A0-A6,-(A7)
           MOVE.B     COUNT,D0
           SUBI.B     #1,D0
           BNE        NEXT
           BSR        ONESEC
           MOVE.B     #60,D0
NEXT       MOVE.B     D0,COUNT
           MOVEM.L    (A7)+,D0-D7/A0-A6
           RTE
```

A few points deserve mention. First, notice that *all* processor registers are saved at the beginning of the handler. Why is this necessary if the only register modified is D0? The answer has to do with ONESEC. We will call this routine once every second. With no information about what registers are used by ONESEC, we must assume that all registers have the possibility of being modified.

   Also, we must use RTE at the end of the handler (instead of RTS) since the routine is servicing an interrupt, and executes in supervisor state with the return address and flags on the stack.

4.9   TROUBLESHOOTING TECHNIQUES

It pays to remember the details of exception processing when troubleshooting a program. Many times the fault of erratic execution in a 68000-based system is poorly written or incomplete exception handlers. Reviewing the basic principles can help eliminate some of the more obvious problems.

- All exception processing is performed in the supervisor state.
- A typical exception pushes the current program counter and flags.
- Vector addresses are equal to four times the vector number.
- The first vector is reserved for the initial supervisor stack pointer and program counter.
- It may be necessary to save and restore registers (via MOVEM) in the exception handler.

- Use RTE (return from exception) to return from an exception handler. RTS does not work properly with exceptions.
- For an exception to work, its vector must be loaded with the starting address of the handler, and the handler code must be in place as well.
- The $\overline{IPL}$ inputs must all be high when not requesting a hardware interrupt.
- The interrupt level (7 is highest) is selected when all three $\overline{IPL}$ inputs are low.

These points are a minimal set of programming tips. You will discover others as you begin writing your own exception handlers.

SUMMARY

We have seen that there are three processing states in the 68000: normal, halted, and exception. In addition, the 68000 supports two privilege states—user and supervisor—with all exception processing being done in the supervisor state. Some instructions are privileged and may only be executed in the supervisor state. Any attempt to utilize them while in the user state results in a privilege violation. Furthermore, the only way to go from the user state to the supervisor state is through exception processing.

We also saw that there is a fixed process used by the 68000 to implement an exception. The CPU enters the supervisor state (by setting the S bit in the status register), saves the current program counter and status register on the supervisor stack, and loads the exception routine address from a special vector address table. The vector address table occupies locations 000 through 3FF in memory and contains pairs of words that represent execution addresses for the 255 allowable exceptions. The vector number used to calculate the vector table address may be internally generated by the 68000, or may be supplied by external hardware during an interrupt acknowledge cycle. The 68000 supports seven levels of external, prioritized interrupts.

Exceptions are either caused by an external hardware failure, by an external hardware request, or through software. The software exceptions may be generated intentionally by the user, or by accident, via a run-time error (such as division by 0). All exceptions (except for reset and double bus faults) allow the program to resume execution where it was interrupted.

STUDY QUESTIONS

1. What is the processor's context? Why is it important to save the context during exception processing?
2. How does the 68000 prevent a user program from entering the supervisor state, by any method other than exception processing?
3. Explain what happens if the 68000 encounters the STOP instruction while in the user state.

4. Explain the different uses of address register A7 in both user and supervisor states.
5. How is FC_2 used to distinguish between the user and supervisor states?
6. Why must an exception handler terminate with an RTE instruction? What kinds of stack problems would occur if RTS were used instead?
7. In Example 4.2, an ANDI instruction was used to clear the S bit and thus select the user state. Modify the operand so that the trace bit is also cleared.
8. How may the status register be changed so that the processor goes from the supervisor state to the user state, with no other bits affected, if an EORI to SR instruction is used? Show the exact instruction needed.
9. Show the memory locations that are altered, and the new data in them, if exception processing begins with the following register values: SSP = 0081FA, PC = 002578, SR = 0314.
10. Why are there only 255 exception vectors, and not 256?
11. From what locations will the exception routine address be fetched, if vector 70 is used?
12. Figure 4.19 shows the contents of a few locations within the vector table address space. What will the new program counter be when a TRAP   #3 exception occurs?
13. Why is there no need for a priority scheme within group 2 (see Table 4.2)?
14. What happens if an external interrupt arrives during execution of a TRAP instruction?
15. Assume that each exception handler implemented in a particular system saves all data registers and all address registers (except for A7). How many levels of nested exceptions are possible with a stack size of 1,024 words?
16. How is the trace exception useful for examining the operation of a program?
17. What hardware event signals a bus error condition?
18. What happens if a bus error is followed by an address error?
19. During an interrupt acknowledge cycle, the value 7C is placed on D_0–D_7. From what locations does the processor fetch the exception routine starting address?
20. What autovector is used if the $\overline{IPL}$ inputs are at 101?
21. Design a prioritized interrupt circuit that has two inputs: $\overline{INT}_a$ and $\overline{INT}_b$. $\overline{INT}_a$ has higher priority and should generate a level-6 interrupt when active. $\overline{INT}_b$ should generate a level-3 interrupt.
22. Refer to Figure 4.10. What vector number is supplied to the 68000 if $\overline{INT}_{14}$ is active? Repeat if $\overline{INT}_2$ is active.
23. What vector number is generated if $\overline{INT}_{10}$ and $\overline{INT}_6$ are active at the same time, in the circuit of Figure 4.10?

FIGURE 4.19 For Question 4.12

Address	Data
08A	518B
08C	002A
08E	639C
090	0047

24. What happens if the 68000 encounters the opcode 4AFC during an instruction fetch? What happens if A300 is encountered?

25. The status register contains the value 0C in the user byte. Will TRAPV generate an exception?

26. Data register D2 contains the value 1FB3009A. What occurs when CHK.W #80,D2 is encountered?

27. An exception handler changes the values of registers D1, D4, and D5. How might a MOVEM instruction be used to save these registers?

28. A divide-by-zero exception handler routine is placed at location 30A50. Show the required memory contents in the appropriate locations within the vector address table.

29. Suppose that bits 10, 9, and 8 in the status register are set to 100, and a level-2 interrupt arrives. What happens?

30. Is an RTE needed at the end of the reset exception?

31. How are the lower three address lines used during an interrupt acknowledge cycle?

32. What is the most probable exception generated if the 68000 is mistakenly directed to fetch an instruction from an area of stack memory?

33. Design an external interrupt circuit that will bring $\overline{\text{VPA}}$ low whenever an interrupt acknowledge cycle occurs.

34. Repeat Question 33, but limit autovectoring to only level-4 and level-6 interrupts.

35. Design an external circuit that places vector 9E on the data bus during an interrupt acknowledge cycle.

36. Modify the design of Figure 4.10 so that vectors C0 through CF are generated.

37. Write an exception handler that will add the contents of registers D0 through D6 to D7 and return the new value of D7. The exception handler should respond to TRAPV.

38. Modify the keyboard echo program of Example 4.12 so that:
    a) the program exits if a lowercase q is entered (in addition to uppercase Q)
    b) two codes, $0D and $0A, are output to the display whenever $0D is received from the keyboard.

39. Design a circuit that enables an EPROM at power-on in the address range 000000–001FFF, but disables the EPROM and enables a RAM after the initial PC and SSP have been read.

40. Show the instructions required to store three parameters into the stack frame prior to calling the subroutine ASUB. The following registers must be saved: D0 (.L), D1 (.L), and D2 (.W). Use the LINK instruction to set up a stack frame using register A4.

PART 3

Programming

Flowcharts allow a graphical
approach to program design
and analysis.

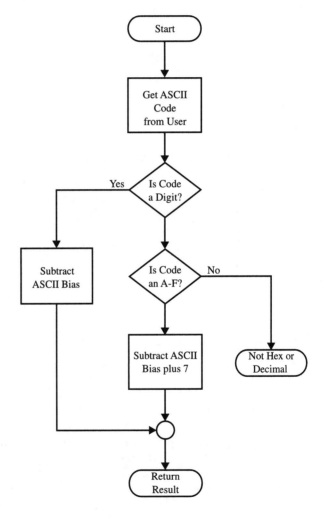

CHAPTER 5

An Introduction to Data Structures

OBJECTIVES

In this chapter you will learn about:

- The method for defining constants and variables
- The storage and access requirements of arrays and strings
- Two-dimensional arrays
- Different techniques for accessing stored information
- The difference between static and dynamic storage allocation
- The operation of linked-lists and binary trees
- The operation of stacks and queues

5.1   INTRODUCTION

A programmer must have a good grasp of a processor's instruction set before any real programming challenges can be undertaken. Just as important to the programmer are the various ways of representing data. Data comes in a wide variety of forms, such as bits, bytes, words, and longwords. Let us take the time to review what we have seen about these types of data. Table 5.1 shows how we access data stored in the 68000's data registers. The lower 8 bits of a data register are used for 8-bit operations. The lower 16 bits are used for word operations. Longwords require a total of 32 bits (4 bytes) for storage. We can easily specify the size of a data value through the use of the required size extension, as in:

TABLE 5.1 Byte, word, and longword storage

Data type	Range	Data register bits used
Byte	$00–$FF	0–7
Word	$0000–$FFFF	0–15
Longword	$00000000–$FFFFFFFF	0–31

```
MOVE.B      #32,D5
MOVE.W      #5000,D6
MOVE.L      #186000,D7
```

Note that the C language typically reserves the same amounts of storage for three important data types: char, int, and long. Table 5.2 shows the correspondence.

Another data type is BCD. Each BCD digit requires 4 bits to represent its value of 0 to 9. Table 5.3 shows the range of BCD digits possible with the three common data sizes. Both the C data types and each range of BCD numbers may be stored in any data register (or group of memory locations). In Chapter 6 we will examine how to work with these data types in more detail. We also group large chunks of data together into **structures,** and call them arrays, matrices, or linked-lists. In this chapter we will examine the different ways data can be represented and accessed.

Section 5.2 shows how constants are defined and used. Section 5.3 presents two methods of defining variables for storage in memory. One-dimensional arrays are the subject of Section 5.4, which is followed by two-dimensional arrays and ASCII character strings in Sections 5.5 and 5.6, respectively. An explanation of the difference between

TABLE 5.2 Sample C data types

C data type	Bits	Range	
		Unsigned	Signed
char	8	0–255	−128–127
int	16	0–65535	−32768–32767
long	32	0–4294967295	−2147483648–2147483647

TABLE 5.3 BCD ranges

Data size	BCD* range
Byte	$00–$99
Word	$0000–$9999
Longword	$00000000–$99999999

*Note: $99 means 99 BCD, not 99 hexadecimal, although they are indistinguishable.

static and dynamic storage allocation is given in Section 5.7. Sections 5.8 through 5.10 provide coverage of linked-lists, binary trees, and stacks and queues, respectively. Troubleshooting techniques are presented in Section 5.11.

5.2   DEFINING CONSTANTS WITH EQU

Constants are data values that do not change during execution of a program. For example, the instruction:

```
MOVE.W    #250,D4
```

contains the constant value 250 decimal. This value does not change during execution. A constant is defined through the **EQU** (equate) assembler directive, as in:

```
<label>    EQU    <value>
```

The assembler equates the symbol *<label>* with *<value>*, as in *<label>* = *<value>*. Whenever *<label>* is seen in a source statement, the assembler will substitute *<value>*. Constants are very useful when a program must initialize registers to predefined values, as shown in the following example.

Example 5.1:   Suppose that the 1,470-line source file of a new program contains the instruction:

```
MOVE.B    #7,D0
```

in many different places. After assembling and testing the program, it is discovered that we really need to put 8 into D0 instead of 7. Unfortunately, someone will have to spend a good deal of time editing the source file, changing all the 7s to 8s. This can all be avoided if a constant is used. For instance, if a different version of the program contained the following source statement at the beginning of the source file:

```
XVAL      EQU    7
```

and every MOVE instruction requiring this value was coded like this:

```
MOVE.B    #XVAL,D0
```

then only *one* line of the source file (the EQU statement) would have to be changed to make the modification from 7 to 8 in the MOVE instructions. Do not forget that the source file must then be reassembled for the changes to take effect.

It is important to note that an EQU statement does not create any data; instead, it causes the assembler to make an entry for the EQU label and value in its **symbol table.** The assembler's symbol table is a storage structure that it uses to keep track of the value of all user-defined symbols (such as labels) in the source file.

5.3 DEFINING VARIABLES WITH DC AND DS

There are times when a programmer needs to temporarily save a value, such as the information in a data or address register, in memory. After all, the number of data and address registers is limited to a total of 16 (allowing 64 bytes of storage). Memory, on the other hand, allows over 16 *million* bytes of storage, and is therefore better suited for long-term data storage.

There are also times when a memory location must be initialized to a particular value (such as the number of loops to perform), and other times when a memory location already contains data and the programmer must read it. Let us refer to any memory location used to store a single data value as a **variable.** The assembler directives **DC** (define constant) and **DS** (define storage) are used to create storage locations for variables. The DC directive has the following syntax:

```
<label>    DC.<size>    <value>
```

where *<label>* is the variable name, *<size>* is B, W, or L, and *<value>* is the data that will be stored. When the assembler encounters the DC directive, it assigns the current program counter address to *<label>*, uses 1, 2, or 4 bytes of memory to represent *<value>*, and then advances the program counter by 1, 2, or 4.

Example 5.2:   Examine the following lines of output from a list file:

```
008400                        ORG   $8400
008400   07            ABC   DC.B   7
008401   41            DEF   DC.B   'A'
008402   03E8          GHI   DC.W   1000
008404   0002 D690     JKL   DC.L   186000
008408
```

Notice that the two DC.B statements each generate a single byte of data. One byte is stored at address 8400 and the other at address 8401. When single quotation marks are used in the *<value>* field of the DC statement, the assembler looks up the ASCII code of the quoted character and places the appropriate byte into memory.

The value of the ABC and DEF variable labels are 8400 and 8401, respectively. The byte values stored at these two locations can be accessed using these instructions:

```
MOVE.B    ABC,D0     ;read 8400
MOVE.B    D0,ABC     ;write 8400
```

and

```
MOVE.B    DEF,D0     ;read 8401
MOVE.B    D0,DEF     ;write 8401
```

This is a form of absolute addressing. Similar instructions are used for the GHI and JKL labels.

Also notice that the assembler converts the decimal values 1000 and 186000 into their respective hexadecimal equivalents 3E8 and 2D690. The assembler's program counter is advanced from 8400 to 8408 by the four DC statements.

If necessary, multiple data items can be declared with a single DC statement. This is useful when we must initialize an **array** of data, a collection of identical data types.

Example 5.3:   The following five output lines from a list file show examples of the DC directive with multiple operands.

```
008408  0102 0304      MNO  DC.B  1,2,3,4
00840C  0064 0100 03E8  PQR  DC.W  100,$100,1000,$1000
008412  1000
008414  4865 6C6C 6F21  STU  DC.B  'Hello!'
00841A                  END
```

Label MNO points to a 4-byte array, PQR to a four-word array, and STU to a 6-byte string of ASCII characters. Examine the difference between 100 and $100 and between 1000 and $1000 in the DC.W statement. One number is converted and the other is not.

The assembler's program counter has been advanced from 8408 to 841A.

When we simply want to *reserve* room for variables and *not* initialize them, we use the DS directive. The syntax of the DS directive is as follows:

```
<label>   DS.<size>   <number>
```

where *<number>* indicates the number of bytes, words, or longwords we wish to reserve. The assembler assigns the current program counter to *<label>* and then advances the program counter by an amount equal to *<size>***<number>*.

The statement:

```
COUNT    DS.B    1
```

reserves 1 byte of storage for the variable COUNT. It does *not* set the value of COUNT to 1. The assembler's program counter is advanced by 1. The statements:

```
SIZE      DS.W    1
LENGTH    DS.L    1
```

reserve 2 and 4 bytes, respectively, for variables SIZE and LENGTH. The assembler's program counter is advanced by 2 after processing DS.W and by 4 after processing DS.L.

Example 5.4:   How many bytes of storage are reserved by these three DS statements?

```
          ORG     $8500
SVAL      DS.B    6
TVAL      DS.W    2
UVAL      DS.L    3
```

Solution: The resulting list file output shows the following information:

```
008500                          ORG     $8500
008500 ??           SVAL        DS.B    6
008506 ????         TVAL        DS.W    2
00850A ???? ????    UVAL        DS.L    3
008516
```

The question marks indicate that the assembler has not placed any data into the reserved locations. A total of 22 bytes are reserved by the three statements, 6 bytes for SVAL, 4 bytes for the two words in TVAL, and 12 bytes for the three longwords in UVAL. Notice how the assembler's program counter has advanced after each statement.

5.4   ACCESSING ONE-DIMENSIONAL ARRAYS

As shown in previous examples, multiple data items can be associated with a label with a single DC or DS statement, as in:

```
CARDS   DC.B    0,0,0,0,0
SUITS   DC.B    0,0,0,0,0
```

or

```
CARDS   DS.B    5
SUITS   DS.B    5
```

where CARDS and SUITS are 5-byte *one-dimensional arrays* representing the cards and suits of a poker hand. Each array represents one row of five values (or *elements*). In a high-level language such as C, we use statements like:

```
card = cards[i];
```

to read the value of one card out of the CARDS array and save it. The specific element accessed depends on the value of the variable *i*, which is called the **index.** For a five-element array, the index has a range of 0 to 4 (although some prefer to use 1 to 5 for various reasons). Thus, cards[0] represents the first element of the array, and cards[4] the last.

In 68000 assembly language, we can use address register indirect with index addressing to both read and write individual elements of an array. Recall that the syntax of the memory operand for this addressing mode is (A*n*,R*n*), where A*n* is any address register, and R*n* is any address or data register. Let us assume that A0 has been initialized to the starting address of the CARDS array (via MOVEA.L   #CARDS,A0 or LEA   CARDS,A0). If D7 is used as the index, we can access any byte in the CARDS array as follows:

```
MOVE.B  (A0,D7),D0      ;read CARD value
MOVE.B  D0,(A0,D7)      ;write CARD value
```

where D0 contains the CARD value.

Word and longword arrays are handled slightly differently, due to the fact that each new word or longword begins at an address that is a product of two or four times the index value. For instance, consider these two array definitions:

TABLE 5.4 Addresses of array elements in WORDS and LONGS

Index value	WORDS address	LONGS address
0	8620	8640
1	8622	8644
2	8624	8648
3	8626	864C
4	8628	8650
5	862A	8654
6	862C	8658

```
              ORG     $8620
WORDS         DS.W    7
              ORG     $8640
LONGS         DS.L    7
```

The address of each word within the WORDS array is found by adding twice the index to the starting address of 8620. The address of each longword within the LONGS array is found by adding 4 times the index to the starting address of 8640. In both cases the index ranges from 0 to 6.

The addresses of each element in both arrays are shown in Table 5.4. If the index value resides in D7 (and A0 points to the starting address of the WORDS array), we can use these instructions to read an element from the WORDS array:

```
MOVE.L   D7,D6         ;make copy of index
ADD.L    D6,D6         ;double index value
MOVE.W   (A0,D6),D0    ;read array value
```

A copy of the index register (D6) is used to compute the offset into the array, since we do not want to alter the index value (D7) during an access.

In a similar fashion, the LONGS array can be written as follows:

```
MOVE.L   D7,D6         ;make copy of index
ADD.L    D6,D6         ;double index value
ADD.L    D6,D6         ;double index value again
MOVE.L   D0,(A0,D6)    ;write array value
```

Example 5.5: The starting address of a word array is 830C. What is the address of the word whose index value is 5?

Solution: Adding 5 to itself gives 10, which equals 0A hexadecimal. Adding this to the starting address gives

830C + 0A = 8316

This is illustrated in Figure 5.1. Notice that you must count up by twos in hexadecimal to generate the successive element addresses within the array.

FIGURE 5.1 Word-size array addressing

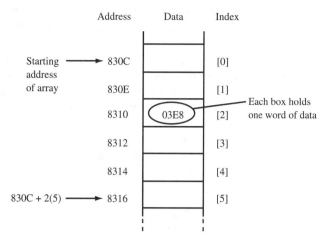

In general, we can derive a formula for calculating the address of any element in a byte, word, or longword array. Let us define some terms first:

EA	Effective address. The actual element address.
Base	Base address. The starting address of the array.
Index	Element index.
Size	Element size. Byte = 1, Word = 2, Longword = 4.
Offset	Offset into array to element's position.

The effective address is calculated as follows:

$$EA = Base + Offset$$

where

$$Offset = Size * Index$$

Thus, we have:

$$EA = Base + Size * Index$$

as a general formula.

Suppose that D6 contains the size value (1, 2, or 4) as a word value, and that D7 is the index value (as a longword). Then we can use:

```
MULU    D7,D6
```

to generate the correct offset into an array of any size. If register A0 contains the base address of the array, any element can be accessed through the (A0,D6) addressing mode.

5.5   ACCESSING TWO-DIMENSIONAL ARRAYS

A two-dimensional array (also called a *matrix*) contains a certain number of rows and columns of elements, such as the 3-by-3 array shown in Figure 5.2. Each row/column can be thought of as a one-dimensional array, so we are partially familiar with the addressing scheme involved. If we know the starting address of the first element in any row/column, we can access any column/row element with the formula from the previous section. The trick now with a matrix is to know how it has been stored in memory.

One way a matrix can be stored is through **row-major** order. This is illustrated in Figure 5.3. Do you see how each row is stored, one after another?

FIGURE 5.2 Sample two-dimensional array

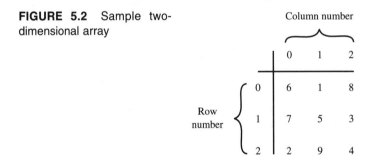

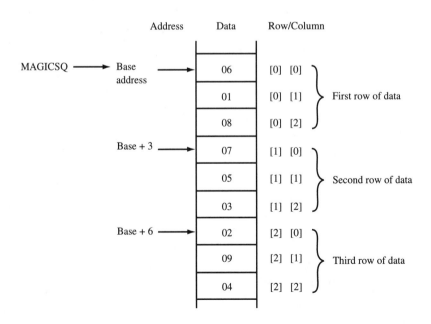

FIGURE 5.3 Two-dimensional array stored in row-major order

It is very easy to define a matrix within a source file. The matrix of Figure 5.2 is defined and stored in row-major order (as shown in Figure 5.3) as follows:

```
MAGICSQ  DC.B   6,1,8   ;first row
         DC.B   7,5,3   ;second row
         DC.B   2,9,4   ;third row
```

where MAGICSQ (for *magic square*) represents the base address of the matrix.

To access any element of the matrix, we must be supplied with a row number and a column number. In C, this translates to a statement like:

```
val = magicsq[row][col];
```

The row number is used to jump over groups of locations (first row of data, second row of data, etc.). The column number is used to calculate an offset within a row. Consider the following definitions:

EA	Effective address. The actual element address.
Base	Base address. Starting address of matrix.
Row	Row number.
Col	Column number.
NCols	Number of columns in matrix.
Size	Element size. Byte = 1, Word = 2, Longword = 4.
RowOffset	Offset into matrix to element's row.
ColumnOffset	Offset into row to element's position.

The effective address is calculated as follows:

$$EA = Base + RowOffset + ColumnOffset$$

where

$$RowOffset = Row * Size * NCols$$

and

$$ColumnOffset = Size * Col$$

Thus, we have:

$$EA = Base + Row * Size * NCols + Size * Col$$

which simplifies to

$$EA = Base + Size * (Row * NCols + Col)$$

as a general formula.

Example 5.6:   A 5-by-7 matrix is defined and stored in row-major order. The matrix has a base address of 83E0 and is composed of longwords. What is the address of element <3><5>?

Solution: We are looking for the address of the sixth element in the fourth row (remember that we begin counting at 0). Each row contains seven longwords, or a total of 28 bytes. We must skip over three groups of 28 bytes in order to get to the beginning of the fourth row of data. This is 84 bytes worth of data (54 hexadecimal). Then we have to skip over the first five longwords in the row to get to the selected element. This represents 20 more bytes (14 hexadecimal). The address of element <3><5> equals these two values plus the base address of 83E0, giving 8448.

Using the simplified formula, we have:

$$EA = 83E0 + 4 * (3 * 7 + 5)$$
$$EA = 83E0 + 4 * (21 + 5)$$
$$EA = 83E0 + 4 * 26$$
$$EA = 83E0 + 104$$

Now convert the decimal offset 104 into hexadecimal:

$$EA = 83E0 + 68$$
$$EA = 8448$$

Matrices may also be stored in **column-major** order. Figure 5.4 shows the same 3-by-3 matrix of Figure 5.2, stored in memory in column-major format. Each column is stored, one after another. This requires some changes to the access formula, and you are encouraged to determine these changes on your own.

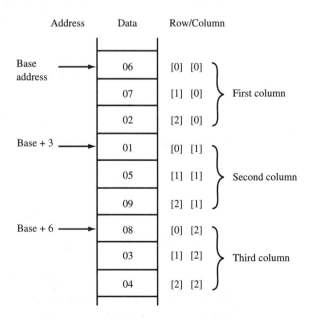

FIGURE 5.4 Column-major order storage format

5.6   CHARACTER STRINGS

For our purposes, a character string is a collection of ASCII character codes terminated by the byte value 00. ASCII codes (see Appendix G) are 7-bit codes that uniquely represent all uppercase and lowercase letters, the digits 0 through 9, punctuation symbols, and other, special, control codes. We have already seen an example of a character string in the HELLO program of Chapter 1, where the statements:

```
HMSG    DC.B     'Hello!'           ;message characters
        DC.B     0                  ;end-of-message marker
```

were used to define a character string of length 6. Figure 5.5 shows how the string is stored in memory. Using 00 as the end-of-string marker allows us to specify zero-length strings simply by placing 00 into the first element position.

It is easy to determine the length of a string. Starting at the beginning of the string, we examine each string element until the zero is found. Each element that is not zero increments a counter. This is illustrated in the following example.

Example 5.7:   The starting address of a string has been loaded into A0. The length of the string, in bytes, will be saved in the lower word of D0 when the following instructions complete execution:

```
        CLR.W    D0          ;clear element counter
MORE    CMPI.B   #0,(A0)+    ;check element for zero and advance A0
        BEQ      DONE        ;branch if element is 00
        ADDI.B   #1,D0       ;increment element counter
        BRA      MORE        ;and continue with next element
DONE    ---
```

FIGURE 5.5   String storage

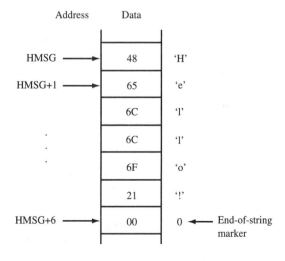

Using post-increment addressing in the CMPI instruction saves us the trouble of advancing A0 ourselves every time the string element is not zero. If the very first element is zero, the code branches to DONE with 0000 in the lower word of D0 (due to the CLR instruction). Otherwise, the code loops until it locates the 00 end-of-string marker and then branches to DONE with the proper string length in D0. What do you think happens when the end-of-string marker is missing?

There are many operations that we can perform on strings, some of which we will examine in the next chapter. These operations include searching a string for a character or substring, catenation of two strings, comparison of two strings, and string conversions. One useful string conversion takes any lowercase string characters and converts them into uppercase. So, a string stored as "Hello!" would be converted into "HELLO!" by changing the four lowercase characters "ello" into uppercase. Let us see how this is done.

Example 5.8:   The following instructions assume that the starting address of the string has already been placed into A0:

```
NEXT     MOVE.B   (A0),D0      ;get string character
         CMPI.B   #0,D0        ;is it the end-of-string marker?
         BEQ      DONE         ;branch if true
         CMPI.B   #'a',D0      ;check for lowercase ASCII range
         BCS      NOTLC        ;branch if below range
         CMPI.B   #'z',D0
         BHI      NOTLC        ;branch if above range
         ANDI.B   #$DF,D0      ;convert into uppercase
         MOVE.B   D0,(A0)      ;and replace character
NOTLC    ADDA.L   #1,A0        ;advance pointer to next character
         BRA      NEXT         ;and repeat
DONE     ---
```

The lowercase to uppercase conversion is accomplished by clearing the sixth bit in the ASCII code byte. For instance, a lowercase 'a' has an ASCII code of $61, or 01100001 binary. An uppercase 'A' has $41 as its ASCII code, which is 01000001 binary. The only difference is in the sixth bit, which is low for uppercase characters and high for lowercase characters. However, it is important to first ensure that the ASCII code to be converted is in the 'a' to 'z' range. Otherwise, we might clear the sixth bit in a non-alphabetic code, such as '{', and convert it into some other ASCII character, like '['.

5.7   STATIC VERSUS DYNAMIC STORAGE ALLOCATION

All of the data structures we have examined so far fall under the category of **static** storage allocation. Variables, arrays, and strings all have predefined, fixed lengths that do not change during execution of a program. We only allocate storage once to these types of data structures. This simplifies the addressing scheme used to access the static structure, and allows quick access to the stored information, since the data does not have to be searched for. However, since the size of the structure is fixed at assembly time, we cannot increase its size (i.e., add more elements to an array or string) without overwriting other important storage locations. Furthermore, if a variable is used a few times at the beginning of a program, and then never again, we do not give the variable's storage space back to the system. It simply sits unused until the program completes execution. This results in poor utilization of available memory.

Dynamic storage allocation operates differently. Space required for data storage is assigned *as requested,* not in advance. A program might ask for a single byte of storage space, or 4,000 bytes, as many times as it needs to. The system the program is running on must supply and manage a block of free memory (a **storage pool**) for the user program to work with. For example, a 68000-based system might reserve an 8KB block of memory for dynamic storage allocation. A system call (a specific TRAP or predefined subroutine address) is provided so that user programs may request memory from the 8KB storage pool.

To make better use of memory, the user program can return (deallocate) memory to the storage pool. Thus, the size of a dynamic data structure is allowed to grow and shrink. As a result, different addressing schemes are needed, and a search method must be used to access data, which increases the access time. For instance, it takes longer to read data out of a linked-list than an array.

So, there are advantages and disadvantages to each type of allocation. In the next few sections we will examine new data structures that make use of dynamic storage allocation.

5.8 LINKED-LISTS

A linked-list is a collection of data elements called **nodes** that is created dynamically. Dynamic creation means that the size of the linked-list is not fixed when it is first created. As a matter of fact, it is empty when first created. As an example, if we want to reserve enough room in memory for 100 integer bytes, we use

```
DATA    DS.B    100
```

This assembler directive is utilized because we know beforehand how many numbers are going to be used. The beauty of a linked-list is that its size can be changed as necessary, either increased or decreased, with a maximum size limited only by the amount of free memory available in the system. This method actually saves space in memory, since it does not dedicate entire blocks of RAM for storing numbers. Rather, a small piece of memory is allocated each time a node is added to the linked-list. A node is most commonly represented by a pair of items. The first item is usually used for storing a piece of data. The second item is called a **pointer,** and is used to point to the next node in the linked-list. Figure 5.6 contains an example of a three-node linked-list. Each node in the list stores a single ASCII character. The beginning of the linked-list, the first node, is pointed to by P. The nodes are linked via pointers from one node to another. The last node in the list, node 3, contains 0 in its pointer field. We will interpret this as a pointer to nowhere (and thus the last node). The pointer to 0 is commonly called a **null** pointer.

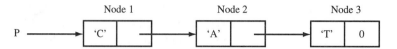

FIGURE 5.6   A sample linked-list

The actual representation of the node on a particular system can take many forms. Since the linked-list must reside in memory, it makes sense to assign one or more locations for the data part (or data field), and four locations for the pointer part (also called the pointer field). Why four locations for the pointer field? Because all nodes reside in memory. To point to a certain node, we must know its address, and addresses in the 68000 occupy 32 bits.

For this discussion, assume that all nodes consist of four data bytes and four address bytes. Consider a subroutine called GETNODE that can be called every time a new node is added to the linked-list. GETNODE must find 8 bytes of contiguous (sequential) memory to allocate the node. When it finds them, it will return the starting address of the 8-byte block in address register A5. Let us take another look at our example linked-list; this time addresses have been added to each node. Figure 5.7 shows how the pointer field of each node contains the address of the next node in the list. The address in the pointer field of node 3 indicates the end of the list. The pointer P may be an address register containing 00010000, the address of the first node in the list. To generate this list, GETNODE has been called three times. GETNODE returned different addresses each time it was called. First came 00010000, then 00F78000, and finally 00020520. Linked-lists do not have to occupy a single area of memory. Rather, they may be spread out all over the processor's address space and still be connected by the various pointer fields.

To add a node to the linked-list, a simple procedure is followed. First, a new node is allocated by calling GETNODE. Address register A5 holds the address of this new

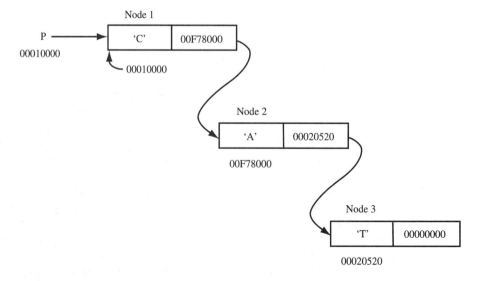

FIGURE 5.7   A linked-list with address assignments

node. The pointer field address of this new node, which will have to be modified to add it to the list, starts at A5 + 4 (because the data field occupies the first 4 bytes). To add the new node to the existing list, a copy of the pointer P is written into the pointer field of the new node. Then, to make the new node the first node in the list, P is changed to the address of the new node. Figure 5.8 shows this step-by-step process, assuming that address register A4 is used to store P. Once the new node has been inserted, its data field, now pointed to by A4, may be loaded with new data. Assume that the new data comes from data register D0. Note from Figure 5.8 that insertion of the new node into the beginning of the linked-list has changed its contents from 'CAT' to 'SCAT'. The code to perform the insertion described in Figure 5.8 is as follows:

```
INSERT    BSR       GETNODE          ;get a new node from storage pool
          MOVE.L    A4,4(A5)         ;load pointer field of new node
          MOVEA.L   A5,A4            ;pointer P to list is updated
          MOVE.L    D0,(A4)          ;data field of new node is loaded
          RTS
```

1. GETNODE returns new node.

2. Pointer field of new node is loaded with P.

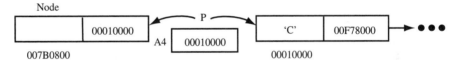

3. Pointer P to list is changed.

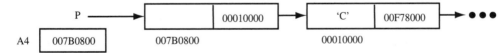

4. Data field of new node is loaded.

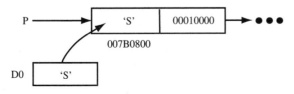

ASCII code for 'S' in lower byte of D0

FIGURE 5.8   Adding a node to a linked-list

Inserting a new node at the end of a linked-list requires that we first search for the end of the list (by looking for 00000000 in the pointer field), and then add on a new node. You are encouraged to figure this technique out for yourself, along with methods for deleting a node from the beginning or end of a linked-list.

Linked-lists are ordinarily used to represent arrays in memory. The data field may be used to store ASCII characters (as in this example), integers, Boolean data, and even pointers to other linked-lists. Linked-lists are very useful tools employed in the functions of operation systems. They are also supported by computer languages such as C.

5.9   BINARY TREES

A binary tree is a specific data structure composed of nodes containing a data field and two link fields, as shown in Figure 5.9. The link fields are called **left child** and **right child.** The term *binary* comes from the fact that each node has the capability of pointing to exactly two other nodes. The first node in the tree is commonly called the **root** node, and is located at the top of the tree's diagram. Figure 5.10 shows a sample binary tree. The root node of the tree is the node containing the asterisk (*). The binary tree in Figure 5.10 was constructed in such a way that each link field points to one of three places: a node containing a math operation, a node containing a variable name, or to NULL (a pointer value of 0). It is important to be able to traverse the binary tree and access the information contained within it. During a traversal, the data stored at each node is displayed or accessed.

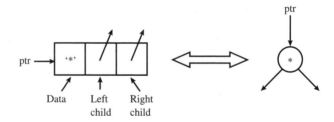

FIGURE 5.9   A sample node in a binary tree

FIGURE 5.10   A sample binary tree

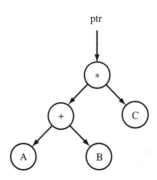

FIGURE 5.11 Binary tree traversal

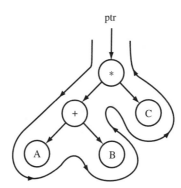

There are three common forms of traversal: pre-order, in-order, and post-order. In all traversal methods, we attempt to go down the tree to the left as far as possible before going to the right. Our trip down the tree continues until we encounter NULL in a link field. The results of each traversal method are different, since each displays/accesses the node's data field at a different time. For example, in a pre-order traversal, the following steps are performed at each node:

1. Access data field.
2. Access left child.
3. Access right child.

An in-order traversal accesses the node fields like this:

1. Access left child.
2. Access data field.
3. Access right child.

A post-order traversal does the following:

1. Access left child.
2. Access right child.
3. Access data field.

Figure 5.11 shows the order in which the nodes in our sample binary tree are accessed by all three techniques. The results of each traversal are as follows:

Pre-order: *+ABC

In-order: A+B*C

Post-order: AB+C*

These traversal techniques are easily implemented through the use of recursive subroutine calls, as the following example illustrates.

Example 5.9: Assume that the binary tree of Figure 5.10 has been dynamically created and stored in memory, and that address register A6 contains the address of the root node. Furthermore, suppose that each node in the tree is represented by a 10-byte block of

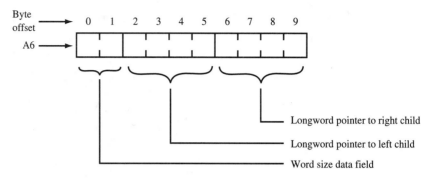

FIGURE 5.12   Structure of binary tree node

storage, partitioned as shown in Figure 5.12. The data field consists of 2 bytes (bytes 0 and 1), and each link field consists of 4 bytes (at byte offsets 2 and 6).

The POSTFIX subroutine presented here calls itself many times during the traversal process. A routine that calls itself employs a technique called **recursion,** and requires that certain information be saved in order for the recursion to work properly. In our case, the address within A6 must be saved each time POSTFIX calls itself, so that, upon return, A6 still contains a valid address.

As a reminder, recall that the EMU68K emulator program uses TRAP   #1 to output a single ASCII character to the PC's display screen. The ASCII code must be in the lower byte of D1.

```
POSTFIX   MOVE.L    A6,-(A7)     ;save A6 on stack
          CMPA.L    #0,A6        ;check for NULL pointer
          BEQ       EXIT         ;branch if NULL found
          MOVE.L    A6,-(A7)     ;save A6 again
          MOVEA.L   2(A6),A6     ;get left child pointer
          BSR       POSTFIX      ;do recursive call
          MOVEA.L   (A7)+,A6     ;read A6 from stack
          MOVEA.L   6(A6),A6     ;get right child pointer
          BSR       POSTFIX      ;do recursive call
          MOVEA.L   (A7)+,A6     ;read A6 from stack again
          MOVE.W    (A6),D1      ;read data field
          TRAP      #1           ;output character to display
          RTS
EXIT      MOVEA.L   (A7)+,A6     ;restore A6 to original value
          RTS
```

During execution of POSTFIX the following output is generated:

```
AB+C*
```

As a matter of interest, the binary tree in Figure 5.10 represents a mathematical expression whose original form was:

```
(A + B)  *  C
```

The parentheses force the addition operation to be performed before the multiplication. This is indicated in the POSTFIX output by the fact that + is output before * is.

5.10   STACKS AND QUEUES

Many computations are greatly simplified by the use of a software-controlled stack or queue. Expression evaluation and round-robin selection algorithms are just two examples of where a stack and a queue are used. The method of implementation is not critical; either may be designed as a special form of a linked-list or simply as fixed-size memory structures. The latter approach will be used here, with address registers pointing to the stack and queue memory structures.

Stacks

A stack is an area in memory reserved for reading and writing special data items such as return addresses and register values. For example, a BSR instruction automatically pushes a return address onto the stack (using A7 with predecrement addressing). Registers may be pushed onto the stack (written into stack memory) with a MOVE.L instruction, as in:

```
MOVE.L     D3,-(A7)
```

where the entire contents of D3 are written into the stack area pointed to by A7. A7 is automatically decremented by 4 during execution.

Items previously pushed onto the stack can be popped off the stack (read out of memory) in a similar fashion, as in:

```
MOVE.L     (A7)+,D3
```

where stack memory is read out into D3 and A7 is automatically incremented by 4 during execution. Thus, we see that using a stack requires manipulation of a stack-pointer register.

One characteristic of a stack is that the last item pushed is always the first item popped. For this reason, stacks are commonly referred to as a **LIFO** (last in, first out) structure.

It is possible, and often necessary, for a programmer to design a custom stack area for use within a program. For instance, suppose that a programmer requires a stack that allows only eight longwords to be pushed onto it. The 68000 has no mechanism for limiting the amount of pushes (or pops) onto a stack. If this is necessary, a set of stack procedures must be written. The following routines implement a stack that allows a maximum of eight pushes. The PUSH routine is used to place data onto the stack. The data pushed must be in D0. If the push is successful, a success code of 00 will be returned in the lower byte of D1. If more than eight pushes are attempted, the routine returns with error code $80 in the lower byte of D1, without pushing any data. The POP routine is used to remove an item from the stack. If a pop is attempted on an empty stack, error code $FF is returned in D1. Successful pops return data in D0. The stack-pointer register is A0 and is assigned the address of a free block of memory from the storage pool by a routine called MAKESTACK. MAKESTACK must be called before the stack can be used. MAKESTACK also assigns addresses to A1 and A2, which are used by PUSH and POP to determine when a stack operation is possible. The structure of the stack is indicated in Figure 5.13. A2 points to the bottom of the stack structure and A1 points to the top. A0 points to the stack location that will be used for the next push or pop.

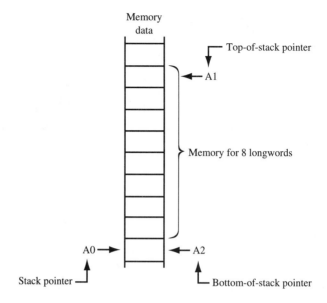

FIGURE 5.13   Software-controlled stack structure

Examine the following routines to see how A1 and A2 used to check for legal pushes and pops.

```
PUSH        CMPA.L    A0,A1              ;ok to push?
            BEQ       STKFULL            ;no, go return error code
            MOVE.L    D0,-(A0)           ;push D0 onto stack
            CLR.B     D1                 ;indicate a successful push
            RTS
STKFULL     MOVE.B    #$80,D1            ;stack full error code
            RTS
POP         CMPA.L    A0,A2              ;ok to pop?
            BEQ       STKEMPTY           ;no, go return error code
            MOVE.L    (A0)+,D0           ;pop D0 off stack
            CLR.B     D1                 ;indicate a successful pop
            RTS
STKEMPTY    MOVE.L    #$FF,D1            ;stack empty error code
            RTS
```

Note that multiple stacks can be maintained by saving the contents of A0, A1, and A2 and loading new addresses into each register.

Queues

Queues are also memory-based structures, but their operation is functionally different from that of a stack. In a queue, the first item loaded is the first item to be removed. For this reason, queues are referred to as **FIFO** (first in, first out) structures. Figure 5.14 shows a diagram of a queue that has had the data items 'A', 'B', and 'C' loaded into it. 'A' was loaded first, 'C' last. When we begin removing items from the queue, the 'A' will come out first (unlike the stack structure, which would have popped 'C' first).

FIGURE 5.14 Software-controlled queue structure

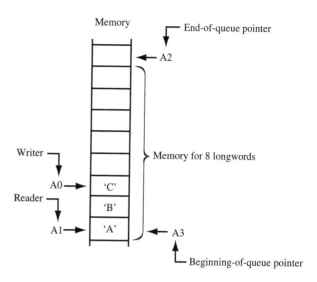

A routine called MAKEQUEUE is used to assign a block of memory from the storage pool. MAKEQUEUE initializes four address registers, whose use and meanings are as follows:

Register	Use/Meaning
A0	Pointer for write operation
A1	Pointer for read operation
A2	Contains end-of-queue address
A3	Contains beginning-of-queue address

Initially, A0, A1, and A3 are all loaded with the same value. The address placed into A2 is determined by the desired size of the queue.

Before an item can be written into the queue, A0 must be compared with A2. If A0 equals A2, it is necessary to reload A0 with A3's address. This allows A0 to *wrap around* the end of the queue (a similar technique is used to wrap A1 around when reading). Next, A0 is used to write the data item into memory. Then A0 is decremented by 4 to prepare for the next write.

Data is removed from the queue by the read-pointer register A1. After data is read, A1 is decremented by 4. Note that serious data errors result when A1 reads a location that has not been written into by A0 yet. For this reason it is necessary to pay special attention to the positions of A0 and A1 within the queue. This part of the queue software is left for you to devise on your own. The routines presented here perform write and read operations with wraparound, but with no error checking.

```
INQUEUE    CMPA.L    A0,A2          ;need to wrap around?
           BNE       NOADJA0        ;no
           MOVEA.L   A3,A0          ;yes, reload A0
NOADJA0    MOVE.L    D0,(A0)        ;write data into queue memory
```

```
                SUBA.L    #4,A0              ;adjust write pointer
                RTS
OUTQUEUE        CMPA.L    A1,A2              ;need to wrap around?
                BNE       NOADJA1            ;no
                MOVEA.L   A3,A1              ;yes, reload A1
NOADJA1         MOVE.L    (A1),D0            ;read data out of queue memory
                SUBA.L    #4,A1              ;adjust read pointer
                RTS
```

In both routines, D0 is used as the queue data register.

5.11   TROUBLESHOOTING TECHNIQUES

When working with data tables and different types of data, it is easy to get lost in all the details. Let us examine a few techniques we can use to help avoid some of the unfortunate consequences of working with data.

- Know your data sizes. Be sure to reserve enough room for all of the variables and structures you are using. This requires careful counting and adding.
- Plan carefully. Draw a diagram of the required data structure. Think about the routines, or instructions, you will need to access and modify it.
- Review the addressing modes you are using. Are you using them correctly? Is there an easier way?
- Be familiar with the various signed and unsigned data ranges.
- Structure your data so that word and longword values are stored at even address boundaries to avoid bus errors.
- Work through the code instruction by instruction with actual values when a difficult bug cannot be identified. Perhaps by watching the flags and individual register contents, the error may be discovered. Single-stepping with EMU68K could help in this area.

Overall, the most important things about an item of data are its size and its address. Keep this in mind as you study the programming examples in the next chapter.

SUMMARY

In this chapter we examined methods for defining constants and variables, data structures that contain a single item. We then covered a number of different data structures, all of which are used to store multiple data items. One- and two-dimensional arrays, ASCII character strings, binary trees, and linked-lists were discussed, as were the various addressing modes and techniques involved in data access for each structure. The advantages and disadvantages of static and dynamic storage allocation were also examined. The chapter concluded with examples of software-controlled stacks and queues, and some important troubleshooting techniques.

STUDY QUESTIONS

1. Rewrite the following instructions using the EQU directive to define all constants:

```
MOVEA.L  #$8700,A0
MOVE.W   (A0),D4
ADDI.W   #20,D4
CMPI.W   #20,D4
BEQ      EXIT
```

2. How does the use of a constant declaration in a source file simplify editing when the constant needs to be changed?

3. What happens when the following instructions are assembled? Are any errors generated?

```
CVAL    EQU      $1000

        MOVE.B   CVAL,D0
        MOVE.W   CVAL,D1
        MOVE.L   CVAL,D2
```

4. How many bytes of code are generated by these DC statements?

```
        ORG     $8408
X1      DC.B    25,100,212
X2      DC.W    1000
X3      DC.L    $1000,7,$12345678
```

5. What are the actual bytes generated by the assembler when processing the statements in Question 4?

6. What addresses are associated with the labels X1, X2, and X3 in Question 4?

7. Show the instruction (or instructions) needed to read the longword 7 out of the data area defined in Question 4.

8. How many bytes of code are reserved by these DS statements?

```
        ORG     $8C00
Y1      DS.B    6
Y2      DS.W    10
Y3      DS.L    5
```

9. What are the addresses associated with labels Y1, Y2, and Y3 in Question 8?

10. What is the difference between these two statements?

```
VAL     EQU      20
VAL     DC.B     20
```

11. Show the instruction (or instructions) required to read or write any element of the following array:

```
ARRAY    DC.W     1,2,3,5,7,11,13,17,19,0
```

Assume that register D2 is used as an index pointer and that ARRAY[0] is the first element.

12. Suppose that an array is composed of elements that require six bytes each for storage. Show the instruction (or instructions) required to read or write any element using D5 as an index register. The base address of the array is associated with the label BIGNUMS.

CHAPTER 6

An Introduction to Programming the 68000

OBJECTIVES

In this chapter you will learn about:

- Breaking a large program down into small tasks
- Driver programs
- Parameter passing
- Code optimization
- Reading character strings from a keyboard
- Packing BCD digits into a byte
- Item search and lookup in a data table
- String operations
- Sorting
- The use of condition flags to return results from a routine
- Binary and BCD math
- Number conversions
- Writing a routine to perform a complex mathematical function
- Open- and closed-loop control systems (simplified theory)
- Determining program execution time and the meaning of overhead
- Simple exception handling
- The operation of a simple calculator

6.1 INTRODUCTION

Getting the most use out of your microprocessor requires expertise, both in designing the hardware around it and in writing the code that it will execute. The purpose of this chapter is to familiarize you with some of the standard programming principles as they apply to the

68000. We will limit ourselves to writing straight code in this chapter, using only the power of the 68000's instruction set. You will see that many complex tasks may be performed in this way, without the use of external peripherals, which will be covered in Chapter 10.

Section 6.2 explains how large programming jobs are broken down into smaller tasks. Section 6.3 shows how a software driver program is written. This is followed by discussions of parameter passing and equivalent instructions in Sections 6.4 and 6.5, respectively. Gathering data is the subject of Section 6.6, which leads to techniques for searching data tables in Section 6.7. The next two sections, 6.8 and 6.9, deal respectively with string operations and sorting techniques, respectively. These are followed by coverage of binary and BCD math routines in Section 6.10. Number conversions, control applications, and instruction execution times are discussed in Sections 6.11 through 6.13, respectively. Another look at exception handling is presented in Section 6.14, followed by the presentation of a working four-function calculator in Section 6.15. Sections 6.16 and 6.17 complete the chapter with coverage of EMU68K's DOS interface and troubleshooting techniques, respectively.

6.2   TACKLING A LARGE PROGRAMMING ASSIGNMENT

Writing a large, complex program from scratch is a difficult job, even for the most seasoned programmers. Even if this could be done easily, other considerations exist to complicate matters. The final program must be tested, to ensure correct operation. It is a rare occurrence for a new program to work perfectly the first time.

For these reasons, a more sensible approach is to break the large program down into smaller tasks. Each task may be thought of as a subroutine, to be called, when needed, by the main program. The subroutines will each perform a single task, and thus will be easier to individually test and correct, as necessary. The technique for writing a large program in this way is often referred to as **structured programming.** We will not concern ourselves with all the details of structured programming. Instead, we will study a sample programming assignment and use the techniques previously mentioned to break the assignment down into smaller jobs.

The Assignment

The programming assignment is presented to us in the form of a **specification.** The specification describes the job that must be performed by the program. It also contains information concerning any input and output that may need to be performed, and sometimes a limit on the amount of time the program may take to execute.

Consider the following specification:

Specification: Subroutine WORDCOUNT

Purpose: To generate a data table of all different words contained in a paragraph of text, and a second data table containing the frequency of occurrence of each word.

Restrictions: Do not distinguish between uppercase and lowercase characters. Ignore punctuation, except where it defines the end of a word. No words will appear more than 255 times.

Inputs:	A data table headed by symbol TEXT that contains the paragraph to be analyzed, represented by ASCII codes. The length of the paragraph text is undefined, but the last character in the text will always be '$'. This character will not appear anywhere else in the text.
Outputs:	A data table, headed by symbol WORDS, that contains a list of all different words encountered in the paragraph text. Each word ends with '.', and the entire table ends with '$'. A second data table, headed by symbol COUNTS, containing the frequency counts for each entry in WORDS.

There is sufficient detail in the specification for us to determine what must be done. How to do it is another matter.

Breaking the Program Down into Modules

Once we understand what is required of the program, through information presented in the specification, the next step is to break the program down into smaller modules. This means that subroutine WORDCOUNT will actually become a main subroutine, which calls other subroutines. We must identify the other subroutines needed. This step of the process requires skill and practice. When you have given it enough thought, you might agree that these subroutines are required:

INITIALIZE	Initialize all pointers, counters, and tables needed.
GETWORD	Get the next word from the paragraph text.
LOOKUP	Search WORDS to see if it contains the present word.
INSERT	Insert new word into WORDS.
MAKECOUNT	Make a new entry in COUNTS.
INCREASE	Increase frequency count in COUNTS for a word found by LOOKUP.

There may, of course, be other required routines, depending on who is writing the code. The idea is to create a subroutine to accomplish only *one* task. None of the identified routines performs more than one task.

Once the inputs and outputs for each subroutine are identified, the code can be written for each one, and the subroutines tested.

Testing the Modules

Testing of each subroutine module is done separately, through a special program called a **driver.** The driver supplies the subroutine with sample input data and examines the subroutine's output for correctness. It is up to the programmer to select the type and quantity of the sample data. We will look at an example of a software driver in Section 6.3.

When all modules have been tested and verified for proper operation, they can be combined into one large module—WORDCOUNT in our example—and this module can be tested also.

Creating the Final Module

WORDCOUNT, as mentioned before, will consist of calls to the subroutines identified in the section on breaking the program down into modules. **Pseudocode,** a generic programming language, can be used to determine the structure of WORDCOUNT (and of the other subroutines as well). The following pseudocode is one way WORDCOUNT may be implemented:

```
subroutine WORDCOUNT
  INITIALIZE
  repeat
    GETWORD
    if no word found then
      return
    LOOKUP
    if word found then
      INCREASE
      else
        INSERT
        MAKECOUNT
  forever
end WORDCOUNT
```

WORDCOUNT is implemented as an infinite loop, since the length of the paragraph text is unknown. The only way out of the loop is to have GETWORD fail to find a new word in the text (that is, by reaching the '$'). This approach satisfies another requirement of structured programming: Routines should contain one entry point and one exit point. Many of the routine examples that we will study in this chapter will be written in this fashion. It is up to the programmer how the repeat-forever and if-then-else statements are implemented.

The IF-THEN statement can be coded in many different ways. The actual structure is IF <condition> THEN <action>. The *condition* must be satisfied for the *action* to take place. In the WORDCOUNT example, the first IF statement causes the subroutine to return if GETWORD does not find a new word in the paragraph text. Let us assume that GETWORD returns a 00 in the lower byte of D5 if it does find a word, and $FF if it does not. One way to code the IF statement might look like this:

```
        BSR     GETWORD
        CMPI.B  #0,D5
        BNE     NEXT
        RTS
NEXT    ---
```

The BSR to GETWORD will adjust the value of D5 accordingly. The CMP instruction is used to determine whether D5 contains 0. If it does not, a branch to NEXT is performed. This will cause execution to continue (with a BSR to LOOKUP as the next instruction). If D5 does contain 0, the BNE will not take place, and the RTS instruction will execute instead.

IF-THEN-ELSE statements are very similar, with coding somewhat like this:

```
          BSR     LOOKUP
          CMPI.B  #0,D5
          BEQ     THENCODE
ELSECODE: ---
```

Other pseudocode structures include the REPEAT-UNTIL and WHILE-DO. The REPEAT-UNTIL structure looks like this:

```
repeat
  <statements>
until <condition>
```

Coding the REPEAT-UNTIL structure depends on the type of condition being tested. A sample structure and its associated code may look like this:

```
initialize counter to 100          MOVE.B    #100,D2
repeat
  GETDATA                  AGAIN    BSR       GETDATA
  PROCESSDATA                       BSR       PROCESSDATA
  decrement counter                 SUBI.B    #1,D2
until counter = 0                   BNE       AGAIN
```

One important point about using loop counters is that the loop-count register (D2 in this example) must be altered during execution of the statements within the loop.

The WHILE-DO structure is slightly different, performing the condition test at the *beginning* of the loop instead of the end. One example of a WHILE-DO is:

```
while char <> 'A' do
    <statements>
end-while
```

The corresponding machine instructions for this loop might look like this:

```
WHILE    CMPI.B #'A',D1
         BEQ    NEXT
         <loop instructions>
         BRA    WHILE
NEXT     ---
```

Here it is important to modify the loop variable (D1 in this case) somewhere within the loop, to avoid getting stuck inside it.

Another programming structure that is useful is the CASE statement. The structure of the CASE statement may look like this:

```
case <item> of
   <item 1> : <statement 1>
   <item 2> : <statement 2>
      .
      .
   <item x> : <statement x>
```

Where each item $(1 - x)$ is checked for a match with the item at the beginning of the CASE statement. Only the statement associated with the matching item is executed. The number of items to match is not limited. An example of a CASE statement with three choices is as follows:

```
case selvalue of
   0 : clear counter
   1 : increment counter
   2 : decrement counter
```

In this example, the *selvalue* variable must contain a value from 0 to 2 to select one of the three statements. The corresponding assembly language for this CASE statement is:

```
        CMPI.B  #0,D1      ;is it 0?
        BNE     C1
        CLR.B   D2         ;clear counter
        BRA     NEXT
C1      CMPI.B  #1,D1      ;is it 1?
        BNE     C2
        ADDI.B  #1,D2      ;increment counter
        BRA     NEXT
C2      CMPI.B  #2,D1      ;is it 2?
        BNE     NEXT       ;no matches
        SUBI.B  #1,D2      ;decrement counter
NEXT: ---
```

In this example the *selvalue* variable is stored in D1 and the counter is represented by D2.

Some programmers use a modified form of the CASE structure that allows execution of a statement when no match is made with any item. The pseudocode for this structure is:

```
case <item> of
   <item 1>  : <statement 1>
   <item 2>  : <statement 2>
     .
     .
   <item x>  : <statement x>
   otherwise : <otherwise-statement>
```

It is a simple matter to execute the otherwise-statement when no matches are found by performing a BNE   OTHER after the last CMPI instruction.

Remember that there are no fixed methods for converting pseudocode into machine instructions. Use your imagination and come up with your own techniques.

6.3   WRITING A SOFTWARE DRIVER

The programs presented in the remaining sections of this chapter are written as subroutines (or procedures) that must be called to perform their function. As we saw in the previous section, there may be many subroutines combined together in a single application, with each subroutine possibly written by a different person. It is thus the responsibility of each programmer to test (and correct if necessary) the subroutine he or she has written. In this section we see how a new programming module is tested with a *software driver* program. The driver executes the new module with data supplied by the programmer and verifies that the module performs the associated task correctly.

The following subroutine was written to solve the quadratic equation

$$y = 5x^2 - 2x + 6$$

where the word value for x is stored in D0 and the result of the equation (y) is returned in the lower word of D1:

```
QUAD    MOVE.W  D0,D1      ;save copy of input value
        MULS    D0,D1      ;compute X^2
        MULS    #5,D1      ;compute 5X^2
        MULS    #2,D0      ;compute 2X
        SUB.W   D0,D1      ;compute 5X^2 - 2X
        ADD.W   #6,D1      ;compute 5X^2 - 2X + 6
        RTS
```

The driver program must pass an X value into the procedure and check the returned value for accuracy. Multiple test cases are preferable, since they will show how the routine performs over a range of input values. This requires the programmer to first determine what the correct results should be. Consider these input and output pairs:

```
x-input     y-output

     0            6
     1            9
    10          486
   100        49806
```

The software driver presented here will send each *x*-input value to the procedure one at a time and check for a match with the expected *y*-output value each time. If all four tests pass, the driver assumes the new routine is acceptable. If any one test fails, an error message is output.

```
        ORG      $8000
X1      DC.W     0              ;test case 1
Y1      DC.W     6
X2      DC.W     1              ;test case 2
Y2      DC.W     9
X3      DC.W     10             ;test case 3
Y3      DC.W     486
X4      DC.W     100            ;test case 4
Y4      DC.W     49806
PASS    DC.B     'Subroutine passes.',0
FAIL    DC.B     'Subroutine fails.',0

        ORG      $8100
TEST    MOVE.W   X1,D0          ;load first test value
        BSR      QUAD           ;compute result
        CMP.W    Y1,D1          ;look for match
        BNE      BAD
        MOVE.W   X2,D0          ;load second test value
        BSR      QUAD           ;compute result
        CMP.W    Y2,D1          ;look for match
        BNE      BAD
        MOVE.W   X3,D0          ;load third test value
        BSR      QUAD           ;compute result
        CMP.W    Y3,D1          ;look for match
        BNE      BAD
        MOVE.W   X4,D0          ;load fourth test value
        BSR      QUAD           ;compute result
        CMP.W    Y4,D1          ;look for match
        BNE      BAD
        MOVEA.L  #PASS,A3       ;set up pointer to pass message
SEND    TRAP     #3             ;EMU68K display string function
        TRAP     #9             ;terminate program
BAD     MOVEA.L  #FAIL,A3       ;set up pointer to fail message
        BRA      SEND           ;go output message

QUAD    MOVE.W   D0,D1          ;save copy of input value
        MULS     D0,D1          ;compute X^2
        MULS     #5,D1          ;compute 5X^2
        MULS     #2,D0          ;compute 2X
        SUB.W    D0,D1          ;compute 5X^2 - 2X
        ADD.W    #6,D1          ;compute 5X^2 - 2X + 6
        RTS

        END      TEST
```

If more test cases are needed, the test data should be arranged as a data table so that a loop can be used to step through each test case.

Some subroutines may only require a single test case to determine whether they function correctly or not. In any case, if the new procedure should fail, it may be necessary to examine instructions one-by-one (using EMU68K) in the program until the error is found.

Writing driver programs for the routines presented in the remaining sections should be a rewarding programming experience.

6.4 PARAMETER PASSING

In the previous section data registers D0 and D1 were used to check the operation of the QUAD subroutine. We refer to the contents of these registers as **parameters.** A parameter may be an actual data value, or it may be an address where data can be found. Register D0 contains the input parameter to the QUAD subroutine, and D1 contains the output parameter. We could say that D0 passes the input parameter into QUAD, and D1 passes the output parameter back from QUAD, as shown in Figure 6.1(a). This is one very common technique used for parameter passing. This technique is also referred to as "passing parameters by value." Using data and address registers (or even a single flag, such as the zero flag) to pass parameters is quick and easy. However, only a limited number of parameters can be passed with this method (since we only have a handful of address and data registers).

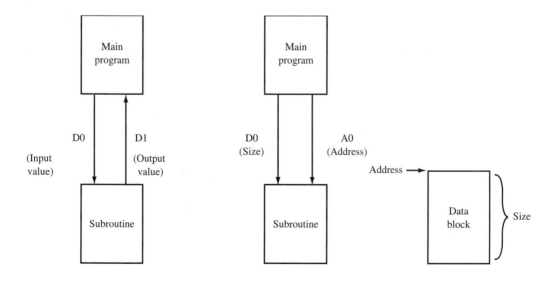

(a) Passing by value   (b) Passing by value (D0) and reference (A0)

FIGURE 6.1   Parameter passing

When large amounts of data must be passed between routines, we only need to supply information about where the data begins and how large the data area is, as shown in Figure 6.1(b). For example, access to a 5000-byte data block beginning at address 8600 can be passed to a subroutine in the following way:

```
MOVE.W     #5000,D0
MOVEA.L    #$8600,A0
BSR        ROUTINE
```

We are still using registers to pass information. The instructions within the subroutine must be written in such a way that they use A0 to access the data. Any addressing mode that uses A0 indirectly will work, as in:

```
MOVE.B     (A0),D1
```

Since there are many more memory locations than registers, we do not have to worry as much about running out of storage space for our data. This method is also called "passing parameters by reference."

Note that we use memory for storage of single values (variables) also. For example, if a word variable XYZ is declared as follows:

```
XYZ    DC.W    1000
```

it is passed either as a value (via MOVE.W   XYZ,D0) or as an address (via MOVEA.L #XYZ,A0).

A third method for parameter passing involves the processor's stack. Remember that we must have a proper stack setup in order to use subroutines, since the 68000 automatically uses the stack (accessed through A7) to save and retrieve return addresses. In stack-based parameter passing, the main program pushes parameters onto the stack before executing the subroutine that uses them. The subroutine skips over the return address to get at the parameter information passed to it by the main program. Stack values (with the exception of the return address) may be modified by the subroutine before it returns. For example, suppose the main program contains these instructions:

```
0080AA  203C 1234 5678   MOVE.L     #$12345678,D0  ;load D0
0080B0  223C 1111 2222   MOVE.L     #$11112222,D1  ;load D1
0080B6  2F00             MOVE.L     D0,-(A7)       ;push D0 onto stack
0080B8  2F01             MOVE.L     D1,-(A7)       ;push D1 onto stack
0080BA  4EB9 001E 4D00   JSR        ADDER          ;execute subroutine
0080C0                   ---
```

The contents of the stack after these instructions execute are shown in Figure 6.2. Notice that the processor has also pushed the return address (0000 80C0) onto the stack.

The ADDER subroutine must use instructions such as:

```
MOVE.L    4(A7),D1    ;read D1 from stack
MOVE.L    8(A7),D0    ;read D0 from stack
```

to retrieve the input parameters. The displacements of 4 and 8 are required to skip over the return address longword.

Suppose that ADDER does the following:

```
ADDER    MOVE.L    4(A7),D1    ;read D1 from stack
         MOVE.L    8(A7),D0    ;read D0 from stack
```

FIGURE 6.2 Stack-based parameter passing

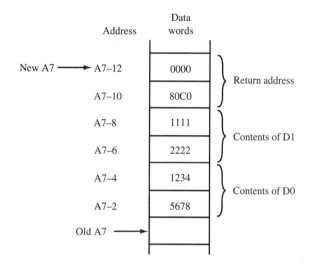

```
ADD.L    D0,D1       ;add registers
MOVE.L   D1,4(A7)    ;save result on stack
RTS
```

The sum of the two registers (2345789A) is written back into stack memory before return. This is shown in Figure 6.3. When the main program gets control again, the stack still contains two longword values. The first is the sum produced by ADDER. The second is the old D0 parameter that can be discarded now. These longwords are processed as follows (by the main program):

```
JSR      ADDER       ;execute subroutine
MOVE.L   (A7)+,D1    ;read result from stack
ADDA.L   #4,A7       ;adjust stack pointer
```

FIGURE 6.3 Updated stack parameters after return

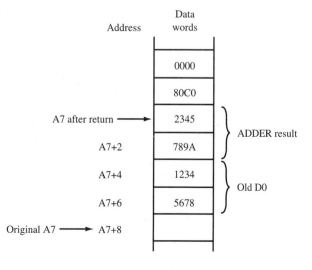

Since the size of the MOVE instruction is .L, the processor will automatically add 4 to A7 after the MOVE completes. Adding another 4 to A7 with ADDA skips over the original D0 parameter and sets the stack pointer back to its original address.

This method of parameter passing has been used in specific ways to implement procedures and functions in high-level programming languages such as C. For instance, the C statement:

```
adder(X,Y);
```

causes the value of the *y* variable to be pushed onto the stack, followed by the value of the *x* variable, and then the return address. The C compiler knows how to generate the correct displacements because it builds the required **stack frame** for the subroutine according to predefined rules. We have already examined the operation of stack frames in Chapter 4. Let us take a deeper look. The stack frame contains reserved locations for each variable used in the subroutine and each variable passed to the subroutine, as well as a pointer (called a **frame pointer**) that indicates the address of the previous stack frame. Two instructions are provided to assist in the management of stack frames. They are LINK and UNLK. For instance, to use A0 as the frame pointer and allocate 32 bytes of stack space, we use:

```
LINK   A0,#-32
```

The negative value on the stack size (–32) is required to allocate memory for stacks, since a push operation will use predecrement addressing (via –(A7) as the destination).

Consider the adder() function shown here:

```
adder(int X, int Y)
{
     int Z;
     Z = X + Y;
}
```

The stack frame created by a call to adder() is illustrated in Figure 6.4. Addresses are included to show how A0 and A7 are modified. With A0 as the frame pointer, the *x, y,* and *z* variables can be accessed by using the appropriate displacements of 8(A0), 12(A0), and –4(A0), respectively. Notice that variables are now referenced by the frame pointer (A0) instead of the stack pointer (A7). The instructions needed to build the stack frame are:

```
In main():            In adder():

MOVE.L  Y,-(A7)       LINK   A0,#-32
MOVE.L  X,-(A7)
JSR     ADDER
```

The three instructions in main() decrement the stack pointer (A7) to address 8E26. The LINK instruction in adder() then does the following:

*   Pushes the contents of A0 (00008F00) onto the stack
*   Loads the new stack pointer (8E22) into A0
*   Adds the signed displacement to the stack pointer

Code for the adder() routine might look like this:

```
LINK    A0,#-32       ;allocate frame
MOVE.L  8(A0),D0      ;read X value
ADD.L   12(A0),D0     ;read and add Y value
```

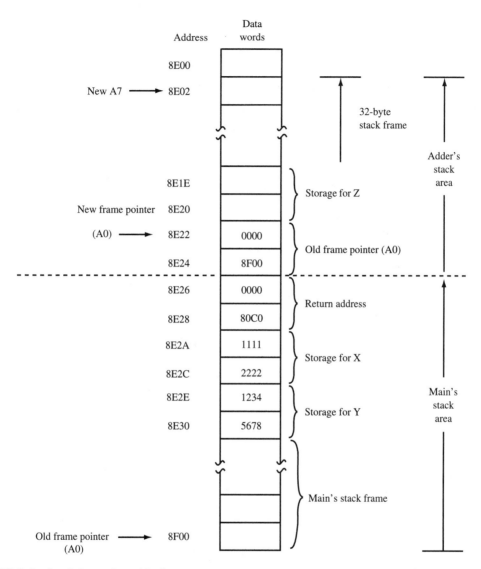

FIGURE 6.4 Stack frame for adder()

```
MOVE.L   D0,-4(A0)      ;save Z
UNLK     A0             ;deallocate frame
RTS
```

where the UNLK   A0 instruction does the following:

- Puts a copy of A0 (00008E22) into the stack pointer (A7)
- Pops longword off stack (00008F00) into A0

The final value of the stack pointer is 8E26.

Stack Frames and Recursive Subroutines

A subroutine that calls itself is called a *recursive* subroutine. This may be the result of writing a recursive function in C such as one that computes the factorial of a number:

```
int fact(int k)
{
    if (k == 0)
        return(1);
    else
        return(k * fact(k-1));
}
```

The `else` statement contains the recursive call to `fact()`.

To support recursion it is necessary to build a new stack frame for each call to `fact()`. Figure 6.5 illustrates several recursive levels of `fact()` in effect. The stack frames are all connected via the various frame pointers. This is commonly called a *dynamic chain*. Each stack frame, when released by UNLK, activates the previous stack frame by updating the frame pointer. The level of recursion possible is limited by the total stack space available.

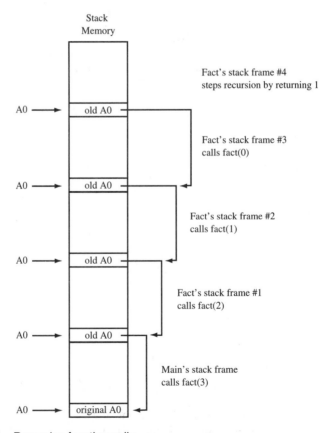

FIGURE 6.5   Recursive function calls

Any registers used by the recursive subroutine must be preserved to guarantee that recursion does not destroy important values needed by `main()`. They are saved in a portion of the stack frame's reserved block of stack space.

So, we have seen that registers and memory (especially stack memory) can both be used to pass parameters to and from individual routines. Knowing different ways to transfer data is a valuable addition to our programming skills.

6.5   EQUIVALENT INSTRUCTIONS AND CODE OPTIMIZATION

The programming examples in the rest of this chapter all use a mix of instructions to accomplish their respective tasks. It is usually possible, however, to use more than one type of instruction to perform a function. Therefore, the instructions that you see in the example programs represent only one of many possible solutions. In this section we will examine a number of pairs of *equivalent* instructions. Both instructions in the pair will do the same thing (i.e., fill a register with zeros, alter a bit in a register), but with different amounts of machine code. In general, it takes less time to fetch a one-word instruction than a two- or three-word instruction. So, we can improve the execution speed of our program by using the appropriate equivalent instruction.

The first pair of equivalent instructions loads an address into A0:

```
207C 0000 1234    MOVEA.L    #$1234,A0
41F8 1234         LEA        $1234,A0
```

The MOVEA instruction requires three words of machine code, one word (207C) for the opcode and two words (0000 1234) for the longword address. The LEA instruction performs the same function, but only requires two words. Thus, we can *optimize* our code by using the LEA instruction instead of the MOVEA instruction. If the MOVEA instruction is used in 200 different places, we can save 200 words of storage by replacing it with the LEA instruction.

The next group of instructions shows three equivalent pairs, all of which load zero into D0. Byte, word, and longword zeros are loaded as follows:

```
4200              CLR.B      D0
103C 0000         MOVE.B     #0,D0

4240              CLR.W      D0
303C 0000         MOVE.W     #0,D0

4280              CLR.L      D0
203C 0000 0000    MOVE.L     #0,D0
```

Verify for yourself that the condition codes are affected the same way in these instructions.

When non-zero values must be loaded, we may use either of the next two instructions to load a byte-size value as a longword:

```
203C 0000 0006    MOVE.L     #6,D0
7006              MOVEQ      #6,D0
```

The .L extension is not required on the MOVEQ instruction since it is designed to automatically load a longword value (0000 0006) by sign extending the supplied byte value (06).

The next group of equivalent instructions perform Boolean operations on individual bits within a register.

```
0880 0007          BCLR.L      #7,D0
0280 FFFF FF7F     ANDI.L      #$FFFFFF7F,D0

08C0 000D          BSET.L      #13,D0
0080 0000 2000     ORI.L       #$2000,D0

0840 0000          BCHG.L      #0,D0
0A80 0000 0001     EORI.L      #$01,D0
```

Note the use of different bit **masks** in the logical instructions. There is more work involved in determining the correct bit mask than there is in simply specifying the bit number.

The last pair of equivalent instructions is used to pass control to a subroutine. The subroutine address is $8150, not too far ahead of the instruction address ($8100). This allows the short form of the BSR instruction to be used.

```
                   ORG         $8100
614E               BSR.S       $8150
4EB9 0000 8150     JSR         $8150
```

Recall that the short form (614E) uses a single byte (4E) to indicate the signed displacement. The JSR instruction does not use a displacement value or relative addressing. The entire longword subroutine address (0000 8150) is used in the machine code. However, if a program is very large or distributed over many different parts of memory, the JSR instruction is more useful than BSR, since it does not have a range restriction.

You will discover many different ways to optimize the programming examples that follow. You are encouraged to experiment with your own optimizations also. For example, can you determine the equivalence between SWAP and ROL (or ROR)? Is there an equivalence between CLR and ANDI? You will find that the 68000's instruction set is full of equivalent instructions.

6.6   DATA GATHERING

When a microprocessor is used in a control application, one of its most important tasks is to gather data from the external process. This data may be composed of inputs from different types of sensors, parallel or serial information transmitted to the system from a separate source, or simply keystrokes from the user's keyboard.

Usually a section of memory is set aside for the storage of the accumulated data, so the processor can alter or examine it at a later time. The rate at which new data arrives, as in keystrokes from a keyboard, may be very slow, with a new item arriving every few milliseconds or so. When the data rate is slow, the processor will waste valuable execution time waiting for the next new data item. Therefore, an efficient solution is to store the data as it arrives, and only process it when all items have been stored. We will examine two examples of gathering data in this section. The first one deals with keyboard buffering and the second with packing BCD numbers.

The Keyboard Buffer

One of the first things anyone involved with computers learns is that nothing happens until you hit Return. All keystrokes up to Return must be saved for processing after return is hit. The subroutine presented here, KEYBUFF, is used to store these keystrokes in a buffer until Return is hit. The processor will then be free to examine the contents of the keyboard buffer at a later time. KEYBUFF makes use of a subroutine called GETKEY, which is used to get a keystroke from the keyboard. The ASCII code for the key is returned by GETKEY in the lower byte of D0. GETKEY takes care of echoing the key back to the user's display. The code for GETKEY can be written with the SBC or EMU68K   TRAPs that perform character I/O. The required instructions are:

```
GETKEY  MOVE.L  A0,-(A7)   ;save A0 on stack
        TRAP    #0         ;read PC keyboard
        TRAP    #1         ;echo character to display
        MOVE.B  D1,D0      ;return ASCII code in D0
        MOVEA.L (A7)+,A0   ;reload A0 from stack
        RTS
```

The MOVE.B   D1,D0 instruction is required since both TRAPs use D1 as the location for the ASCII character code. Also, since these two TRAPs destroy the contents of A0, the stack is used to hold a copy so that A0 has the same value upon exit of GETKEY that it does on entry.

An important point to keep in mind is that GETKEY will not return a value in D0 until a key is struck.

The ASCII codes for the keys entered are saved in a buffer called KEYS, which is limited to 128 characters. No code is provided to prevent more than this number of keystrokes. Can you imagine what problems occur when the 129th key is entered?

```
        ORG     $8000
CR      EQU     $0D                ;ASCII code for return
KEYS    DS.B    128                ;key buffer

        ORG     $8100
KEYBUFF MOVEA.L #KEYS,A0           ;A0 points to start of buffer
NEXTKEY BSR     GETKEY             ;get key from user
        MOVE.B  D0,(A0)+           ;save key in buffer
        CMPI.B  #CR,D0             ;continue until return is seen
        BNE     NEXTKEY
        RTS
```

An important feature missing in this example is the use of special codes for editing. No means are provided for editing mistaken keys entered by the user. At the very least, the user should be able to enter a backspace to correct a previous error. You are encouraged to solve this problem, and the other one dealing with limiting the number of keystrokes, yourself.

Packing BCD Numbers

Any program that deals with numbers must use one of two approaches to numeric processing. The program must either treat the numbers as binary values or as BCD values.

The use of binary operations provides for large numbers with a small number of bits (integers over 16 million can be represented with only 24 bits), but is limited in accuracy when it comes to dealing with fractions. The use of BCD provides for greater accuracy but requires software to support the mathematical routines, and this software greatly increases the execution time required to get a result. Even so, BCD numbers have found many uses, especially in smaller computing systems. The example we will study here is used to accept a multidigit BCD number from a keyboard and store it in a buffer called BCDNUM. The trick is to take the ASCII codes that represent the numbers 0 through 9 and convert them into BCD numbers. Two BCD numbers at a time are packed into a byte as shown in Figure 6.6. So, if the user enters 53297, three of the bytes in BCDNUM will be 05 32 97. BCDNUM will be limited to 6 bytes, thus making 12-digit BCD numbers possible. The subroutine PACKBCD will take care of packing the received BCD numbers into bytes and storing them in BCDNUM. No error checking is provided to ensure that no more than 12 digits are entered, or that the user has entered a valid digit. If the number entered is less than 12 digits long, the user enters Return to complete the entry. All numbers will be right-justified when saved in BCDNUM. This means that numbers less than 12 digits long will be filled with leading zeros. For example, 53297 will become 00 00 00 05 32 97, with the first 00 byte being the beginning of the buffer.

```
                ORG       $8000
CR              EQU       $0D
BCDNUM          DS.B      6                    ;storage for 12 BCD digits

                ORG       $8100
PACKBCD         MOVEA.L   #BCDNUM,A0           ;point to beginning of buffer
                MOVEQ     #5,D0                ;init loop counter
CLEARBUFF       CLR.B     (A0)+                ;clear all bytes in BCDNUM
                DBRA      D0,CLEARBUFF         ;with this loop
                SUBA.L    #1,A0                ;move A0 back to last pair
GETDIGIT        BSR       GETKEY               ;get a number from the user
                CMPI.B    #CR,D0               ;done?
                BEQ       DONE
                SUBI.B    #$30,D0              ;remove ASCII bias
                BSR       SHIFTBUFF            ;shift BCDNUM left one digit
                OR.B      (A0),D0              ;pack new digit into D0
```

FIGURE 6.6 Packing two BCD digits together

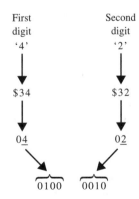

$42

```
               MOVE.B      D0,(A0)           ;save new digit in buffer
               BRA         GETDIGIT
DONE           RTS

SHIFTBUFF      MOVE.W      #4,D3             ;set up loop counter (5 passes)
               MOVEA.L     #BCDNUM,A1        ;set up pointer to BCDNUM
DOPAIR         MOVE.B      (A1),D1           ;read digit pair
               LSL.B       #4,D1             ;shift left one digit
               MOVE.B      1(A1),D2          ;read next digit pair
               LSR.B       #4,D2             ;shift right one digit
               OR.B        D1,D2             ;pack new pair of digits
               MOVE.B      D2,(A1)+          ;save in buffer
               DBRA        D3,DOPAIR         ;repeat for next pair
               MOVE.B      (A1),D1           ;get last buffer pair
               LSL.B       #4,D1             ;shift left one digit
               MOVE.B      D1,(A1)           ;and save
               RTS
```

The loop at the beginning of PACKBCD writes zeros into all 6 bytes of BCDNUM. This is done to automatically place all leading zeros into the buffer before any digits are accepted. Notice also that A0 has been advanced to the end of the buffer when the loop has finished. We need A0 to start at the end of BCDNUM because this is where we store the digits as they are entered. GETKEY is used to get a BCD number from the user (assuming that no invalid digits are entered). Subtracting $30 from the ASCII values returned by GETKEY converts the ASCII character code ($35 for '5') into the correct BCD value. The digit buffer (BCDNUM) is shifted one digit to the left each time a new number is returned by GETKEY. The process of shifting the buffer left is shown in Figure 6.7. Assume that the user has already entered the digits 1, 2, 3, 4, and 5. If another digit is entered, the buffer must be shifted left one digit, to become 123450. As Figure 6.7 shows, the first pair of digits (01) is shifted left, giving 10, and the second pair (23) is shifted right, giving 02. The 10 and 02 values are ORed together to produce 12. This process is repeated five times to shift the entire buffer left one digit. By using this shifting technique we can always load the new BCD digit into the last byte of BCDNUM.

FIGURE 6.7 Shifting BCDNUM left one digit

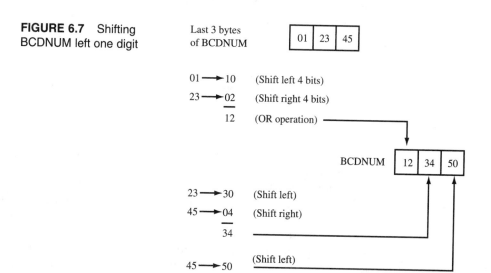

Programming Exercise 6.1:   Modify the keyboard buffer routine KEYBUFF so that the user may not enter more than 128 keys. KEYBUFF should automatically return if 128 keys are entered.

Programming Exercise 6.2:   Modify KEYBUFF to allow for two simple editing features. If a backspace key is entered (ASCII code $08), the last key entered should be deleted. (What problem occurs, though, when backspace is the first key entered?) The second editing feature is used to cancel an entire line. If the user enters a Control-C (ASCII code $03), the contents of the entire buffer are deleted.

Programming Exercise 6.3:   Modify KEYBUFF to include a count of the number of keys entered, including the final return key. This number should be stored in COUNT on return from KEYBUFF.

Programming Exercise 6.4:   Modify KEYBUFF so that all lowercase letters (a–z) are converted to uppercase (A–Z) before being placed in the buffer. All other ASCII codes should remain unchanged.

Programming Exercise 6.5:   Modify KEYBUFF so that the contents of the buffer are displayed (using TRAP   #1 for character output) if Control-R is entered, and the buffer is cleared if Control-C is entered.

Programming Exercise 6.6:   Modify the PACKBCD routine so that a maximum of 12 digits may be entered. PACKBCD should automatically return after processing the 12th digit.

Programming Exercise 6.7:   Modify the PACKBCD routine to check for illegal digits returned from GETKEY. Only ASCII codes '0' through '9' are valid.

6.7   SEARCHING DATA TABLES

In this section we will see a few examples of how a block of data may be searched for single or multibyte items. This technique is a valuable tool and has many applications. In a large database, information about many individuals may be stored. Their names, addresses, social security numbers, phone numbers, and many other items of importance may be saved. Finding out if a person is in the database by searching for any of the items just mentioned requires an extensive search of the database. In an operating system, information about users may be stored in a special access table. Their user names, account numbers, and passwords might be included in this table. When users desire to gain access to the system, their entries in the table must be located by account number or name and their passwords checked and verified. Once on the system, a user will begin entering commands. The commands entered must be checked against an internal list to see whether they exist before processing can take place. In a word processing program, a special feature might exist that allows a search of the entire document for a desired string. Every occurrence of this string must be replaced by a second string. For example, the user may notice that every occurrence of "apples" must be changed to "oranges." If only one or two of these strings exist, the user will edit them accordingly. But if "apples" occurs in 50 different places, it becomes very time consuming and inefficient to do this manually. Let us now look at a few examples of how a data table may be searched.

Searching for a Single Item

The first search technique we will examine involves searching for a single item. This item might be a byte, word, or longword value. The following subroutine searches a 100-element data table for a particular byte value. Upon entry to the subroutine, the byte to be searched for is stored in ITEM. The item may or may not exist within the data table. To account for these two conditions, we will need to return an indication of the result of the search. The carry flag is used to do this. If the search is successful, we will return with the carry flag set. If the search fails, we return with the carry flag cleared.

```
                ORG       $8000
DATA            DS.B      100
ITEM            DS.B      1

                ORG       $8100
FINDBYTE        MOVEA.L   #DATA,A0        ;init data pointer
                MOVE.W    #99,D0          ;init loop counter
                MOVE.B    ITEM,D1         ;load D1 with search item
COMPARE         CMP.B     (A0)+,D1        ;compare item with data in table
                BEQ       FOUND
                DBRA      D0,COMPARE      ;compare next item
                ANDI      #$FE,CCR        ;clear carry flag
                RTS
FOUND           ORI       #$01,CCR        ;set carry flag
                RTS
```

Notice how logical operations have been used to directly modify the carry flag, depending on the results of the search. Using the carry flag in this manner allows the programmer to write much simpler code. For example, only two instructions are needed to determine the result of the search:

```
BSR    FINDBYTE
BCS    SUCCESS
```

Of course, other techniques may be used to indicate the results. The nice thing about using the flags is that they require no external storage and can be used whenever a binary condition (true/false) is the result.

Searching for the Highest Integer

When working with data it often becomes necessary to find the largest value in a given set of numbers. This is useful in finding the range of the given set and also has an application in sorting. MAXVAL is a subroutine that will search an array called NUMBERS for the largest positive byte integer. No negative numbers are allowed at this time. The result of the search is passed back to the caller in the lower byte of D4.

```
          ORG       $8000
NUMBERS   DS.B      128

          ORG       $8100
MAXVAL    MOVEA.L   #NUMBERS,A2      ;init data pointer
          CLR.B     D4              ;assume 0 is largest to begin with
          MOVE.W    #127,D2         ;init loop counter
CHECKIT   CMP.B     (A2)+,D4        ;compare current value with new data
          BCC       NOCHANGE        ;branch if new value is not greater
          MOVE.B    -(A2),D4        ;load new maximum value
          ADDA.L    #1,A2           ;point to next location
NOCHANGE  DBRA      D2,CHECKIT      ;continue until all locations checked
          RTS
```

Since postincrement addressing is used during the compare operation, it is necessary to use predecrement addressing when loading the new maximum value. The ADDA instruction is also necessary to ensure that we do not get into an infinite loop when we encounter a new maximum value. Using BCC after the compare operation treats all bytes as unsigned integers. Other forms of the conditional branch will allow signed numbers to be detected as well.

A Command Recognizer

Consider a small single-board computer system that allows you to do all of the following:

- Examine/alter memory (EXAM)
- Display memory (DUMP)
- Execute a program (RUN)
- Terminate program execution (STOP)
- Load a program into memory (LOAD)

Each of the five example commands has a specific routine address within the memory map of the system. For example, the DUMP command is processed by the code beginning at address 0004A2C. The **command recognizer** within the operating system of the small computer must recognize that the user has entered the DUMP command, and jump to address 0004A2C. This requires that both a string compare operation and a table lookup be performed. The following routine is one way this may be accomplished.

```
          ORG       $8000
COMMANDS  DC.B      'EXAM'
          DC.B      'DUMP'
          DC.B      'RUN '
          DC.B      'STOP'
          DC.B      'LOAD'
JUMPTABLE DC.L      DOEXAM
          DC.L      DODUMP
          DC.L      DORUN
          DC.L      DOSTOP
          DC.L      DOLOAD
COMBUFF   DS.L      1

          ORG       $8100
RECOGNIZE MOVEA.L   #COMMANDS,A0    ;point to command table
          MOVEA.L   #JUMPTABLE,A4   ;point to routine address table
```

```
              MOVE.W      #4,D0                  ;init loop counter
              MOVE.L      COMBUFF,D2             ;load D2 with command text
NEXTCOM       CMP.L       (A0)+,D2               ;compare command text
              BEQ         GETADDR                ;go get jump address if match
              ADDA.L      #4,A4                  ;move to next routine address
              DBRA        D0,NEXTCOM
              JMP         COMERROR               ;command not found
GETADDR       MOVEA.L     (A4),A2                ;load A2 with routine address
              JMP         (A2)                   ;jump to command routine
```

The set of valid commands begins at COMMANDS. The addresses for each command
routine begin at JUMPTABLE. The command entered by the user is saved in the 4 bytes
beginning at COMBUFF. The purpose of RECOGNIZE is to compare entries in COM-
MANDS with COMBUFF. Every time a match is not found, a pointer (A4) is advanced
to point to the next routine address in JUMPTABLE. When a match is found, A4 will
point to the start of the routine address saved in memory. This routine address is then
loaded into A2 for use by JMP. If none of the commands match the user's, a jump is made
to COMERROR (possibly a routine that will output an error message saying "Illegal
command").

Programming Exercise 6.8:   Modify the FINDBYTE data search subroutine so that
the length, in bytes, of the data table is passed via LENGTH. The maximum length of the
data is 1,024 items.

Programming Exercise 6.9:   Modify FINDBYTE so that the position of ITEM
within VALUES is returned in POSITION, if the search is successful. For example,
if ITEM is the first element, POSITION should be 0. If ITEM is the 11th element,
POSITION should be 000A.

Programming Exercise 6.10:   Modify the MAXVAL subroutine so that negative
numbers (represented in 2's complement notation) may be included in the data.

Programming Exercise 6.11:   Modify the MAXVAL subroutine in two ways: (1)
The memory address of the maximum value is returned in A0, and (2) comparison of positive *word* values is performed if D0 equals 00 upon entry to MAXVAL, and comparison
of byte values is performed otherwise.

Programming Exercise 6.12:   The command recognizer RECOGNIZE works only
with uppercase commands. Rewrite the code so that lowercase commands are also accepted. For example, "DUMP" and "dump" should both be recognized.

Programming Exercise 6.13:   Write a command recognizer that will recognize
single-letter commands. The commands may be either uppercase or lowercase, and have
the following addresses associated with them:

A: 20BE

B: 3000

C: 589C

D: 2900

6.8   STRING OPERATIONS

As previously covered, a string is a collection of bytes that represent ASCII characters. For example, the TRAP   #3 exception handler provided on the SBC and EMU68K requires strings of the form:

```
ANYSTRING         DC.B    'This is a text string.',0
```

in which the end of the string is indicated by the 0 byte. What is required to process a display string? Assume that address register A3 has been loaded with the starting address of ANYSTRING (via LEA   ANYSTRING,A3). The SENDOUT routine shown here reads string characters one at a time and outputs them to the display until the end-of-string marker is seen.

```
SENDOUT   MOVE.B   (A3)+,D1     ;read a string character
          CMPI.B   #0,D1        ;end of string?
          BEQ      EXIT         ;jump if match
          TRAP     #1           ;display character function
          BRA      SENDOUT      ;and repeat
EXIT      RTS
```

There are many other uses for text strings. They can specify a program's file name, as in:

```
SYSFILE  DC.B     'IODAT.SYS',0
```

or a list of abbreviated days of the week:

```
DAYS    DC.B      'MonTueWedThuFriSatSun',0
```

In this section we will examine a number of techniques that utilize text strings to perform useful operations.

Searching a String for a Single Character

This string operation examines every character in the string to see if it matches the character supplied by the user. If a match is found, the address of the character is returned. If no match is found, an address of 00000000 is returned. For example, consider the string:

```
HEYSTR   DC.B     'Hey!',0
```

If we search HEYSTR for the '!' character, the address of the '!' character will be returned. If we instead search HEYSTR for the '?' character, 00000000 will be returned. Note that the only way to determine whether a string contains a character is to look at every element of the string until a match is found, or until the end-of-string marker is found. Both cases are handled by the SEARCHSTR routine.

```
SEARCHSTR   MOVE.B   (A3),D0       ;read string character
            CMPI.B   #0,D0         ;end of string found?
            BEQ      NOMATCH       ;branch if true
            CMP.B    D0,D1         ;search character found?
            BEQ      MATCH         ;branch if true
            ADDA.L   #1,A3         ;advance character pointer
            BRA      SEARCHSTR     ;and repeat
MATCH       RTS                    ;A3 contains character address
NOMATCH     LEA      0,A3          ;A3 contains 00000000
            RTS
```

The search character must be in D1 and the starting address of the string in A3 before SEARCHSTR is executed.

Comparing Strings

A very important part of any program that deals with input from a user involves recognizing the input data. Consider the password required by most users of large computing systems. The user must enter a correct password or be denied access to the system. Since the password may be thought of as a string of ASCII characters, some kind of string comparison operation is needed to see if the user's password matches the one expected by the system. The following subroutine compares two strings, returning with the carry flag set if the strings are exactly the same. If you think of one string as the password entered by the user and the other as the password stored within the system, you will see how they are compared.

```
            ORG      $8000
STRINGA     DC.B     'alphabetic',0
STRINGB     DC.B     'alphabet',0

            ORG      $8100
CMPSTR      MOVEA.L  #STRINGA,A0   ;init pointer to first string
            MOVEA.L  #STRINGB,A1   ;init pointer to second string
CHECKCHAR   CMPM.B   (A0)+,(A1)+   ;compare item from each string
            BNE      NOMATCH       ;even one difference causes failure
            CMPI.B   #0,-1(A0)     ;was last element the end marker?
            BNE      CHECKCHAR     ;branch if false
            ORI      #$01,CCR      ;strings are identical, set carry
            RTS
NOMATCH     ANDI     #$FE,CCR      ;strings are different, clear carry
            RTS
```

The two strings used in the example are not identical because they have different lengths.

Concatenating Two Strings

This string operation is used to combine two strings into a single string, whose length equals the sum of the individual string lengths. For example, consider these two strings:

```
STR1   DC.B   'This is ',0
STR2   DC.B   'only a test...',0
```

Concatenating these two strings results in a third string equivalent to this definition:

```
DC.B    'This is only a test...',0
```

Now, if we wish to concatenate STR2 onto STR1, there must be storage available at the end of STR1 for the new string characters. Otherwise, we risk the danger of overwriting other data that might happen to be stored directly after the string data. One way to reserve room at the end of a string is as follows:

```
STR1    DC.B    'This is ',0,0,0,0,0
```

which is fine if we only need a few extra character positions. But if we expect larger strings to be concatenated, it is better to use the DS directive to reserve space, as in:

```
STR1    DC.B    'This is ',0
        DS.B    16
```

where we easily reserve room for 16 additional string characters.

The CATSTR subroutine shown here concatenates two strings by searching to the end of the first string (pointed to by A0), and then copying all elements from the second string (pointed to by A1) to the end of the first string until the end-of-string marker is found. Upon entry, A0 and A1 must contain the starting addresses of each string.

```
CATSTR    CMPI.B   #0,(A0)+      ;check for end marker
          BNE      CATSTR        ;keep looking until found
          SUBA.L   #1,A0         ;back up to end marker address
COPYCHAR  MOVE.B   (A1)+,(A0)+   ;copy character from 2nd string
          CMPI.B   #0,-1(A1)     ;check for end marker
          BNE      COPYCHAR      ;branch if not found
          RTS
```

Since postincrement addressing is used to perform the copying, it is necessary to look at the *previous* address pointed to by A1 to check for the end marker. This is done in the CMPI.B   #0,–1(A1) instruction.

Numeric String Conversion

This last string operation is used to convert a string containing only numerical digits into an actual value. For example, the string:

```
'123',0
```

which is composed of the characters '1', '2', and '3', is converted into the hexadecimal value 7B (which is 123 decimal). Figure 6.8 shows one way this can be done. The ASCII codes for '1', '2', and '3' are converted into their binary equivalents. The '1' value is multiplied by 100, and the '2' value by 10. All three values are added together to get the result. Although this technique works, it is not easily coded to handle numeric strings of different lengths.

A different technique that works with strings of different lengths is shown in Figure 6.9. The conversion method is simple. We begin with a temporary result of zero. For each new digit we pull out of the string, we multiply the temporary result by 10 and then add the new digit value to the temporary result. This continues until we reach the end marker. The NUMSTR routine presented here performs this type of conversion, but is only accurate for strings whose numeric value is in the range 0 to 65535. This is because the MULU instruction used to multiply the temporary value by 10 only works with word-size operands.

FIGURE 6.8 Converting a numeric string into an actual number

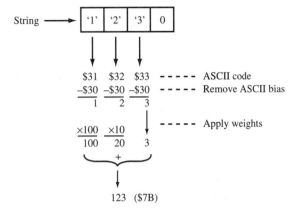

FIGURE 6.9 Simplified numeric string conversion

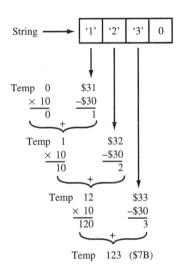

```
NUMSTR    CLR.W     D0              ;clear temp result
NEXT      MOVE.B    (A0)+,D1        ;get string element
          BEQ       EXIT            ;branch if end marker found
          SUBI.B    #$30,D1         ;remove ASCII bias
          EXT.W     D1              ;adjust to word size
          MULU      #10,D0          ;multiply temp result by 10
          ADD.W     D1,D0           ;add new digit to temp result
          BRA       NEXT            ;and repeat
EXIT      RTS
```

Upon entry to NUMSTR, A0 must contain the starting address of the numeric string. The result is returned in the lower word of D0.

Programming Exercise 6.14: Modify the SEARCHSTR routine so that the string is searched for a *substring* provided by the user. For example, the string 'abcde' contains the substring 'bc' but does not contain the substring 'bd'.

Programming Exercise 6.15:   Modify the CATSTR routine so that only the number of characters specified by the value in D0 are concatenated.

Programming Exercise 6.16:   Modify the NUMSTR routine so that hexadecimal number strings are also converted. For example, if the input string is '$10' we get 16 decimal as the result.

6.9   SORTING

It is often necessary to sort a group of data items into ascending (increasing) or descending (decreasing) order. On average, the search time for a sorted list of numbers is smaller than that of an unsorted list, so it pays to examine different sorting techniques, so that we may in turn write efficient search algorithms. Many different sorting algorithms exist, with some more efficient than others. In this section we will examine two simple sorting techniques and compare their operation.

Bubble Sort

This sorting technique derives its name from the fact that a data element within the array "bubbles" its way into its correct position. A bubble sort consists of many passes over the elements being sorted, with comparisons and swaps of numbers being made during each pass. The range of the input numbers is not limited. A short example will serve to introduce you to the bubble-sort technique. Consider this group of numbers:

```
7  10  6  3  9
```

It is necessary to perform only four comparisons to determine the highest number in the group. We will repeatedly compare one element in the group with the next element, starting with the first. If the second element is larger than the first, the two numbers will be swapped. By this method we guarantee that after four comparisons, the largest number is at the end of the array. Check this for yourself. Initially, 7 and 10 are compared and not swapped. Then 10 and 6 are compared and swapped because 10 is greater than 6. The new array looks like this:

```
7  6  10  3  9
```

Next 10 and 3 are compared and swapped. Then 10 and 9 are compared and swapped. At the end of the first pass, the array is:

```
7  6  3  9  10
```

It is no longer necessary to compare any of the elements in the array with the last one, since we know it to be the largest. The next pass will only compare the first four numbers, giving this array at the end of the second pass:

```
6  3  7  9  10
```

The third pass will produce:

```
3  6  7  9  10
```

and you may notice now that the array is sorted. However, this is due to the original arrangement of the numbers, and for completeness a final pass must be performed on the first two numbers. It is interesting to note that the five numbers being sorted required four passes. In general, N numbers will require N − 1 passes. The subroutine BUBSORT presented here implements a bubble sort. D1 is used as the pass counter, registers D2 and D3 are used for swapping elements, and D4 is used as a loop counter. The number of elements to be sorted is saved as a word count in NVAL. The appropriately sized DS statement is needed for VALUES, with only 16 bytes reserved in this example. Also, only positive integers may be sorted (because of the use of BCC in the comparison).

```
              ORG       $8000
VALUES        DS.B      16
NVALS         DC.W      16

              ORG       $8100
BUBSORT       MOVE.W    NVALS,D1      ;get number of data items
              SUBI.W    #1,D1         ;adjust count for looping
DOPASS        MOVE.W    D1,D4         ;init loop counter
              MOVEA.L   #VALUES,A0    ;init data pointer
CHECK         MOVE.B    (A0),D2       ;get first element
              MOVE.B    1(A0),D3      ;get second element
              CMP.B     D2,D3         ;compare elements
              BCC       NOSWAP        ;no swap if D3 > D2
              MOVE.B    D2,1(A0)      ;swap elements
              MOVE.B    D3,(A0)
NOSWAP        ADDA.L    #1,A0         ;point to next pair
              SUBI.W    #1,D4         ;decrement loop counter
              BNE       CHECK
              SUBI.W    #1,D1         ;decrement pass counter
              BNE       DOPASS
              RTS
```

By advancing A0 in steps of 1, memory references (A0) and 1(A0) always access the next two elements in the VALUES array. When it is necessary to swap them, the registers are written back into memory in a swapped fashion. The use of different conditional jump instructions will allow for negative numbers to be sorted as well.

Bucket Sort

Unlike the bubble sort, which can sort numbers of any size, a bucket sort requires the numbers to be within a predetermined range. For example, consider the same numbers used in the bubble sort illustration:

```
7  10  6  3  9
```

None of the numbers are larger than 10. In addition, none of the numbers are duplicated. Knowing in advance that we will be sorting unduplicated numbers less than or equal to 10

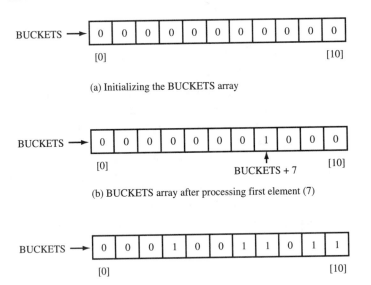

(a) Initializing the BUCKETS array

(b) BUCKETS array after processing first element (7)

(c) Final BUCKETS array

FIGURE 6.10   Bucket sort array: (a) initializing the BUCKETS array; (b) BUCKETS array after processing first element (7); (c) final BUCKETS array

allows us to initialize an 11-element **bucket array** as shown in Figure 6.10(a). As you can see, each element of the array is initialized to 0, which is used to represent an **empty bucket.**

The process of bucket sorting is now straightforward: For each element in the input, we set the associated element in the bucket array equal to 1 (or any non-zero integer), signifying a full bucket. For example, Figure 6.10(b) shows the bucket array after the first input element (7) has been processed. Notice the 1-value stored in element position [7].

After all five input elements have been processed, the bucket array contains the element values shown in Figure 6.10(c). The zeros remaining in the array indicate that we did not see the values 0, 1, 2, 4, 5, and 8 in our input list.

To store the sorted array, it is only necessary to start at the beginning of the bucket array and store each index number that contains a value of 1.

The BKTSORT routine shown here performs a bucket sort on a five-element array of numbers. The largest input number equals 10, which requires an 11-element bucket array.

```
            ORG     $8000
NUMBERS     DS.B    5                   ;room for 5 input values
BUCKETS     DS.B    11                  ;room for 11 buckets

            ORG     $8100
BKTSORT     MOVEA.L #BUCKETS,A0         ;init array pointer
            MOVE.W  #10,D0              ;init loop counter
CLRBKT      CLR.B   (A0)+               ;clear bucket
            DBRA    D0,CLRBKT           ;do this 11 times
            MOVEA.L #NUMBERS,A1         ;init pointer to input data
            MOVEA.L #BUCKETS,A0         ;init pointer to buckets
            MOVE.W  #4,D0               ;init loop counter
```

```
SETBKT    MOVE.B    (A1)+,D1          ;read input value
          ANDI.L    #$000000FF,D1     ;convert value to longword
          MOVE.B    #1,(A0,D1)        ;set bucket to 1
          DBRA      D0,SETBKT         ;do this 5 times
          MOVEA.L   #NUMBERS,A1       ;init pointer to input array
          MOVEA.L   #BUCKETS,A0       ;init pointer to buckets
          CLR.L     D1                ;clear index pointer
          MOVE.W    #10,D0            ;init loop counter
READBKT   CMPI.B    #1,(A0,D1)        ;read bucket value
          BNE       NOSAVE            ;branch if bucket = 0
          MOVE.B    D1,(A1)+          ;save index value
NOSAVE    ADDI.B    #1,D1             ;advance index value
          DBRA      D0,READBKT        ;do this 11 times
          RTS
```

Register D1 is used as an index pointer (together with A0) whenever the BUCKETS array must be accessed. This allows the input number to be converted into index values easily through the ANDI instruction.

Programming Exercise 6.17: Modify the BUBSORT routine so that positive and negative numbers can be sorted.

Programming Exercise 6.18: Modify the BUBSORT routine so that it terminates if no swaps are performed in a single pass (the array is already sorted).

Programming Exercise 6.19: The BKTSORT routine can be altered to allow duplicate numbers in the input. Instead of setting a bucket to 1, *increment* the bucket instead. So, if the number 3 is seen four times, the bucket for element 3 should contain 4. Modify the BKTSORT routine accordingly.

6.10   COMPUTATIONAL ROUTINES

This section covers examples of how the 68000 performs standard mathematical functions. Since the processor has specific instructions for both binary and BCD operations, we will examine sample routines written around those instructions. Math processing is a major part of most high-level languages and the backbone of specialized application programs, such as spreadsheets and statistical analysis packages. Most processors, however, are limited in their ability to perform complicated math. When complex functions such as SIN(X) or LOG(Y) are needed, the programmer is faced with a very difficult task of writing the code to support them. Even after the code is written and judged to be correct, it will most likely be very lengthy and slow in execution speed. For this reason, some systems are designed with math coprocessor chips. These chips are actually microprocessors whose instruction sets contain only mathematical instructions. Adding a coprocessor eliminates the need to write code to perform the math function. SIN(X) is now an instruction executed by the coprocessor. The main CPU simply reads the result from the coprocessor. The coprocessor available for the 68000 is the 68881 floating-point coprocessor, which we will examine in Chapter 10.

The examples we will see in this section deal only with addition, subtraction, multiplication, and division. We will, however, also look at a few ways these simple operations can be applied to simulate more complex ones.

Binary Addition

Binary addition is accomplished with any of the following instructions: ADD, ADDI, ADDQ, or ADDX. All four perform addition on data registers and/or memory locations. To add data to an address register, use ADDA. The example presented here is used to find the signed sum of a set of data. The data consists of signed 8-bit numbers. Since it is possible for the sum to exceed 127, we use 16 bits to represent the result.

```
        ORG       $8000
SCORES  DS.B      200
SUM     DS.W      1

        ORG       $8100
TOTAL   MOVEA.L   #SCORES,A0      ;init pointer to data
        MOVE.W    #199,D6         ;init loop counter
        CLR.W     D1              ;clear result
ADDEM   MOVE.B    (A0)+,D0        ;load D0 with value
        EXT.W     D0              ;sign extend into 16 bits
        ADD.W     D0,D1           ;add new value to result
        DBRA      D6,ADDEM        ;do all values
        MOVE.W    D1,SUM          ;save result in memory
        RTS
```

Even though the data consists of signed 8-bit numbers, we can perform 16-bit additions if we first use EXT to extend the signs of the input numbers (from 8 to 16 bits).

Binary Subtraction

Binary subtraction is implemented by SUB, SUBI, SUBQ, and SUBX. All of these instructions work with memory locations and/or data registers. SUBA is also included when subtraction involves an address register. The example presented here shows how two blocks of memory may be subtracted from each other. One application in which this technique is useful involves digitally encoded waveforms. Suppose that two analog signals, sampled at an identical rate, must be compared. If the difference is computed by subtracting the binary representation of each waveform and the resultant waveform displayed by sending the new data to an analog-to-digital converter, we will see a straight line at the output if the waveforms are identical. WAVE1 and WAVE2 are labels associated with the 2K word blocks of memory that must be subtracted. Because of the addressing mode used, the resulting data will overwrite the data saved in WAVE2's area.

```
        ORG       $8000
WAVE1   DS.W      2048
WAVE2   DS.W      2048

        ORG       $A000
SUBWAVE MOVEA.L   #WAVE1,A0       ;init pointer to beginning of WAVE1
        ADDA.L    #2048,A0        ;advance pointer to end of data table
        MOVEA.L   #WAVE2,A1       ;repeat for second table
        ADDA.L    #2048,A1
        MOVE.W    #2047,D0        ;init loop counter
SUBEM   ANDI      #$EF,CCR        ;clear X flag
```

```
SUBX.W     -(A0),-(A1)    ;subtract (WAVE1-WAVE2)
DBRA       D0,SUBEM
RTS
```

For the subtraction to work properly, we must ensure that the X bit in the condition code register is cleared prior to each SUBX. Why does ANDI need to be inside the loop?

Binary Multiplication

Two instructions are available for performing binary multiplication. MULU (unsigned multiply) is used to multiply two 16-bit operands, one of them contained in a data register. MULS (signed multiply) generates a signed result using signed operands of 16 bits each. In both cases, the result is 32 bits long. When 32-bit precision is not enough, we must turn to an alternative method to perform the math. One solution is to add a math coprocessor chip. The benefit of doing this is reflected in a decrease in time needed to perform the math. Also, many complex functions are available in the coprocessor. The disadvantage is in the added cost of the hardware. Coprocessors tend to be expensive.

If the hardware cost is excessive, the only other solution is to use software. The example presented here is used to perform 32-bit by 16-bit multiplication on unsigned integers. The 48-bit result represents a significant increase over the 32 bits the processor is limited to. The method used to perform the multiplication is diagrammed in Figure 6.11. The 32-bit operand is represented by two 16-bit halves, A and B. The 16-bit operand is represented by C. Multiplying B by C will yield a 32-bit result. The same is true for A and C, except that A is effectively shifted 16 bits to the left, making its actual value much larger. To accommodate this, 16 zeros are placed into the summing area in such a way that they shift the result of A times C the same number of positions to the left. This is analogous to writing down a zero during decimal multiplication by hand. The lower 16 bits of the result are the same as the lower 16 bits of the BC product. The middle 16 bits of the result are found by adding the upper 16 bits of the BC product to

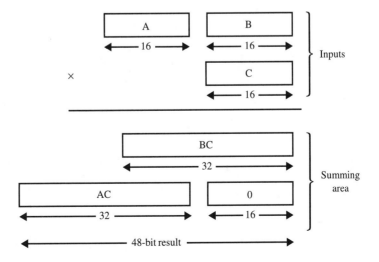

FIGURE 6.11 Diagram of 32-bit by 16-bit multiplication

the lower 16 bits of the AC product. The upper 16 bits of the result equal the upper 16 bits of the AC product, plus any carry out of the middle 16 bits. In the following routine, data register D0 contains the 32-bit value we know as AB. Data register D1 contains the 16-bit multiplier C.

```
                ORG      $8000
LOWER           DS.W     1                ;lower 16-bits of result
MIDDLE          DS.W     1                ;middle 16-bits of result
UPPER           DS.W     1                ;upper 16-bits of result

                ORG      $8100
MULTIPLY        MOVE.W   D1,D2            ;let D2 equal D1 (C)
                MULU     D0,D2            ;do B times C
                MOVE.W   D2,LOWER         ;save partial result
                MOVE.W   D1,D3            ;let D3 equal D1 (C)
                SWAP     D0               ;exchange bits 0-15 with 16-31
                MULU     D0,D3            ;do A times C
                SWAP     D2               ;switch 16-bit halves of BC product
                ADD.W    D3,D2            ;generate middle 16-bits of result
                MOVE.W   D2,MIDDLE        ;and save them
                SWAP     D3               ;get upper bits of AC product
                CLR.W    D2               ;clear lower word of D2
                ADDX.W   D2,D3            ;form upper 16-bits of result
                MOVE.W   D3,UPPER         ;and save them
                RTS
```

It should be possible to relate the code of this example to Figure 6.11. Generating the individual AC and BC products is easily done via MULU. Adding the upper 16 bits of the BC product to the lower 16 bits of the AC product is accomplished by using SWAP before the ADD instruction. Any overflow out of the middle 16 bits will be placed into the X flag. This carry is then added to the upper 16 bits of the AC product to complete the operation. It is necessary to clear a data register for use with the ADDX instruction in this case, because ADDX does not support immediate data operands.

Binary Division

The 68000 microprocessor supports binary division with its DIVU and DIVS (unsigned and signed division) instructions. Both instructions divide a 32-bit quantity by a 16-bit quantity. The 32-bit result is composed of a 16-bit quotient and a 16-bit remainder. When division by 0 is attempted, an exception will be generated on completion of the instruction. Many applications exist for the division operation. It can be used to find averages, probabilities, factors, and many other items that are useful when we are working with sets of data. The following subroutine is used to find a factor of a given number, when supplied with another factor. For example, FACTOR will return 50 as a factor, when 6 and 300 are supplied as input (because 300 divided by 6 equals 50 exactly). FACTOR will return 0 if no factor exists (that is, 300 divided by 7 gives 42.857143, which is not an integer; thus, both numbers cannot be factors).

```
                ORG      $8000
NUMBER          DS.L     1                ;32-bit input number
FACTOR1         DS.W     1                ;16-bit input factor
FACTOR2         DS.W     1                ;16-bit output factor

                ORG      $8100
FACTOR          MOVE.L   NUMBER,D0        ;load D0 with number
                DIVU     FACTOR1,D0       ;divide D0 by input factor
                SWAP     D0               ;test remainder for zero
```

```
                CMPI.W      #0,D0
                BEQ         SAVED0      ;branch if factor is integer
                CLR.L       D0          ;clear entire result
SAVED0          SWAP        D0          ;get result back
                MOVE.W      D0,FACTOR2  ;and save it
                RTS
```

Since the remainder appears in the upper 16 bits of D0, we use SWAP to move the remainder into the lower 16 bits for testing. It then becomes necessary to do a second SWAP, in case the result is a valid factor.

BCD Addition

In the binary number system, we use 8 bits to represent integer numbers in the range 0 to 255 (00 to FF hexadecimal). When the same 8 bits are used to store a binary coded decimal (BCD) number, the range changes. Integers from 0 to 99 may now be represented, with the 10s and 1s digits using 4 bits each. If we expand this reasoning to 16 bits, we get a 0–65535 binary integer range, and a 0–9999 BCD range. Notice that the binary range has increased significantly. This is always the case and represents one of the major differences between binary and BCD numbers. Even so, we use BCD to solve a nasty problem encountered when we try to represent some numbers using binary. Consider the fractional value 0.7. It is impossible to exactly represent this number using a binary string. We end up with 0.101100110011. . . . The last 4 bits (0011) keep repeating. So, we can get very close to 0.7 this way (0.699999 . . .), but never actually get 0.7. When we use this binary representation in a calculation, we will automatically generate a round-off error. The purpose of BCD is to eliminate the round-off error (at the cost of a slower computational routine).

For the purposes of this discussion we will use a BCD representation that consists of 4 bytes stored in consecutive memory locations. The first byte is the most significant byte. The fourth byte is least significant. All BCD numbers stored this way (0 to 99999999) will be right justified. Examine the following two numbers and their memory representations to see what is meant by right justification:

34298: 00 03 42 98

7571364: 07 57 13 64

Notice the leading zeros in each number when it has been stored. It is necessary to use leading zeros with small numbers to make them fit correctly into memory. We can increase the range of numbers by adding more bytes of storage per number. Each new byte gives two additional BCD digits. Furthermore, we could also add an additional byte to store the exponent of the number. A single byte could represent exponents from 127 to –128 if we use signed binary numbers. Standards exist that define the format of a BCD number (and of binary numbers as well, for use with math coprocessors), but we will not cover them at this time.

The example presented here shows how two BCD numbers (each stored in memory at NUMA and NUMB) can be added together. The ABCD (add decimal with extend) instruction is used to perform the BCD addition. ABCD will add 2 bytes, each containing two BCD digits. The result will also be in BCD, with the X flag containing any carry out of the most significant digit. For example, if 37 and 85 are added, X will equal 1 and the result operand will contain 22. Because we have defined the 4-byte storage array for a BCD number to be right justified, it is necessary to begin adding with the least significant byte in the array. The result is stored in NUMB, overwriting the BCD number already saved.

```
                ORG         $8000
NUMA            DS.B        4
NUMB            DS.B        4

                ORG         $8100
ADDBCD          MOVE.W      #3,D0          ;init loop counter
                MOVEA.L     #NUMA,A0       ;init pointer to first BCD number
                ADDA.L      #4,A0          ;and adjust pointer to LSB
                MOVEA.L     #NUMB,A1       ;repeat for second BCD number
                ADDA.L      #4,A1
                ANDI        #$EF,CCR       ;clear X flag
DECIADD         ABCD        -(A0),-(A1)    ;decrement pointers and add decimal
                DBRA        D0,DECIADD
                RTS
```

Upon return from the subroutine, the X flag will contain any carry out of the MSB.

BCD Subtraction

BCD subtraction is implemented in much the same way as BCD addition, and the subroutine presented here uses the same 4-byte BCD number definition covered in the previous section. The difference in this routine is that the addresses of the two BCD numbers are assumed to be contained in address registers A0 and A1 upon entry. Assuming that A0 points to NUMA and A1 to NUMB, two different subtractions are possible.

```
AMINUSB         EXG         A0,A1          ;swap pointers
BMINUSA         MOVE.W      #3,D0          ;init loop counter
                ANDI        #$EF,CCR       ;clear X flag
DECISUB         SBCD        -(A0),-(A1)    ;adjust pointers and subtract decimal
                DBRA        D0,DECISUB
                RTS
```

Upon return, the X flag will indicate any borrow from the MSB. If X is set upon return, the result of the subtraction is negative. The result will replace the contents of NUMB when BMINUSA is the entry point to the subroutine. Entering at AMINUSB will cause the result to replace NUMA.

BCD Multiplication

Since BCD multiplication is not directly implemented on the 68000, there are at least two ways it can be simulated. One method is to convert both BCD numbers into their binary equivalents and then use MULS to find the result. Of course, the binary result will have to be converted back into BCD. A procedure to accomplish this method is as follows:

```
BCDMUL1         MOVEA.L     A0,A6          ;convert first number
                BSR         TOBINARY
                MOVE.L      D6,D0          ;save it in D0
                MOVEA.L     A1,A6          ;convert second number
                BSR         TOBINARY
                MOVE.L      D6,D1          ;save it in D1
                MULS        D0,D1          ;multiply numbers
                MOVE.L      D1,D6          ;convert back into BCD
                BSR         TOBCD
                RTS
```

The two conversion routines, TOBINARY and TOBCD, operate as follows: TOBINARY converts the BCD number pointed to by A6 into a signed binary number and returns it in D6. TOBCD converts the signed binary number in D6 into a BCD number and saves it in memory at BCDNUM.

The advantage of BCDMUL1 is its speed of execution. Its disadvantage lies in the size of the numbers that may be multiplied. Since MULS will only multiply signed 16-bit numbers, the range of inputs is limited (32767 to −32768).

A second approach is to do all the math in BCD. This will require a number of repetitive additions to generate the answer. The need for this looping will unfortunately slow down the execution speed. This disadvantage is overcome by the ability to multiply larger numbers than BCDMUL1. BCDMUL2 will multiply two 2-digit BCD numbers stored in the lower byte of registers D0 and D1. The BCD result will be placed in D6. Further programming easily extends the input numbers into additional digits.

The multiplication performed by BCDMUL2 is detailed in Figure 6.12. As the figure shows, the product resulting from the 10s digits of the multiplier is shifted left one BCD digit, to simulate the result of multiplying by 10.

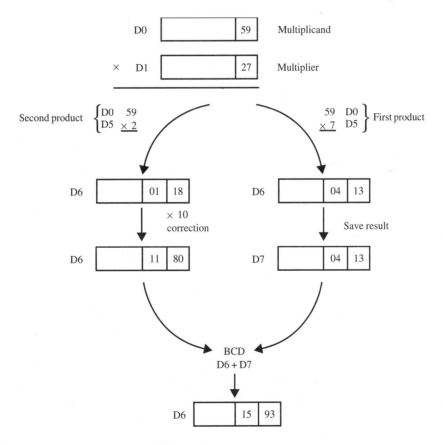

FIGURE 6.12 Multiplying two BCD numbers

```
            ORG       $8100
BCDMUL2     MOVE.W    D1,D5          ;find the 1s product
            ANDI.W    #$0F,D5
            BSR       D0TIMESD5
            MOVE.W    D6,D7          ;save result in D7
            MOVE.W    D1,D5          ;find the 10s product
            LSR.B     #4,D5
            ANDI.W    #$0F,D5
            BSR       D0TIMESD5
            LSL.W     #4,D6          ;multiply D6 by 10 (one digit shift left)
            BSR       D6PLUSD7       ;find final result
            RTS

D0TIMESD5   CLR.W     D6             ;init result
            BRA       NEXT
AGAIN       ANDI      #$EF,CCR       ;clear X
            ABCD      D0,D6          ;add to result
            BCC       NEXT           ;test for 100s carry
            ADD.W     #$100,D6
NEXT        DBRA      D5,AGAIN
            RTS                      ;return result in D6

D6PLUSD7    ANDI      #$EF,CCR       ;clear X
            ABCD      D7,D6          ;add lower 2 digits
            ROR.W     #8,D6
            ROR.W     #8,D7
            ABCD      D7,D6          ;add upper 2 digits
            ROL.W     #8,D6          ;correct result
            RTS
```

Notice how subroutines are used to make the overall process easier to code and read. Although we have not shown it in this example or in previous ones, it is assumed that a valid stack pointer has been assigned to save the subroutine return addresses.

BCD Division

All of the BCD operations we have examined so far have ignored treatment of exponents. A collection of subroutines that perform BCD math must have methods of dealing with exponents or be very limited in its applications. As previously mentioned, we can add a single byte to our BCD format to include exponents in the calculations. A single byte gives a signed integer range from –128 to 127. This slightly changes the format of the BCD numbers represented, and requires **normalization** of the numbers before conversion. Normalization is necessary because we have no way of storing a decimal point within the binary data we use to represent a number. Through normalization, we end up with a standard representation by altering the mantissa and adjusting the exponent accordingly. For example, 576.4 and 5.764E2 are equal, as are 23497.28 and 2.349728E4. In these examples, both numbers have been normalized so that the first digit of the mantissa is always between 1 and 9. This method works for fractional numbers as well. Here we have 0.0035 equaling 3.5E–3. The addition of an exponent byte to our format, together with the new technique of normalization, will require that we now left justify our BCD numbers. Representing these numbers in our standard format gives

576.4:   02 57 64 00 00

23497.28:   04 23 49 72 80

0.0035:   FD 35 00 00 00

where the first byte is used to represent the signed binary exponent.

Notice the 2's complement representation of the exponent −3 in the third set of data bytes.

Adding exponent capability to our BCD format complicates the routines we have already seen. The addition routine (as well as subtraction) will only give valid results when we are adding two numbers whose exponents are equal. Since this is rarely the case, we need to adjust the exponent of one number before doing the addition. For instance, if we wish to add 5027 and 394, we must first normalize both numbers:

5027: 03 50 27 00 00

394: 02 39 40 00 00

Because the exponents are different, we have to adjust one of the numbers to correctly add them. If we adjust the number with the higher exponent, we may lose accuracy in our answer. It is much safer to adjust the smaller number. This gives us

5027: 03 50 27 00 00

394: 03 03 94 00 00

It is clear now that BCD addition of the four trailing bytes will give the correct answer. Notice that we have not changed the value of the second number, only its representation.

BCD multiplication and division also require the use of exponents for best results. Unfortunately, it is not a simple matter of adding exponents for multiplication and subtracting them for division. Special rules are invoked when we multiply or divide two negative numbers. In any case, we must take all rules into account when writing a routine that will handle exponents.

The BCD division routine presented here keeps track of exponents during its calculations. The subroutine ALIGN adjusts the dividend so that it is always 1 to 9 times greater than the divisor. ALIGN modifies the exponent of the dividend as well. Subroutine MSUB performs multiple subtractions. The number of times (0 to 9) the divisor is subtracted from the dividend is returned in the lower 4 bits of D6. Both routines utilize A0 and A1 as pointers to the memory locations containing the BCD representations of the dividend and divisor. Data register D4 accumulates the individual results from MSUB into an eight-digit BCD result. The exponent is generated by the EXPONENT subroutine, which uses the initial exponent values plus the results of ALIGN to calculate the final exponent, which is returned in the lower byte of D5.

```
             ORG      $8000
DIVIDEND     DS.B     5              ;reserve 5 bytes (one byte for exponent)
DIVISOR      DS.B     5

             ORG      $8100
BCDDIV       MOVEA.L  #DIVIDEND,A0   ;init first pointer
             MOVEA.L  #DIVISOR,A1    ;init second pointer
             CLR.B    D5             ;clear exponent accumulator
             MOVE.W   #8,D0          ;init loop counter
DIVIDE       BSR      ALIGN          ;align numbers
             BSR      MSUB           ;perform multiple subtractions
             ROXL.L   #4,D4          ;shift result one digit left
             ANDI.B   #$0F,D6        ;mask out result from MSUB
             OR.B     D6,D4          ;save result in D4
             DBRA     D0,DIVIDE      ;continue for more precision
             BSR      EXPONENT       ;generate final exponent
             RTS
```

BCDDIV does not check for division by zero, but this test could be added easily with a few instructions.

Deriving Other Mathematical Functions

Once subroutines exist for performing the basic mathematical functions (addition, subtraction, multiplication, and division), they may be used to derive more complex functions. Although all forms of high-level math operations are available with the addition of a math coprocessor, there are times when we must get by without one. The examples presented here show how existing routines can be combined to simulate higher-level operations. All of the examples to be presented assume that the following multiprecision subroutines exist:

Routine	Operation
ADD	(A2) = (A0) + (A1)
SUBTRACT	(A2) = (A0) − (A1)
MULTIPLY	(A2) = (A0) * (A1)
DIVIDE	(A2) = (A0) / (A1)

In all cases, A0 and A1 point to the two input numbers upon entry to the subroutine, and A2 points to the result. Thus, (A0) means the number pointed to by A0, not the contents of A0. By defining the routines in this way, we can avoid discussion about whether the numbers are binary or BCD.

The first routine examined is used to raise a number to a specified power (for example, 5 raised to the third power is 125). This routine uses the binary number in D0 as the power. The number raised to this power is pointed to by A0. The final result is pointed to by A1.

```
POWER       BSR     COPY                    ;make a copy of the input num-
ber
MAKEPOW     BSR     MULTIPLY                ;compute next power result
            EXG     A1,A2                   ;use result as next input
            DBRA    D0,MAKEPOW
            RTS
```

POWER is written such that the power must be 2 or more. Negative powers and powers equal to 0 or 1 are not implemented in this routine (and are left as an exercise). COPY is a subroutine that makes a copy of the input number pointed to by A0. The copied number is pointed to by A1.

The next routine is used to generate factorials. A factorial of a number (for example, 5! or 10! or 37!) is found by multiplying all integers up to and including the input number. For instance, 5! equals 1 * 2 * 3 * 4 * 5. This results in 5! equaling 120. Do a few factorial calculations yourself, and you will see that the result gets very large, very quickly! FACTORIAL will compute the factorial of the integer value stored in D0. The result is pointed to by A1.

```
FACTORIAL   MOVEA.L #ONE,A0                 ;init counter
            MOVEA.L #ONE,A1                 ;init result
NEXTNUM     BSR     MULTIPLY                ;compute partial factorial
            EXG     A1,A2                   ;use result as next input
            BSR     INCREMENT               ;increment counter
            DBRA    D0,NEXTNUM
            RTS
```

INCREMENT is a subroutine that performs a specific task: Add 1 to the number pointed to by A0. We use INCREMENT to generate the sequence of integers that get multiplied together. The symbol ONE refers to a predefined storage area in memory that contains the value 1.

   The next routine, ROOT, computes square roots. The formula, and an example of how it works, is presented in Figure 6.13. This type of formula is **iterative.** This means that we must run through the formula a number of times before getting the desired result. Notice in the figure how each new application of the square root formula brings the estimate of the answer closer to the correct value. After applying the formula only five times, we have a result that comes very close to the square root. A few more iterations will increase the accuracy of the result even more. Fewer iterations are needed when the initial estimate is close to the desired value. For instance, if the original estimate used in Figure 6.13 was 7 instead of 21, fewer iterations would have been needed to get to 6.4807. The routine presented here implements the formula of Figure 6.13.

```
            ORG         $8000
NUMBER      DS.B        5
ESTIMATE    DS.B        5

            ORG         $8100
ROOT        MOVEA.L     #NUMBER,A0       ;input number
            MOVEA.L     #TWO,A1          ;predefined constant 2
            BSR         DIVIDE           ;calculate original estimate
            MOVEA.L     #ESTIMATE,A3     ;save estimate
            BSR         SAVE
            MOVE.W      #9,D0            ;init loop counter for 10 iterations
ITERATE     MOVEA.L     #NUMBER,A0
            MOVEA.L     #ESTIMATE,A1
            BSR         DIVIDE           ;number / estimate
            EXG         A0,A2            ;use result in following addition
            BSR         ADD              ;(number / estimate) + estimate
            EXG         A0,A2            ;use this result in following division
            MOVEA.L     #TWO,A1
```

FIGURE 6.13 Finding
square roots by iteration

$$\text{Estimate} = \frac{\dfrac{number}{estimate} + estimate}{2}$$

Example: Find square root of 42
Initial estimate: 21

Number of iterations	Estimate
0	21
1	11.5
2	7.57608
3	6.55992
4	6.48121
5	6.4807

$(6.4807)^2 = 41.999$

```
BSR          DIVIDE            ;entire formula implemented now
MOVEA.L      #ESTIMATE,A3      ;save new estimate
BSR          SAVE
DBRA         D0,ITERATE
RTS
```

The subroutine SAVE is used to make a copy of the number pointed to by A2. The copy is stored in memory starting at the location pointed to by A3. The EXG instruction is used to swap pointers, thus making the results of ADD and DIVIDE available for the next operation.

If only an integer approximation of a square root is required, the formula shown in Figure 6.13 can be implemented with a shorter set of instructions. The SQROOT routine shown here finds the integer square root of the word-size input number found in register D0. The result is returned in the lower word of D2.

```
            ORG          $8100
SQROOT      MOVE.W       D0,D2       ;first estimate equals original number
            MOVE.W       #9,D3       ;prepare for 10 iterations
ITERATE     MOVE.W       D0,D1       ;make a copy of number
            ANDI.L       #$FFFF,D1   ;convert into unsigned longword
            DIVU         D2,D1       ;number / estimate
            ADD.W        D1,D2       ;(number / estimate) + estimate
            LSR.W        #1,D2       ;((number / estimate) + estimate) / 2
            DBRA         D3,ITERATE
            RTS
```

The ANDI instruction is used to perform an unsigned conversion by setting the upper word of D1 equal to zero. This is necessary, since DIVU uses all 32 bits of D1 as the source operand.

The last example we will examine is used to compute powers of **base e.** From calculus, it can be shown that an infinite series of terms can be used to generate the result of raising e (2.7182818) to any power, as Figure 6.14 illustrates. Notice that only the first seven terms are needed to get a reasonable amount of accuracy. Many complex functions can be represented by an infinite series, which we can then implement in software using a loop operation. The following routine generates the first 10 terms of the exponential series, using the POWER and FACTORIAL routines already discussed. We assume, however, that POWER and FACTORIAL give valid results for all input values (including 0 and 1).

```
            ORG          $8000
X           DS.B         5
TEMP        DS.B         5
ETOX        DS.B         5


            ORG          $8100
EPOWER      MOVE.W       #9,D1        ;init loop counter
NEXTERM     MOVEA.L      #X,A0        ;compute numerator
            MOVE.W       D1,D0
            BSR          POWER
            MOVEA.L      #TEMP,A3     ;save numerator
            EXG          A1,A2
            BSR          SAVE
            MOVE.W       #D1,D0       ;compute denominator
            BSR          FACTORIAL
            MOVEA.L      #TEMP,A0     ;divide to generate term
            BSR          DIVIDE
            EXG          A1,A2        ;add current term to result
            MOVEA.L      #ETOX,A0
            BSR          ADD
            MOVEA.L      #ETOX,A3     ;save result
```

FIGURE 6.14 Generation of e^x by infinite series

$$e^x = \sum_{n=0}^{\infty} \frac{x^n}{n!}$$

$$= \frac{x^0}{0!} + \frac{x^1}{1!} + \frac{x^2}{2!} + \frac{x^3}{3!} + \frac{x^4}{4!} + \dots$$

$$= 1 + x + \frac{x^2}{2} + \frac{x^3}{6} + \frac{x^4}{24} + \dots$$

Example: Find e^1

Number of terms	Result
1	1
2	2
3	2.5
4	2.66666
5	2.70833
6	2.71666
7	2.71805

($e^1 = 2.7182818$)

```
        BSR         SAVE
        DBF         D0,NEXTERM
        RTS
```

Again, EXG is used to redirect output results back into the math routines. EXG is also used to swap pointers for storing results in memory. ETOX contains the final result when EPOWER finishes execution.

These examples should serve to illustrate the point that complex mathematical functions can be implemented with a small amount of software. Once a library of these routines has been defined and tested, even more complex equations and functions may be implemented. All that is needed is a BSR to the appropriate subroutine (or collection of subroutines).

Programming Exercise 6.20: Modify the TOTAL routine so that 32-bit numbers are added together.

Programming Exercise 6.21: Write a subroutine called BIGMUL that will compute the 64-bit result obtained by multiplying two 32-bit integers. The two input numbers should be in registers D6 and D7 on entry to BIGMUL. Use the MULTIPLY subroutine in your code to implement a process similar to that shown in Figure 6.12.

Programming Exercise 6.22: Use the FACTOR subroutine to find all factors of the number saved in a new variable, INVALUE. Place the factors into a data array called FACTORS.

Programming Exercise 6.23:   Write a subroutine called BIGADD that will perform a BCD addition of all 32 bits in registers D5 and D6. Place the result in D7.

Programming Exercise 6.24:   Write a subroutine called TOBINARY that will convert the six-digit BCD number pointed to by A0 into an unsigned binary number. The result should be returned in D0.

Programming Exercise 6.25:   Write a subroutine called TOBCD that converts the unsigned 16-bit binary number in D7 into a BCD number. The BCD result should be returned in D7 also.

6.11   NUMBER CONVERSIONS

In the preceding section we were introduced to both binary and BCD operations. It was mentioned that routines were needed to convert from BCD into binary and from binary to BCD. In this section we will examine a few of the techniques used to perform these conversions.

BCD to Binary Conversion

A BCD number consists of any number of digits that can take on the values 0 to 9. These digits may be individually entered from a keyboard or keypad, or they may all reside together as a group (when stored in memory or processor registers). BCD numbers are different from hexadecimal numbers (a shorthand way for representing binary), which utilize 0 through 9 and the letters A through F. For this reason, 57 BCD and $57 do not have the same values. As a matter of fact, 57 BCD equals $39. To convert 57 BCD into binary, we multiply the value of the 10s digit by 10 and add it to the value of the 1s digit. This gives 5 * 10 = 50 ($32) plus 7 (which gives $39). The following subroutine uses this technique to convert the lower byte of register D2 into binary, returning the result in D3. Remember that any 68000 data register is capable of storing eight BCD digits.

```
MAKEBIN   MOVE.B   D2,D4          ;save a copy of initial data
          ANDI.B   $0F,D2         ;get the 1s digit
          EXT.W    D2             ;extend into 16 bits
          MOVE.W   D2,D3          ;initialize result to 1s digit
          LSR.B    #4,D4          ;shift 10s digit into 1s place
          EXT.W    D4             ;extend into 16 bits
          MOVE.W   #10,D2         ;multiply digit by 10
          MULU     D2,D4
          ADD.W    D4,D3          ;add to result register
          RTS
```

The MAKEBIN subroutine can be used to convert larger BCD numbers into binary. For example, to convert a four-digit BCD number into binary, put the lower two digits into D2 and call MAKEBIN. Save the result, put the upper two digits into D2, and call MAKEBIN again. Multiply the value returned in D3 by 100, and add the result of the first MAKEBIN call. The following routine will accomplish this, with the BCD input number in D0 and the binary result in D5 upon return.

```
UPTO9999  MOVE.W  D0,D1           ;save a copy of initial data
          MOVE.W  D0,D2           ;load D2 for MAKEBIN
          BSR     MAKEBIN         ;convert lower 2 digits
          MOVE.W  D3,D5           ;save temp result
          LSR.W   #8,D1           ;shift upper 2 digits down
          MOVE.W  D1,D2           ;load D2 for MAKEBIN
          BSR     MAKEBIN         ;convert upper 2 digits
          MOVE.W  #100,D1         ;multiply digits by 100
          MULU    D1,D3
          ADD.W   D3,D5           ;add to result register
          RTS
```

Another technique can be used to convert BCD into binary *on the fly*. This involves multiplying the result by 10 each time a new digit is received, and then adding the digit to the result. For example, to convert 349 BCD into binary we start with a zero result. When the 3 is received, we multiply the result by 10 (giving 0) and add 3 to it. Next, the 4 is received. Multiplying the result by 10 gives 30. Adding 4 to this gives 34. When the last digit comes in, we multiply the result by 10 a last time (giving us 340) and get the final value when we add 7 to it. This technique is favorable when the BCD number contains many digits. You are encouraged to write this routine on your own.

Binary to BCD Conversion

This type of conversion is needed when a number must be displayed on a terminal screen, which requires ASCII characters. For example, the value $7B must be converted into the three ASCII characters '1', '2', and '3' (since $7B equals 123 decimal). One way to do this is to divide the input number by successively smaller powers of 10, beginning with 100. Each time we do this we get a result whose value is between 0 and 9. This result is converted into its displayable ASCII equivalent by adding the ASCII **digit bias** of $30 to it. Then the digit is output to the display. The remainder of the division is then used to get the next digit until the remainder itself is between 0 and 9. So, for our example number of 123 we do the following:

Division	Result	Remainder
123 ÷ 100	1	23
23 ÷ 10	2	3
3	3	

The following routine performs this conversion on input numbers that can be as large as 65535. The input number must be in the lower word of D7 when SHOWDECI is called. Once again, TRAP   #1 takes care of outputting the ASCII number in D1 to the terminal screen.

```
          ORG     $8100
SHOWDECI  MOVE.W  D7,D6           ;make copy of input number
          MOVE.W  #10000,D5       ;do 10,000s digit
```

```
              BSR       DODIGIT
              MOVE.W    #1000,D5      ;do 1,000s digit
              BSR       DODIGIT
              MOVE.W    #100,D5       ;do 100s digit
              BSR       DODIGIT
              MOVE.W    #10,D5        ;do 10s digit
              BSR       DODIGIT
              MOVE.B    D6,D1         ;load 1s digit
              ADDI.B    #$30,D1       ;add ASCII bias
              TRAP      #1            ;and display
              RTS

DODIGIT       ANDI.L    #$FFFF,D6     ;clear upper word of D6
              DIVU      D5,D6         ;divide D6 by D5
              MOVE.B    D6,D1         ;load result digit
              ADDI.B    #$30,D1       ;add ASCII bias
              TRAP      #1            ;display digit
              SWAP      D6            ;get remainder
              RTS
```

Note that SHOWDECI outputs leading zeros whenever the input number is less than 10,000. Elimination of leading zeros can be performed in at least two different ways. First, a series of CMPI instructions can be used to determine what power of 10 to begin dividing by, as in:

```
SHOWDECI      CMPI.W    #10000,D7     ;check 10,000s digit
              BCS       LT10000
              MOVE.W    #10000,D5
              BSR       DODIGIT
              BRA       GET1000
LT10000       CMPI.W    #1000,D7      ;check 1,000s digit
              BCS       LT1000
GET1000       MOVE.W    #1000,D5
              BSR       DODIGIT
              BRA       GET100
LT1000        CMPI.W    #100,D7       ;check 100s digit
              BCS       LT100
              etc.
```

Second, instead of outputting the individual digits to the display, they could be written into an output array. Then the array could be examined for leading zeros. Any that are found are simply skipped over. Assuming that the five ASCII digits have been stored in the array DIGITS, the following code will perform this type of leading-zero suppression:

```
MOVEA.L   #DIGITS,A0
CMPI.B    #'0',(A0)     ;check 10,000s digit
BNE       DISPLAY
ADDA.L    #1,A0
CMPI.B    #'0',(A0)     ;check 1,000s digit
BNE       DISPLAY
ADDA.L    #1,A0
CMPI.B    #'0',(A0)     ;check 100s digit
etc.
```

Clearly, this method can be simplified with the use of a loop.

Binary to Hexadecimal Conversion

This conversion method takes a binary number and converts it into the appropriate hexa-decimal ASCII characters. Thus, an input value of $7B will be converted into the two

ASCII characters, '7' and 'B'. The DISPBYTE routine shown here uses a lookup table to select the equivalent ASCII character for display using PC relative addressing.

```
            ORG     $8100
HEXTAB      DC.B    '0123456789ABCDEF'

DISPBYTE    MOVE.B  D7,D6                ;load number
            LSR.B   #4,D6                ;index equals upper digit
            EXT.W   D6                   ;extend index to word size
            EXT.L   D6                   ;extend index to longword size
            MOVE.B  HEXTAB(PC,D6),D1     ;read ASCII equivalent
            TRAP    #1                   ;output digit to display
            MOVE.B  D7,D6                ;load number again
            ANDI.B  #$0F,D6              ;index equals lower digit
            EXT.W   D6                   ;sign-extend index
            EXT.L   D6
            MOVE.B  HEXTAB(PC,D6),D1     ;read ASCII equivalent
            TRAP    #1                   ;and output
            RTS
```

The data table HEXTAB contains the valid ASCII hexadecimal digits listed in increasing order. Digit '0' has an index of 0, digit '9' has an index of 9, digit 'A' has an index of 10 ($0A), and digit 'F' has an index of 15 ($0F). The index is formed from the 4-bit value used to represent a single hex digit, and is used to read the corresponding ASCII equivalent out of HEXTAB. Since the lower byte of D7 contains a pair of 4-bit values, the LSR instruction is used to obtain the bits of the most significant digit in the byte, which is displayed first. All work is performed on register D6 so that the input value in D7 remains unchanged.

Displaying word and longword data values can be accomplished by executing DISPBYTE more than once, as shown in the DISPWORD and DISPLONG routines.

```
DISPWORD    ROR.W   #8,D7        ;do upper hexadecimal pair
            BSR     DISPBYTE
            ROR.W   #8,D7        ;do lower hexadecimal pair
            BSR     DISPBYTE
            RTS

DISPLONG    SWAP    D7           ;do upper word first
            BSR     DISPWORD
            SWAP    D7           ;and then lower word
            BSR     DISPWORD
            RTS
```

In both cases, the value of D7 is preserved. This is important, since we do not want our value destroyed just because we need to display it.

Programming Exercise 6.26: Write a routine called GETDECI that reads ASCII numeric characters '0' through '9' from the keyboard and converts them into the corresponding decimal equivalent. The user must enter a carriage return to complete the number. Return the result in D7. Use TRAP   #0 to read the keyboard (ASCII code is returned in D1).

Programming Exercise 6.27: Repeat Exercise 6.26 for hexadecimal input numbers.

Programming Exercise 6.28: Write a routine called BIGBCD that will display the eight-digit BCD number stored in D7. Do not output the leading eight zeros.

Programming Exercise 6.29:   Write a routine called BIGDECI that will display the decimal equivalent of the number stored in D7. The range of D7 is 0 to 10,000,000. *Hint:* First divide D7 by 10,000 to get at the upper digits.

6.12   CONTROL APPLICATIONS

In this section we will examine two examples of how the 68000 may be used in control applications. Control systems are designed in two different ways: open-loop and closed-loop systems. Figure 6.15 shows two simple block diagrams outlining the main difference between these two types of control systems. An open-loop control system uses its input data to effect changes in its outputs. A closed-loop system contains a feedback path, where data concerning the present output conditions is sampled and supplied along with the external inputs. A burglar alarm is an example of an open-loop control system. The system may be designed to monitor sensors at various windows and doors. It may also include circuitry to digitize readings from temperature sensors. When any of the sensors detects an abnormal condition (for example, a window opening), the computer may be directed to dial an emergency phone number and play a recorded help message.

A typical application of a closed-loop control system involves the operation of a motor. Suppose that we want to control the speed of the motor by making adjustments to an input voltage to the system. The speed of the motor is proportional to the input voltage and increases as the input voltage increases. We cannot simply apply the input voltage to the motor's windings, for it may not be large enough to operate the motor. Usually an amplifier is involved that is capable of driving the motor. But a problem occurs when the motor encounters a load (for example, by connecting the motor shaft to a pump). The increased load on the motor will cause the motor speed to decrease. To maintain a constant

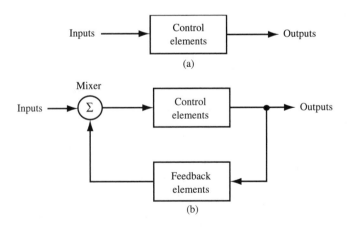

FIGURE 6.15   Control system block diagram: (a) open loop and (b) closed loop

speed in the motor at this point, we need an increase in the input voltage. We cannot hope or expect the operator to constantly watch the motor and adjust the input voltage accordingly. For this reason, we add a feedback loop, which is used to sample the motor speed and generate an equivalent voltage. An error voltage is generated by comparing the actual speed of the motor (the voltage generated by the feedback circuit) with the desired speed (set by the input voltage). The motor speed voltage may be generated by a tachometer connected to the output of the motor. The error voltage is used to increase or decrease the speed of the motor until it is operating at the proper speed.

Let us look at how the 68000 might be used to implement the two control systems just described.

A Computerized Burglar Alarm

In this section, we will use the 68000 to monitor activity on 100 windows and doors in a small office building. The office building consists of four floors, with 15 doors and 10 windows on each floor. The alarm console consists of an electronic display containing a labeled light-emitting diode for each window and door and a serial data terminal capable of displaying ASCII information. The operation of the system consists of two tasks: (1) illuminating the appropriate LED for all open doors and windows, and (2) sending a message to the terminal whenever a door or window opens or closes. It is necessary to continuously scan all of the windows and doors to detect any changes. The circuitry used to monitor the doors and windows and drive the LED displays is connected to the processor's system bus so that all I/O can be done by reading and writing to memory. Figure 6.16 shows the assignments of all input and output devices for the first floor of the office building.

As the figure shows, 15 door and 10 window inputs are assigned for the first floor. Whenever a door or window is open, its associated bit will be low. To sample the bits, the processor must do a memory read from the indicated addresses (7000 to 7003). Floors 2, 3, and 4 are assigned the same way, with the following addresses:

Floor 2:   7004–7007

Floor 3:   7008–700B

Floor 4:   700C–700F

The door and window LEDs for the first floor are illuminated when their respective bits are high. The processor must do a memory write to locations 7800 through 7803 to activate LEDs for the first floor. The other-floor LEDs work the same way, with these addresses assigned to them:

Floor 2:   7804–7807

Floor 3:   7808–780B

Floor 4:   780C–780F

The serial device used by the system to communicate with the ASCII terminal is driven by a subroutine called CONSOLE. The 7-bit ASCII code in the lower byte of data register D0 is sent to the terminal when CONSOLE is called. The operation of CONSOLE is similar to that of TRAP   #1 on the SBC and in EMU68K.

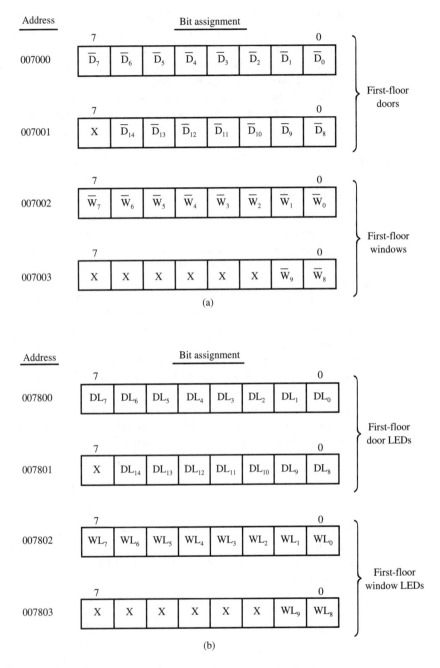

FIGURE 6.16   Burglar alarm I/O assignments: (a) system inputs and (b) system outputs

Knowing these definitions, we can design a system to constantly monitor all 100 doors and windows. The technique we will use is called **polling.** Each input address will be read and examined for any changes. If a door or window has changed state since the last time it was read, a message will be sent to the terminal, via CONSOLE, indicating the floor and door/window number. Since we need to remember the last state of each door and window, their states must be saved. A block of memory, called STATUS, will be used for this purpose. STATUS points to a 16-byte block of memory, which we will think of as four blocks of 4 bytes each. Each 4-byte block will store the bits for all doors and windows on a single floor.

When the program first begins operation, the state of each door and window is unknown. For this reason, we *initialize* STATUS by reading all system inputs when the program starts up. The code to perform the initialization is contained in a subroutine called INIT, and is as follows:

```
            ORG       $8000
STATUS      DS.B      16
INPUTS      EQU       $7000
DISPLAY     EQU       $7800

            ORG       $8100
INIT        MOVEA.L   #STATUS,A0    ;init pointer to STATUS
            MOVEA.L   #INPUTS,A1    ;init pointer to system input data
            MOVEA.L   #DISPLAY,A2   ;init pointer to LED display
            MOVE.W    #15,D0        ;init loop counter
SYSREAD     MOVE.B    (A1)+,D1      ;read system information
            MOVE.B    D1,(A0)+      ;save it in memory
            NOT.B     D1            ;complement input data
            MOVE.B    D1,(A2)+      ;update display
            DBRA      D0,SYSREAD
            RTS
```

When INIT completes execution, the display has been updated to show the state of all 100 doors and windows, and STATUS has been loaded with the same information.

Once the initial states are known, future changes can be detected by using an exclusive OR operation. Exclusive OR produces a 1 only when both inputs are different. Figure 6.17 shows how state changes can be detected with exclusive OR. To incorporate this into the program, EOR is used during updates to detect changes. Note that up to 16 changes at once can be detected by EORing entire words. It is then a matter of scanning the individual bits to determine if any state changes occurred. A subroutine called DETECT will do this for us. DETECT will sense any state changes and send the appropriate message (for example, first floor: door 12 opened) to the terminal. When DETECT is called, data registers D4, D5, and D6 will be interpreted as follows:

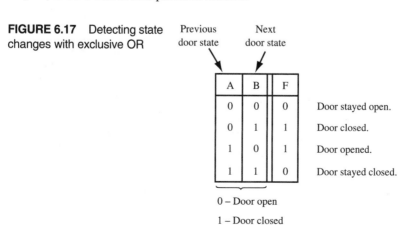

FIGURE 6.17 Detecting state changes with exclusive OR

	Previous door state	Next door state	
A	B	F	
0	0	0	Door stayed open.
0	1	1	Door closed.
1	0	1	Door opened.
1	1	0	Door stayed closed.

0 – Door open

1 – Door closed

D4:   Lower 16 bits contain current door or window states.

D5:   Lower 16 bits contain door or window state changes.

D6:   Bit 8 cleared means D4 contains door information

Bit 8 set means D4 contains window information.

Bits 0 and 1 contain the floor number 00—first, 01—second, 10—third, 11—fourth)

We will not cover the code involved in getting DETECT to do its job. You are encouraged to write this routine yourself, preferably using a rotate or shift instruction to do the bit testing. Since DETECT will have to output ASCII text strings (for example, 'First floor', 'Second floor', 'opened'), the following code may come in handy:

```
(This follows STATUS data area...)
STAR     EQU       $2A                 ;ASCII code for '*'
MSG1     DC.B      'First floor *'
MSG2     DC.B      'Second floor *'
MSG3     DC.B      'Third floor *'
MSG4     DC.B      'Fourth floor *'
MSG5     DC.B      'door *'
MSG6     DC.B      'window *'
MSG7     DC.B      'opened *'
MSG8     DC.B      'closed *'


(This follows INIT routine...)
SEND     MOVE.B    (A3)+,D0            ;get a message character
         CMPI.B    #STAR,D0            ;end of message character?
         BEQ       EXIT
         BSR       CONSOLE             ;send character to terminal
         BRA       SEND                ;get next character
EXIT     RTS
```

The SEND subroutine must be entered with address register A3 pointing to the address of the first character in the text string to be sent. SEND could be a subroutine called by DETECT during its analysis of the door and window states.

When we use DETECT, the code to poll all doors and windows in the office building becomes

```
         ORG       $8200
         BSR       INIT                ;get initial states and update display
BEGIN    CLR.W     D6                  ;start with first floor doors
         MOVE.W    #3,D7               ;init loop counter
         MOVEA.L   #INPUTS,A0          ;init pointer to first floor doors
         MOVEA.L   #STATUS,A1          ;init pointer to STATUS information
         MOVEA.L   #DISPLAY,A2         ;init pointer to display
NEWFLOOR MOVE.W    (A0)+,D4            ;get floor door data
         NOT.W     D4
         MOVE.W    D4,(A2)+            ;update display
         NOT.W     D4
         MOVE.W    (A1),D5             ;get past door status
         EOR.W     D4,D5               ;compute state changes
         MOVE.W    D4,(A1)+            ;update status
         BSR       DETECT              ;find doors that have changed
         MOVE.W    (A0)+,D4            ;get floor window data
         NOT.W     D4
         MOVE.W    D4,(A2)+            ;update display
```

```
NOT.W     D4
MOVE.W    (A1),D5         ;get past window status
EOR.W     D4,D5           ;compute state changes
MOVE.W    D4,(A1)+        ;update status
ORI.W     #$400,D6        ;select window detection
BSR       DETECT          ;find windows that have changed
ANDI.W    #$FF,D6         ;select door detection again
ADDQ.B    #1,D6           ;go to next floor
DBRA      D7,NEWFLOOR
JMP       BEGIN
```

While you write DETECT, do not forget that the main routine uses a number of registers (A0, A1, A2, D4 through D7) and that these registers should not be altered. The stack would be a good place to store them for safekeeping.

A Constant-Speed Motor Controller

In this section we will see how the 68000 may be used in a closed-loop control system to maintain constant speed in a motor. The schematic of the system is shown in Figure 6.18. The speed control is a potentiometer whose output voltage varies from 0 to some positive voltage. This voltage is digitized by an 8-bit analog-to-digital converter, such that 0 volts is $00 and the most positive voltage is $FF. The processor reads this data from location 8000. For a purely digital speed control system, this circuitry is eliminated and the speed set directly by software.

The motor speed is controlled by the output of an 8-bit digital-to-analog converter (with appropriate output amplifier, capable of driving the motor). The motor's minimum speed, 0 RPM, occurs when the computer outputs 00 to the D/A (by writing to location 8020). The motor's maximum speed occurs when FF is sent to the D/A. A tachometer is connected to the motor shaft through a mechanical coupling. The output of the tachometer

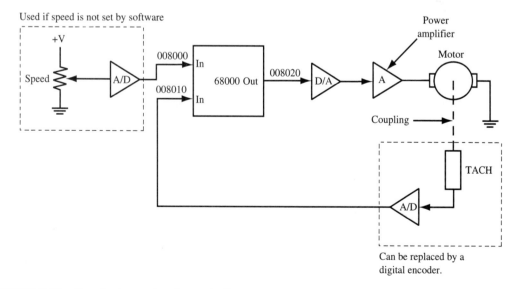

FIGURE 6.18 Constant-speed motor controller

is digitized also. Again, the minimum and maximum tachometer readings correspond to 00 and FF. A purely digital system would use a digital shaft encoder instead of a tachometer and A/D.

Each converter is calibrated with respect to a common reference. In theory, a 17 from the SPEED A/D causes a 17 to be sent to the MOTOR D/A, which in turn causes the TACH A/D to read 17 when at the proper speed. In practice the relationship is not so linear, due to external effects of deadband, friction, and other losses in the motor.

The purpose of the program is to operate the motor at a constant speed, by comparing the SPEED data with the TACH data. When SPEED equals TACH, the motor is turning at the desired speed. When SPEED is less than TACH, the motor is spinning too fast. When SPEED is greater than TACH, the motor is rotating too slowly. The idea is to subtract the TACH value from the SPEED value. The difference determines how much the motor speed should be increased or decreased.

```
                ORG       $8100
SPEED           EQU       $8000
MOTOR           EQU       $8020
TACH            EQU       $8010
SERVO           MOVE.B    #0,D0         ;initial motor speed is 0 RPM
                MOVE.B    D0,MOTOR
GETSPEED        MOVE.B    SPEED,D0      ;read new speed value
                MOVE.B    TACH,D1       ;and new tachometer value
                CMP.B     D1,D0         ;compare SPEED and TACH
                BCS       INCREASE      ;branch if SPEED < TACH
                BEQ       GETSPEED      ;no change
                EXG       D0,D1         ;SPEED > TACH, switch for SUB
INCREASE        SUB.B     D1,D0         ;adjust motor speed
                BSR       GAIN
                MOVE.B    D0,MOTOR
                BRA       GETSPEED
```

Since the motor's speed will not change instantly from very slow to very fast, or vice versa, the program will loop many times before the motor gets to the proper speed. For safety or functional reasons, it may not be desirable to try to change the motor speed from slow to fast instantly. Instead, the program should *ramp up* to speed gradually by limiting the size of the error voltage presented to the D/A during speed increases. Subroutine GAIN is used for this purpose, to alter the contents of D0, before D0 is output to the motor D/A. The ramp up/down speed of the motor, and therefore the response of the closed-loop system, will be a function of the operation of GAIN. Note that if GAIN contains the single instruction RTS, D0 remains unchanged and the entire difference is used to change the motor speed.

Programming Exercise 6.30:   Write the DETECT subroutine used by the computerized burglar alarm.

Programming Exercise 6.31:   An office complex consisting of 64 offices and 16 hallways is to have its lighting controlled by a computer. Each office has one switch to control its light. Hallways have a switch at each end. Each switch is assigned a bit position in a particular memory location that can be read by the computer, and a closed switch represents a zero. Each light (think of all lights in a hallway as a single light) is also assigned a certain bit in a memory location that the computer can write to. A logic 1 is needed to turn on any light. How many byte locations are needed for all I/O? Write a program that will constantly monitor all switches and adjust the complex lighting as necessary.

Programming Exercise 6.32: Modify the SERVO program so that the motor's speed will ramp up/down during periods where a large speed increase/decrease is desired.

6.13   INSTRUCTION EXECUTION TIMES

An important topic in the study of any microprocessor involves analysis of the execution time of programs, subroutines, or short sections of code. The most direct application of this study is in the design of programs that function under a time constraint. For example, high-resolution graphics operations, such as image rotation, filtering, and motion simulation, require all processing to be completed within a very short period of time (usually a few milliseconds or less). If analysis of the total instruction execution time for the graphics routine exceeds the allowed time of the system, a loss in image quality will most likely result. We will not get this involved with our analysis of the 68000's instruction times. Instead, we will look at three examples and how their total execution times may be determined.

Our first example is a subroutine called RANDOM that contains only four instructions. Table 6.1 shows the instructions and their respective clock cycles. Our goal is to determine how many clock cycles are required to execute the entire subroutine, including the RTS instruction. Each of the instructions in Table 6.1 can be found in Appendix C. The first instruction, ADDI.B   #17,D0, requires eight clock cycles. Four of the clock cycles are for fetching and decoding the ADDI instruction. The other four are for reading the immediate data out of memory. This implies that the addressing mode used in an instruction affects its execution time. This is especially true for instructions that access memory.

The second instruction, LSL.B   #3,D0, requires 12 clock cycles. The formula shown in Appendix C for this instruction is $6 + 2n$, where n is the shift count used in the instruction. Since the shift count is 3, we get $6 + 2(3) = 12$ clock cycles. Larger shift counts require additional clock cycles.

The third instruction, NOT.B   D0, requires only four clock cycles. This is the advantage we get when using register direct addressing mode. Since the registers are stored *inside* the processor, their contents are quickly accessible.

So far we have a total of $8 + 12 + 4 = 24$ clock cycles. Note that all three instructions operated on byte-size operands. In general, byte and word operations usually require the

TABLE 6.1   A four-instruction subroutine

Instruction			Clock cycles
RANDOM ADDI.B	#17, D0		8
LSL.B	#3, D0		12
NOT.B	D0		4
RTS			16

TABLE 6.2   Instructions in a loop

	Instructions		Clock cycles	
			Overhead	Loop
	MOVE.B	#10,D0	8	
READ	ADD.W	(A0)+,D1		8
	SUBQ.B	#1,D0		4
	BNE	READ		10 (8)

same number of clock cycles, and longwords require additional clock cycles (since the processor needs to access memory *twice* to read or write a longword).

The last instruction, RTS, requires 16 clock cycles. This gives a total of 40 clock cycles for the RANDOM subroutine. For a 68000 running at 8 MHz, each clock cycle is 125 ns long. So, these 40 clock cycles will take 5 μs to execute.

When some form of a loop is used in a section of code, the number of clock cycles gets multiplied. Consider the instructions shown in Table 6.2. Of the four instructions shown, only one instruction (MOVE.B   #10,D0) is outside the loop. Since this instruction only executes once, while the other three instructions each execute 10 times, its contribution to the overall execution time is small. This contribution is often called **overhead.** Thus, we have eight clock cycles of overhead.

The three loop instructions require a total of 22 clock cycles to execute once. Let us see where these 22 clock cycles come from. The ADD.W   (A0)+,D1 instruction takes eight clock cycles due to the fact that the (A0)+ source addressing mode must access memory. The SUBQ.B   #1,D0 instruction requires just four clock cycles, even though there is immediate data within the instruction. But, unlike SUBI.B   #1,D0, the SUBQ instruction has a limited range of immediate data (1 to 8 only) and thus codes the immediate data right into the instruction word. There is no need for a second memory access to read the immediate data (as there is in SUBI).

The BNE instruction comes in two forms. One allows the displacement to be specified as a single byte whenever the range is +127 to –128. This is certainly the case here, since the READ label is only two instructions behind BNE   READ. This means that the BNE READ instruction will take 10 clock cycles each time it is taken. So, a single pass through the loop takes 22 cycles. At the end of the last pass (10th) through the loop the BNE   READ will *not* be taken, and will only take eight clock cycles to execute. This means that the last pass through the loop only requires 20 clock cycles. The total number of clock cycles becomes:

     8   overhead
+ 9(22)   nine passes through loop
  + 20   last pass through loop

which equals 226 clock cycles. How large can this number get? In general, for this loop we have $28 + 22(N - 1)$ clock cycles, where N is the number loaded into D0 at the beginning of the loop. The largest unsigned integer we can use for N is 255, giving a maximum of 5,616 clock cycles.

TOBIN is a subroutine that will convert a four-digit BCD number in the lower word of data register D0 into a binary number. The result is returned by TOBIN in D2. TOBIN

TABLE 6.3 Required instruction clock cycles in a nested loop

Instructions			Overhead cycles	Outer loop cycles	Inner loop cycles
TOBIN	CLR.L	D2	6		
	MOVEQ	#3,D6	4		
NEXTDIGIT	MOVEQ	#3,D5		4	
	CLR.W	D1		4	
GETNUM	LSL.W	#1,D0			8
	ROXL.W	#1,D1			8
	DBRA	D5,GETNUM			10
	MULU	#10,D2		42	
	ADD.W	D1,D2		4	
	DBRA	D6,NEXTDIGIT		10	
	RTS		16		

is a good example to use for execution time determination because it contains two nested loops. Examine Table 6.3. Once again the number of clock cycles an instruction takes to execute depends on a number of factors. Operand size is the first variable. Look at the clock cycles required by both CLR instructions. CLR.W requires four clock cycles, and CLR.L requires six. In this case, the 32-bit clear operation takes an additional two clock cycles to complete.

The addressing mode used by an instruction also affects the number of clock cycles required. Immediate operand addressing, as in MOVEQ and MULU, takes fewer clock cycles than an instruction that uses a memory location as its operand. Memory references, be they read or write operations, take four clock cycles or more, depending on how many wait states are inserted by the hardware. The instruction times in Appendix C assume four clock cycles for all memory references.

The operand size is also a determining factor in some instructions. LSL and ROXL both take a minimum of six clock cycles for byte and word operands, plus another two cycles for each shift or rotate that must be performed. Since the shift count for both instructions in TOBIN is 1, both LSL and ROXL require eight clock cycles each to execute.

MULU is another example of an instruction whose execution time is a function of the operand size. MULU takes 38 cycles, plus two additional ones for each bit in the source operand that is a 1. In MULU #10,D2, the source operand, in hexadecimal, is 000A. This number contains 2 bits that are 1. Thus, the instruction takes 42 clock cycles to execute. If all bits in the source operand were high, MULU would take a maximum of 70 cycles to execute.

The fastest instructions are those that use register direct addressing. An example of this is TOBIN's ADD instruction. The number of clock cycles is minimized because no external memory references (outside of the instruction fetch) are needed.

The RTS instruction takes a long time to execute due to its need to read information from the stack. This requires a number of memory read operations, which always slow down execution.

The last instruction in TOBIN that has two execution times is DBRA, which takes 10 cycles to execute when the branch is taken and 14 when it is not. The additional four cycles may not be ignored, particularly when nested loops are involved. Notice how the clock cycles for each instruction are split up into three columns. The first column contains instructions that only execute once. These instructions, CLR.L, MOVEQ, and RTS, contribute a total of 26 clock cycles to the overall execution time. We will see that this overhead is a small amount compared to what the nested loops will generate.

The second column of numbers composes the main, or outer, loop of the subroutine. The third column represents the inner loop. The inner loop takes a total of 108 clock cycles to execute. Where does this number come from? The inner loop takes 26 clock cycles to execute *once*. Since four loops are requested, via D5, this number (26) must be multiplied by 4, giving 104 cycles. The last time DBRA   D5,GETNUM executes, it will take an additional four clock cycles, and that is where the 108 comes from.

The execution time of the outer loop will contain the execution time of the inner loop, plus the additional cycles required by the outer loop instructions. This gives 64 plus 108, or 172 cycles for the outer loop to execute just once. The subroutine calls for the outer loop to execute four times. This gives a total of 688 clock cycles. To this number we add four more cycles for the last execution of DBRA, giving us 692 clock cycles. This number is added to our overhead (do you see why the overhead is not substantial?) of 26, giving a grand total of 718 clock cycles for the TOBIN subroutine.

How does this number translate into an execution time? Suppose the clock frequency is 8 MHz. Each clock cycle will then have a period of 125 ns. Multiplying 125 ns by 718 gives 89.75 µs! This is the execution time of TOBIN.

In conclusion, it is interesting to note that TOBIN can convert over 11,000 BCD numbers to binary in one second.

Programming Exercise 6.33:   What is the execution time of this code:

```
        MOVE.W    #$1000,D0
NEXT    ADDQ.B    #2,D2
        MOVE.B    D2,(A0)
        DBRA      D0,NEXT
```

The processor is running at 8 MHz.

Programming Exercise 6.34:   Repeat Programming Exercise 6.33 assuming all memory references take an additional two clock cycles.

6.14   EXCEPTION HANDLING

In Chapter 4, various kinds of exceptions supported by the 68000 were examined. In this section we will see two examples of how exceptions may be handled by the programmer. The first example deals with two exceptions that, we hope, never occur: divide-by-zero and illegal instruction. The purpose of the exception handler for these two routines will be to send an error message to the user's terminal, indicating that the error has occurred. Execution will then resume at the address following the instruction that caused the exception. Remember: All exceptions cause the 68000 to reference its exception vector table (located in memory from 000000 to 0003FF). In this table, the processor will expect to find the address of the routine that will handle the designated exception. The following routines, DIV0 and BADINST, are ORGed at $1000 and $2000, respectively. For the 68000 to find its way to them during exception processing, the starting addresses of these routines must be placed into the exception vector table at the proper addresses. Divide-by-zero, vector 5, requires its exception handler address to be stored in locations 000014 through 000017, with 000014 and 000015 containing the upper word of the routine address and 000016 and 000017 the lower word. Illegal instruction, vector 4, operates in a similar manner but uses locations 000010 through 000013. Figure 6.19 shows how these locations should be loaded with the addresses of the exception handlers.

Once the CPU has loaded the program counter with the address of the exception handler, it will begin executing instructions at a new location. The exception handlers, to be on the safe side, should save all processor registers, so that execution may resume with the proper information after exception processing has completed. This can be done very easily with the MOVEM instruction. Once this is done, DIV0 or BADINST may perform their associated tasks. In this example they will simply load address register A3 with the address of an ASCII text message that will be output to the user's terminal via TRAP #3.

FIGURE 6.19 Exception vector table address assignments for divide-by-zero and illegal instruction

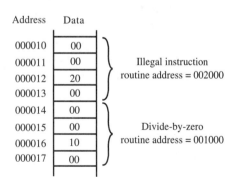

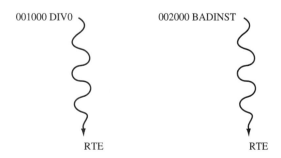

```
              ORG        $1000
DIV0          MOVEM.L    D0-D7/A0-A6,-(A7)    ;save all registers on stack
              MOVEA.L    #DIVMSG,A3           ;point to error message
PEXCEP        TRAP       #3                   ;send message to user
              MOVEM.L    (A7)+,D0-D7/A0-A6    ;get processor registers back
              RTE                             ;return from exception
DIVMSG        DC.B       'Warning: Divide by zero has occurred!',0


              ORG        $2000
BADINST       MOVEM.L    D0-D7/A0-A6,-(A7)    ;save all registers on stack
              MOVEA.L    #BADMSG,A3           ;point to error message
              JMP        PEXCEP               ;go process exception
BADMSG        DC.B       'Illegal instruction encountered!',0
```

Note that we save some code in BADINST's handler by using a portion of DIV0's routine. Also, the final instruction in any exception handler must be RTE (not RTS). This is required to load the proper information from the stack when returning to the main routine (where the exception was initiated).

In addition to sending the error message, it may be desirable to output a list of register contents as well. The user would then be able to see the state of the processor registers at the instant the exception occurred. This requires a routine capable of converting the hexadecimal register contents into a corresponding string of ASCII characters to send to the user's terminal. This routine must be called *after* all registers have already been saved. Otherwise we end up destroying the contents of whatever registers we use in the conversion routine.

Our second example, a routine called TIMER, is used to keep accurate time. This example assumes that a stable 60-Hz clock is used to interrupt the processor 60 times a second, generating a level-7 interrupt each time. Again, the address of TIMER must be placed into the proper locations (00007C through 00007F) in the processor's exception vector table.

TIMER simply increments SECONDS, MINUTES, and HOURS (all byte locations) as necessary. These locations may be read by other routines that need to make use of timing functions. Why is the MOVEM instruction absent from this exception handler?

```
              ORG        $8000
SECONDS       DC.B       0
MINUTES       DC.B       0
HOURS         DC.B       0

              ORG        $8100
TIMER         ADDI.B     #1,SECONDS           ;increment seconds
              CMPI.B     #60,SECONDS          ;has a minute passed?
              BNE        EXITIMER
              CLR.B      SECONDS              ;reset seconds count
              ADDI.B     #1,MINUTES           ;increment minutes
              CMPI.B     #60,MINUTES          ;has an hour passed?
              BNE        EXITIMER
              CLR.B      MINUTES              ;reset minutes count
              ADDI.B     #1,HOURS             ;increment hours
              CMPI.B     #24,HOURS            ;has a day passed?
              BNE        EXITIMER
              CLR.B      HOURS                ;reset hour count
EXITIMER      RTE
```

Since a level-7 interrupt was used, we never have to worry about losing time. Level-7 interrupts are nonmaskable, so we never miss a 60-cycle interrupt signal.

Programming Exercise 6.35:   Write an exception handler for TRAPV that outputs the message:

```
Warning! Overflow occurred.
```

Programming Exercise 6.36:   Modify the TIMER exception handler so that a 12-hour clock is implemented.

6.15   A WORKING CALCULATOR

The entire source file of a working calculator is presented in this section. Many of the programming principles and techniques covered in previous sections of this chapter are used here to implement a simple 68000-based calculator. The specifications of the calculator are as follows:

*   Numbers are represented as unsigned 16-bit integers. This gives a range of 0 to 65535 on all input and output numbers.
*   Only the basic four math operations +, –, *, and / are provided.
*   Division by zero is not allowed.
*   The user may exit the calculator program by entering 'Q' in place of a math operation.
*   The user must enter '=' to perform a computation.
*   The last result displayed is used in the next computation. A sample execution of the calculator looks like this (user input in **boldface**):

```
68000 Calculator, Ver2.0

100<cr>
/
4<cr>
=
00025
+
17<cr>
=
00042
*
100<cr>
=
04200
-
99999
Overflow
25<cr>
=
04175
```

Notice that the user must enter a carriage return at the end of a number, whereas the four math symbols and the equal sign do not require a carriage return. Also notice that the calculator does not suppress leading zeros on results that are less than 10000. In addition, the calculator checks for input numbers greater than 65535 and outputs the *Overflow* error message when the 99999 value is entered.

One type of overflow that the calculator does not check for regards the size of the result of a computation. For example, adding 65500 and 500 gives the following result:

```
65500<cr>
+
500<cr>
=
00464
```

Which is not correct, but is all we can get with 16-bit integers. However, since 70000 minus 65536 *is* 464, the calculator at least got the first 16 bits of the result correct!

Let us examine the individual routines utilized by the calculator to see how the whole thing works. The first part of the calculator program contains the reserved data area:

```
        ORG     $8000
HELLO   DC.B    $0D,$0A,$0D,$0A
        DC.B    '68000 Calculator, Ver2.0'
        DC.B    $0D,$0A
CRLF    DC.B    $0D,$0A,0
NUM1    DS.W    1                       ;first user input
NUM2    DS.W    1                       ;second user input
MATHOP  DS.B    1
EQUST   DC.B    '=',$0D,$0A,0
BADDIG  DC.B    $0D,$0A,'Illegal digit',$0D,$0A,0
DIVBY0  DC.B    'Divide by 0',$0D,$0A,0
TOOBIG  DC.B    $0D,$0A,'Overflow',$0D,$0A,0
```

ASCII text strings are used to store error messages and other screen-control strings. Storage for user numbers is also provided.

The main part of the calculator program calls a number of subroutines to perform its job. These subroutines are:

GETNUM   Get a number from the user.

GETOP   Get a math operation from the user.

MATH   Perform the desired math operation.

SHOWDECI   Display the result.

When the calculator begins execution, the user must enter two numbers to produce the first result. Then, each result becomes the first number used in the next computation.

```
        ORG     $8100
        MOVEA.L #$8700,A7       ;init stack
TOP     MOVEA.L #HELLO,A3       ;send greeting
        TRAP    #3
        BSR     GETNUM          ;get first number
NEXTOP  MOVE.W  D7,NUM1         ;save it
        BSR     GETOP           ;get math operation
        BSR     GETNUM          ;get second number
        MOVE.W  D7,NUM2         ;save it
WAITEQ  TRAP    #0              ;read keyboard
        CMPI.B  #'=',D1         ;wait for equal sign
        BNE     WAITEQ
        MOVEA.L #EQUST,A3       ;display equal sign
        TRAP    #3
```

```
        BSR      MATH               ;perform computation
        BSR      SHOWDECI           ;show result
        MOVEA.L  #CRLF,A3           ;fix display
        TRAP     #3
        BRA      NEXTOP
```

GETNUM returns the user input number in the lower word of D7. The NUM1 and NUM2
storage variables in the data area are used to save these input numbers. GETOP waits for
the user to enter a valid math operation and then stores it in the MATHOP location.
MATHOP, NUM1, and NUM2 are all used by MATH to calculate the result, which is
returned in the lower word of D7 also. The SHOWDECI routine (covered in Section 6.11)
then uses D7 to display the result.

GETNUM converts the user input number into binary on-the-fly, checking for
values greater than 65535 as it runs. The user may exit the calculator program from inside
GETNUM at any time by entering 'Q'. Otherwise, GETNUM keeps looking for new input
digits until a carriage return is entered. Any illegal input characters generate an error message and start GETNUM over from the beginning.

```
GETNUM  CLR.W    D7                 ;init temp result
GETDIG  TRAP     #0                 ;read keyboard
        TRAP     #1                 ;echo to display
        CMPI.B   #$0D,D1            ;return?
        BEQ      GOTNUM
        CMPI.B   #'Q',D1            ;quit?
        BNE      CHKDIG
        TRAP     #9
CHKDIG  CMPI.B   #'0',D1            ;test for numeric
        BCS      SORRY
        CMPI.B   #':',D1
        BCC      SORRY
        MULU     #10,D7             ;adjust temp result
        SWAP     D7                 ;check upper half for overflow
        CMPI.W   #0,D7
        BNE      OVER
        SWAP     D7
        SUBI.B   #$30,D1            ;remove ASCII bias
        EXT.W    D1                 ;sign-extend
        ADD.W    D1,D7              ;fix result
        BCC      GETDIG
OVER    MOVEA.L  #TOOBIG,A3         ;output overflow message
        BRA      FIXNUM
SORRY   MOVEA.L  #BADDIG,A3         ;output illegal-digit message
FIXNUM  TRAP     #3
        BRA      GETNUM             ;start over
GOTNUM  MOVEA.L  #CRLF,A3           ;fix display
        TRAP     #3
        RTS
```

The GETOP routine also provides an exit from the calculator program if the user enters 'Q' instead of one of the four math operations. Only valid input characters are echoed
to the display. The math operation entered by the user is stored in MATHOP. Users do not
have to enter a carriage return to enter their choice.

```
GETOP   TRAP     #0                 ;read keyboard
        CMPI.B   #'Q',D1
        BNE      CHKOP
        TRAP     #9
```

```
CHKOP       CMPI.B    #'+',D1       ;check for valid math operations
            BEQ       OKOP
            CMPI.B    #'-',D1
            BEQ       OKOP
            CMPI.B    #'*',D1
            BEQ       OKOP
            CMPI.B    #'/',D1
            BEQ       OKOP
            BRA       GETOP
OKOP        MOVE.B    D1,MATHOP     ;save operation
            TRAP      #1            ;echo to display
            MOVEA.L   #CRLF,A3      ;fix display
            TRAP      #3
            RTS
```

Finally, the MATH routine examines the symbol saved in MATHOP, and then selects the corresponding arithmetic instruction to complete the calculation. No overflow
checking is provided. This means that the add, subtract, and multiply operations may not
display the correct answer for some input values, since we ignore any result that is larger
than 16 bits or that may be interpreted as a negative value.

```
MATH        MOVE.W    NUM1,D7       ;load first number
            ANDI.L    #$FFFF,D7     ;sign extend it
            CMPI.B    #'+',MATHOP   ;add?
            BEQ       MATH1
            CMPI.B    #'-',MATHOP   ;subtract?
            BEQ       MATH2
            CMPI.B    #'*',MATHOP   ;multiply?
            BEQ       MATH3
            BRA       MATH4         ;must be divide
MATH1       ADD.W     NUM2,D7
            RTS
MATH2       SUB.W     NUM2,D7
            RTS
MATH3       MULU      NUM2,D7
            RTS
MATH4       CMPI.W    #0,NUM2       ;check for division by 0
            BEQ       MATH5
            DIVU      NUM2,D7
            RTS
MATH5       MOVEA.L   #DIVBY0,A3    ;output division-by-0 message
            TRAP      #3
            CLR.W     D7            ;set result to zero
            RTS
```

The calculator program was written in this way to show the basic activities involved in performing calculations. You are encouraged to experiment with improvements and changes.

Programming Exercise 6.37:   Modify the calculator so that 32-bit *signed* numbers
are allowed.

Programming Exercise 6.38:   Integrate the integer SQROOT routine into the calculator.
Perform the square root operation when the user enters 'R' (for root) as the math operation.

Programming Exercise 6.39:   Modify the calculator so that all numbers are right justified as they are displayed. For example, if the user enters 12345, the display will look like
this after each digit is entered:

```
    1
   12
  123
 1234
12345
```

6.16   CONNECTING TO DOS

The EMU68K emulator has a new feature that allows your 68000 program to interface with the personal computer's operating system. This is done through the use of 8086 interrupt functions. The C compiler used to create EMU68K has a built-in function that generates an 8086 interrupt. This function is accessed through the use of EMU68K's TRAP #10. Recall that many of the SBC TRAPs are emulated in EMU68K. TRAP #10 is *not* part of the SBC. It is included within EMU68K to allow you to write more interesting and useful 68000 programs for simulation. Imagine that the emulator is really a 68000 machine that has the same capabilities and operating system functions as the personal computer.

In order to use the DOS interrupt, it is necessary to establish a correspondence between the 68000's data registers and the 8086 general purpose registers. Figure 6.20 shows this relationship. Registers D0 through D5 are mapped to the 8086's AX, BX, CX, DX, SI, and DI registers. When an interrupt is generated, the contents of D0 through D5 are transferred to the 8086 registers. When the interrupt completes, D0 through D5 are updated with the new values of the 8086 registers.

The lower byte of D7 specifies the interrupt number ($00 to $FF).

The following instructions display a single character (the letter A) on the screen through DOS INT 21H function 02H:

```
MOVE.B   #'A',D3       ;ASCII code in DL (lower byte of D3)
MOVE.W   #$0200,D0     ;put 02H in AH (second byte of D0)
MOVE.B   #$21,D7       ;select DOS INT 21H
TRAP     #10           ;DOS call
```

Many DOS functions can be used in this way to enhance your program. Functions to read/set the time and date, control the printer, manipulate the video display, access disks,

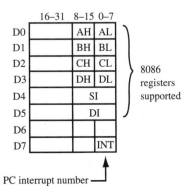

FIGURE 6.20   8086 register assignments for EMU68K

and more are available through the 8086 interrupt mechanism on the personal computer. The following program reads the system time from the PC's clock and displays it. DOS INT 21H function 2CH is used to read the time. Hours, minutes, and seconds are returned in CH, CL, and DH, respectively. Appendix H gives additional details.

```
            ORG      $8000
TMSG        DC.B     'The time is '
            DC.B     0               ;end-of-message marker
CRLF        DC.B     13,10,0     ;CR,LF

            ORG      $8100          ;starting address of program
START       MOVEA.L  #TMSG,A3       ;load A3 with message address
NEXT        MOVE.B   (A3)+,D3       ;load character to display
            BEQ      RDTIME
            MOVE.W   #$0200,D0       ;select function 2 (display character)
            MOVE.B   #$21,D7        ;select DOS INT 21H
            TRAP     #10            ;DOS call
            BRA      NEXT
RDTIME      MOVE.W   #$2C00,D0       ;select function 2CH (read time)
            MOVE.B   #$21,D7        ;of INT 21H
            TRAP     #10
            MOVE.W   D2,D6          ;D2 contains hours and minutes
            LSR.W    #8,D6          ;look at the hours byte first
            BSR      BASE10         ;display it
            MOVE.B   #':',D1        ;and a colon
            TRAP     #1
            MOVE.B   D2,D6          ;now look at the minutes
            ANDI.W   #$00FF,D6
            BSR      BASE10
            MOVEA.L  #CRLF,A3       ;finish with a new line
            TRAP     #2
            TRAP     #9             ;return to command processor

BASE10      MOVE.W   #10,D5         ;do 10's digit
            ANDI.L   #$FFFF,D6       ;clear upper word of D6
            DIVU     D5,D6          ;divide D6 by D5
            MOVE.B   D6,D1          ;load result digit
            ADDI.B   #$30,D1        ;add ASCII bias
            TRAP     #1             ;display digit
            SWAP     D6             ;get remainder
            MOVE.B   D6,D1          ;load 1's digit
            ADDI.B   #$30,D1        ;add ASCII bias
            TRAP     #1             ;and display
            RTS

            END      START          ;end of source file
```

Note the mixture of DOS calls and emulator TRAPs used in the program. Instead of sending the initial message to the display with TRAP   #2, DOS's display character function is used.

When the time program is executed via EMU68K we get

```
The time is 12:22
```

Tapping into DOS should add a little spice to your programming. There are some limitations, however. Interrupts that utilize segment registers are not implemented. Using them anyway will result in unpredictable behavior. Calling an interrupt that does not exist (has not been initialized or is invalid or reserved) will cause problems as well. Use TRAP   #10 with caution. Appendix H provides information on several useful interrupt functions that do not require segment registers. See what is possible with them before trying your own.

6.17 TROUBLESHOOTING TECHNIQUES

This chapter dealt mainly with individual programming modules that performed a single chore. In each case, the subroutine was already developed and tested for you, and limitations of the routines were also mentioned.

How exactly is a subroutine created from scratch? The answer is: lots of different ways. Every programmer will use his or her own individual techniques to create a new set of instructions to solve a problem. Some may prefer to use flowcharts or pseudocodes, others feel comfortable writing the instructions from scratch. Let us look at one method to create a new subroutine.

First comes the problem. It may be something like this:

"Write a subroutine called FIND7 that counts the number of times the value 7 appears in an array of 100 word-size integers. The starting address of the array is the label VALUES. Return the result in D7."

This specification is very clear about what needs to be done. We begin by framing the subroutine:

```
FIND7
    .
    .
    .
        RTS
```

Note that RTS is used and not RTE, which is reserved for exceptions.

Next, we add the instructions that will allow 100 passes through the loop.

```
FIND7     MOVE.W   #99,D6
AGAIN
    .
    .
    .
          DBF      D6,AGAIN
          RTS
```

Using register D6 does not interfere with the requirement that we return the result in register D7, and has the added advantage of being automatically used by the DBF instruction to control the number of passes. Recall that the DBF instruction stops when D6 equals −1, not 0.

Now, in addition to initializing D6, we must initialize D7 and a pointer to the VALUES array. D7 must be initialized to 0, since it represents a count (there may really be zero 7s in the data).

```
FIND7     MOVE.W   #99,D6
          SUB.L    D7,D7
          MOVEA.L  #VALUES,A6
AGAIN
```

.
.
.

```
         DBF       D6,AGAIN
         RTS
```

All that remains is the actual loop code. What we want to do, 100 times, is examine a data item from the VALUES array, compare it with 7, and, if equal, increment the D7 register. This is done as follows:

```
FIND7    MOVE.W    #99,D6
         SUB.L     D7,D7
         MOVEA.L   #VALUES,A6
AGAIN    CMPI.B    #7,(A6)+
         BNE       NOTEQ
         ADDI.B    #1,D7
NOTEQ    DBF       D6,AGAIN
         RTS
```

It is very important to adjust the pointer register A6 with each pass through the loop, so that the CMPI instruction always reads a new data item from the VALUES array. This is done automatically by the postincrement addressing used on A6. FIND7 is now ready to be combined with a driver program for testing.

This step-by-step approach, or one you develop on your own, will become very automatic with practice. It requires familiarity with the most basic programming chores, initialization, counting, looping, and comparing. It is worthwhile to invent your own programming problems as well, and then try to write the instructions to solve them.

SUMMARY

In this chapter we examined a number of different applications the 68000 is capable of performing. These applications find widespread use in industrial and commercial situations. In addition, we covered many different techniques, such as numeric conversion, data table searching, and mathematical processing with both binary and BCD numbers. The overall idea is to get a sense of how the 68000's instructions can be combined to perform any task that we can imagine. Many more applications are possible, and we have barely scratched the surface here. But the routines presented in this chapter should serve as a foundation on which to build when you write an application of your own.

STUDY QUESTIONS

1. What is a program specification? What does it contain?
2. How much test data should be used in a software driver program?
3. When are processor registers better suited for parameter passing than parameters stored in memory?
4. When the stack is used to pass parameters to a subroutine, what must be done by the subroutine to correctly access the parameters?

5. What are the advantages of using MOVEQ instead of MOVE.L? What are the limitations?

6. Show that the following instructions are all equivalent:
   a) SWAP D0 and ROL.L   #16,D0 (or ROR.L #16,D0)
   b) CLR.B D1 and ANDI.B   #0,D1
   c) MULU #2,D2 and LSL.L   #1,D2

7. What determines the length of a text string?

8. The bubble sort technique requires approximately N^2 comparisons for N input values. Where do the N^2 comparisons come from?

9. How must the storage format for BCD numbers be changed so that negative numbers can be used? How about real numbers, such as 0.725 and 431.5?

10. What does it mean to normalize a number? Why is normalization needed?

11. Why is it necessary to clear the X flag before beginning a BCD operation such as ABCD?

12. How are roundoff errors eliminated by using BCD?

13. Why is it necessary to read the initial status information in the computerized burglar alarm?

14. What is overhead? Can overhead be ignored?

15. What are the various reasons for one instruction taking longer to execute than another?

16. Which of these two equivalent instructions takes longer to execute?

```
MULU   #8,D0     or     LSL.L    #3,D0
```

17. Why is it necessary to preserve the processor registers during exception processing?

18. Discuss the advantages and disadvantages of using BCD and binary arithmetic in the calculator program.

19. What type of parameter passing is used in the calculator program?

20. How do stack frames support recursion?

ADDITIONAL PROGRAMMING EXERCISES

1. Write a software driver for the following subroutine. Use at least five test values.

```
XYZ     MOVE.W   D0,D1
        ADD.W    D1,D1
        ADD.W    D1,D1
        ADD.W    D0,D1
        ADD.W    D1,D1
        RTS
```

2. Two large data items containing 10 words and 25 words, respectively, are pushed onto the stack prior to a JSR instruction. What instructions are needed to access the first word of each data item from within the subroutine?

3. Add instructions to the KEYBUFF routine so that leading blanks are removed from the buffer. That is, if the buffer looks like:

```
'     abcd'
```

it is converted to 'abcd'.

4. Write a routine that reverses the contents of a text string. For example, if the original string is 'Hello', the reversed string is 'olleH'.

5. Write a routine that determines whether a text string is a *palindrome*. A palindrome is a sequence of symbols or characters that reads the same forwards and backwards, as in *radar*.

6. Assume that two arrays of numbers, NUMS1 and NUMS2, are already sorted in increasing order and have the same length of 12 bytes. Write a routine called MERGE that will produce a third array containing all 24 bytes in increasing order.

7. What instructions are needed to left justify a right justified BCD number stored in D3? For example, if D3 initially contains 00012345, how is it converted into 12345000?

8. Write a routine called PMADD that adds two *signed* BCD numbers. Use a sign byte to represent the sign of the number (0 = positive, 1 = negative).

9. Modify the SHOWDECI routine so that leading zeros are eliminated.

10. Write a routine called DISPBIN that displays the 8-bit number stored in the lower byte of D7.

11. A small-scale burglar alarm uses a single memory location to read 8 status bits and write 4 control bits. Each control bit is the exclusive OR of 2 status bits. Write a routine that will constantly read the status byte and then write the new control byte.

12. Write an exception handler for a level-2 interrupt. Whenever 25 interrupts have occurred, the handler should perform a JSR   TIMEOUT.

13. Modify the calculator program so that the current display is saved in memory when 'S' is entered and loaded from memory when 'M' is entered.

14. Add a clear operation to the calculator (triggered by 'C').

15. Rewrite the calculator so that it uses stack frames.

16. Write a program that displays a countdown from 10 to 0, with 1 second between each count.

17. Write a program that displays the PC's date, as in December 25, 1997.

18. Write a program that accepts a secret password from the keyboard, without echoing it as it is entered.

19. Write and test the recursive function fact().

20. Add recursion to the calculator to support parentheses, as in 5 * (4 + 3).

PART 4

Hardware Architecture

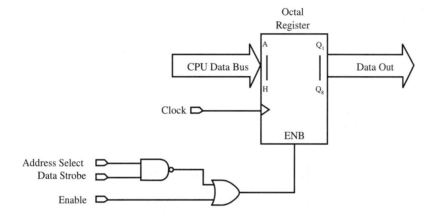

Basic logic gates play an important role in hardware interfacing.

CHAPTER 7

Hardware Details of the 68000

OBJECTIVES

In this chapter you will learn about:

- The general specifications of the 68000 microprocessor
- The processor's control signal names and functions
- General signal relationships and timing
- Methods in which the 68000 may interface with external devices
- The external interrupt signals and their operations
- The 68000's bus arbitration scheme
- The method used to access a peripheral
- How hardware expansion is provided by a bus connector

7.1   INTRODUCTION

Before using any microprocessor, we must necessarily have an understanding of both its hardware requirements and its software functions. In this chapter we will examine all 64 pins of the 68000's package, learning what their use may be in a larger system employing the 68000 as its CPU. We will not concentrate on interfacing, since this important topic is covered in Chapters 8, 9, and 10. Upon completion of this chapter, we should, however, know about the various signals of the processor in order to begin interfacing it with support circuitry, which includes memories, I/O devices, and coprocessors.

Section 7.2 gives a quick overview of the capabilities of the 68000, its memory addressing capabilities, available clock speeds, and various other functions. Section 7.3 covers all 64 pins of the 68000 in detail. The pins are separated into eight functional groups, such as interrupt control, system control, processor status, and so on. Block diagrams and timing waveforms are given, where applicable, except where they might apply to interfacing. Section 7.4 describes two system timing diagrams: HALT timing and bus arbitration timing. Other system timing diagrams will be covered in Chapter 8. Section 7.5 discusses the use of processor signals in a typical hardware calculator system. Section 7.6 describes bus connectors found on 68000 motherboards. The chapter concludes with troubleshooting techniques in Section 7.7.

7.2 CPU SPECIFICATIONS

The 68000 is a 16-bit microprocessor that communicates with the outside world via a 16-bit bidirectional data bus. The immediate advantage of a 16-bit bus over an 8-bit bus is that twice as much data may be exchanged at once, which has the effect of decreasing the time required for memory accesses and program execution. The 68000's 24-bit address bus can address over 16 *million* bytes of memory (or over 8 million words of 2 bytes each). Control signals are provided that enable external circuitry to take over the 68000's buses (a must for DMA operations), and three interrupt lines provide seven levels of external hardware interrupts. Three control outputs may be used to decode any of eight internal CPU states, and there are also control signals provided that enable the 68000 to be interfaced with existing 6800 peripherals (for those users who may wish to upgrade their existing hardware with a minimum of fuss). The 68000 is available with maximum clock speeds from 4 to 16 MHz as of this writing, and has been on the market long enough to be purchased at a reasonable cost. But the power of the 68000 can be tapped only if we know how to use it, so on to the next section, where we will examine the 68000's signals and their usage.

7.3 CPU PIN DESCRIPTIONS

Figure 7.1 details the input and output signals of the 68000 CPU. There are eight groups of pins that we will examine. Each group of pins (or signal lines) performs a specific function, necessary to the proper operation of the 68000.

V_{CC}, GND, and CLK

This group deals with the processor power and clock inputs. Note that there are two pins each for V_{CC} and GND as shown in Figure 7.2. Each pair must be connected for proper operation. The 68000 operates on a single supply voltage of 5 volts, plus or minus 5 percent, and will dissipate 1.5 watts of power (with an 8 MHz clock speed) at this voltage. Maybe this is why the designers used two power pins each for V_{CC} and ground. A rough calculation yields 300 mA as the required supply current. As for the CLK input, the rise and fall times are limited to 10 ns (on all versions except the 12.5 and 16 MHz, which limits them to 5 ns), and must be a TTL-compatible waveform with a 50 percent duty cycle. Even though the clock input is internally buffered, the clock signal should be kept at a constant frequency (via an external crystal oscillator) for best operation.

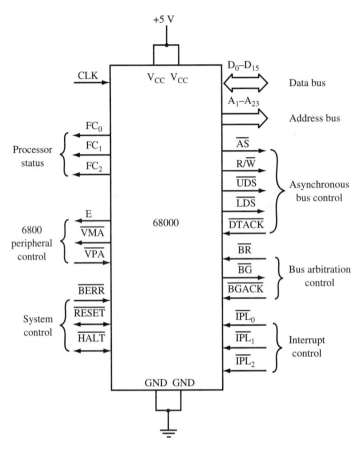

FIGURE 7.1   68000 CPU input and output signals

FIGURE 7.2 V_{CC}, GND, and CLK pins

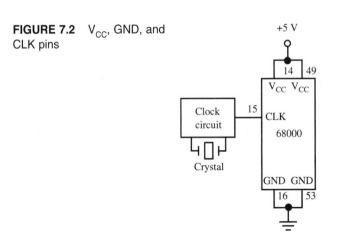

FC_0, FC_1, and FC_2

This group of signals is used to output the encoded processor status. Outputs FC_0, FC_1, and FC_2 are the function code outputs, and they indicate the current internal processing state of the 68000 (what type of cycle it is currently executing). Note that the function code outputs are only valid when the processor's $\overline{AS}$ signal is active. Table 7.1 illustrates how these three outputs are decoded. It shows the division of processing states into user data, user program, supervisor data, supervisor program, and interrupt acknowledge. In some systems, the function code outputs are used to restrict memory accesses by employing them in the memory address decoding circuitry. This is necessary when users must be kept out of supervisor memory space.

An easy way to decode all eight processor states is to use a three- to eight-line decoder such as the 74LS138; this is left for you to do as a homework problem. Normally we have no use for most of the decoded cycle states in a small system, except for the interrupt acknowledge state (indicated by all three function code outputs high), assuming we require interrupt capability. Decoding this state is accomplished with a four-input NAND gate, whose output will go low only during the interrupt acknowledge cycle. Figure 7.3 shows this connection. Remember: The function code outputs are valid only when the $\overline{AS}$ signal is active ($\overline{AS}$ is low).

TABLE 7.1   Function code outputs

FC_2	FC_1	FC_0	Cycle type
0	0	0	Reserved*
0	0	1	User data
0	1	0	User program
0	1	1	Reserved*
1	0	0	Reserved*
1	0	1	Supervisor data
1	1	0	Supervisor program
1	1	1	Interrupt acknowledge

*By Motorola, for future use.

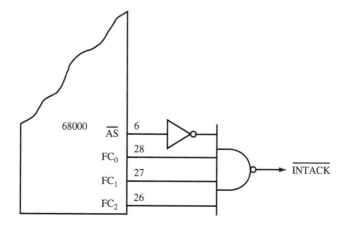

FIGURE 7.3   Interrupt acknowledge cycle decoder

E, $\overline{\text{VMA}}$, and $\overline{\text{VPA}}$

This group of signals is provided to give the 68000 the capability to control older 6800 peripherals. Many of these devices were designed for use with this earlier processor, and it would be a shame to have to replace working, and possibly expensive, hardware just because an increase in computing power is desired. These signals were provided on the 68000 to allow designers to transition from 8-bit designs up to 16-bit designs. They are not found on the later, more advanced 680x0 CPUs. The three signals used for 6800 peripheral control are the E clock, $\overline{\text{VMA}}$ (valid memory address), and $\overline{\text{VPA}}$ (valid peripheral address). Of these three signals, only $\overline{\text{VPA}}$ is an input as shown in Figure 7.4. The E clock is used to generate the proper timing signals for 6800 peripherals, and is derived from the 68000's clock by dividing it by 10, with the resulting waveform having a 40 percent duty cycle (high for four CLK cycles, low for six). Figure 7.5 indicates this timing relationship.

The $\overline{\text{VPA}}$ input is used to inform the 68000 that it has addressed a 6800 peripheral and that the data transfer should be synchronized with the E clock. This synchronization is indicated by the $\overline{\text{VMA}}$ output, which goes low when the processor synchronizes with the E clock. The entire 6800 peripheral control sequence is as follows: The 68000 outputs the address of a 6800 peripheral on A_1 through A_{23}. The peripheral circuitry responds by pulling $\overline{\text{VPA}}$ low. When the processor has synchronized with the E clock, it outputs a low on $\overline{\text{VMA}}$ and the data transfer takes place. Figure 7.6(a) shows this timing sequence.

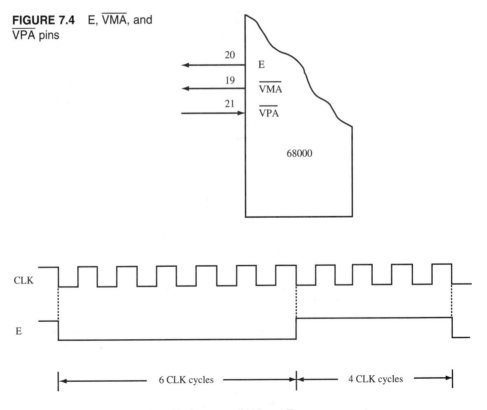

FIGURE 7.4   E, $\overline{\text{VMA}}$, and $\overline{\text{VPA}}$ pins

FIGURE 7.5   Timing relationship between CLK and E

Figure 7.6(b) shows the block diagram connections for the 6800 peripheral interface. Of
course, the length of the bus cycle depends on how long it takes the 68000 to internally
synchronize with the E clock. Figure 7.6(a) shows the *best case* timing for the peripheral
access. It may very well be that more *w* states (wait states) are required before synchronization occurs. This would therefore increase the entire bus cycle time.

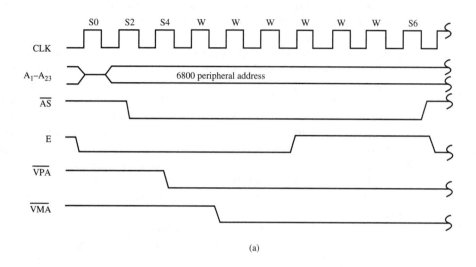

(a)

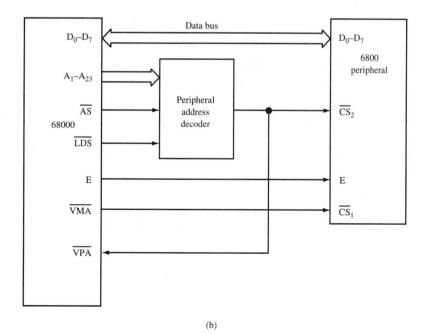

(b)

FIGURE 7.6   (a) 6800 peripheral timing and (b) 6800 peripheral interface

RESET, HALT, and BERR

This group of signals is used for system control. $\overline{\text{BERR}}$ (bus error), $\overline{\text{RESET}}$, and $\overline{\text{HALT}}$ are, in effect, the panic buttons of the 68000 and are shown in Figure 7.7. The $\overline{\text{BERR}}$ input is used to inform the processor that the cycle currently executing has a problem. Maybe the processor has addressed an illegal memory location (a location that is not mapped to EPROM, RAM, or a peripheral), or some other kind of bus error has occurred. By asserting $\overline{\text{BERR}}$, we inform the 68000 that an error has occurred during execution of the current bus cycle, and that it should take appropriate action. The 68000 will then choose between performing a bus error exception or rerunning the bus cycle. If the $\overline{\text{HALT}}$ line is not asserted when the bus error occurs, the 68000 starts bus error exception processing, terminating the currently failed cycle. If the $\overline{\text{HALT}}$ line was asserted before or at the same time as the $\overline{\text{BERR}}$ signal, the 68000 will instead rerun the bus cycle. It does this by terminating the cycle and placing the data and address buses in the high-impedance state. The 68000 then enters a "do nothing" state until there is activity on $\overline{\text{HALT}}$. When the $\overline{\text{HALT}}$ line is deactivated (returned high), the processor will rerun the previous cycle (using the same data, address, and control codes). The design of the 68000 requires that the $\overline{\text{BERR}}$ signal be deactivated at least one clock cycle before the $\overline{\text{HALT}}$ signal. The processor will not rerun a read-modify-write cycle (used during execution of the TAS instruction) to ensure data integrity.

Both $\overline{\text{RESET}}$ and $\overline{\text{HALT}}$ are bidirectional signals. This means that there are times when they act as inputs and others when they act as outputs. For example, $\overline{\text{RESET}}$, together with $\overline{\text{HALT}}$, must initially (after a power-on) be taken low for a minimum of 100 ms. This is accomplished with the use of a timer (such as the 555 timer) designed to output a 100-ms pulse at power-on. Among other things, this ensures that V_{CC} has stabilized. This type of reset sequence results in a total processor reset, meaning that the 68000 loads the

FIGURE 7.7 $\overline{\text{RESET}}$,
$\overline{\text{HALT}}$, and $\overline{\text{BERR}}$ pins

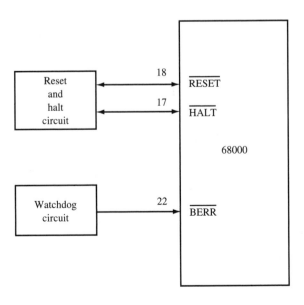

supervisor stack pointer from vector 0 (at address 000000) and the program counter from vector 1 (at address 000004). All other registers remain unaffected except for the status register, which is adjusted to indicate a level-7 interrupt priority. In addition, the trace bit is cleared and the supervisor bit is set. $\overline{\text{RESET}}$ and $\overline{\text{HALT}}$ may be used again at a later time (long after power-on, during normal execution) to reset the processor, by asserting them for at least 10 clock cycles.

The 68000 instruction set includes RESET, an instruction that, when executed, will cause the 68000 to output a low level on $\overline{\text{RESET}}$ for 124 clock cycles. In this case the $\overline{\text{RESET}}$ line acts as an output signal. This has the effect of resetting external circuitry connected to $\overline{\text{RESET}}$, without affecting the state of the processor at all.

The $\overline{\text{HALT}}$ line may be driven low at any time by an external device. When this happens, the processor completes the current bus cycle and halts. When halted, all tristate signals are put into the high-impedance state, and all control signals are inactive. Recall that a tristate signal may be in one of three states at any time: low (0) state, high (1) state, and high-impedance (open) state, as indicated in Figure 7.8. Execution will resume when $\overline{\text{HALT}}$ is returned to a high level. If an internal event such as execution of the 68000's STOP instruction triggers the halt condition, the 68000 will output a low level on $\overline{\text{HALT}}$. Care must be taken when designing circuitry to control the $\overline{\text{RESET}}$ and $\overline{\text{HALT}}$ lines, to ensure that they both continue to work properly when in either state (input or output). Figure 7.9 shows a typical power-on operation and its related timing. As the figure indicates, processing does not begin for at least 100 ms *after* V_{CC} has stabilized at 5 volts.

$\overline{\text{IPL}}_0$, $\overline{\text{IPL}}_1$, and $\overline{\text{IPL}}_2$

This group is used for interrupt control. The three inputs $\overline{\text{IPL}}_0$, $\overline{\text{IPL}}_1$, and $\overline{\text{IPL}}_2$ are used by external circuitry to request the encoded priority level of the hardware interrupt. Level 7 has the highest priority. Level 0 indicates that no interrupts are present (and thus all three

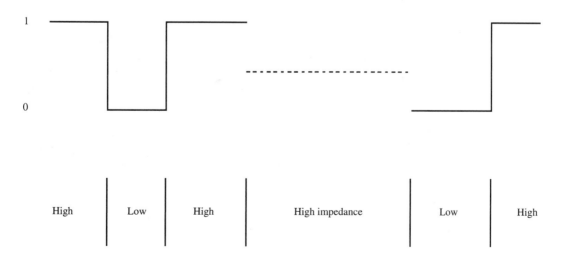

FIGURE 7.8   Tristate signal levels

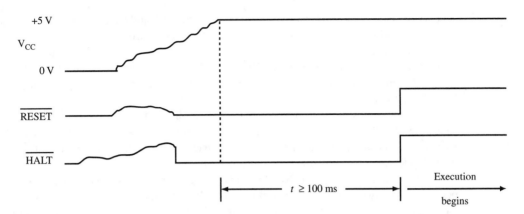

FIGURE 7.9   Power-on $\overline{\text{RESET}}$ and $\overline{\text{HALT}}$ timing

inputs are high). Table 7.2 indicates the logic levels needed on these three inputs to achieve a certain interrupt. As shown, $\overline{\text{IPL}_2}$ is the most significant of the three interrupt inputs. Note that the level numbers (0–7) and the actual binary codes needed on $\overline{\text{IPL}_0}$ through $\overline{\text{IPL}_2}$ are inverted. Both vectored and autovectored interrupts may be generated via $\overline{\text{IPL}_0}$ through $\overline{\text{IPL}_2}$. Figure 7.10 shows how a pushbutton can be used to generate a level-7 interrupt.

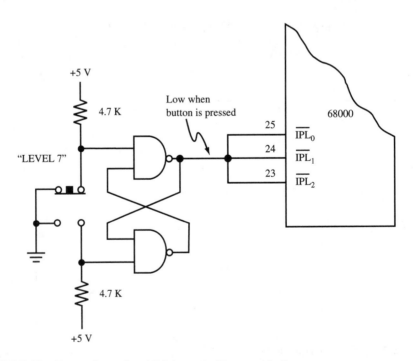

FIGURE 7.10   Generating a level-7 interrupt with a pushbutton

TABLE 7.2 Interrupt-level encoding

Interrupt			
$\overline{IPL_2}$	$\overline{IPL_1}$	$\overline{IPL_0}$	Interrupt level*
1	1	1	0 (Lowest, none)
1	1	0	1
1	0	1	2
1	0	0	3
0	1	1	4
0	1	0	5
0	0	1	6
0	0	0	7 (Highest, nonmaskable)

*Note the inversion of the binary bits needed on $\overline{IPL_2}$–$\overline{IPL_0}$.

$\overline{BR}$, $\overline{BG}$, and $\overline{BGACK}$

This group is used for bus arbitration control. This is a fancy way of saying that we will use these signals to place the 68000 in a wait state while we make use of the hardware connected to its buses. The circuit that takes over control of the bus may be a DMA controller or another CPU in the system. The bus takeover must be done in an orderly fashion, by predetermined rules, or bus contention will occur. The most likely result: Our program will fail and we may lose all our data (these problems may be the least of our worries if the processor is being used to fly an airplane or rocket).

The bus arbitration protocol works like this: The requesting device, called a **bus master,** requests use of the 68000's buses by activating the $\overline{BR}$ (bus request) input. The 68000 will respond to the bus master by taking the $\overline{BG}$ (bus grant) output low, indicating that it will release control of its buses at the end of the current cycle. When the new bus master wants to take control, it asserts $\overline{BGACK}$ (bus grant acknowledge). There are four conditions that must be met before the new bus master may activate $\overline{BGACK}$:

- $\overline{BG}$ must be active.
- $\overline{AS}$ must be inactive, to show that the processor is not using the bus.
- $\overline{DTACK}$ must be inactive, to show that no external devices are using the bus.
- $\overline{BGACK}$ must be inactive, to ensure that another bus master is not already using the bus.

Thus, taking over control of the 68000's buses is a tricky matter and is best approached with caution.

Releasing control of the bus is an easier matter. Once $\overline{BGACK}$ has been asserted, the new bus master can release $\overline{BR}$ (bring $\overline{BR}$ to a high level). When the bus operation is completed and the bus master gives control back to the 68000, it does so by negating $\overline{BGACK}$. Figure 7.11 shows the connections for bus arbitration.

Motorola uses three signals to implement bus arbitration. Many other processors (the 68008 is one) use only two: $\overline{BR}$ and $\overline{BG}$. Why the need for $\overline{BGACK}$? The answer is that Motorola wanted to provide a way to eliminate false takeovers of the bus due to *noise* on the $\overline{BR}$ input. If $\overline{BGACK}$ is not active when $\overline{BR}$ goes inactive (the noise goes away), the 68000 will not release the bus and continues normal execution.

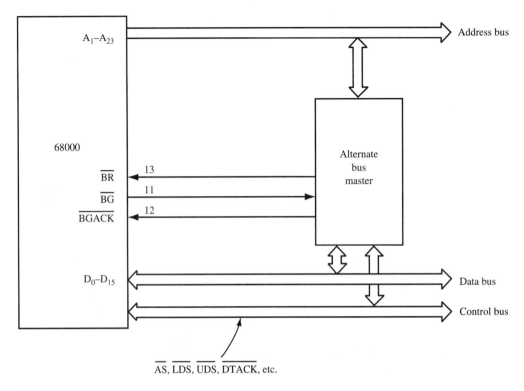

FIGURE 7.11   Bus arbitration logic block diagram

$\overline{\text{AS}}$, R/W, $\overline{\text{UDS}}$, $\overline{\text{LDS}}$, and $\overline{\text{DTACK}}$

This group, asynchronous bus control, contains five signals that are essential to the proper operation of external hardware. Four of the signals—$\overline{\text{AS}}$, R/$\overline{\text{W}}$, $\overline{\text{UDS}}$, and $\overline{\text{LDS}}$—are outputs, and the fifth one ($\overline{\text{DTACK}}$) is an input. The $\overline{\text{AS}}$ (address strobe) signal is used to indicate that a valid memory address exists on the address bus. The R/$\overline{\text{W}}$ (read/write) signal determines whether the current cycle is a read or write cycle. $\overline{\text{UDS}}$ (upper data strobe) and $\overline{\text{LDS}}$ (lower data strobe) are used to gate 8 bits of information to/from selected portions of the 68000's data bus. These signals are described more fully in Chapter 8, but for now all we need to know is that $\overline{\text{UDS}}$ controls data on bits 8 to 15 of the data bus (the *even* bytes), and $\overline{\text{LDS}}$ does the same for bits 0 to 7 (the *odd* bytes). Thus, to transfer only 8 bits of data, we need only activate either $\overline{\text{UDS}}$ or $\overline{\text{LDS}}$, whichever is appropriate. To transfer 16 bits at once, both $\overline{\text{UDS}}$ and $\overline{\text{LDS}}$ must be low. The last signal, $\overline{\text{DTACK}}$ (data transfer acknowledge), is used by external circuitry to perform **asynchronous** data transfers (we perform **synchronous** transfers using E, $\overline{\text{VMA}}$, and $\overline{\text{VPA}}$). When $\overline{\text{DTACK}}$ is activated by external hardware, the processor recognizes that the current bus cycle can be completed. For example, data present on the data bus is latched by the 68000 when $\overline{\text{DTACK}}$ goes active during a read cycle, or data is placed on the data bus and kept there for writing until $\overline{\text{DTACK}}$ is asserted by external hardware (a memory or I/O device). It is the job of the external hardware (either memory circuitry or peripheral) to activate/deactivate $\overline{\text{DTACK}}$ at the proper times. When the 68000 is operated at full speed, the timing of $\overline{\text{DTACK}}$ becomes important (to ensure the correct cap-

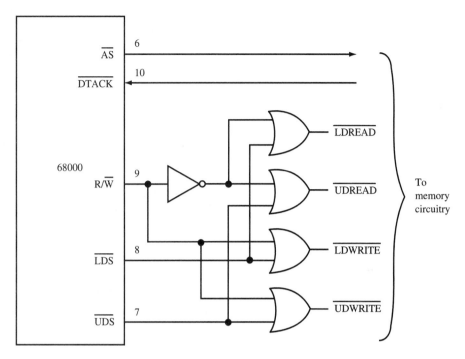

FIGURE 7.12   Decoding memory read/write signals

ture of data), but we will not go into this topic here. Figure 7.12 shows how OR gates can be used to decode memory read and write signals using the asynchronous control signals.

A_1 Through A_{23}, D_0 Through D_{15}

The last group contains the address and data buses. As mentioned before, the address bus, containing $\overline{\text{UDS}}$ and $\overline{\text{LDS}}$, can be used to access more than 16 million bytes of memory. The ability to directly address this many locations provides the programmer with a very flexible programming environment. Older systems required that special software be used to manage memory in "pages" that were some portion of the system's main address space. This software is not even needed now, in some applications, because of the great increase in address lines and hence memory locations. One important note: A_1 through A_{23} are **unidirectional** (output only), and are always used to transmit a memory address, except during an interrupt acknowledge cycle. In this case, A_1, A_2, and A_3 contain the interrupt level and all other address lines are high.

The bidirectional data bus on lines D_0 through D_{15} is used to transfer information between the processor and the outside world. D_0 through D_7 serve double duty during an interrupt acknowledge cycle, where they then transmit the interrupt vector number (supplied by an external device). A very important note here is that most 68000-based systems, even small ones, will require extensive use of the address and data lines (at least four EPROMs/RAMs requiring use of the buses), and it will be very easy to exceed the maximum drive capability of these signal lines, which is limited to a few milliamperes each.

For this reason (see the detailed coverage of this topic in Chapter 8), we usually use special buffering circuits between the address/data lines and the external hardware. The buffers give us the ability to drive many more devices than the processor's address/data lines do. Figure 7.13 shows an example of how the address and data buses can be buffered.

FIGURE 7.13 Buffering the address and data buses

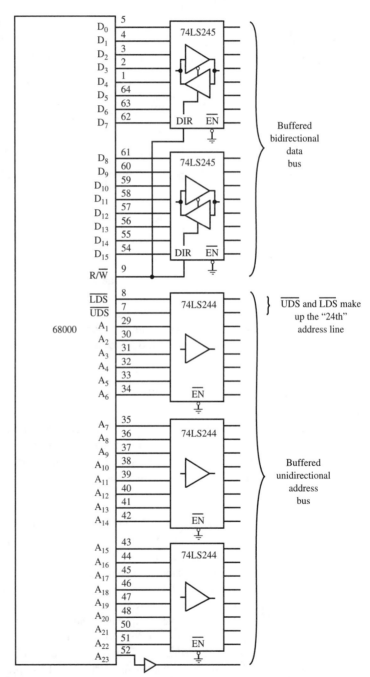

TABLE 7.3 Summary of 68000 signals

Signal	Input	Output	Tristate
CLK	✓		
FC_0–FC_2		✓	✓
E		✓	
$\overline{VMA}$		✓	✓
$\overline{VPA}$	✓		
$\overline{BERR}$	✓		
$\overline{RESET}$	✓	✓	
$\overline{HALT}$	✓	✓	
$\overline{IPL_0}$–$\overline{IPL_2}$	✓		
$\overline{BR}$	✓		
$\overline{BG}$		✓	
$\overline{BGACK}$	✓		
$\overline{AS}$		✓	✓
$R/\overline{W}$		✓	✓
$\overline{UDS}$		✓	✓
$\overline{LDS}$		✓	✓
$\overline{DTACK}$	✓		
A_1–A_{23}		✓	✓
D_0–D_{15}	✓	✓	✓

SIGNAL SUMMARY

Since this concludes the pin description section, it would be good to end with a summary of all the CPU signals, whether they are inputs or outputs and whether they have tristate capability. Table 7.3 does this for us. In the next section we will examine some of the processor signals again, and see their timing relationships to one another in some selected bus cycles.

7.4 SYSTEM TIMING DIAGRAMS

When timing is a critical issue in the design of a new 68000-based system, it pays to know how to interpret the CPU timing diagrams supplied by the manufacturer. This section will provide more details on the processor's control and timing signals by analyzing two timing scenarios.

Since the timing of memory read and write cycles is covered in Chapter 8, we will not study them here. Instead, we will look at some special cases, the first being the processor's HALT timing and the second being the processor's timing during bus arbitration.

Processor HALT Timing

Figure 7.14 shows the timing diagram that results when the 68000 enters the halt mode. The left portion of the timing diagram represents the processor timing during its current read cycle. States S0 (CLK is high) through S7 (CLK is low) are the eight states needed to complete this cycle. A state, sometimes called a **cycle state** or **state time,** represents a half cycle of the CLK input to the processor. For example, a 10-MHz 68000 would have a state time equal to 50 ns. A state is the smallest amount of time in which the processor can perform *any* function. Since all instructions are composed of a number of these cycle states, instruction execution time depends on the CLK frequency. During the eight states of the current instruction, we see that there are times when the address bus is valid (S1 through S7), and others when it is tristated (S0). The activity on other processor signal lines, such as $\overline{AS}$, $\overline{DTACK}$, or D_0 through D_{15}, is also related to the processor's state time. Notice that signal levels do not change except during transitions between states.

When the processor finishes the current cycle, its control signals revert to their inactive state (high levels for $\overline{AS}$, $\overline{UDS}$, and $\overline{LDS}$). Since the $\overline{HALT}$ line was asserted during the current cycle, the processor does not begin a new bus cycle. It instead leaves its control lines in the inactive state and tristates its address and data buses. The 68000 is really in a "wait" state now, where the length of time that it "waits" before beginning a new instruction cycle is a function of how long the external hardware keeps the $\overline{HALT}$ line low (although there are exceptions to this rule). When the circuitry releases the $\overline{HALT}$ line, the processor uses a few states to do some internal housekeeping, and then begins execution of the next instruction. Exceptions to this rule are the arrival of either a $\overline{RESET}$ or an interrupt during the halt state. Both of these cases cause the halt state to terminate.

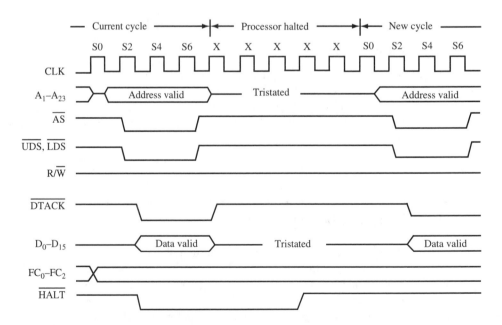

FIGURE 7.14   Processor HALT timing

While there do not seem to be any useful applications for the halt state, it is actually quite helpful when debugging programs. A simple circuit may be designed to assert the $\overline{\text{HALT}}$ line during execution of *every* instruction, which results in a single-step mode of operation. This is left for you to design as a homework problem.

Bus Arbitration Timing

This section deals with the timing involved when another bus master wishes to take over the 68000's buses. The timing signals in Figure 7.15 show what happens when the processor receives a bus request during execution of an instruction. We see that the processor finishes the current bus cycle, even though it has received a bus request ($\overline{\text{BR}}$ active), and has granted it ($\overline{\text{BG}}$ active). When the current bus cycle finishes, the processor's control lines ($\overline{\text{AS}}$, $\overline{\text{UDS}}$, $\overline{\text{LDS}}$, R/$\overline{\text{W}}$, and FC_0–FC_2), as well as the address and data buses, all go into the high-impedance state. This permits a parallel connection of these signal lines with other signal lines (from another processor or a DMA device), which themselves must be tristated when the 68000 has control, or we would have a major bus contention problem. While the new bus master has control of the 68000's buses, it may do whatever it wishes with them. The 68000 will not respond to activity on the buses while $\overline{\text{BGACK}}$ is active. Only when the bus master finishes, and releases $\overline{\text{BGACK}}$, does the 68000 come back to life and begin execution of the next instruction. One attractive application of this process is in the use of dynamic RAMs, which need to be continually refreshed in order to retain their data. Since the RAMs have to be refreshed when the processor is not accessing them, one solution is to employ refresh circuitry designed to implement the bus arbitration scheme of the 68000. The refresh circuit actually becomes the bus master.

While these two timing examples fall short of completely describing the timing relationships between the numerous signal lines of the 68000, they should serve as a starting point for further exploration. A very good exercise at this point would be to hook a 68000-based system up to a logic analyzer and examine the states of $\overline{\text{AS}}$, $\overline{\text{DTACK}}$, R/$\overline{\text{W}}$, $\overline{\text{UDS}}$, $\overline{\text{LDS}}$, the address and data lines, and any other signals you can think of. Watch their activity as the 68000 accesses memory and I/O devices and handles external interrupts.

7.5   HARDWARE ASPECTS OF THE CALCULATOR PROJECT

In the last chapter we examined the software portion of a working four-function calculator. We were not concerned at the time with the specific details of how the calculator's display and keyboard were implemented. TRAP numbers 0 and 1 were provided to handle this chore for us. In this section it is appropriate to ask "How *does* the calculator perform I/O?" and discuss the various options available.

Since the calculator must accept numbers and operations from the user, an input device is needed. This is usually in the form of a keyboard (or keypad). The keyboard may generate either parallel or serial data for each keystroke. When a key is pressed, a unique code is generated and read by the calculator's input hardware.

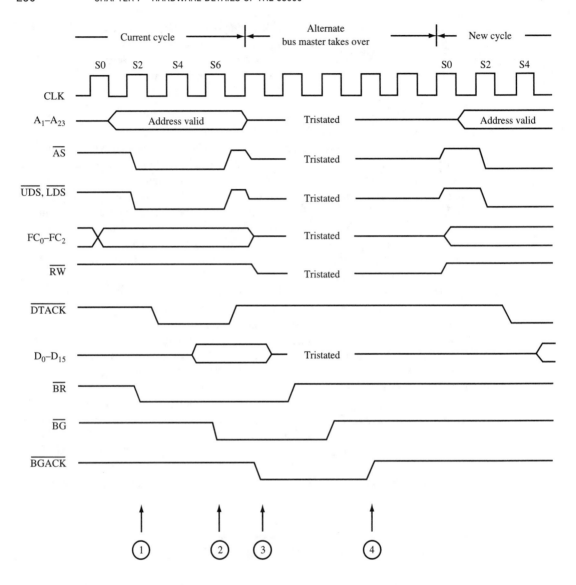

1. Alternate bus master requests use of bus.

2. 68000 grants request.

3. Alternate bus master takes control of bus.

4. Bus returned to 68000.

FIGURE 7.15   Bus arbitration timing

The results of calculations must then be displayed. They may be output to a serial ASCII data terminal, to a seven-segment LED or LCD display, or to an alphanumeric display. A printer may also be attached for hard copy of results.

The overall operation of the calculator is controlled by a program stored in EPROM. Since EPROMs store only 8 bits of data in each location, an EPROM is needed for each half of the processor's data bus. Likewise, two RAMs are needed for storage of temporary results and stack operations.

If an automatic power-down option is provided on the calculator, an external timing circuit will be needed to tell the 68000 when to shut off. The circuit usually times out when no keys have been pressed for a predetermined length of time (such as 5 minutes). The timing circuit may use an interrupt to inform the processor to shut down. Other interrupts, possibly from the keyboard or display, may also be used. So, interrupt logic may be needed as well.

In terms of processor signals, practically all of them are needed. The address and data buses must be connected to the EPROMs, RAMs, and I/O sections. Control signals $\overline{AS}$, $\overline{LDS}$, $\overline{UDS}$, R/$\overline{W}$, and $\overline{DTACK}$ will also be needed to control memory accesses and I/O operations. The interrupt circuitry will require the use of the FC and $\overline{IPL}$ signals (and $\overline{VPA}$ if autovectoring is used).

$\overline{RESET}$, $\overline{HALT}$, and CLK must also be used, to properly start the processor at power-on and control system timing.

In the next few chapters we will see examples of how these various circuits may be implemented.

7.6 PROVIDING ACCESS TO THE PROCESSOR SIGNALS

It is often cost effective to design a main logic board around a host processor, where the main components of the system are implemented. This includes memory and I/O interfaces, as well as the required timing circuitry. Then, to make the board as useful as possible, an *expansion connector* is provided to allow users to customize the system to suit their own needs by plugging their own logic boards into the expansion connector. The expansion connector provides access to all processor signals, all of which might be needed by the expansion logic. This is what is done in the Macintosh SE personal computer, where a 96-pin connector called the **SE Bus** is provided on the motherboard to allow hardware expansion. The connector consists of three rows of 32 pins each. Examine the pin assignments shown in Table 7.4. As you can see, all processor signals are provided. This gives complete access to the processor and all other hardware (EPROM, RAM, etc.) on the motherboard. Any expansion card plugged into the SE Bus connector must conform to these pin assignments. When the Macintosh II came out, a different expansion connector called the **NuBus** was used. This connector (also 96 pins) provided access to the additional address and data lines found on the 68020 processor used in the Macintosh II. So, a designer must be familiar with many different bus connectors to guarantee an expansion card will be properly interfaced.

TABLE 7.4 Macintosh SE expansion connector

Pin	Signal description		
	Row A	Row B	Row C
1	FC_2	GND	$\overline{VPA}$
2	FC_1	GND	$\overline{VMA}$
3	FC_0	GND	$\overline{BR}$
4	A_1	GND	$\overline{BGACK}$
5	A_2	GND	$\overline{BG}$
6	A_3	GND	$\overline{DTACK}$
7	A_4	GND	$R/\overline{W}$
8	A_5	GND	$\overline{LDS}$
9	A_6	GND	$\overline{UDS}$
10	A_7	Reserved	$\overline{AS}$
11	A_8	Reserved	$\overline{PMCYC}$**
12	A_9	$\overline{HALT}$	$\overline{RESET}$
13	A_{10}	+5 V	+5 V
14	A_{11}	+5 V	D_0
15	A_{12}	+5 V	D_1
16	A_{13}	+5 V	D_2
17	A_{14}	+5 V	D_3
18	A_{15}	$\overline{IPL_0}$	D_4
19	A_{16}	$\overline{IPL_1}$	D_5
20	A_{17}	$\overline{IPL_2}$	D_6
21	A_{18}	$\overline{BERR}$	D_7
22	A_{19}	Spare	D_8
23	A_{20}	Reserved	D_9
24	A_{21}	Reserved	D_{10}
25	A_{22}	Reserved	D_{11}
26	A_{23}	Reserved	D_{12}
27	E	Reserved	D_{13}
28	8 MHz	$\overline{EXTDTACK}$*	D_{14}
29	16 MHz	GND	D_{15}
30	GND	+12 V	GND
31	+12 V	+12 V	Spare
32	+12 V	−5 V	−12 V

*External $\overline{DTACK}$

**Processor memory cycle

| 7.7 | TROUBLESHOOTING TECHNIQUES |

The hardware operation of the 68000 is complex and requires great patience during design or troubleshooting sessions. Your basic knowledge of 68000 hardware architecture should include operation of all its signals and their functional groupings:

- System signals CLK, $\overline{\text{RESET}}$, $\overline{\text{HALT}}$, and $\overline{\text{BERR}}$.
- The signals involved with memory accesses. These are data lines D_0 through D_{15}, address lines A_1 through A_{23}, $\overline{\text{AS}}$, R/$\overline{\text{W}}$, $\overline{\text{UDS}}$, $\overline{\text{LDS}}$, and $\overline{\text{DTACK}}$.
- The signals involved with I/O (to peripherals). These are E, $\overline{\text{VMA}}$, and $\overline{\text{VPA}}$, and all memory signals except $\overline{\text{DTACK}}$.
- The signals involved with interrupts. These are $\overline{\text{IPL}}_0$ through $\overline{\text{IPL}}_2$, FC_0 through FC_2 (which indicate an interrupt acknowledge cycle), and A_1 through A_3 (which indicate the interrupt level being serviced).
- Bus arbitration signals $\overline{\text{BR}}$, $\overline{\text{BG}}$, and $\overline{\text{BGACK}}$.

The groupings are important because they point the way at the beginning of a design or while troubleshooting. Knowing the relationships between signals allows you to make informed decisions about what to do next, such as how to design a memory address decoder or determine why the $\overline{\text{HALT}}$ input stays low in a faulty system.

SUMMARY

In this chapter we examined the signal lines of the 68000, and saw how they were separated into eight functional groups. Those groups were power and clock, processor status, 6800 peripheral control, system control, interrupt control, bus arbitration control, asynchronous bus control, and finally the address and data buses. Some of the signals, such as $\overline{\text{RESET}}$ and $\overline{\text{HALT}}$, are bidirectional, while others are simply inputs (CLK, $\overline{\text{VPA}}$, $\overline{\text{BERR}}$) or outputs ($\overline{\text{BG}}$, $\overline{\text{VMA}}$, E). All signals are summarized in Table 7.3.

We also looked at two examples of the 68000's timing: one while it entered HALT mode and the other while it released its bus to an external bus master. We will examine timing waveforms again in Chapter 8. The chapter concluded with another look at the calculator project and a brief example of a bus connector.

STUDY QUESTIONS

1. Design a 4-MHz clock oscillator for the 68000's CLK input. Use an 8-MHz crystal oscillator and a flip-flop in your design.
2. Draw one complete cycle of the E clock waveform, assuming an 8-MHz CLK frequency.

3. Design a function code decoder using a three- to eight-line decoder. Make sure that $\overline{AS}$ is used to enable/disable the decoder.

4. Redesign the circuit of Figure 7.3 so that $\overline{AS}$ is used with an OR gate to generate $\overline{INTACK}$.

5. Design a circuit that will turn on an LED during the 68000's interrupt acknowledge cycle. The LED should be off at all other times.

6. Explain how $\overline{VPA}$, $\overline{VMA}$, and E might be used to control a non-6800-type peripheral (such as the Intel 8251 or 8255).

7. Design a circuit using a seven-segment display so that the display shows a lowercase *r* while the processor is running (segments e and g on), and a lowercase *h* while it is halted (segments c, e, f, and g on).

8. Design a circuit that will reset the processor after 16 CLK cycles, unless a word write to address C8000 is performed.

9. Design a circuit that uses a single-pole, single-throw switch to switch the processor from running to halted, and vice versa. *Note:* Do not directly ground the $\overline{HALT}$ line with the switch!

10. Design a circuit that will encode seven different interrupt signals ($\overline{INTA}$ through $\overline{INTG}$) into 3 binary bits that may be connected to $\overline{IPL_0}$ through $\overline{IPL_2}$. $\overline{INTA}$ is least significant.

11. Four 68000 microprocessors are wired together in a system with one address and data bus. Show how one 68000 can be wired as the master processor, with the other three having the capability of becoming bus masters if they wish.

12. What do you suppose happens to program execution with frequent, alternate bus master activity?

13. What is the state of signal lines $\overline{AS}$, $\overline{UDS}$, $\overline{LDS}$, and R/$\overline{W}$, when the 68000 is writing data from D_8 to D_{15}?

14. Repeat Question 13 if the processor is reading data from D_0 to D_7.

15. Repeat Question 13 if the processor is writing data from D_0 to D_{15}.

16. Repeat Question 13 if the processor is in a halt state.

17. What special function is performed by A_1 to A_3 during an interrupt acknowledge cycle?

18. What special function is performed by D_0 to D_7 during an interrupt acknowledge cycle?

19. Three signals must all have the capability of generating a level-7 interrupt. The signals are $L7_a$, $L7_b$, and $L7_c$, with $L7_a$ having the highest priority. Design a circuit that will generate a level-7 interrupt if any of the three signals go low. The circuit should also have outputs I_0 and I_1, which indicate which of the three L7 inputs caused the interrupt.

20. Examine the proposed single-step circuit of Figure 7.16. Will it work? If not, what changes need to be made? The processor should execute one instruction every time the button is pushed.

21. Sketch the timing diagram for the single-step circuit. Be sure to show expected activity on $\overline{HALT}$, $\overline{AS}$, and $\overline{DTACK}$.

22. Figure 7.17 shows a proposed $\overline{BERR}$ time-out circuit. How long is a bus cycle allowed to take?

23. Modify the design of Figure 7.17 so that the timer can be disabled by throwing an external switch.

FIGURE 7.16 For Question
7.20

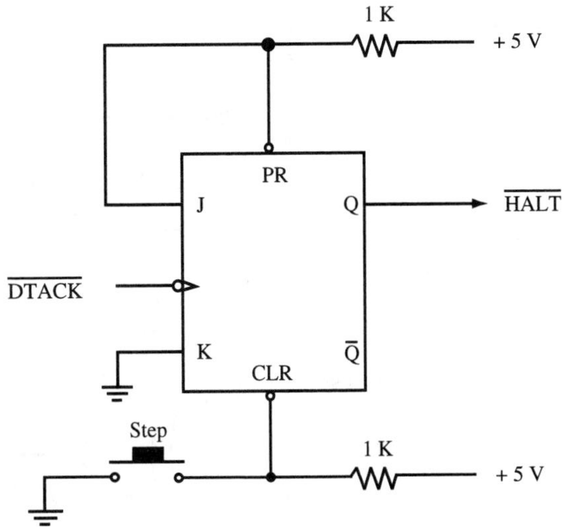

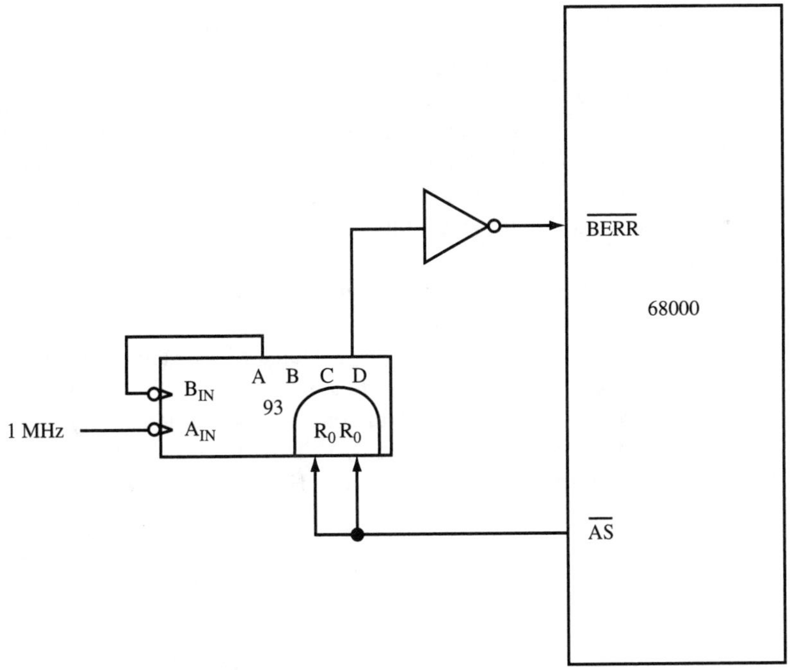

FIGURE 7.17 For Question 7.22

24. Which two interrupt levels are possible with the circuit of Figure 7.18?
25. What is the advantage of tristating the processor's signal lines during a bus request?
26. Explain how $\overline{BR}$ (and $\overline{BG}$ and $\overline{BGACK}$) can be used to single-step the 68000. What other uses can you find for $\overline{BR}$?

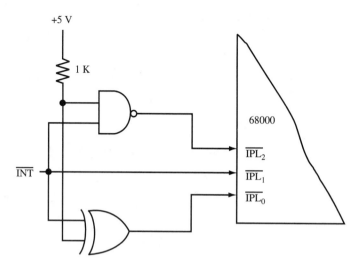

FIGURE 7.18   For Question 7.24

27. Explain how the user/supervisor data/program processor states, as indicated by the function code outputs, could be used to restrict accesses to certain blocks of memory.
28. Design a circuit that will generate a bus error ($\overline{BERR}$) whenever a user program reference or user data reference is made to any address above 80000.
29. The 68000 does not have an external A_0 address bit but does have an internal A_0 bit. Speculate on the state of the internal A_0 bit as the processor accesses even and odd locations in memory.
30. Use $\overline{LDS}$ and $\overline{UDS}$ to generate a synthetic A_0 bit outside the processor. Will A_0 be valid during word operations?
31. Draw a block diagram of the calculator hardware described in Section 7.5. Label all signals. Use an eight-digit LCD display and a 20-key keypad for the I/O sections.
32. Another standard bus used in 68000-based systems is the VME bus. Find some information on this bus and discuss its basic features.
33. Why is the $\overline{BGACK}$ signal used? How does it work?

CHAPTER 8

Memory System Design

OBJECTIVES

In this chapter you will learn about:

- The importance of bus buffering
- How the 68000 addresses (accesses) memory
- The design of custom memory address decoders
- The difference between full- and partial-address decoding
- How wait states may be inserted into memory read/write cycles
- The differences between static and dynamic RAMs
- How a dynamic RAM is addressed and what refresh cycles are for
- DMA (direct memory access)

8.1   INTRODUCTION

The internal memory capacity of any microprocessor, with the exception of single-chip microprocessors, is severely limited. The 68000 itself has only a handful of 32-bit locations in which it can store numbers, and these locations are the actual data and address registers available to the programmer. The need for larger, external memories quickly becomes apparent, especially if an application involves number crunching or word processing. The purpose of this chapter is to explore ways of adding external memory to 68000-based systems. We will examine how the 68000's various control signals ($\overline{AS}$, $\overline{DTACK}$, $\overline{UDS}$, $\overline{LDS}$, and R/$\overline{W}$) are used to supply memory read and write signals to read-only memories and both static and dynamic random access memories.

In addition we will see how an external device called a **bus master** takes over control of the 68000's memory system during a process called **direct memory access.**

The information provided in this chapter should enable you to design future memory systems from scratch.

Section 8.2 explains the 68000's address and data buses. The importance of bus buffering is discussed in Section 8.3. Section 8.4 shows how the 68000 accesses memory, Section 8.5 covers the design of a memory address decoder, and Section 8.6 introduces the partial-address decoder. This is followed by a discussion of actual RAMs and EPROMs in Section 8.7. Section 8.8 explores the use of a shift register to generate the $\overline{\text{DTACK}}$ signal. In Section 8.9, we show how dynamic RAM can be used with the 68000. Section 8.10 explains how DMA works. Some tips on testing address decoders are presented in Section 8.11.

8.2   THE 68000 ADDRESS AND DATA BUSES

The 68000 microprocessor has a 16-bit-wide data bus capable of reading or writing 8 and 16 bits of information at a time, and a 24-bit address bus that can address 16MB of external memory. Only 23 of these address lines, A_1 through A_{23}, are available for use. Address line A_0 is used *inside* the processor to control two other signals: $\overline{\text{UDS}}$ and $\overline{\text{LDS}}$. Byte and word memory transfers are controlled by these signals. The $\overline{\text{UDS}}$ and $\overline{\text{LDS}}$ address lines indicate which part of the CPU's data bus is transferring information. Table 8.1 details the operation of the $\overline{\text{UDS}}$ and $\overline{\text{LDS}}$ signals.

The levels of the $\overline{\text{UDS}}$ and $\overline{\text{LDS}}$ lines are controlled by the state of the 68000's *internal* A_0 bit. If A_0 is low during a byte operation, $\overline{\text{UDS}}$ is activated. If A_0 is high during a byte operation, $\overline{\text{LDS}}$ is asserted. Word addressing ignores the state of the internal A_0 bit; both $\overline{\text{UDS}}$ and $\overline{\text{LDS}}$ are made active.

When designing a memory system to be both byte and word addressable, good use is made of $\overline{\text{UDS}}$ and $\overline{\text{LDS}}$ to enable individual odd/even RAMs or EPROMs.

TABLE 8.1 $\overline{\text{UDS}}$ and $\overline{\text{LDS}}$ functions

$\overline{\text{UDS}}$	$\overline{\text{LDS}}$	$R/\overline{W}$	D_8–D_{15} *(even byte)*	D_0–D_7 *(odd byte)*
High	High	—	No valid data	No valid data
Low	Low	High	Valid data bits 8–15	Valid data bits 0–7
High	Low	High	No valid data	Valid data bits 0–7
Low	High	High	Valid data bits 8–15	No valid data
Low	Low	Low	Valid data bits 8–15	Valid data bits 0–7
High	Low	Low	Valid data bits 0–7	Valid data bits 0–7
Low	High	Low	Valid data bits 8–15	Valid data bits 8–15*

*These conditions are a result of current implementation and may not appear on future devices.

8.3 BUS BUFFERING

Every microprocessor-based memory system, whether EPROM or RAM, will have standard buses connecting it to the microprocessor, whose functions are to direct the flow of information to and from the memory system. These buses are generally called the **control** bus, the **data** bus, and the **address** bus. Figure 8.1 shows the relationship between the CPU, the buses, and the memory system. Note that the address bus is unidirectional, which means that data on the address bus goes one way, from the CPU to the memory system. The data and control buses, on the other hand, are bidirectional. Data may be written to or read from memory, hence the need for a bidirectional data bus. We will soon see why the control bus is also bidirectional.

Whether they are bidirectional or not, some care must be taken when the buses are connected to the memory section. It is possible to overload an address or data line by forcing it to drive too many loads. As always, it is important to *not* exceed the fanout of a digital output. If, for example, a certain output is capable of sinking 2 mA, how many 0.4 mA inputs can it drive? The answer is five, which we get by dividing the output sink current by the required input current. If more than five inputs are connected, the output is overloaded and its ability to function properly is diminished. Clearly, the possibility of overloading the 68000's address or data buses exists when they are connected to external memory. For this reason, we will **buffer** the address and data buses.

Figure 8.2 shows how the address lines are buffered by connecting them to a standard high-current buffer, the 74LS244 octal line driver/receiver. An address line on the 68000 is capable of sinking 3.2 mA all by itself. When the output of the 74LS244 is used instead, the address line has an effective sink current of 24 mA. This means that six times as many gates can be driven. Buffering the address lines allows the CPU to drive all the devices in our memory system, without the added worry of overloading the address line.

Buffering the data bus is a little trickier because the data bus is bidirectional. Data must now be buffered in both directions. Figure 8.3 shows how this bidirectional buffering is accomplished. The 74LS245 is an octal bus transceiver. Data flow through this device is controlled by the DIR signal, which tells the buffer to pass data from left to right, or from right to left. Left-to-right data is CPU output data. Right-to-left data is

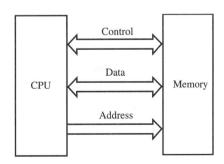

FIGURE 8.1 Memory bus structure

FIGURE 8.2 Address bus buffering

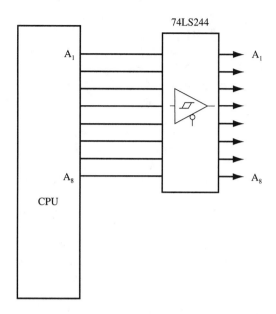

considered CPU input data. The natural choice for controlling the direction of the 74LS245 is the 68000's R/$\overline{\text{W}}$ line, which always indicates the direction of data on the 68000's data bus.

FIGURE 8.3 Data bus buffering

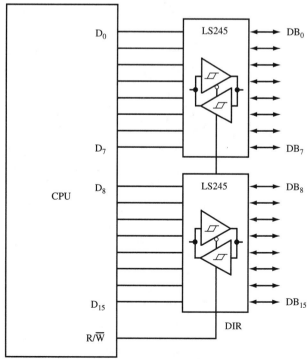

In conclusion, then, remember that address and data buses should be buffered so that many gates can be connected to them instead of the few that can be directly driven by the unbuffered address or data line. All designs presented in this chapter will assume that the buses are already well buffered.

8.4   ACCESSING MEMORY

In addition to well-buffered address and data buses, a control bus must also be used to control the operation of the memory circuitry. The three operations we have to consider are the following:

- Read data from memory
- Write data to memory
- Do not access memory

The first two cases represent data (bytes or words) that gets transferred between the 68000 and memory. The third case occurs when the 68000 is performing some other duty, internal instruction execution perhaps, and has no need for the memory section. Thus, it appears that the 68000 either accesses memory or does not access it. Does a processor signal exist that tells external circuitry the 68000 needs to use its memory? Yes, the $\overline{AS}$ output. A low on $\overline{AS}$ indicates that the 68000 wishes to access memory. A high shows that the 68000 is busy doing something else. When the memory circuitry sees a low on $\overline{AS}$, it will assume that the 68000 is outputting a valid memory address on the address bus. If the memory address falls inside the range of allowable memory addresses, the memory circuitry will tell the 68000 to proceed with the data transfer by pulling the $\overline{DTACK}$ line low. $\overline{DTACK}$ is used to synchronize the 68000 with memory devices. After issuing a valid memory address and $\overline{AS}$ signal, the 68000 will enter a wait loop until it sees a low on $\overline{DTACK}$. It is the responsibility of the memory circuitry to activate $\overline{DTACK}$. During normal operation, the 68000 will complete the data transfer shortly after receiving $\overline{DTACK}$.

Figure 8.4 shows the timing signals for a memory read cycle. The cycle is divided into a minimum of eight states, S0 through S7. Every two states equal one cycle of the system clock (therefore, the state times are equal to one-half the period of the clock).

During S0, the beginning of the read cycle, the address bus and the function code outputs are in a high-impedance state, and the processor's R/$\overline{W}$ line is set to a logic 1 (to indicate the read operation).

At the start of state S1, the function codes become valid (to indicate the type of bus cycle being performed), and the address bus leaves its high-impedance state to point to the desired memory location.

One-half clock cycle later, the beginning of state S2, the $\overline{AS}$, $\overline{UDS}$, and $\overline{LDS}$ signals become active (the last two depending on the type of read being performed, byte or word). At this point, the external hardware should be in the process of decoding the memory address and activating the selected memory component.

During states S3 and S4, no new signals are issued by the processor. It is during these states that the external hardware will most likely inform the processor to continue with the cycle. It will do this by placing data onto the processor's data bus and by asserting

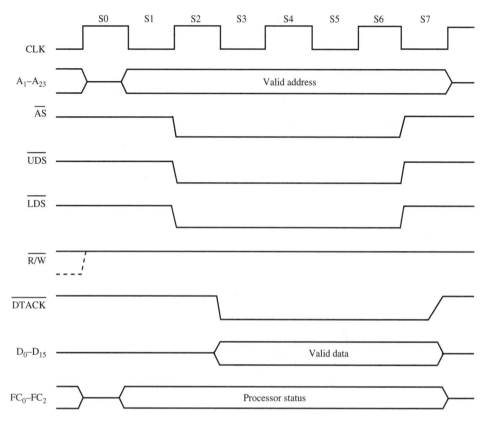

FIGURE 8.4   Memory read cycle timing

$\overline{\text{DTACK}}$. The processor will look for $\overline{\text{DTACK}}$ at the beginning of state S5. If $\overline{\text{DTACK}}$ is not asserted (into a logic 0) by this time, the processor will insert wait states into the read cycle, two states at a time, before trying to continue with state S5. Two wait states equal one CLK period. If fast RAM and EPROM are used, $\overline{\text{DTACK}}$ may be asserted in S5. Otherwise, it may be necessary to wait one or more CLK cycles to assert $\overline{\text{DTACK}}$.

If $\overline{\text{DTACK}}$ is asserted, the processor will continue through to state S7, where the data on the data bus becomes internally latched, and control signals $\overline{\text{AS}}$, $\overline{\text{UDS}}$, and $\overline{\text{LDS}}$ are negated. This should cause the external hardware to negate $\overline{\text{DTACK}}$ and also remove data from the data bus. It is necessary that this happen within one clock cycle after the completion of state S7 (somewhere in S0 or S1 at the latest), to ensure proper operation of the processor.

In Figure 8.5, the details of the processor's memory write cycle are shown. This cycle also requires eight states (S0 through S7). During state S0, the function code outputs and the address bus are tristated. At the beginning of state S1, these signals become active. At the start of state S2, the processor asserts $\overline{\text{AS}}$, and takes R/$\overline{\text{W}}$ low, to indicate a memory write cycle. Notice that the issuing of $\overline{\text{UDS}}$ and $\overline{\text{LDS}}$ do not occur at this time, nor is any data placed onto the processor's data bus. This has the double advantage of letting the external hardware prepare the memories for the write operation, as well as giving the bidirectional data bus drivers time to switch directions without bus conflicts.

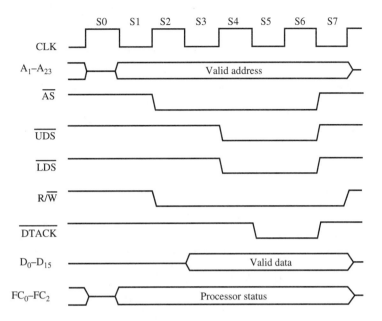

FIGURE 8.5 Memory write cycle timing

If $\overline{\text{DTACK}}$ is asserted by the time state S5 begins, the cycle will continue normally. If $\overline{\text{DTACK}}$ is not asserted, the processor will insert wait states until it is. Signals $\overline{\text{AS}}$, $\overline{\text{UDS}}$, and $\overline{\text{LDS}}$ are all negated at the start of S7, the last state in the cycle.

In both cases (read and write), the actual time between negations of $\overline{\text{AS}}$ and $\overline{\text{DTACK}}$ will vary, depending on the external hardware used to implement the design, but is not allowed to exceed the time of one clock period.

In the next section we will see how a memory address decoder uses the address bus and $\overline{\text{AS}}$ to enable RAM and EPROM memories.

8.5 DESIGNING A MEMORY ADDRESS DECODER

The sole function of a memory address decoder is to monitor the state of the address bus and determine when the memory chips should be enabled. But what is meant by *memory chips?* These are the actual RAMs or EPROMs the designer wants to use in the computer. So, before the designing begins, it must be decided how much memory is needed. If 8K words of EPROM is enough, then the designer knows that 13 address lines are needed to address a specific location inside the EPROM (because 2 raised to the 13th power is 8,192). How many address lines are needed to select a specific location in a 32K memory? The answer is 15, because 2 to the 15th power is 32,768. The first step in designing a memory address decoder is determining how many address lines are needed just for the memory device itself. Any address lines remaining are used in the address decoder.

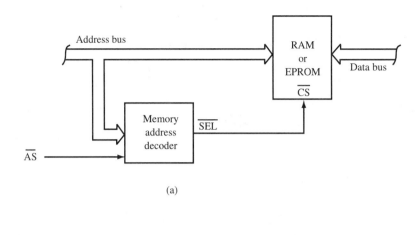

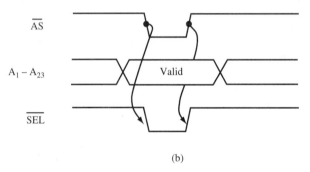

(b)

FIGURE 8.6   Simple memory address decoder: (a) block diagram and (b) timing

Figure 8.6(a) shows a block diagram of a memory address decoder connected to a memory chip. Figure 8.6(b) shows a simplified timing diagram representing the activity on the address bus and the $\overline{AS}$ output. The memory address decoder waits for a particular pattern on the address lines and a low on $\overline{AS}$ before making $\overline{SEL}$ low. When these conditions are satisfied, the low on $\overline{SEL}$ causes the $\overline{CS}$ (chip select) input on the memory chip to go low, which enables its internal circuitry, thus connecting the RAM or EPROM to the processor's data bus. When the address bus contains an address different from the one the address decoder expects to see, or if $\overline{AS}$ is high, the output of the decoder will remain high, disabling the memory chip and causing its internal buffers to tristate themselves. Thus, the RAM or EPROM is effectively disconnected from the data bus.

The challenge presented to us, the designers of the memory address decoder, is to chip-enable the memory device at the correct time. The following example illustrates the steps involved in the design of a memory address decoder.

Example 8.1:   A circuit containing 64K words (128KB) of RAM is to be interfaced to a 68000-based system, so that the first address of the RAM (also called the **base** address) is at 480000. What is the entire range of RAM addresses? How is the address bus used to enable the RAMs? What address lines should be used?

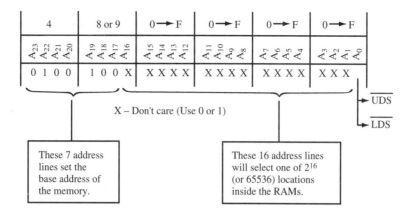

4	8 or 9	0 ➛ F	0 ➛ F	0 ➛ F	0 ➛ F
A_{23} A_{22} A_{21} A_{20}	A_{19} A_{18} A_{17} A_{16}	A_{15} A_{14} A_{13} A_{12}	A_{11} A_{10} A_9 A_8	A_7 A_6 A_5 A_4	A_3 A_2 A_1 A_0
0 1 0 0	1 0 0 X	X X X X	X X X X	X X X X	X X X

X – Don't care (Use 0 or 1)

➛ $\overline{UDS}$
➛ $\overline{LDS}$

These 7 address lines set the base address of the memory.

These 16 address lines will select one of 2^{16} (or 65536) locations inside the RAMs.

FIGURE 8.7   Memory address range decoding

Solution: Figure 8.7 shows how the memory lines are assigned. Two 64K RAM devices must be used in order to get the 16-bit-wide data bus required by the 68000. $\overline{UDS}$ and $\overline{LDS}$ will be used to select one of the devices, with each device containing 64K locations.

Since we are using 64K locations per device, we need 16 address lines to select one of 64K possible addresses. We always use the lowest-numbered address lines first (the least significant ones). We start with A_1 (because A_0 is not available), and use the next 15 just for the RAM. This means that A_1 through A_{16} go directly to the RAM circuitry, where they will be used to select a location inside the RAM. The remaining seven address lines, A_{17} through A_{23}, are used to *select* the specific 64K word bank located at address 480000.

To determine the entire range of addresses, first make all the "don't cares" (the Xs in Figure 8.7) zeros. That gives address 480000, the first address in the range of addresses. Next, make all "don't cares" high to generate the last address, which becomes 49FFFF.

Note that A_{22} and A_{19} are high when the 64K word RAM bank is being accessed, while the other 5 upper address bits are low. This particular pattern of 1s and 0s is one of 128 possible binary combinations that may occur on the upper 7 address bits. We need to detect a *single* pattern, so that the RAM circuit responds only to the address range 480000 to 49FFFF. The circuit of Figure 8.8 is one way to do this. The output of the eight-input NAND gate is the output of the memory address decoder, which in turn gets connected to the chip-enable inputs of the 64K word RAM bank. The only time the output of the NAND gate will go low is when *all* of its inputs are high. Since some of the upper address bits are low when the desired memory range is present on the address bus, they must be inverted before they reach the NAND gate. Even though A_{23}, A_{21}, A_{20}, A_{18}, and A_{17} are low, the NAND gate receives five 1s from them, via the inverters. Since A_{22} and A_{19} are already high, there are now seven 1s present on the input of the NAND gate. When $\overline{AS}$ goes low, indicating a valid memory address, the last required logic 1 is presented to the NAND gate, and its output goes low, enabling the 64K word RAM bank.

In general, a memory address decoder is used to reduce many inputs to a single output. The inputs are address lines and control signals. The single output is usually an enable signal sent to the memory section. Various TTL gates are used depending on the

FIGURE 8.8 Memory address decoder for 480000–49FFFF range

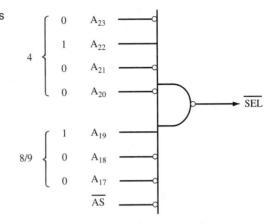

addressing requirements, and the following examples present only a few of the hundreds of ways we can design memory address decoders to suit our needs.

Example 8.2:   A 16K word EPROM section, with a starting address of 300000, is to be added to an existing memory system. The following circuitry will properly decode the entire address range, 300000 to 307FFF. Note that the address range indicates 32KB of memory. The eight-input NAND gate in Figure 8.9 is used to detect the 30 pattern on the upper 8 address bits, and the three-input OR gate is used to detect the last address bit (A_{15}) and $\overline{AS}$ signal. Nine address lines are used in this decoder, because the other 14 are needed to address one of 16K possible byte locations in each EPROM. Two EPROMs are needed for the same reason two RAMs were needed in Example 8.1, and these EPROMs are directly addressed by A_1 through A_{14}.

FIGURE 8.9 Memory address decoder for 300000–307FFF range

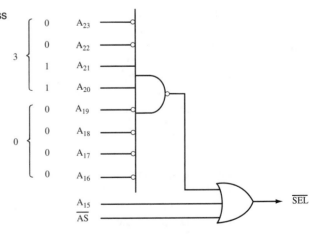

Example 8.3: Two 16K word memories, an EPROM with a starting address of 600000 and a RAM with a starting address of 700000, are needed for a new memory system. Figure 8.10(a) shows how the EPROM is enabled, and Figure 8.10(b) shows how the RAM section is enabled.

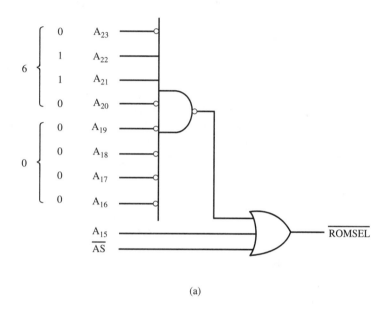

(a)

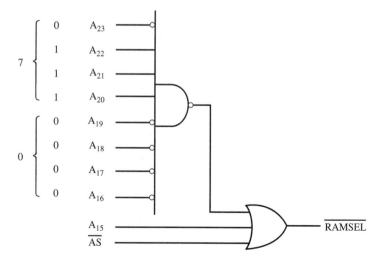

FIGURE 8.10 Memory address decoders for two different ranges: (a) EPROM bank at 600000 and (b) RAM bank at 700000

It is useful to perform a timing analysis on the address decoder to determine exactly how long it takes the output ($\overline{\text{ROMSEL}}$ or $\overline{\text{RAMSEL}}$) to become active once the address bus is stable. The time it takes the address decoder to activate its output is a function of the number of cascaded gate propagation delays in the circuit. In Figure 8.3, there is an initial delay of 15 ns due to the inverters at the inputs of the eight-input NAND gate. The NAND gate itself introduces another 20 ns of delay. Finally, the three-input OR gate adds an additional 15 ns. The total propagation delay of the circuit is 50 ns. This is how long it takes the output to become active once the address bus is stable. For a high-speed 68000 system it may be necessary to redesign the address decoder if the delay introduces wait states in the bus cycle.

The propagation delay times used here are worst-case times for typical 74LSxxx logic gates.

Experienced digital designers can detect binary patterns, and the reward in finding a pattern is generally a reduction in the digital circuitry needed to implement a desired function. Did you notice that the address ranges for the RAM and EPROM in the previous example are very similar? In fact, they are identical, except for the A_{20} address bit. Let us look at another example to see how we can use pattern detection to simplify the required hardware.

Example 8.4:   The EPROM and RAM sections of Example 8.3 are enabled by the simplified decoder presented in Figure 8.11. Do you see how the NAND gate is used to detect the 60 pattern, and how the A_{20} address line is used to enable the RAM *or* the EPROM?

In this case, we were able to eliminate one integrated circuit, an eight-input NAND gate. The next example shows how to design a decoder to respond to *eight* different address ranges.

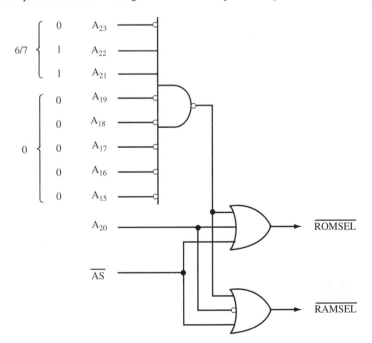

FIGURE 8.11   Combined RAM/EPROM address decoder

Example 8.5: A 256KB RAM memory is composed of sixteen 16KB RAMs. The address ranges for the RAMs are as follows:

- 00000 to 07FFF
- 08000 to 0FFFF
- 10000 to 17FFF
- 18000 to 1FFFF
- 20000 to 27FFF
- 28000 to 2FFFF
- 30000 to 37FFF
- 38000 to 3FFFF

How might all eight RAM pairs be selectively enabled by one device?

Solution: Our first thought may be to use eight individual memory address decoders, one for each address range and RAM pair. But this would be an unnecessary waste of circuitry. If we instead look for a pattern, we see that address lines A_{23} through A_{18} (the upper six address lines) are always low in the memory range 00000 to 3FFFF. In addition to this important piece of information, each RAM requires 14 address lines, A_1 through A_{14}, to select one of 16K locations within the RAM. This leaves us with address lines A_{15}, A_{16}, and A_{17} actually indicating a specific 32KB memory range. When these three address lines are all low, address range 00000 to 07FFF is selected. When A_{15} is high, and A_{16} and A_{17} are low, address range 08000 to 0FFFF is selected. The last range, 38000 to 3FFFF, is selected when A_{15}, A_{16}, and A_{17} are all high. What we need then is a circuit that can decode these eight possible conditions by using only the three address lines. Figure 8.12 shows the required circuitry.

FIGURE 8.12 Multibank address decoder

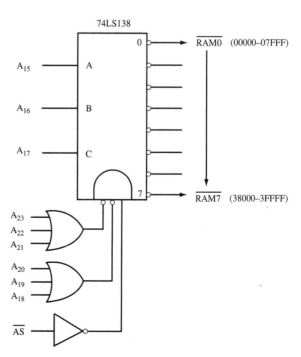

In this circuit, a 74LS138 three- to eight-line decoder is used to decode the different memory ranges. The 74LS138 has three select inputs and three control inputs. The select inputs are connected to address lines A_{15}, A_{16}, and A_{17}. The 3-bit binary number present on the select inputs will pull the selected output of the 74LS138 low (assuming that the 74LS138 is enabled), thus activating a specific RAM pair. To enable the 74LS138, two lows and a high must be placed on its control inputs. The two lows are generated by three-input OR gates, whose outputs are low only when the upper six address lines are low. The $\overline{AS}$ signal is inverted to generate the last control input.

By using special integrated circuits like the 74LS138 and a simple pattern recognition technique, we are able to greatly simplify the hardware required to generate all of our memory enables.

The last four examples have shown how we can decode a specific range of memory addresses using the full address bus of the 68000. In the next section, we will see how to further simplify our decoder, by using a technique called **partial-address decoding.**

8.6   PARTIAL-ADDRESS DECODING

Although the 68000 is capable of addressing over 16 million bytes of memory, it would be safe to assume that most applications would require much smaller memories. A good example might be an educational 68000 single-board computer, much like the one presented in Chapter 11, which requires only 2K words of EPROM and 2K words of RAM. This type of system needs only 12 address lines. The first 11, A_1 through A_{11}, go directly to the EPROM and RAM, and the last address line, A_{12}, is used to select either the EPROM or the RAM. Figure 8.13 details this example system.

FIGURE 8.13   Partial-address decoding for RAM/EPROM

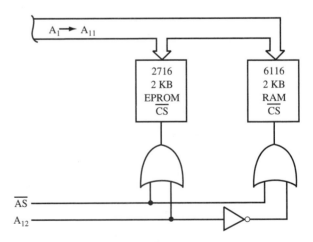

In this figure, A_{12} is connected directly to the OR gate driving the $\overline{CS}$ input of the EPROM, and is *inverted* before it gets to the $\overline{CS}$ input of the RAM. So, whenever A_{12} is low, the EPROM is enabled, and whenever A_{12} is high, the RAM is enabled. We only have to use an inverter to do all the decoding in our memory section!

But what about the other address lines, A_{13} through A_{23}? They are ignored, and here is why: When the 68000 is powered up, a reset exception is generated, and the processor looks first at memory location 000000. The 68000 is looking at memory location 000000 to get part of its initial supervisor stack pointer. We had better make sure good data is in that location at power-on. If we use a RAM with a starting memory address of 000000, its contents are random at power-on, and we have only a 1-in-256 chance of seeing the correct data in any location. The chances of fetching a correct long word is 1 in 4,294,967,296! If we instead use an EPROM at 000000, we can be assured that the correct information will be present.

Going back to Figure 8.13, it is clear that the EPROM will be enabled at power-on, since A_{12} will be low when the processor tries to access location 000000.

But we still do not know why the other 11 upper address lines, A_{13} through A_{23} in this case, can be ignored. The answer lies in Figure 8.13. Do you see any address lines other than A_{12} being used to enable or disable the EPROM or RAM memories? No! Since we ignore these address lines, it does not matter whether they are high or low. In this fashion we can read from memory locations 000000, 006000, 3E7000, or FF2000 and get the same data each time. The upper address bits have no effect on our memory circuitry, since we are only using the lower 12 address lines.

Partial-address decoding gives us a way to get the job done with a minimum of hardware. Since fewer address lines have to be decoded, less hardware is needed. This is its greatest advantage. A major disadvantage is that future expansion of memory is difficult, and usually requires a redesign of the memory address decoder. This may turn out to be a difficult, or even impossible, job. The difficulty lies in having to add hardware to the system. If the system has been manufactured by some company, and distributed to a number of users, making changes to all systems becomes a challenge. Furthermore, individuals wanting to make changes themselves may mistakenly place a new memory device into a partially decoded area. This will unfortunately result in two memories being accessed at the same time, probably resulting in invalid data during reads.

As long as these dangers and limitations are understood, partial-address decoding is a suitable compromise and acceptable in small systems.

Two more examples are presented to further show the simplicity of partial-address decoding.

Example 8.6: Two 8KB EPROMs are to have a starting address of 4000. Figure 8.14(a) shows the required partial-address decoding circuitry needed. Whenever A_{14} is high and $\overline{AS}$ is low, the output of the NAND gate will be low, which in turn will enable the EPROM pair. Address lines A_1 through A_{13} are connected directly to the EPROMs, together with $\overline{UDS}$ and $\overline{LDS}$. Figure 8.14(b) shows how the two EPROMs are connected to the system. It is necessary to always implement two EPROMs at a time, to allow for the 68000's word size instruction fetches.

FIGURE 8.14 8K word EPROM storage: (a) partial-address decoder and (b) EPROM circuitry

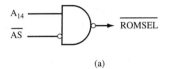

(a)

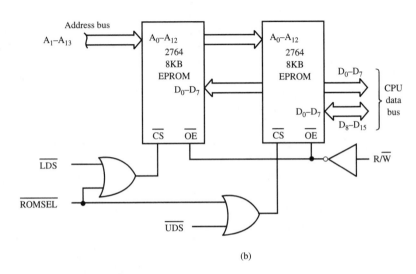

(b)

Example 8.7: A 32KB EPROM pair needs a starting address of 000000, and a 32KB RAM pair needs a starting address of FF8000. The circuitry in Figure 8.15 shows how these addresses are partially decoded. In this example, A_{23} is used to select the EPROM when low, and the RAM when high. This allows absolute short addressing to be used. Address lines A_1 through A_{14}, and $\overline{LDS}$ and $\overline{UDS}$ are used to read/write any of 32KB locations.

FIGURE 8.15 Partial-address decoding for 32KB RAM/EPROM pair

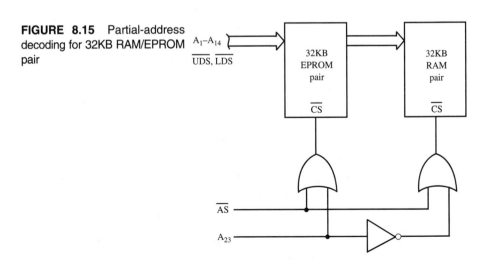

Remember that partial-address decoding has both advantages and disadvantages. Small, dedicated systems are better suited for this design technique than large, general purpose systems.

8.7   DESIGNING WITH AVAILABLE MEMORIES

In the previous examples, the emphasis was on the design of the address decoding logic. Since the address decoder in turn enables the selected memory chips, we need to know what kinds of memories are available.

Since no 68000 system is complete without EPROM, we will examine these memories first. Table 8.2 lists some typical EPROMs. Each EPROM stores 8 bits of data in each location. The number of address lines required to access the internal EPROM locations are also shown. Since all addresses on the address bus are binary numbers, we always end up with a memory size that is a multiple of 2. This helps explain why the size of the EPROMs double each time we add an address line.

How much EPROM should a 68000-based system contain? This depends on the application. The SBC system of Chapter 11 requires only 2K words (4KB) of EPROM to support a useful monitor program. This means two 2716s can be used (one for each half of the 16-bit data bus). A 68000-based communication server might require more EPROM, say 512KB. If the desired EPROM size is not available, multiple EPROMs are used to obtain the required amount of locations. The next example shows how this is done.

Example 8.8:   The communication server just mentioned requires 512KB of EPROM with a base address of 000000. How many 27512 EPROMs are needed? What does the memory circuitry look like?

Solution: Since each 27512 EPROM stores 64KB of data, we will need eight EPROMs to make up the 512KB memory. Figure 8.16(a) shows how the 64KB EPROMs are mapped into the address bus. Each pair of 27512s makes up 64K words of memory (or 128KB). The eight EPROMs are split up into four pairs. Each pair is accessed by a unique 2-bit binary pattern on address lines A_{17} and A_{18}. As Figure 8.16(b) shows, one half of a

TABLE 8.2 Typical EPROM sizes		
Part	Size	Address lines
2716	2KB	11
2732	4KB	12
2764	8KB	13
27128	16KB	14
27256	32KB	15
27512	64KB	16

74LS139 decoder is used to decode the 2-bit pattern and select one pair of 27512s. The 139 is enabled when address lines A_{23} through A_{19} are low, $\overline{AS}$ is low, and $R/\overline{W}$ is high.

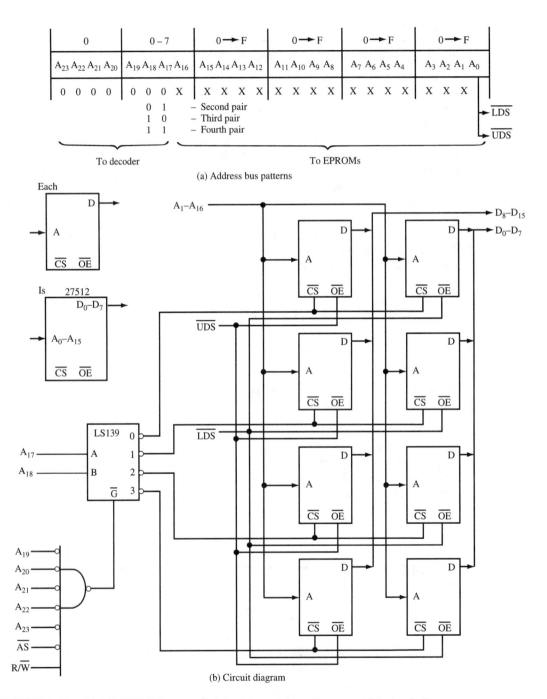

(a) Address bus patterns

(b) Circuit diagram

FIGURE 8.16   512KB EPROM memory: (a) address bus patterns and (b) circuit diagram

The range of addresses for each pair of 27512s is as follows:

000000 to 01FFFF	First pair
020000 to 03FFFF	Second pair
040000 to 05FFFF	Third pair
060000 to 07FFFF	Fourth pair

Systems that require a small amount of RAM might use one of the static RAMs shown in Table 8.3. The SBC of Chapter 11 uses two 6116s to make up a 2K word memory. This is plenty of RAM for small educational test programs. The communication server mentioned previously might require 16KB of RAM per channel, for 32 channels of data. Clearly, we must resort to using multiple RAM chips in some designs to make up the required memory.

Personal computers require even larger amounts of RAM, usually in multiples of 2 or 4MB. It is not economical to design a 4MB memory out of 32KB RAM chips, so an entirely different kind of memory is used. This memory, called **dynamic RAM,** comes in large sizes, as you can see in Table 8.4. The dynamic RAMs are soldered onto a small plug-in memory card called a SIMM (for Single Inline Memory Module) and inserted into a socket on the motherboard. Another common package is called a SIPP, for Single Inline Pin Package, which is also available in megabyte sizes. SIMMs have edge connectors, and SIPPs have actual metal pins for making contact.

The larger the RAM memory, the simpler the address decoder. This is because there are very few upper address lines left over after the lower ones have been used for the SIMM or SIPP itself. For example, a 1MB SIMM requires 20 address lines just for itself. This means that only A_{23}, A_{22}, and A_{21} can be used in the address decoder. Let us look at a detailed example.

TABLE 8.3 Typical static RAM sizes

Part	Size	Address lines
6116	2KB	11
6264	8KB	13
62256	32KB	15

TABLE 8.4 Typical SIMM sizes

Part	Size	Address lines
41256	256KB	18
421000	1MBit	20
9400	4MB	22

Example 8.9:   A small workstation requires 4MB of RAM for color-graphic screen memory. The base address of the RAM is 800000. What does the memory circuitry look like if 1MB SIMMs are used?

Solution: Figure 8.17 shows the 4MB memory system design. Four 1MB SIMMs are used as two pairs. Each pair represents a 2MB block of RAM. Address lines A_1 through A_{20} are used, through an address multiplexer, to access locations within each 1MB SIMM.

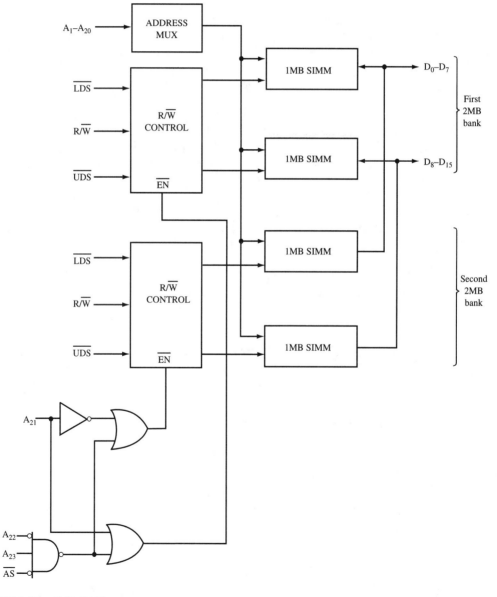

FIGURE 8.17   4MB RAM memory

The address multiplexer sends two groups of 10 bits each (A_1 through A_{10} first, A_{11} through A_{20} second) to the SIMMs to access any location. This is required by the architecture of the dynamic RAM chips.

The remaining address lines, together with $\overline{\text{AS}}$, are used to enable the individual R/$\overline{\text{W}}$ control blocks for each pair. The corresponding address ranges are:

800000 to 9FFFFF	First 2MB pair
A00000 to BFFFFF	Second 2MB pair

So, we rarely can design any large type of memory system without the need to combine a group of memory chips together.

In the next section we will see how the speed of the memories influences the design of the memory system.

8.8   GENERATING $\overline{\text{DTACK}}$

$\overline{\text{DTACK}}$ (data transfer acknowledge) is a signal that tells the 68000 CPU that data may be read from or written into memory. We have ignored this signal so far, so that we could develop an understanding of how memory address decoders work. The first function of the memory address decoder is to monitor the address bus and activate the RAMs or EPROMs when a specific address, or range of addresses, is seen. The second function of the decoder is to tell the CPU to *wait,* if necessary, until the memories have been given enough time to become completely active. A typical RAM might require 100 ns to become active after it gets enabled. This is due to the time required by the internal RAM circuitry to correctly decode the supplied address and turn on its internal buffers. This 100 ns access time must fit within the time frame of the memory read or write cycle, or else problems such as data loss might arise. The $\overline{\text{DTACK}}$ signal gives us a way to synchronize the 68000 with the memory access time.

In Figure 8.18 we see that the output of the address decoder is connected to the memory device and to a delay circuit. The delay circuit generates the required delay (if any) and asserts $\overline{\text{DTACK}}$ at the appropriate time. $\overline{\text{AS}}$ is connected to the delay circuit to guarantee that $\overline{\text{DTACK}}$ goes high when $\overline{\text{AS}}$ goes high (at the end of the bus cycle). The R/$\overline{\text{W}}$ signal is connected to the delay circuit because read cycles and write cycles do not usually take the same amount of time, and the delay time (if any) may need to be adjusted depending on the type of bus cycle.

How do we determine if no delay is needed? This can only be determined if we know the following:

- The state time of the processor
- The propagation time of the address decoder
- The access time of the memory

The state time of the processor is the amount of time the CLK signal is low or high. Thus, one-half of the CLK period equals the state time. The period of the CLK signal is the inverse of its frequency. So, a 68000 running at 8 MHz has a period of $1 \div 8$ MHz, or 125 ns. This gives a state time of 62.5 ns.

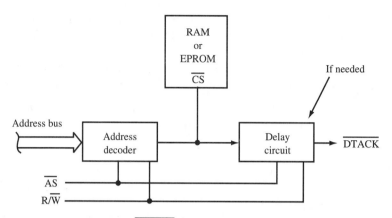

FIGURE 8.18   Block diagram of $\overline{\text{DTACK}}$ circuit

The propagation time of the address decoder equals the sum of the propagation times of the individual gates used in the decoder. If the decoder consists of a few inverters connected to a multi-input NAND gate, the propagation time of the decoder will be around 35 ns. The more circuitry used in the decoder, the longer its propagation time will be.

Typical EPROM access times are in the 150 to 200 ns range. Static RAM typically requires at least 100 ns, and dynamic RAM around 70 ns. Now, during a memory read cycle, $\overline{\text{AS}}$ does not become active until state S2 (refer to Figure 8.4). Since data will be read by the processor at the end of state S6, this means that the decoder and the memory device must complete their operation within five state times. With a state time of 62.5 ns, this allows the memory system up to 312.5 ns for a read operation. With 35 ns for the decoder, and another 150 ns for a fast EPROM, we only require 185 ns for the read operation to complete after state S2. This is well within the 312.5 ns timeframe, and thus $\overline{\text{DTACK}}$ may be asserted as soon as possible (during state S3).

On the other hand, suppose that an older EPROM must be used, whose access time is 350 ns. In this case, data will not be available for 385 ns, which is longer than the 312.5 ns window allowed. In this case, we must *delay* the generation of $\overline{\text{DTACK}}$, so that it is *not* active during state S5. This will force the 68000 to insert a complete CLK cycle (two state times) into the memory read cycle, and give us the extra 72.5 ns needed.

How do we delay the generation of $\overline{\text{DTACK}}$? A very simple solution is to use a shift register. Consider the 2-bit shift register of Figure 8.19. It will take two clock pulses for information at the first D input ($\overline{\text{SEL}}$) to get to the second Q output ($\overline{\text{DTACK}}$). So, if $\overline{\text{SEL}}$ goes low, we expect $\overline{\text{DTACK}}$ to go low two clock pulses later. If an 8-MHz clock is used, the flip-flops are clocked every 125 ns, which results in $\overline{\text{DTACK}}$ being asserted 250 ns after $\overline{\text{AS}}$! This causes $\overline{\text{DTACK}}$ to be asserted *after* state S5, thus extending the bus cycle.

But we are not finished yet. Remember that we want $\overline{\text{DTACK}}$ to go high when $\overline{\text{AS}}$ goes high. Since $\overline{\text{AS}}$ controls the address decoder, and hence the $\overline{\text{SEL}}$ signal, $\overline{\text{SEL}}$ will go high when $\overline{\text{AS}}$ goes high, but $\overline{\text{DTACK}}$ will not go high for another 250 ns. Since we want $\overline{\text{DTACK}}$ to go high when $\overline{\text{AS}}$ does, we must modify our shift register.

In Figure 8.20, we see that connecting $\overline{\text{AS}}$ to the flip-flops through an inverter presets the flip-flops whenever $\overline{\text{AS}}$ is high. The high on $\overline{\text{AS}}$ causes a low on the preset inputs,

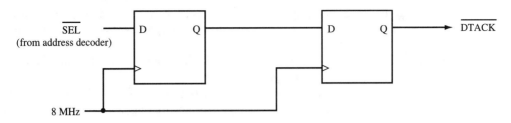

FIGURE 8.19 $\overline{\text{DTACK}}$ delay generator composed of 2-bit shift register

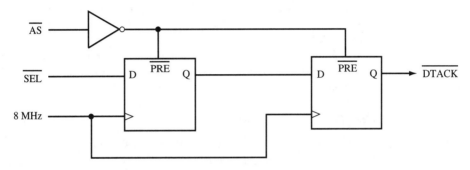

FIGURE 8.20 Improved $\overline{\text{DTACK}}$ delay generator

which force the Q outputs, and thus $\overline{\text{DTACK}}$, to be high. When $\overline{\text{AS}}$ goes low at the start of a memory reference, the preset inputs are pulled high and data may be clocked through the shift register. The instant $\overline{\text{AS}}$ goes high (indicating the completion of the memory cycle), the flip-flops will be preset, causing $\overline{\text{DTACK}}$ to go high.

So we have a circuit that will assert $\overline{\text{DTACK}}$ at the appropriate time during a memory cycle.

8.9 DYNAMIC RAM INTERFACING

What Is Dynamic RAM?

Dynamic RAM is a special type of RAM memory that is currently the most popular form of memory used in large memory systems for microprocessors. It is important to discuss a few of the specific differences between static RAMs and dynamic RAMs. Static RAMs use digital flip-flops to store the required binary information, whereas dynamic RAMs use MOS capacitors. Because of the capacitive nature of the storage element, dynamic RAMs require less space per chip, per bit, and thus have larger densities.

In addition, static RAMs draw more power per bit. Dynamic RAMs employ MOS capacitors that retain their charges (stored information) for short periods of time, whereas static RAMs must saturate transistors within the flip-flop to retain the stored binary information, and saturated transistors dissipate maximum power.

A disadvantage of the dynamic RAM stems from the usage of the MOS capacitor as the storage element. Left alone, the capacitor will eventually discharge, thus losing the stored binary information. For this reason the dynamic RAM must be constantly **refreshed** to avoid data loss. During a refresh operation, all of the capacitors within the dynamic RAM (called DRAM from now on) are recharged.

This leads to a second disadvantage. The refresh operation takes time to complete, and the DRAM is unavailable for use by the processor during this time.

Older DRAMs required that all storage elements inside the chip were refreshed every 2 ms. Newer DRAMs have an extended 4-ms refresh time, but the overall refresh operation ties up an average of 3 percent of the total available DRAM time, which implies that the CPU only has access to the DRAM 97 percent of the time. Since static RAMs require no refresh, they are available to the CPU 100 percent of the time, a slight improvement over DRAMs.

In summary, we have static RAMs that are fast, require no refresh, and have low bit densities. DRAMs are slower and require extra logic for refresh and other timing controls, but are cheaper, consume less power, and have very large bit densities.

Accessing Dynamic RAM

A major difference in the usage of DRAMs lies in the way in which the DRAM is addressed. A 1MBit DRAM requires 20 address bits to select one of 1,048,576 possible locations, but its circuitry contains only 10 address lines. A study of Figure 8.21 will show how these 10 address lines are expanded into 20 address lines with the help of two additional control lines: $\overline{RAS}$ and $\overline{CAS}$.

The 10 address lines are presented to row and column address buffers, and latched accordingly by the application of the $\overline{RAS}$ and $\overline{CAS}$ signals. To load a 20-bit address into the DRAM, 10 bits of the address are first latched by pulling $\overline{RAS}$ low.

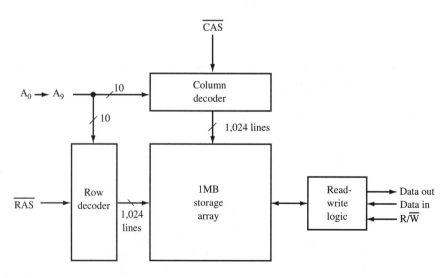

FIGURE 8.21   Internal block diagram of a 1MBit dynamic RAM

Then the other 10 address bits are presented to A_0 through A_9, and $\overline{CAS}$ is pulled low. By adding just one more address line to the DRAM, the addressing capability is increased by a factor of 4, since one extra address line signifies an extra row and column address bit. This explains why DRAMs tend to quadruple in size with each new release.

The actual method for addressing the DRAM is presented in Figure 8.22. First the 10 row address bits are applied to A_0 through A_9, and $\overline{RAS}$ is pulled low. Then A_0 through A_9 receive column address information, and $\overline{CAS}$ is pulled low. After a short delay, the circuitry inside the DRAM will have decoded the full 20-bit address, and reading or writing may commence.

External logic is needed to generate the $\overline{RAS}$ and $\overline{CAS}$ signals, and also take care of presenting the right address bits to the DRAMs. The circuit of Figure 8.23 shows an example of the required logic.

The operation of this circuit is as follows: The address decoder monitors the address bus for an address in the desired DRAM range, and outputs a logic 0 when it sees one. Normally the three Q outputs of the shift register are all high. The first clock pulse will shift the logic 0 from the address decoder to the output of the first flip-flop, causing $\overline{RAS}$ to go low. Since the output of the second flip-flop is still high, the 74LS157s (quad two-line to one-line multiplexers) are told to pass processor address lines A_1 through A_{10}. This is how we load the ROW address bits into the DRAM.

The second clock pulse will shift the logic 0 to the second Q output (the first is still low also), which causes the 74LS157s to select the processor address lines A_{11} through A_{20}. These address bits are recognized and latched by the DRAM when the third clock pulse occurs, because the logic 0 has been shifted to the third Q output, which causes $\overline{CAS}$ to go low. The DRAM has been loaded with a full 20-bit address, and reading or writing may commence. At the end of the read or write cycle, $\overline{AS}$ will go high, presetting all three flip-flops via the preset line, and the shift register reverts back to its original state.

This sequence will repeat every time the address decoder detects a valid address.

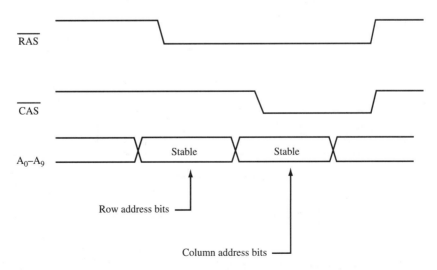

FIGURE 8.22 DRAM cycle timing

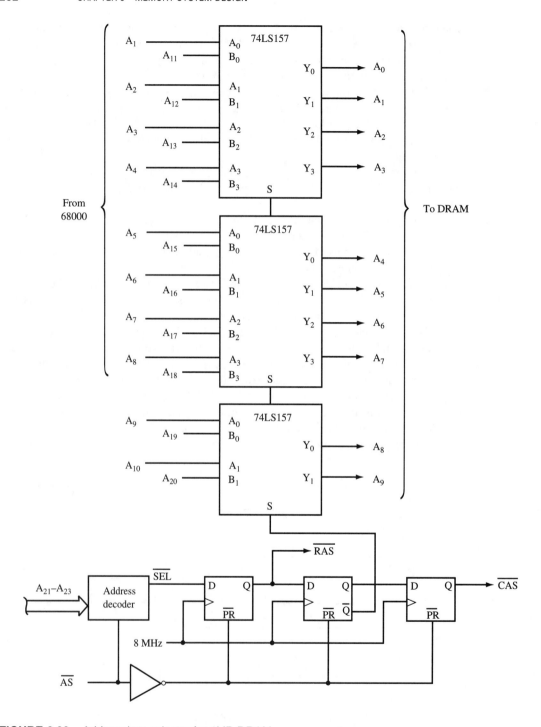

FIGURE 8.23   Address bus selector for 1MB DRAM

Refereshing Dynamic RAM

Previously we learned that DRAMs need to be refreshed, or the MOS capacitors that retain the binary information will discharge and data will be lost. Older DRAMs required that all cells (storage elements) be refreshed within 2 ms. Although the process of reading or writing a DRAM cell is a form of refresh, it is possible that entire banks of DRAM remain inactive while the CPU addresses other memories or I/O devices, so a safe designer will include a refresh circuit in the new DRAM system.

Newer DRAMs (such as the MCM6664) contain a single control line called $\overline{REF}$ that automatically refreshes the DRAM whenever it is pulled low. We will instead look at the process that is used to refresh a DRAM and the circuitry needed to control the process.

DRAMs are internally designed as a grid of memory cells arranged as a matrix, with an equal number of rows and columns (hence the $\overline{RAS}$ and $\overline{CAS}$ control signals). A 1MBit DRAM would need 20 address lines: 10 for the row decoder and 10 for the column decoder. Each decoder would pick one row and column out of a possible 1,024. During a refresh operation, all 1,024 column cells would be refreshed by the application of a single $\overline{RAS}$ signal; this is called **RAS-only refresh.** To refresh all locations, it is only necessary to $\overline{RAS}$ select all 64 rows. The easiest way to ensure that all rows get selected during a refresh operation is to use a binary counter and connect the output of the counter to the DRAM address lines during a refresh. To ensure that the DRAMs get refreshed periodically, a timer is needed to generate a $\overline{REFRESH}$ signal. The $\overline{REFRESH}$ signal will suspend processor activity while the DRAM is refreshed. Figure 8.24 shows how a 555 timer can be used to generate a $\overline{REFRESH}$ signal every 100 μs. The 555 timer clocks a D-type flip-flop, whose output is $\overline{REFRESH}$. When the refresh cycle is completed, $\overline{DONE}$ is used to preset the flip-flop and remove the $\overline{REFRESH}$ request, until the 555 times out again. Figure 8.25 shows how the refresh timer, together with the $\overline{RAS}$ refresh circuitry, is used

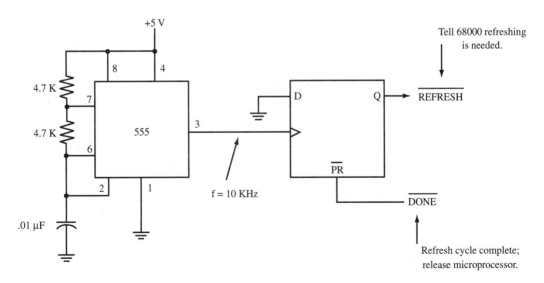

FIGURE 8.24 555 timer generates refresh signals every 100 μs

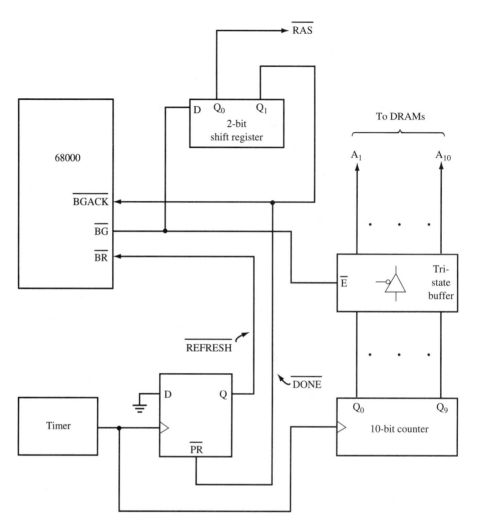

FIGURE 8.25 DRAM refresh generator

to refresh the DRAMs. When the 555 timer initiates a refresh cycle, $\overline{\text{REFRESH}}$ will go low, issuing a $\overline{\text{BR}}$ (bus request) to the 68000. The processor will respond by asserting $\overline{\text{BG}}$, which allows a 0 to be clocked into the 2-bit shift register used to control $\overline{\text{RAS}}$ and $\overline{\text{DONE}}$. When $\overline{\text{RAS}}$ is active, bits A_1 through A_{10} of the address bus will contain the 10-bit counter value. When $\overline{\text{DONE}}$ is active, the refresh flip-flop is preset, which removes the $\overline{\text{BR}}$ signal. This causes the processor to release $\overline{\text{BG}}$, which in turn causes the 2-bit shift register to be loaded with 1s. At this point, the bus request is over, and the processor resumes execution. Since the 555 timer also clocks the 10-bit counter, a unique row address is generated each refresh cycle.

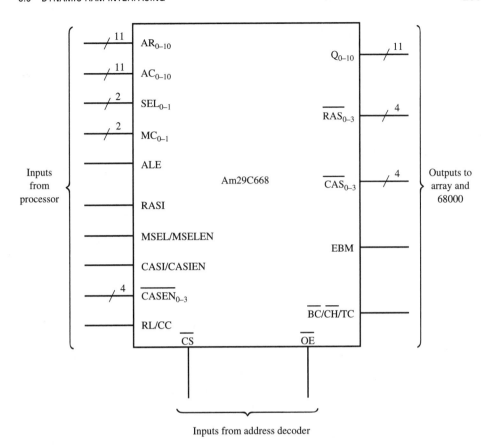

FIGURE 8.26 Am29C668 dynamic RAM controller

A Dynamic RAM Controller

We may conclude that the circuitry required to address, control, and refresh DRAMs is both complicated and extensive (which may translate into *expensive*). There must be a simpler way. There is!

Various companies make DRAM controller devices that take care of all refreshing and timing requirements needed by the dynamic RAMs. Most will support 64KB, 256KB, 1MB, and 4MB dynamic RAMs. All circuitry is contained in a single package in most cases. The DRAM controller does its work independently of the processor. This means that the DRAM controller will issue a wait to the processor when the processor tries to access memory during a refresh cycle. Figure 8.26 shows the signals associated with the Am29C668 DRAM controller. This controller is con-configurable and will handle DRAMs as large as 4MB! The input and output signals are described as follows:

Input Signals

AR_{0-10}	Row Address (from 68000)
AC_{0-10}	Column Address (from 68000)
SEL_{0-1}	Bank Select
MC_{0-1}	Mode Control
ALE	Address Latch Enable
RASI	Row Address Strobe Input
MSEL/MSELEN	Multiplexer Select/Multiplexer Select Enable
CASI/CASIEN	Column Address Strobe/Column Address Strobe Enable
$\overline{CASEN}_{0-3}$	Column Address Strobe Enable
RL/CC	Register Load/Column Clock
$\overline{CS}$	Chip Select
$\overline{OE}$	Output Enable

Output Signals

Q_{0-10}	Address (to DRAM)
$\overline{RAS}_{0-3}$	Row Address Strobe
$\overline{CAS}_{0-3}$	Column Address Strobe
EBM	End Burst/Block Mode
$\overline{BC}/\overline{CH}/TC$	Bank Compare/Cache Hit/Terminal Count

The MC inputs select the mode of operation (read/write, refresh, reset). ALE is used to latch a valid address into the controller. The other input signals are used in many different ways to control how the DRAMs are accessed.

The operation of the Am29C668 is controlled by data written into various internal control registers. These registers are the configuration register, the mask register, and the burst count register. These registers are loaded with data after power-on to configure the controller for the particular memory system being used. Figure 8.27 shows a simplified block diagram of the Am29C668 being used to control a 2MB array of DRAM. The array is composed of eight 256KB DRAMs arranged as four 512KB banks. A minimal amount of addressing logic is needed to chip-select the Am29C668. The base address of the DRAM array is E00000, as decoded by the NAND gate.

Using the Am29C668 (or some other controller) in a design greatly reduces the logic required to support dynamic RAM.

Dynamic RAM Summary

Our study of DRAMs has shown that they require complicated circuitry to get them to work (unless a DRAM controller is used). On the other hand, DRAMs are cheaper, per bit, than static RAM, they consume less power, and have much larger bit densities. With the advance of the microcomputer into the word processing arena, where very

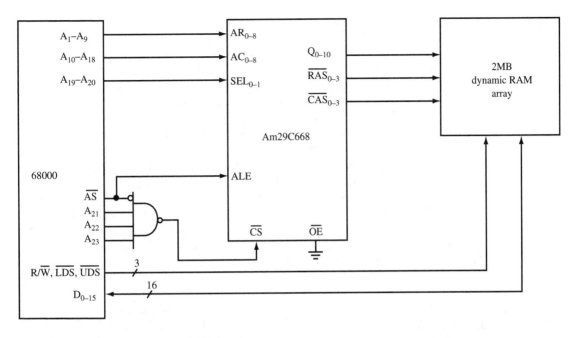

FIGURE 8.27 Simplified Am29C668 interface

large memories are needed to store and manipulate text, dynamic RAM becomes a very economical solution. Image processing, large informational databases, and virtually any large storage system is an ideal choice for the use of dynamic RAMs. Furthermore, interfacing dynamic RAMs is made easier with the use of a DRAM controller.

8.10   DIRECT MEMORY ACCESS

Direct memory access, usually called DMA for short, is a process in which a device external to the CPU requests the use of the CPU's buses (address bus, data bus, and control bus) for its own use. Examples of external circuits that might wish to perform DMA are video pattern generators, which share video RAM with the CPU, and high-speed data transfer circuits such as those used in hard disks.

In general, a DMA process consists of a **slave** device requesting the use of the **master's** buses. In a microprocessor-based system, the master is usually the CPU. Once the slave device has control of the bus, it can read or write to the system memory as necessary. When the slave device is finished, it releases control of the master's buses, and system operation returns to normal.

An example of why DMA is a useful technique can be illustrated in the following way: Suppose that you wish to add a hard disk storage unit to your microcomputer. The hard disk boasts a data transfer rate of 5 million bytes per second. This comes to 1 byte transferred every 200 ns! Many microprocessors would be hard put to execute even *one* instruction in 200 ns, much less the multiple number of instructions that would be required to read the byte from the hard disk, place it in memory, increment a memory pointer, and then test for another byte to read. A DMA controller would be very handy in this example. It would merely take over the CPU's buses, write all the bytes into memory very quickly, and then return control to the CPU.

To perform DMA on the 68000, three signals must be used. They are the $\overline{BR}$ (bus request), $\overline{BG}$ (bus grant), and $\overline{BGACK}$ (bus grant acknowledge) signals. The flowchart in Figure 8.28 shows how an external device requests the 68000's bus, gets it, and finally gives it back. Figure 8.29 shows the typical bus timing during a DMA operation. It is important to note that the device performing the DMA is responsible for maintaining the DRAM refresh requirements, either by performing them itself, or by allowing them to happen normally with existing circuitry.

FIGURE 8.28 Bus arbitration cycle flowchart (Reprinted with permission of Motorola, Inc.)

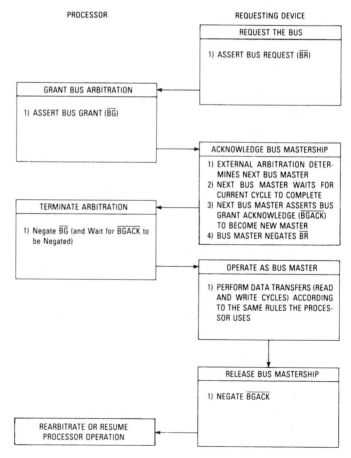

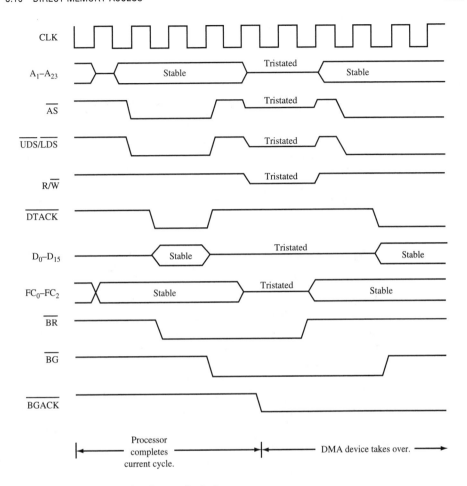

FIGURE 8.29 Bus arbitration cycle timing

Example 8.10: Recall that a memory-read cycle requires a minimum of eight states (four CLK cycles) to read a word from memory. If 1,024 reads are performed, a total of 4,096 CLK cycles are needed. A DMA controller is able to read memory faster, since it uses a different type of bus cycle to access memory. Let us assume that we are using a DMA controller that is capable of reading memory every two CLK cycles (once it controls the buses). Now, suppose that the DMA controller has been programmed to read the same 1,024 locations, in **bursts** of 128 words. Each burst will use 256 CLK cycles, plus some additional CLK cycles for controller overhead, such as bus request/grant protocol. With 16 CLK cycles of overhead, each burst requires 272 CLK cycles. The total number of CLK cycles required for the 1,024-word block is 2,176, a little more than half that required by the 68000 itself. Thus, we see that a DMA controller can be very useful when large blocks of data must be transferred.

Look at the discussion of the 68440 dual DMA controller covered in Chapter 10 for more information.

8.11   TROUBLESHOOTING TECHNIQUES

As we have seen before, a good knowledge of binary numbers is beneficial when working with microprocessors and necessary to the efficient design of address decoders (and other interfacing circuitry). Once the decoders are designed, however, they must be tested to see if they perform as required. Testing the operation of a memory address decoder can be accomplished any number of ways. The circuit can be set up on a breadboard, simulated via software, or just plain stared at on paper until it seems correct.

In situations like the troubleshooting phase of the single-board computer project in Chapter 11, testing the memory address decoder (similar to Figure 8.12) is often necessary. In addition to checking the wiring connections visually (or via a continuity tester or DMM), a logic analyzer is connected so that the waveforms can be examined. Even an old 8-channel logic analyzer can be used to diagnose difficult problems in a microprocessor-based system. Oscilloscopes typically fall short in showing the associated high-speed timing relationships (unless they are storage scopes).

The logic analyzer is connected so that all of the inputs to the address decoder, and as many of the outputs as necessary, are sampled. In addition, the logic analyzer is set up so that it triggers (and begins capturing data) at RESET, so that the initial activity of the processor's address bus can be observed. If the single-board's EPROM is not enabled, the system will not function at all. The logic analyzer will show how the address decoder responds at power-on.

The logic analyzer can also be used to examine the data coming out of the EPROM or RAM as well. Sampling the data and a few of the address lines should be enough to verify if the data is correct. It is sometimes possible to spot switched data or address lines this way.

These techniques also apply to I/O circuitry, which we examine in the next chapter.

SUMMARY

In this chapter we studied some of the most common methods used in the design of memory circuitry for microprocessor-based systems. Bus buffering, full- and partial-address decoding, direct memory access, and the logical requirements for static and dynamic RAMs were covered. A good designer will employ many of these techniques in an effort to construct a new system that is logically simple and elegant and also functional and easy to troubleshoot. The end-of-chapter questions are designed to further test your knowledge of these topics, and you are encouraged to work *all* of them to increase your ability to design memory address decoders, partial-address decoders, and complete memory systems.

In the next chapter you will see how a different technique, **memory-mapped I/O,** is used to implement serial and parallel communication.

STUDY QUESTIONS

1. Explain the different functions of the internal A_0 bit when used for:
   a) byte addressing
   b) word addressing
2. How does the $R/\overline{W}$ line in Figure 8.3 control the direction of the 74LS245?
3. Show the states of $\overline{UDS}$ and $\overline{LDS}$ when the 68000 is involved in the following memory accesses:
   a) a byte write to address 3000
   b) a byte write to address 3001
   c) a word write to address 3000
4. Explain the difference between synchronous and asynchronous data bus transfers, and what processor signals are used in each method.
5. If a state time in a 68000-based system is 250 ns, what is the minimum time spent doing a memory read?
6. When (and why) are wait states inserted into memory accesses?
7. Calculate the worst-case delay time for each address decoder in Figure 8.30.
8. Why do wait states always come in twos?
9. For the state time of Question 5, what is the time spent doing a memory read with two wait states?
10. Two EPROMs are used to make a 2K word memory. How many address lines are needed for the EPROMs? What upper address lines must be used for the decoder?
11. For the memory of Question 10, what is the address of the last memory location, if the starting address of the EPROM is E04000?
12. Design a memory address decoder for the EPROM memory of Question 10, using a circuit similar to that in Figure 8.8.
13. Repeat Questions 10 through 12 for these memory sizes and starting addresses:
    a) 8K words, base address of CD0000
    b) 32K words, base address of 8C0000
    c) 256K words, base address of 180000
14. Explain why it is not possible for an 8K word memory to have a starting address of $1000.
15. What are the decoded address ranges for the circuits in Figure 8.30?
16. What signal (or signals) is missing from the address decoder in Figure 8.30? Modify the decoders to include the missing signal (or signals).
17. What are the address range groups for the decoder in Figure 8.31?

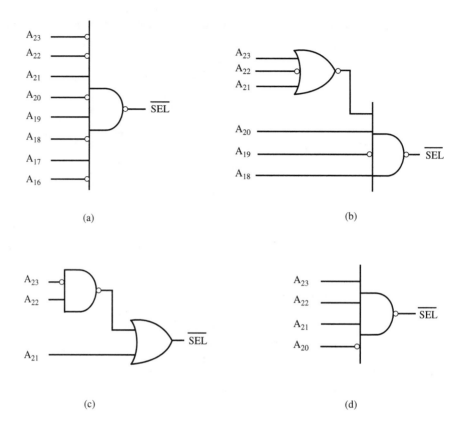

(a)   (b)

(c)   (d)

FIGURE 8.30   For Questions 8.7, 8.15, and 8.16

FIGURE 8.31   For Questions
8.17 and 8.18

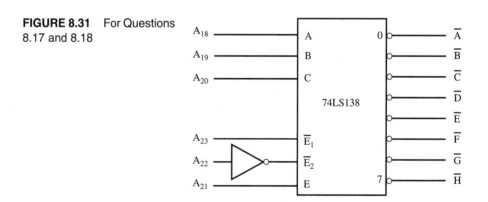

18. Use a circuit similar to that of Figure 8.31 to decode these address ranges:

108000 to 1087FF

108800 to 108FFF

109000 to 1097FF

109800 to 109FFF

10A000 to 10A7FF

10A800 to 10AFFF

10B000 to 10B7FF

10B800 to 10BFFF

19. Why is RAM preferred in memory from 0 to 3FF, instead of EPROM?
20. What are two main advantages gained in using partial-address decoding? Two disadvantages?
21. Give three possible address ranges for each decoder in Figure 8.32. Address lines A_1 through A_{13} are used by the memories.
22. Suppose that three different memory decoders have output signals $\overline{RAMA}$, $\overline{RAMB}$, and $\overline{ROM}$. Design a circuit to generate a $\overline{DTACK}$ delay of 200 ns using a 100-ns-period clock and a circuit similar to that of Figure 8.19. Any of the three signals going low triggers the generator.

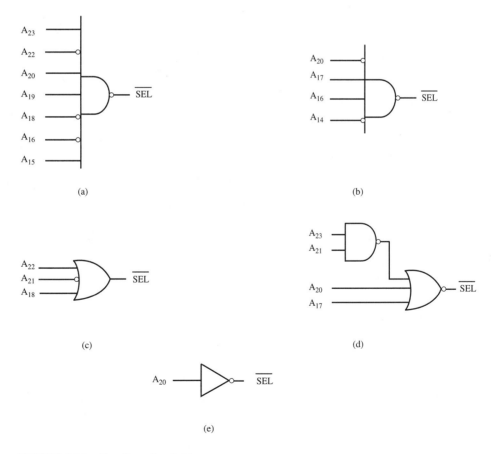

FIGURE 8.32 For Question 8.21

23. Why is the delay circuit eliminated in synchronous bus transfers?
24. Design a 16K word memory using 8KB EPROMs. Show the address and data line connections to all EPROMs and the circuitry needed to switch between the two 8K word sections. The base address is 280000.
25. Design a 256KB RAM memory using 64KB RAMs. The base address is C00000.
26. How do the $\overline{RAS}$ and $\overline{CAS}$ lines on a DRAM eliminate half of the required chip address lines?
27. Why does the size of a DRAM go up by a factor of 4 for each single address line that is added?
28. Why do DRAMs consume less power than static RAMs?
29. Explain how DRAM refreshing could be accomplished using an interrupt service routine.
30. How does program execution change on a system that supports DMA?
31. What is the 68000 doing while its external buses are involved in a DMA transfer?
32. Name at least three operations that might require the high data-transfer rate of DMA.

CHAPTER 9

I/O System Design

OBJECTIVES

In this chapter you will learn about:

- Memory-mapped I/O
- Parallel data transfer
- Serial data transfer
- The basics of memory-mapped video

9.1 INTRODUCTION

In the last chapter we saw how memory systems were designed for use with the 68000. I/O systems are designed in a similar way. I/O peripherals are interfaced with the 68000 through the same kind of address decoders we used in the previous chapter. These I/O address decoders are used to map hardware peripheral reigisters to specific memory locations. When the processor reads or writes into one of these **memory-mappped I/O** locations, some action will be taken by the peripheral device. The techniques involved in performing memory-mapped I/O will be introduced in this chapter through the use of 6800-based peripherals. More I/O interfacing will be shown in Chapter 10.

Section 9.2 describes the way memory-mapped I/O is used to transfer information between the processor and the outside world. Sections 9.3 and 9.4 detail two specific types of memory-mapped I/O: parallel data transfer and serial data transfer, respectively. A special kind of I/O called memory-mapped video is covered in Section 9.5. Troubleshooting I/O techniques are covered in Section 9.6.

9.2   MEMORY-MAPPED I/O

Normally, a memory location, or group of locations, is used to store program data and other important information. Data is written into a particular memory location and read later for use. Through a process called memory-mapped I/O, we remove the storage capability of the memory location and instead use it to communicate with the outside world. Imagine that you have a keyboard that supplies an 8-bit ASCII code (complete with parity) whenever you press a key. Your job is to somehow get this parallel information into your computer. By using memory-mapped I/O, a memory location may be set aside that, when read, will contain the 8-bit code generated by the keyboard. Conversely, data may be sent to the outside world by writing to a memory-mapped output location. The 68000 CPU is capable of performing memory-mapped I/O in either byte or word lengths. All that is required is a memory address decoder, coupled with the appropriate bus circuitry. For a memory-mapped output location, the memory address decoder provides a clock pulse to an octal latch capable of storing the output data. A memory-mapped input location would use the memory address decoder to enable a tristate octal buffer, placing data from the outside world onto the CPU's data bus when active. Figure 9.1 shows the circuitry for an 8-bit memory-mapped I/O location, sometimes referred to as a memory-mapped I/O port. The memory address decoder may be used for both input and output. Study Figure 9.1 and see how the R/$\overline{\text{W}}$ line is used to control the operation of the memory-mapped I/O port. The design of the address decoder is shown in Figure 9.2. Note the number of address lines used in the decoder. Partial-address decoding would greatly simplify the logic needed here.

The address decoded in Figure 9.2 is CFFFFF (since $\overline{\text{LDS}}$ was used to complete the I/O decoding). Reading and writing this memory-mapped location can be accomplished in many different ways. One way is to specify the address directly in the instruction, as in:

```
MOVE.B    $CFFFFF,D0     ;read input data
```

or

```
MOVE.B    D0,$CFFFFF     ;write output data
```

Other methods involve the use of address register indirect addressing. The I/O address must be loaded into an address register prior to any I/O accesses. For example, the following loop uses address register indirect addressing to constantly read the input data and echo it to the output.

```
        MOVEA.L   #$CFFFFF,A0    ;set up I/O address
ECHO    MOVE.B    (A0),D0        ;read input data
        MOVE.B    D0,(A0)        ;write output data
        BRA       ECHO
```

A 16-bit I/O design requires two octal latches and two octal buffers, one latch/buffer for each half of the data bus. The same address decoder may be used, with $\overline{\text{UDS}}$ and $\overline{\text{LDS}}$ controlling the latch/buffer that gets activated.

Many companies now make specialized input and output devices that are easily interfaced with the 68000 (and other CPUs). One useful output device is the NSM-1416, a

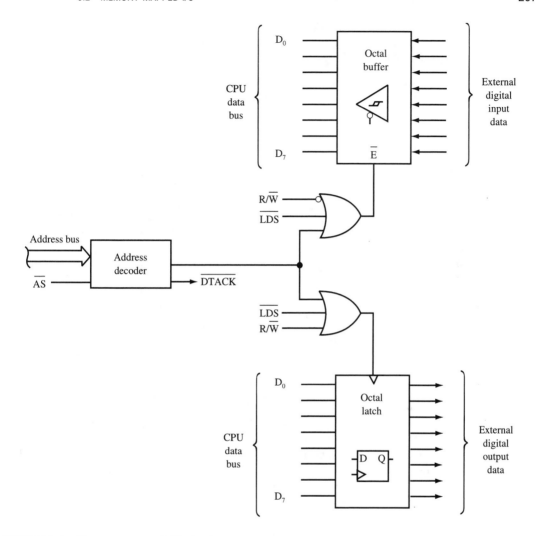

FIGURE 9.1   Memory-mapped I/O circuitry

four-digit alphanumeric display. Each digit is composed of 16 LED segments capable of forming letters, numbers, and punctuation symbols. The 1416 is designed to mount side by side with other 1416s to create multi-digit displays. In Figure 9.3, two 1416s are used to make an 8-digit alphanumeric display. Each 1416 has two address inputs (A_0 and A_1) that are used to select one of four digits (with the rightmost digit being digit 0). The ASCII code for the selected digit is placed on the 1416's data inputs (D_0 through D_6) and stored *inside* the 1416 when $\overline{CE}$ and $\overline{WR}$ are active. The 1416 contains its own internal digit storage and multiplexing logic, reducing the external hardware needed to operate the device.

FIGURE 9.2 Address decoder for memory-mapped I/O device at address CFFFFF

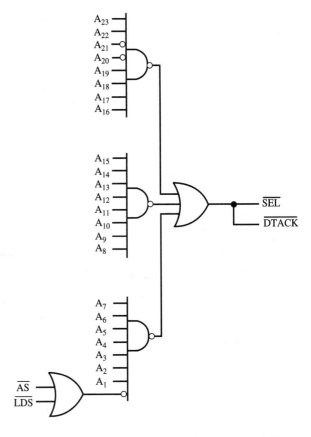

To display an eight-character message on the display, we need to output ASCII information to all eight memory-mapped locations. This is accomplished by the following routine, which uses E00001 as the base address of the display.

```
        ORG     $8000
MSG     DC.B    'MY 68000'
        ORG     $8100
DISP8   MOVEA.L #MSG,A0         ;set up pointer to message
        ADDA.L  #8,A0           ;advance pointer to end of message
        MOVEA.L #$E00001,A1     ;set up pointer to display
        MOVE.W  #7,D0           ;prepare for 8 passes
NEXT    MOVE.B  -(A0),D1        ;read message character
        MOVE.B  D1,(A1)         ;write character to display
        ADDA.L  #2,A1           ;advance to next digit address
        DBRA    D0,NEXT         ;repeat for all digits
        RTS
```

It is necessary to use odd addresses when accessing the display, since $\overline{\text{LDS}}$ was used in the $\overline{\text{WR}}$ logic and the upper half of the data bus is ignored. This is why 2 is added to A1 during each pass of the loop. We also must read the message characters from right to left to match the requirements of the display. This accounts for the predecrement addressing used to read message characters.

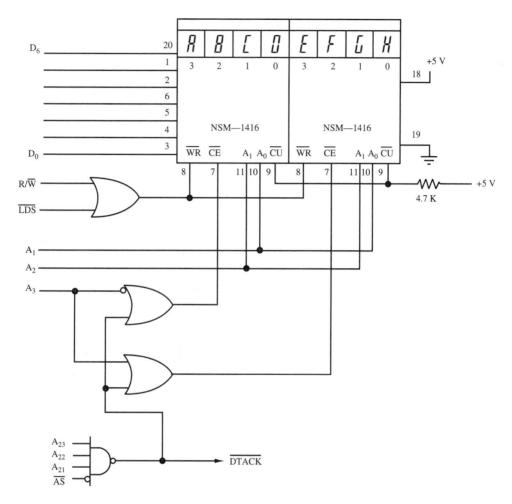

FIGURE 9.3   Eight-digit alphanumeric display

9.3   PARALLEL DATA TRANSFER: THE 6821 PIA

In the previous section we examined the operation of a simple I/O port that used a technique called memory-mapped I/O to send and receive data with the outside world. Although the hardware was simple and dedicated to fixed 8-bit I/O, the message is clear: Memory addresses may be used to perform external CPU I/O. In this section we will look at parallel I/O again, now seeing how a dedicated peripheral, the 6821 peripheral interface adapter (PIA), uses a number of memory-mapped locations to implement parallel I/O with the external system hardware. The 6821 was initially designed for use in 6800-based systems, but we will examine the method used to interface it with the 68000.

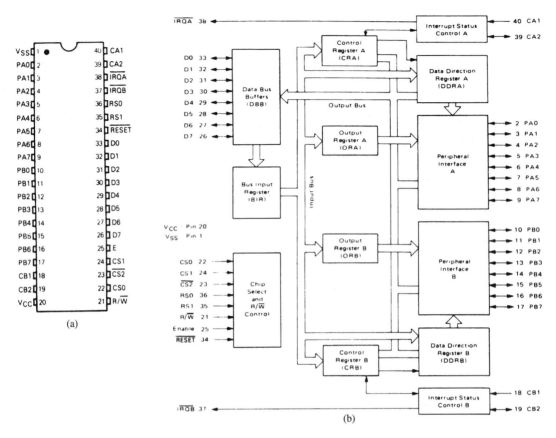

FIGURE 9.4   The 6821 PIA: (a) pin assignments and (b) internal block diagram (Reprinted with permission of Motorola Inc.)

Figure 9.4 shows a simple block diagram of the 40-pin 6821, and also a pinout of the device. The block diagram contains a fair amount of symmetry between the logical functions. That is because the 6821 supports two separate 8-bit ports. These two ports (A and B) are configured by software and may drive two TTL loads. In addition, port A will also drive CMOS loads.

In the next two sections we will examine both the software and hardware requirements for operating this device and the method needed to connect the 6821 to a 68000-based system.

Programming the 6821

Did you notice that the block diagram of the 6821 contained control and data direction registers for both the A and B interfaces? Together, these two registers give the programmer a very flexible I/O system. Each *bit* in the A and B interfaces may be programmed as either an input or an output bit. By writing the appropriate data into the data direction register, the programmer may configure each signal line (PA_0 through PA_7, and PB_0 through PB_7).

This has the great advantage of allowing the software to adjust to new system requirements, without having to make changes to the actual output circuitry. In addition, there are additional signal lines (CA_1, CA_2, CB_1, and CB_2) that may be programmed to allow interrupts, and strobe signals for each port.

Selecting any configuration involves the exchange of data between the CPU and the 6821 control/data direction registers. These registers are shown in Figure 9.5. Accessing any of the internal registers is accomplished by writing the appropriate data to specific memory locations. These locations act like memory locations that we utilize to make up EPROM or RAM memory, except they reside in the 6821 peripheral and therefore we make use of two new processor signals to control the data exchange. We will get to these signals in the next section. For now, all we have to understand is that once the hardware is installed, we may use the 6821 like a handful of memory locations: writing data to the 6821 may output data on the port A or B lines, and reading data from the 6821 may input data from port A or B.

To learn how to configure the 6821, let us look only at the port A side. The programming of port B is identical, except for the memory addresses that select the port B registers.

Figure 9.6 shows the bit assignments within the side A control register. The first bit of importance is bit 2, DDR access. This bit controls the path that output data takes when written to the 6821. If bit 2 is low, the data direction register gets the output data, and when it is high the output register receives the data. When the 6821 receives a $\overline{\text{RESET}}$ signal, all bits in each port are set up for input. To make a particular bit an output bit, a logic 1 must be written into the corresponding bit in the data direction register. For example, writing a $0F$ into the data direction register for side A makes PA_0 through PA_3 outputs and PA_4 through PA_7 inputs.

The short code sequence that follows may be used to program the side A data lines, assuming that no programming of the 6821 has been attempted since power-on:

```
MOVE.B     #0,D3              ;Control word to select DDR
MOVEA.L    #$20002,A5         ;memory address for control register A
MOVE.B     D3,(A5)            ;write data to control register A
```

RS$_1$	RS$_0$	Control register bit		Selected register
		CRA-2	CRB-2	
0	0	1	X	Peripheral register A
0	0	0	X	Data direction register A
0	1	X	X	Control register A
1	0	X	1	Peripheral register B
1	0	X	0	Data direction register B
1	1	X	X	Control register B

X = Don't care

FIGURE 9.5   Programmer's model of the 6821

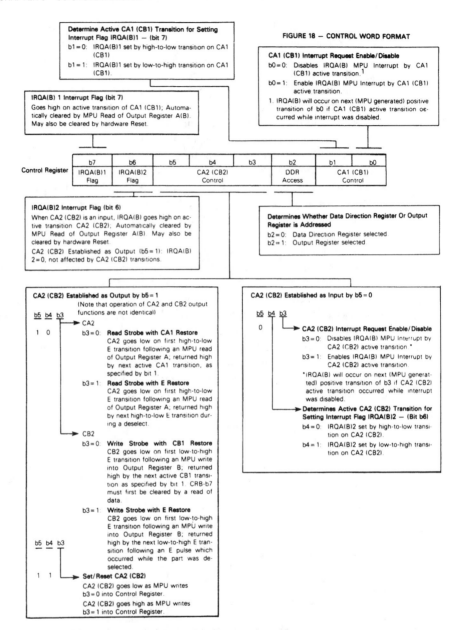

FIGURE 9.6   Control register bit assignments for the 6821 (Reprinted with permission of Motorola Inc.)

```
MOVE.B    $0F,D5          ;I/O configuration
MOVEA.L   #$20000,A5      ;memory address for DDRA
MOVE.B    D5,(A5)         ;write data to DDRA
```

The code sequence assumes that the 6821 has a base address of 20000. This makes 20000 the location for peripheral/data direction registers A, 20002 control register A, 20004 peripheral/data direction registers B, and 20006 control register B.

All other bits in the control register select different configurations that may be used for handshaking with circuitry connected to the 6821. For example, bit 1 selects the activity needed on the CA_1 input (rising-edge or falling-edge) needed to set an interrupt flag. This of course assumes that the interrupt has already been enabled to bit 0 in the control register.

Bits 3, 4, and 5 select whether the CA_2 line is an input or an output, and what type of interrupt/strobing option is used. Bits 6 and 7 are interrupt flags that are set by various actions on the CA_1 and CA_2 lines.

Since simple parallel I/O without handshaking is our requirement here, we will not go into detail on these special functions. We need only concern ourselves with the function of bit 2. If we use zeros for all other control bits, we will always disable the special interrupt/strobing features when we write data into the control register.

To access the 6821 through the previous memory locations, we must have an address decoder that will become activated by the presence of the 6821's memory-mapped addresses on the 68000's address bus. We will see how this is accomplished in the next section.

Interfacing the 6821

You may wish to review the pinout of the 6821 PIA in Figure 9.4(b). Pins 2 through 19, 39, and 40 are all I/O pins. One-half of the CPU data bus connects to D_0 through D_7 (pins 33 through 26). $\overline{IRQA}$ and $\overline{IRQB}$ are interrupt outputs that may be used to interrupt the 68000. RS_0 and RS_1 are the register select inputs. We normally use lower address bits A_1 and A_2 on these lines to control the selection of internal 6821 registers. Review Figure 9.5 to see how RS_0 and RS_1 point to these registers.

The 6821 also contains three chip-select inputs CS_0, CS_1, and $\overline{CS_2}$. When these signals are in their active state (high, high, and low), the 6821 will communicate with the CPU data bus. However, it must do its data transfer at a much slower rate than normal memory. This is because the 6821 is a peripheral designed for use with the slower 6800 CPU. But this does not matter; the designers of the 68000 took this into account and supplied CPU signals that allow us to complete the interfacing. These signals are the E clock, $\overline{VMA}$ (valid memory address), and $\overline{VPA}$ (valid peripheral address).

The E clock is controlled by the main CPU clock, except that its frequency is 10 times smaller and has a 40 percent duty cycle. This means that an 8-MHz 68000 will have an E clock equal to 800 kHz that is high for 500 ns and low for 750 ns. The E clock synchronizes the faster 68000 with the slower 6821 logic circuitry. Data transfers that use the E clock in this fashion are called **synchronous** data transfers, and take additional time to complete because of the need to wait for synchronization to occur.

The $\overline{VPA}$ signal is an input to the 68000 that tells it the address bus contains the address of a slow peripheral and that its bus cycle must be modified to allow synchronization with the E clock. The $\overline{VPA}$ signal is generated by the address decoder for the 6821 (or other peripheral) in much the same way we generated the $\overline{DTACK}$ signal, except there is no need to delay before pulling $\overline{VPA}$ low.

When the processor detects the active $\overline{VPA}$ signal, it will respond by pulling $\overline{VMA}$ low. This output signal may be used to complete the enabling of the peripheral device. Figure 9.7 shows a general interfacing scheme for the 6821.

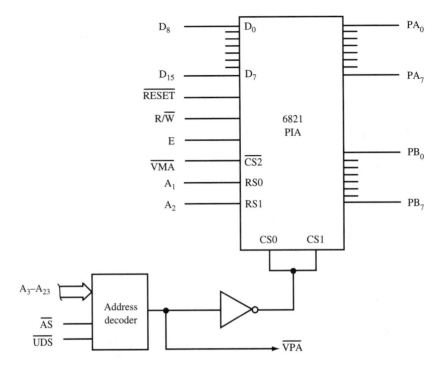

FIGURE 9.7   6821-to-68000 interface

Notice how the remaining address lines, as well as $\overline{AS}$ and $\overline{UDS}$, are used as inputs to the address decoder. These signals determine the base address of the 6821 (we assume that RS_0 and RS_1 are zeros).

Another 16 bits of I/O may be added easily by connecting a second 6821 to the existing circuitry. In this case, the second PIA may communicate over the lower half of the data bus, providing the ability for full 16-bit data transfers if so desired.

A simple I/O application utilizing the 6821 is shown in Figure 9.8. Port A has been programmed for input and is connected to a set of switches. An open switch makes a logic 1 and a closed switch makes a 0. Port B is connected to a set of light-emitting diodes. A logic 1 on any port B output will turn on its associated LED. The following routine reads the switch byte from port A and uses it to load a counter with an initial value. The counter is used to provide some delay between outputs to port B. A binary count is output to port B to flash the LEDs in a specific pattern. The purpose of the switches is to control the speed at which the LEDs count. A circuit such as this is very useful for exploring the operation of the 6821.

The software assumes that the 6821 has already been programmed and that A0 and A1 point to port A and port B, respectively.

```
BINCNT    CLR.B     D1               ;clear display counter
GETSW     MOVE.B    (A0),D0          ;read switches
          ANDI.W    #$FF,D0          ;clear bits 8-15
          ROL.W     #8,D0            ;shift switch data into bits 8-15
          MOVE.B    D1,(A1)          ;output to LEDs
```

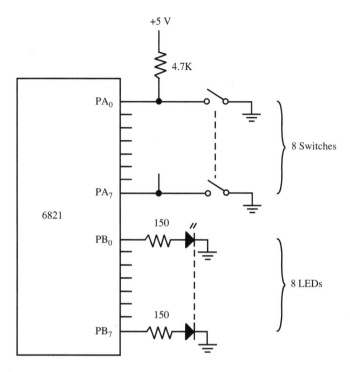

FIGURE 9.8 6821 I/O application

```
          ADDQ.B   #1,D1         ;increment display counter
WAIT      NOP                    ;waste a little time
          NOP
          DBRA     D0,WAIT       ;stay in loop until D0 = -1
          BRA      GETSW         ;go read switches again
```

The routine is written to allow the speed to be changed while the program is running.

When parallel data I/O is not enough, we may wish to add serial I/O. The next section will show how this other standard form of communication may take place.

9.4 SERIAL DATA TRANSFER: THE 6850 ACIA

Serial data transmission offers the convenience of running a small number of wires between two points (three will do the job in most cases), while at the same time being very reliable. Even though we must wait longer to receive our data, since it is transmitted only 1 bit at a time, we are able to place our communication devices (computers, terminals, etc.) far away from each other. Worldwide networks now exist, connected via satellites, based on serial data transmission. The peripheral covered in this section implements serial data transmission in a variety of formats. The standard serial data transmission waveform for any UART is depicted in Figure 9.9. One ASCII character is embedded in the waveform.

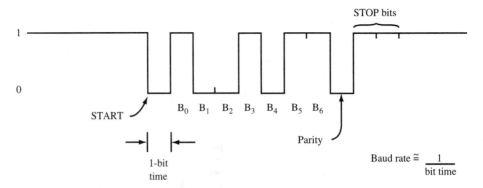

FIGURE 9.9   Standard TTL serial data waveform

The normal state of the serial data line is a logic 1. This level indicates that no activity is present (that is, no data being transmitted). When the line level falls to a logic 0 (the start bit), the receiving UART knows that a new character is being transmitted. The data bits representing the character (or data) being transmitted are clocked out in the order shown, least significant to most significant.

Following the data bits is the parity bit, which will be used by the receiving UART to determine the accuracy of the data it received. The parity bit in Figure 9.9 shows that the data has even parity. The last bits in any transmission are the stop bits, which are always high. This gets the line back into its inactive state. We are able to set the number of data bits, the type of parity used, the number of stop bits, and the bit time through software.

Note that the bit time (seconds/bit) is inversely proportional to the baud* rate (bits/second). In this section we will examine the operation of the Motorola 6850 ACIA (asynchronous communications interface adapter), which is a UART initially designed for the 6800.

Figure 9.10 shows the pinout and block diagram of the 6850, which consists mainly of a control section and data transmitter/receiver sections. The control section allows the 6850 to be programmed by enabling special bits in a control register. Through this register we can modify the internal timing of the 6850, reset it, choose 7 or 8 data bits per character, and also control the parity and stop bits. In order to use the control register, we must be able to write data into it. For this reason we select a memory-mapped address for the 6850, and use it like a regular memory location.

We will look first at the methods used to program the 6850, and then at the hardware requirements that must be satisfied to interface the 6850 with the 68000.

---

* Note that baud rate and bits/second are not identical, but have been loosely accepted to mean the same thing. Baud rate actually refers to the number of discrete signal charges per second.

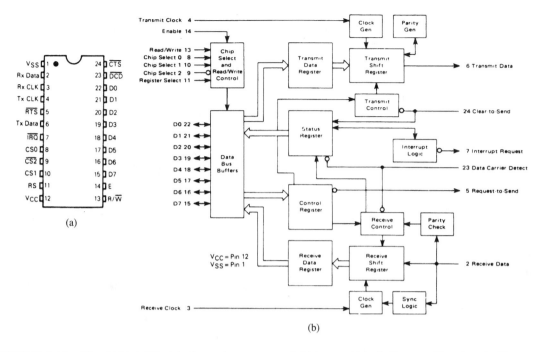

FIGURE 9.10 The 6850 ACIA: (a) pin assignments and (b) internal block diagram (Reprinted with permission of Motorola Inc.)

Programming the 6850

The programmer's model of the 6850 is shown in Figure 9.11. Internal registers are selected by the RS and R/$\overline{\text{W}}$ inputs.

The 6850 is configured by writing a particular binary control word into its control register. Figure 9.12 shows the control bit assignments within the 6850's control register. Before any programming of the 6850 is attempted, it is a good idea to give the device a master reset command. This ensures that the timing signals inside the 6850 are synchronized and that the device is ready for programming. To issue the master reset command, bits 1 and 0 of the control word should be high, with all other bits low. This

FIGURE 9.11 Programmer's model of the 6850

RS	R/$\overline{\text{W}}$	Register selected
0	0	Control register
0	1	Status register
1	0	Transmit data register
1	1	Receive data register

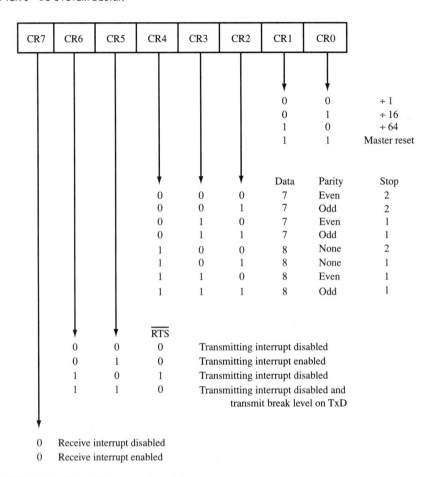

FIGURE 9.12   The 6850 control register

control word (03) is then written into the memory-mapped location that selects the control register.

The baud rate of the 6850 is determined by both the external baud rate clock applied to the RxC and TxC pins and also the setting of bits 1 and 0 in the control register. If these bits are both low, no frequency division occurs. If bit 0 is high with bit 1 low, the 6850 operates on an internal baud rate that is 1/16th that of the external clock frequency. Example 9.1 shows how the divide select bits are used.

Example 9.1:   A 6850 has a 38.4-kHz clock connected to its RxC and TxC pins, and bits 1 and 0 in the control register have been set to 0 and 1, respectively. What is the baud rate?

Solution: Dividing 38.4 kHz by 16 gives 2400 Hz, causing the 6850 to operate at 2400 baud (2400 bits/second).

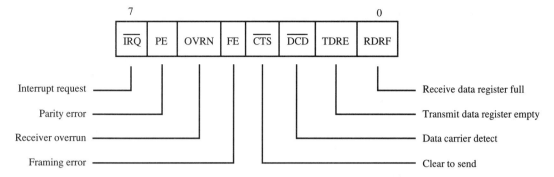

FIGURE 9.13 The 6850 status register

Control register bits 2, 3, and 4 are used to select the number of data bits per character (7 or 8), the type of parity used (odd, even, or none), and the number of stop bits. A 110 code on bits 4, 3, and 2 results in 8 data bits, even parity, and 1 stop bit.

The remaining bits are used in special handshaking and interrupt schemes that we will not go into here. A low on these 3 bits disables their functions.

Once the 6850 is configured, its operational status may be checked by examining various bits in its status register. Figure 9.13 shows the bit assignments in this register.

Bit 0 indicates receive data register full (RDRF) when high. The 6850 has received and decoded a character, which may be read from the 6850's data register.

Bit 1 indicates transmit data register empty (TDRE) when high. A new character may be transmitted by writing it into the 6850's data register.

Bit 2 is data carrier detect ($\overline{DCD}$) when low. When the 6850 is used to communicate with a modem, $\overline{DCD}$ indicates the presence of a carrier via the $\overline{DCD}$ input of the 6850 (pin 23). $\overline{DCD}$ may be used to generate an interrupt request via the receive interrupt enable bit in the control register.

Bit 3 indicates clear to send ($\overline{CTS}$) when low. Again an external input ($\overline{CTS}$ on pin 24), together with a modem, may be used to control this bit. $\overline{CTS}$ will inhibit the TDRE bit when it is high.

Bit 4, framing error (FE), indicates that a problem has occurred with the most recently received character. When high, problems such as a missing stop bit, break character detection, or loss of synchronization may have occurred.

Bit 5 indicates receiver overrun (OVRN) when high. If more than one character is received before the data register is read, data is lost and only the last character received is available.

Bit 6, when high, shows that a parity error (PE) has occurred. This means that the most recently received character's parity did not match the parity selected by the control register.

Bit 7 is interrupt request ($\overline{IRQ}$), and is used to show the state of the $\overline{IRQ}$ output (pin 7). Interrupt conditions enabled via the control register will affect this status bit.

The programmer must utilize the 6850's status bits in order to ensure proper serial data communication. Figure 9.14 shows how the first two status bits are used to implement a simple serial input/output procedure.

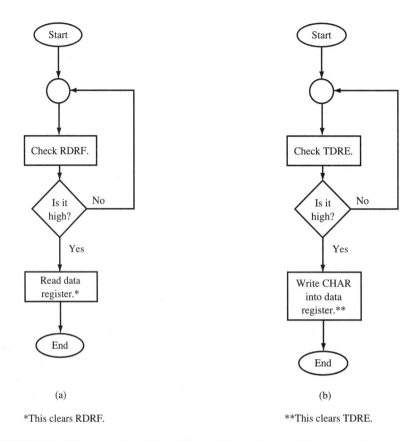

FIGURE 9.14   I/O flowcharts: (a) serial reception and (b) serial transmission

Both flowcharts indicate that repeated testing of the RDRF/TDRE bits may be necessary. An example to show the importance of this repeated testing follows.

Example 9.2:   Suppose that a 6850 is configured to transmit and receive data at 1,200 baud, with 7 data bits, odd parity, and 1 stop bit. How long does it take to fully transmit or receive a character?

Solution: At 1,200 baud, the bit time is just over 833 µs, and the selected word length of 10 bits makes the total time to receive or transmit a single character roughly 8.3 ms.

It is not difficult to imagine how many instructions the 68000 might be able to execute in 8.3 ms. Would a few thousand be unreasonable? Probably not. Therefore, we use the status bits to actually slow down the 68000, so that it does not try to send or receive data from the 6850 faster than the 6850 can handle.

The two short routines that follow show how a character input and a character output routine might be written in 68000 code.

```
CIN       MOVEA.L   #$18000,A4    ;point to ACIA status register
GET_ST    MOVE.B    (A4),D2       ;load status byte into D2
          ANDI.B    $01,D2        ;test RDRF bit
          BEQ       GET_ST        ;loop until high (char rec'd)
          MOVEA.L   #$18002,A4    ;point to ACIA data register
          MOVE.B    (A4),D1       ;load character into D1
          ANDI.B    #$7F,D1       ;clear MSB
          RTS

COUT      MOVEA.L   #$18000,A4    ;point to ACIA status register
GET_ST2   MOVE.B    (A4),D2       ;load status byte into D2
          ANDI.B    #$02,D2       ;test TDRE bit
          BEQ       GET_ST2       ;transmitter busy, try again
          MOVEA.L   #$18002,A4    ;point to ACIA data register
          MOVE.B    D1,(A4)       ;transmit character from D1
          RTS
```

The routines assume a 6850 with decoded register addresses at 18000 (RS = 0) and 18002 (RS = 1). They also assume that the 6850 has been properly initialized. Usually this is done at the beginning of code execution (power-on) but is not limited to a single time. The 6850 may be reconfigured by first issuing a master reset command. The following code may be used to initialize the 6850 for 8 data bits, no parity, 1 stop bit, and an X16 clock.

```
INIT      MOVEA.L   #$18000,A4    ;point to ACIA control register
          MOVE.B    #$03,(A4)     ;Master reset command
          MOVE.B    #$15,(A4)     ;configuration pattern
          RTS
```

Now that we have an understanding of the software needs of the 6850, we will next examine the hardware needs, and see that interfacing the ACIA is a rather simple task.

Interfacing the 6850

Take another look at Figure 9.10(b). The 6850 communicates with the host processor (the 68000 in our case) via its 8-bit data bus, D_7 through D_0 (pins 15 through 22). Reading or writing data is controlled by the $R/\overline{W}$ input (pin 13). The chip itself is enabled when its three chip-select inputs—CS_0, CS_1, and $\overline{CS}_2$—are all active. The internal registers of the 6850 are selected by the $R/\overline{W}$ input and by the RS (register select) input. A low on this input selects the control/status registers, and a high selects the receiver/transmitter data registers. Pin 14 is the E clock input, which controls all internal 6850 timing. Baud rate timing is controlled by the separate RxC and TxC inputs, which means that the 6850 may be configured to transmit and receive at different baud rates. RxD and TxD are the receiver and transmitter serial data lines. The $\overline{IRQ}$ output (when low) may be used to interrupt the 68000 to inform it of special conditions occurring inside the 6850. The $\overline{CTS}$, $\overline{DCD}$, and $\overline{RTS}$ pins (of which only $\overline{RTS}$ is an output) are normally used to interface the 6850 with a modem. In Figure 9.15 we see an example of how the 6850 is interfaced to the 68000.

In this example, the 6850 is connected to the upper half of the data bus (thus requiring the use of $\overline{UDS}$ in the address decoder). When the 68000 outputs an address that

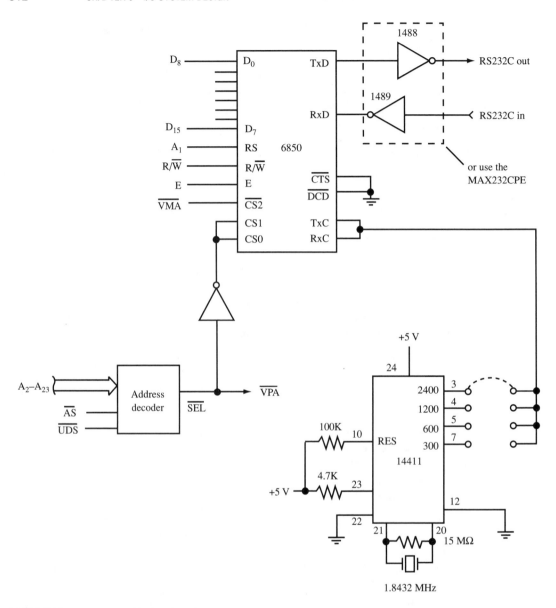

FIGURE 9.15   6850-to-68000 interface

corresponds to the one that will activate the address decoder, the $\overline{\text{SEL}}$ output will go low. This will cause two things to happen. First, two of the 6850's enables (CS_0 and CS_1) will be taken high via the inverter. Second, the $\overline{\text{VPA}}$ (valid peripheral address) signal will go low, indicating to the 68000 that it has addressed a 6800 peripheral. This will cause the 68000 to extend its read or write cycle until it has synchronized with the E clock. When this occurs, the 68000 will proceed with the peripheral read or write by issuing $\overline{\text{VMA}}$ (valid memory address). Since a low on this line will complete the chip-select requirements of the 6850, we will now see data transfer occur. If address line A_1 is low, control/status information will be transferred, depending on the state of the $R/\overline{W}$ line. If A_1 is high, receiver/transmitter data will be transferred.

The baud rate is controlled by the clock supplied by a special baud rate chip, the 14411. This integrated circuit contains internal frequency dividers, and when coupled with a special crystal, will produce the exact clocks needed for many common baud rates. In this example, the 14411 is configured to output clocks that are 16 times the required baud rate. When we select X16 operation inside the 6850, we will get back the original baud rate clock.

The $\overline{\text{CTS}}$ and $\overline{\text{DCD}}$ inputs are grounded to ensure proper operation, since we are not interested in any fancy handshaking with a modem.

The 1488 and 1489 integrated circuits are used to convert the TTL compatible TxD and RxD signals to RS232C levels that swing plus and minus.

This concludes our study of the 6850 ACIA. Together with the 6821 PIA, we have a very powerful and flexible way of communicating with the outside world. In Chapter 10 we will examine the serial and parallel peripherals designed specifically for the 68000.

9.5   MEMORY-MAPPED VIDEO I/O: THE 6845 CRT CONTROLLER

When parallel and serial data communications alone are not enough, the designer/programmer must turn to a different approach to I/O. Since pictures convey a gigantic amount of information at once, it would be nice if we could take an electronic snapshot of the contents of memory and somehow display all the information at once on a single display. This technique is called **memory-mapped video.** A portion of memory is dedicated to performing this video I/O for us. We will soon see how a special peripheral, the 6845 CRT controller, uses this memory to create an image composed of ASCII characters on a television screen.

Ordinary monitors employ a technique called **raster scan** to deflect the electron beam to various parts of the screen in an orderly fashion. The beam is moved rapidly (63 μs for a single trace) from left to right, and slowly down, so that an image composed of 512 lines is created in 1/30th of a second. By varying the intensity of the beam, we can control the brightness of each line as it is traced out. After 1/30th of a second we have an entire image, which will quickly fade if we do not repeat the process. To simulate motion we will slightly change the information we present in the next image; therefore, in one second we may see 30 different images. Since our eyes cannot respond quickly enough to the changing images, they will average the information they see, and there will appear to be smooth transitions from one moment to the next.

Suppose now that we want to display only stationary images. This would mean that we can avoid having to change the image information every scan. To further simplify matters, we also display the information in two shades only, black and white. These two requirements serve to point the way to digital circuitry to store and generate the image. The circuitry would need to accomplish many complex tasks to correctly generate the video image. First, it needs to be able to synchronize its timing with that of the television monitor. Second, it would have to contain circuitry to generate the video image from information stored in system memory (which would also involve circuitry to access that memory). Third, it would have to do all of this very quickly, in order to pack usable information into one scan-line time of 63 µs.

Figure 9.16 shows the pinout and internal block diagram of the 6845 CRT controller. The various counters, sync circuits, and registers within the device all work to implement the raster scan technique. During normal operation, this involves repeated accessing of the system's video RAM. Data read out from the RAM is interpreted as ASCII data, which the internal circuitry uses to create a dot-matrix image of the character in a location on the screen corresponding to the location in memory that is being scanned. Thus, we divide the video memory up into rows and columns, and use sequential memory locations to fill the screen. The 6845 gives the option of choosing the number of rows and columns used. One option gives 24 lines by 80 characters. We would then need 1920 memory locations to store all of the ASCII information we could place on the screen.

The data written into these 1920 locations constitutes what we see on the screen of the display. CRT controllers are designed to interpret the data within screen memory as ASCII characters (in some modes of operation). Thus, by writing the correct ASCII values into memory we actually put the desired characters on the screen. Two useful routines are usually included in any memory-mapped video package. These routines are used for clearing the screen and for scrolling the screen. CLEAR and SCROLL are used for this purpose. Both routines assume that the proper video mode has already been set up and that A4 points to the beginning of a 24-by-80 character screen memory.

```
CLEAR     MOVE.W    #1919,D0      ;init loop counter
BLANK     MOVE.B    #$20,(A4)+    ;write ASCII blank into memory
          DBF       D0,BLANK      ;repeat to clear screen
          RTS


SCROLL    MOVEA.L   A4,A5         ;copy starting address
          ADDA.L    #80,A5        ;advance to 2nd line
          MOVE.W    #1839,D0      ;init loop counter
ROLLUP    MOVE.B    (A5)+,(A4)+   ;copy character from next line
          DBF       D0,ROLLUP     ;repeat for 23 lines
          MOVE.W    #79,D0        ;init loop counter
NEWLYN    MOVE.B    #$20,(A4)+    ;write blanks into last line
          DBF       D0,NEWLYN
          RTS
```

In the CLEAR routine, the initial loop counter value 1919 is one less than the number of screen characters. In the SCROLL routine, the initial loop counter value 1839 is one less than 1920 − 80. To get scrolling to work we have to copy the characters from line 2 to line 1. Then the characters from line 3 are copied to line 2. By the time the ROLLUP loop is finished, the entire screen has been scrolled up, leaving lines 23 and 24 identical. Line 24 is then blanked out to prepare it for the next new line of characters.

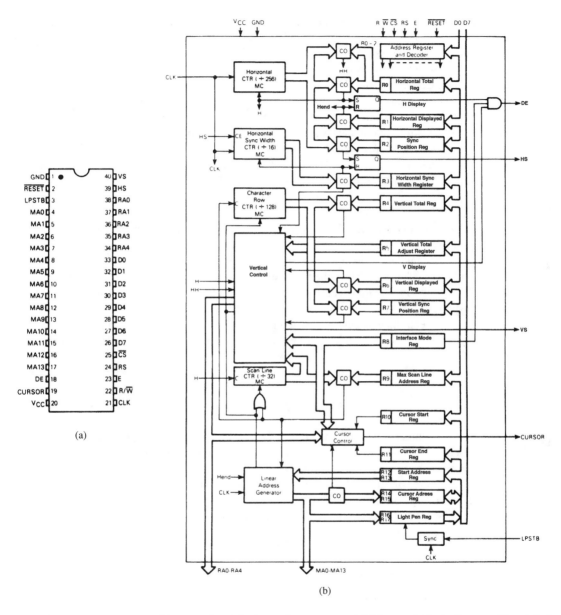

(a)

(b)

FIGURE 9.16   The 6845 CRT controller: (a) pin assignments and (b) internal block diagram (Reprinted with permission of Motorola Inc.)

Since the 6845 will be constantly scanning this video RAM to refresh the screen image, what happens when the CPU also tries to access it? Usually the designer will include circuitry to detect this illegal condition and then do one of two things: (1) the CPU gets priority and blank image data is sent back to the 6845, tricking it into thinking it has read memory, or (2) the 6845 gets priority, forcing the 68000 into a wait state until the 6845 is finished reading the video RAM. The second case is ordinarily used because it generates clearer images.

To move data around on the screen (very useful in screen-oriented text editors), we need only move it around in the video RAM. This function is supplied by software. We must write routines to perform the necessary editing that we wish to perform. The software may also interact with the 6845 registers that control its cursor, light pen, and hardware scrolling circuitry. Since the operation of the 6845 is very complex (the software must control 17 different internal registers), we will not examine the operation of this device further. More information may be found by reading the rather lengthy and comprehensive data sheets for the 6845 and its relative, the 6847 video display generator, which can generate color, bit-mapped images.

The original Macintosh computer used a **bit-mapped** form of memory-mapped video. The display screen was composed of 342 visible scan lines, with each line containing 512 **pixels** (the smallest video dot we can turn on and off). The entire 342-by-512 pixel display required 175,104 bits of storage. This corresponds to 21,888 bytes (or 10,944 words) of video RAM, which the designers of the Macintosh referred to as the **main screen buffer.**

Each word of video RAM sets the intensity (0 = on, 1 = off) of 16 horizontal pixels, with the far left pixel controlled by the MSB of the video word, as shown in Figure 9.17. The 16-bit video word ($1F30) is loaded into a shift register and sent 1 bit at a time to the display.

Screen images, such as the icon for a program or file, were made up of groups of video words. Moving an icon around on the screen was accomplished by moving blocks of video RAM around. Other screen operations, such as graphics, required manipulation of specific bits within selected video RAM locations. For example, to draw a horizontal line between pixels 26 and 215 on line 75 of the display, all bits in the corresponding memory words (or portions of words) had to be cleared. This is shown in Figure 9.18.

A similar technique was used to draw vertical lines. In this case, a single bit in each selected video word was cleared, with each word located 32 words apart from each other (since each scan line used 32 words).

Newer bit-mapped displays assign up to 24-bits for each pixel, 8 bits each for the colors red, green, and blue. This allows 2^{24} different colors for each pixel and results in a very large block of video RAM.

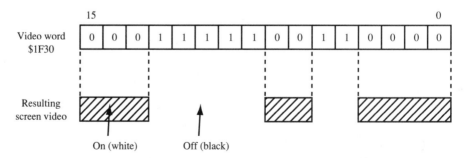

FIGURE 9.17   Format of memory-mapped video word

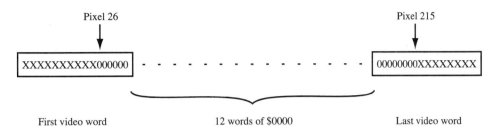

FIGURE 9.18 Video RAM contents for a horizontal line

9.6   TROUBLESHOOTING TECHNIQUES

Finding the cause of a faulty I/O device can be tricky. Here are some suggestions of things to try when you encounter an I/O problem.

- Write a short loop that continually accesses the I/O device. This should allow you to use an oscilloscope to look for a stream of pulses on the output of the address decoder. Use something like this to test an 8-bit output port:

```
        MOVEA.L    #$<I/O address>,A1
        SUB.B      D1,D1
TEST    MOVE.B     D1,(A1)
        ADDI.B     #1,D1
        BRA        TEST
```

In addition to the steady stream of pulses that should appear on the address decoder output, there should be a binary count appearing at the output port. It is easy to see with an oscilloscope if the waveform periods double (or halve) as you step from bit to bit. This is a good way to check for stuck or crossed outputs.

When checking an input port, use these instructions (or something similar):

```
        MOVEA.L    #$<I/O address>,A1
TEST    MOVE.B     (A1),D1
        BRA        TEST
```

This loop is good for checking the operation of the address decoder. If possible, combine both loops so that data read from the input port is echoed to the output port.
- Check for easily overlooked mistakes, such as using D_0 through D_7 to connect to the I/O device, but using UDS in the address decoder.
- Verify that the enable signals on the I/O device all go to their active states when accessed.
- For a serial device, examine the serial output for activity. Check for valid transmitter and receiver clocks. If the serial device is connected to a keyboard, press the keys and watch the serial input of the device. Make sure the TTL-to-RS232 driver is working correctly.

Other I/O devices may require you to test the interrupt system, or write special initialization code to program a peripheral. Keep track of the new software and hardware designs you develop or troubleshoot. They will save you time and effort in the future.

SUMMARY

In this chapter we examined the operation of memory-mapped I/O devices. These devices communicate with the processor through the use of memory read and write cycles, and thus utilize similar address decoding logic.

Two 6800-based peripherals were examined in detail. The hardware and software interfacing requirements of the 6821 peripheral interface adapter and 6850 asynchronous communications interface adapter were examined and examples given to demonstrate their use. A third peripheral, the 6845 CRT controller, also was discussed to introduce the technique of memory-mapped video. This was followed by a brief introduction to the Macintosh's bit-mapped display.

Peripherals designed specifically for the 68000 will be covered in the next chapter.

STUDY QUESTIONS

1. What is the difference between a memory location and a memory-mapped I/O location?
2. Modify the I/O circuitry of Figure 9.1 so that 16 bits of data are transferred.
3. What is the range of the memory-mapped I/O addresses of the decoder shown in Figure 9.19?
4. What is the memory-mapped address range for the decoder of Figure 9.2 if address lines A_1 through A_4 are not used?
5. Why are latches needed for the output section of an I/O device?
6. Design a partial address decoder for a memory-mapped I/O device that will respond to addresses $8000 through $80FF.
7. Write a subroutine that will scroll the eight-digit alphanumeric display left one digit each time it is called. The ASCII code for the new character entering from the right is in D_7.
8. Add two more NSM-1416s to the design of Figure 9.3.

FIGURE 9.19   For Question 9.3

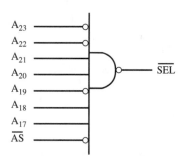

9. A 32-cell LCD alphanumeric display is arranged as two rows of 16 cells each, as shown in Figure 9.20. The display has seven data inputs (D_0 through D_6), five address lines (A_0 through A_4), a write input ($\overline{WR}$), and a chip enable input ($\overline{CE}$). Show how the LCD display is interfaced to the 68000. The base address of the display (cell 0) is $28000.

10. Write the software necessary to output the alphabet to the LCD display of Question 9. Format the output as follows:

```
ABCDEFGHIJKLMNOP
---QRSTUVWXYZ---
```

11. Show how the 6821 is programmed for port A out and port B in. Assume a base I/O address of $40000.

12. Show how an 8-by-8 matrix of LEDs can be controlled with the 6821.

13. What initialization is needed for the LED matrix circuit of Question 12?

14. Rewrite the BINCNT program so that the LEDs generate the following display:

```
XXX-----
-XXX----
--XXX---
---XXX--
----XXX-
-----XXX
----XXX-
---XXX--
--XXX---
-XXX----
XXX-----
```

An X indicates a LED that is on.

15. Show how the 6821 could be used to drive two 7-segment LED displays.

16. Write a BCD counter for the circuit of Question 15.

17. Design an address decoder for a 6821 that will respond to addresses $CE040 through $CE04F.

18. What is needed to program the 6850 for 8 data bits, no parity, and 2 stop bits?

19. In Example 9.1, what is the baud rate if the RxC and TxC clocks are running at 76.8 kHz?

20. Modify the CIN routine so that it checks the receiver overrun and parity error flags. Return an FF in D1 if either error is found.

21. Repeat Example 9.2 for a 6850 running at 9,600 baud.

22. Write a routine that continually calls CIN and COUT (to echo all input characters to the output). If the input character is a carriage return ($0D), output a $0D and a $0A (line feed).

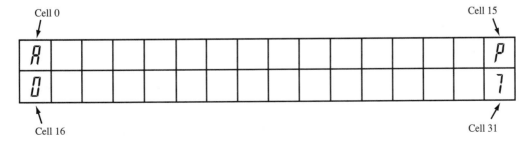

FIGURE 9.20   For Questions 9.9 and 9.10

23. Show how a second 6850 may be added to the circuitry of Figure 9.15.
24. Why is it necessary for the 6850 to utilize synchronous bus transfers?
25. When is it necessary to call the SCROLL routine?
26. Eight 1,920-byte blocks of memory have been loaded with screen characters. To simulate animation, each block is copied, one at a time, to screen memory. A short delay is included between copies. Assume that A4 points to the beginning of screen memory and A5 to the beginning of the eight image blocks. Write an ANIMATE routine to perform the copying.
27. Write a routine called MIRROR that swaps the left and right side of the display screen. Thus, a line that looks like this:

```
The quick brown fox jumps over the lazy cat.
```

becomes:

```
.tac yzal eht revo spmuj xof nworb kciuq ehT
```

28. Write the code needed to copy a 32-by-32 bit image to the bit-mapped video RAM. Assume that A4 points to the image and A5 points to the location within video RAM where the image will be copied.
29. Write a routine called HLINE that draws a horizontal line between the pixels specified in D0 and D1. The scan line is specified in D2.
30. Write a routine called VLINE that draws a vertical line between the scan lines specified in D0 and D1. The pixel number is specified in D2.
31. How many bytes does a 342-by-512 pixel display require if 24 bits are used to specify the pixel color?
32. What addresses will activate the memory-mapped output port in Figure 9.21?
33. Write a routine that will turn the UP LED on in the circuit of Figure 9.21, count to 10 (0 . . . 9), then repeat the count for each of the other directions in this order: UP DOWN LEFT RIGHT. Only one LED may be on at one time.
34. Figure 9.22 shows a Motorola 14499 multiplexed display driver interfaced to four 7-segment displays. The 14499 performs all the work generating the multiplexed display of four BCD digits. If $\overline{\text{ENABLE}}$ is low, DATA is clocked into the 14499 on the falling edge of the clock. Bits are clocked into the 14499 in the following format: DP_1 through DP_4, then the MSB through LSB for digit 1, followed by the bits for the other three digits. A total of 20 bits must be clocked into the 14499 by manipulating the bits of the 8-bit output port. Write a subroutine that outputs the four-digit BCD number stored in register D7, with only the decimal point for digit 4 turned on.

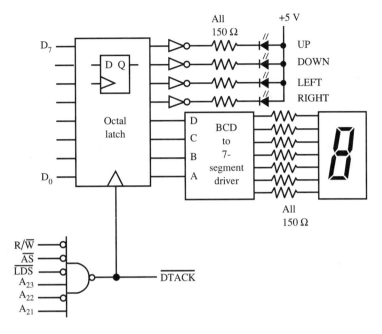

FIGURE 9.21 For Questions 9.32 and 9.33

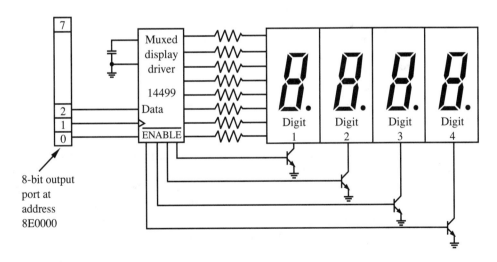

FIGURE 9.22 For Question 9.34

CHAPTER 10

Advanced Programming Using 68000 Peripherals

OBJECTIVES

In this chapter you will learn about:

- Controlling DMA operations
- Managing memory
- Parallel and serial data transmission
- Hardware interrupt handling
- Floating-point coprocessor functions
- Interfacing non-Motorola devices
- Additional 68000-based peripherals

10.1   INTRODUCTION

The power of a microprocessor can be increased by the use of peripherals designed to implement special functions, functions that may be very difficult to implement via software. A good example of this principle would be in the use of a coprocessor. The coprocessor comes equipped with the ability to perform complex mathematical tasks, such as logarithms, exponentials, and trigonometry. The 68000, although powerful, would require extensive programming to implement these functions, and even then would not compute the results with the same speed. Thus, we see that there are times when we have to make a hardware/software trade-off. In this chapter we will concentrate on applications that employ the use of standard peripherals, designed specifically for the 68000. In each case, we will examine the interfacing requirements of the peripheral, and then see how software is used to control it.

Keep in mind that the designers of commercial 68000-based systems have the option of using an **ASIC** (Application Specific Integrated Circuit) that performs the functions of many different peripherals. The ASIC is specifically designed to mimic the operation of each peripheral, from both a hardware and a software viewpoint. This has the double advantage of reducing the hardware chip count *and* supporting the same software. But someone still needs to design and fabricate the ASIC, which may cost more than a company is willing to pay to produce a product. In this case, it is cheaper to use individual peripheral chips. So, it is still worthwhile to study the operation of the 68000-family peripherals.

You are encouraged to refer to Appendix D as you read the chapter, for additional details.

Sections 10.2 through 10.8 cover, respectively, the following peripherals:

- The 68440 Dual DMA Controller
- The 68451 Memory Management Unit
- The 68681 Dual UART (Universal Asynchronous Receiver/Transmitter)
- The 68230 PI/T (Parallel Interface/Timer)
- The 68901 Multifunction Peripheral
- The 68153 Bus Interrupt Module
- The 68881 Floating-Point Coprocessor

Section 10.9 explains how a non-Motorola peripheral can be interfaced to the 68000, and Section 10.10 gives a brief summary of other peripherals designed for use with the 68000. Troubleshooting techniques are the subject of Section 10.11.

10.2   THE 68440 DUAL DMA CONTROLLER

If we stop to consider the operation of a multiuser microprocessor system capable of supporting user terminals, a printer, and a disk storage system, it is clear that the system will spend a great deal of time performing input/output operations. These I/O operations result from transfers between the CPU, memory, and all other devices connected to the system bus. When a single CPU is utilized, the users must all be given a small slice of CPU time in which their respective jobs may be processed. If a user is in the middle of a disk read operation, many time slices will be used simply to transfer data from the disk to memory, as shown in Figure 10.1. This method unfortunately wastes CPU time. Even when multiple CPUs are used, we still have the problem of dedicating large amounts of CPU time for I/O operations. To improve the situation we need to remove the CPU from the I/O operations. This is accomplished with the addition of a **direct memory access** controller.

Figure 10.2 shows the addition of a DMA controller in a microprocessor-based system. The DMA controller is used to supervise high-speed data transfers between the devices connected to the system bus and to the system's memory. It is not uncommon for one of the devices to have a very fast data rate (that is, 5 million bytes per second in a hard disk device). This data rate is much faster than the CPU can handle by itself. It is not possible for the CPU to directly transfer the data from the high-speed device to memory (or vice versa). This leads to the addition of the DMA controller, a hardware chip designed to manipulate the system bus in a way similar to the processor's read and write operations, but at

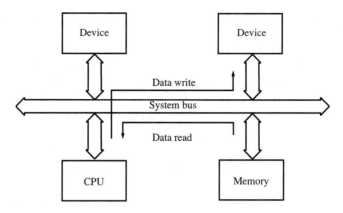

FIGURE 10.1   A system using the CPU to perform all I/O

a much faster speed. The DMA controller will directly address the memory and peripheral devices. The CPU will simply program the DMA controller with information concerning how much data should be transferred, where the data should come from, and where it should go. Once the DMA controller is told to begin the transfer, the CPU is free to perform other processing. The DMA controller allows the processor to spend more time processing information, instead of transferring it.

A device capable of performing DMA in a 68000-based system is the 68440 dual channel DMA controller. A channel is a data path between memory and a peripheral, a peripheral and memory, or memory and memory. Memory-to-peripheral transfers were the subject of our previous discussion. Memory-to-memory transfers are also necessary operations. Consider an operating system that needs to constantly relocate user programs within memory, due to its memory management scheme. We cannot have the CPU performing the data transfers, since this will result in a great loss of performance.

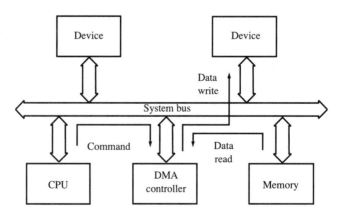

FIGURE 10.2   A system employing a DMA controller to perform I/O

Interfacing the 68440

The 68440 has a number of signals that connect directly to the 68000 and additional signals for communication with peripherals. Figure 10.3 shows the various signal groups that must be properly interfaced to obtain DMA capabilities. The signals in the asynchronous bus control group are already known to us. $\overline{\text{UDS}}$ and $\overline{\text{LDS}}$ (A_0 and $\overline{\text{DS}}$) have two functions, depending on the size of the data bus. $\overline{\text{UDS}}$ and $\overline{\text{LDS}}$ are used with a 16-bit data bus. A_0 and $\overline{\text{DS}}$ are used with an 8-bit data bus (found on the 68008). A_1 through A_7 are used to select the internal 68440 registers. $\overline{\text{BEC}_0}$ through $\overline{\text{BEC}_2}$ (bus exception control) are used to indicate to the 68440 what type of abnormal bus cycle has occurred—that is, bus error or RESET.

The 68440 employs a multiplexed address/data bus (A_8/D_0 through A_{23}/D_{15}) for all communication. When the 68440 is receiving commands from the CPU, the multiplexed bus is a data bus. When the 68440 is transferring data during a DMA operation, the multiplexed bus acts as both an address and data bus. External buffering and latching circuitry are needed, together with control signals from the 68440, to demultiplex the bus. Before we examine the external circuitry required, we must examine the control signals involved. The multiplex con-

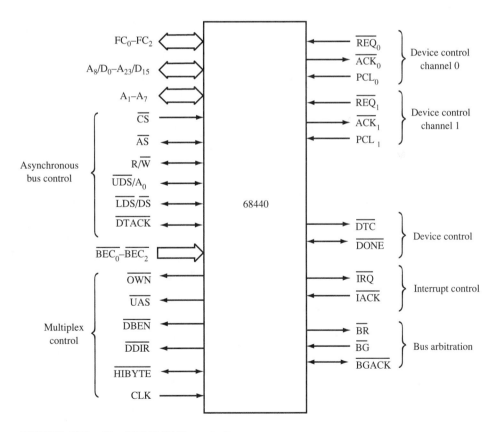

FIGURE 10.3   The 68440 DMA controller

trol signals are used to control the buffering and latching of data and addresses used by the 68440. $\overline{\text{OWN}}$ is an active low signal that indicates when the 68440 controls the system bus. It is used to enable the external address latches and data buffers. $\overline{\text{UAS}}$ (upper address strobe) is used to latch A_8 through A_{23} when they appear on the multiplexed bus. $\overline{\text{DBEN}}$ (data buffer enable) controls the operation of the external data bus buffer. Since the data bus is bidirectional, an additional signal is needed to control its direction during read and write operations. The signal that performs this function is $\overline{\text{DDIR}}$ (data direction). When $\overline{\text{DDIR}}$ is low, data is being read by the 68440. When $\overline{\text{DDIR}}$ is high, the 68440 is outputting data. The last signal in the multiplex control group is $\overline{\text{HIBYTE}}$, which is used in two ways. During a RESET operation, $\overline{\text{HIBYTE}}$ is used as an input, to indicate the size of the data bus (8 bits when low, 16 bits when high). Once the 68440 has been initialized, $\overline{\text{HIBYTE}}$ is used as an output, to gate data onto the correct half of the data bus when the 68440 is accessing an 8-bit device. All of the multiplex control signals work together to control the multiplexed bus. Figure 10.4 shows how external

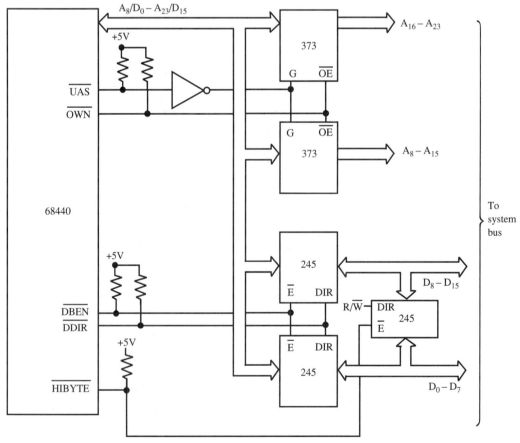

Note: 74LS373 is an octal latch.

74LS245 is a bidirectional buffer.

FIGURE 10.4   External circuitry needed to demultiplex A_8/D_0 through A_{23}/D_{15}

address latches and data buffers can be controlled with these signals. The two 74LS373 octal transparent latches are used to store the 68440's A_8 through A_{23} address information, when it appears on the multiplexed bus. These latches are controlled by $\overline{UAS}$ (to latch the address) and $\overline{OWN}$ (to output enable them during memory/peripheral accesses).

The three 74LS245 bidirectional buffers control the transfer of data between the 68440 and the system bus. $\overline{DBEN}$ and $\overline{DDIR}$ are used to enable the 245s and control their direction (input or output). $\overline{HIBYTE}$ is used with the third 245 to swap the upper and lower halves of the data bus during accesses to 8-bit peripherals.

The bus arbitration signals are used by the 68440 to take control of the system bus during DMA operations. They allow the 68440 to become the bus master, placing the 68000 into a wait state while the system bus is being used. $\overline{BR}$ (bus request) is active when the 68440 is requesting the bus. The processor (or other bus master) will respond to $\overline{BR}$ by asserting $\overline{BG}$ (bus grant). When the 68440 does take control of the bus, it will assert $\overline{BGACK}$ (bus grant acknowledge).

The interrupt control signals $\overline{IRQ}$ and $\overline{IACK}$ are used to support vectored interrupts. The 68440 can be programmed to interrupt the processor when a number of events occur, one of which is the completion of a DMA transfer operation.

The two groups of device control channel signals ($\overline{REQ}$, $\overline{ACK}$, and PCL) are used to control two peripherals that may be connected to the system bus. $\overline{DTC}$ and $\overline{DONE}$ are additional device control signals that are shared by both channels. $\overline{REQ_0}$ and $\overline{REQ_1}$ (request) is used by the peripheral to request data transfer between itself and memory. $\overline{ACK_0}$ and $\overline{ACK_1}$ (acknowledge) is asserted by the 68440 when it is transferring the data. PCL_0 and PCL_1 are peripheral control lines whose functions are programmed for various kinds of status indications. $\overline{DTC}$ (data transfer complete) is asserted by the 68440 whenever a successful data transfer has been completed. $\overline{DONE}$ is a bidirectional signal used to indicate that the last data transfer completed a block operation (that is, all data has been transferred).

Since timing is critical to proper operation, a CLK signal is provided to generate all internal 68440 timing.

Programming the 68440

Seventeen internal registers are used by the 68440 to control a single DMA channel (with a second group of 17 controlling the other channel). These registers control the channel priority, the type of DMA operation used, and the size of the data transfer (in bytes or words). They are also used to hold the source and destination addresses for the block transfers, the interrupt vector numbers, and channel status bits.

The general control register (GCR), located at internal address FF, selects the burst transfer time (the number of clock cycles allowed during a data transfer) and the system bus bandwidth available to the 68440. The bus bandwidth indicates how much activity is found on the system bus. A portion of this activity will be due to the 68440 executing DMA transfers. Figure 10.5 indicates the bit assignments within the GCR, and how they must be coded to select specific functions.

The channel priority register (CPR), located at internal address 2D, sets the channel priority. When both channels have been programmed for transfer and request the bus at the same time, the CPR is used to break the tie and decide which channel gets priority. Priority 00 is higher than 01. When the priority is the same for each channel, a round-robin approach is used, with each channel alternately receiving use of the bus.

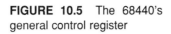

FIGURE 10.5 The 68440's general control register

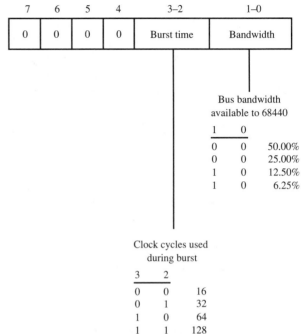

Two interrupt vector registers are used by the 68440. The normal interrupt vector register (NIVR), located at internal address 25, contains the interrupt vector used when the 68440 completes an operation, such as a block transfer. The error interrupt vector register (EIVR), located at internal address 27, contains the interrupt vector used when the 68440 encounters an error during a DMA operation. When both channels request an interrupt, the highest priority channel is serviced first.

Three registers are used to hold addresses used during DMA operations. The memory address register (MAR), located at internal address 0C, holds the address of the memory location to be used during the data transfer. The device address register (DAR), located at internal address 14, contains the address of the peripheral used during a data transfer. The base address register (BAR), located at internal address 1C, is used to access special tables in a **chain** operation. A chain operation is a special type of DMA, in which 68440 registers are loaded from a data table in memory.

Each address register (MAR, DAR, and BAR) is associated with a 4-bit function code register. The memory (MFCR), device (DFCR), and base (BFCR) function code registers, located, respectively, at internal addresses 29, 31, and 39, are used by the 68440 during a DMA operation to provide additional information to external circuitry connected to the system bus. The contents of the function code register involved in the operation are output on the FC_0 through FC_2 lines during the DMA operation.

The size of the block of data to be transferred is contained in the memory transfer count register (MTCR), located internally at address 0A. This 16-bit register is decremented each time data is transferred by the 68440. When the MTCR reaches 0, the channel operation terminates.

The channel error register (CER), located at internal address 01, indicates the type of error that may have occurred during a channel operation. The lower 5 bits are used to code the error. If all bits are low, no error has occurred. The CER indicates configuration, operation timing, address, bus, and counter errors. Software and external abort errors are also indicated. Figure 10.6 shows how these errors are coded within the CER.

The 68440's channel status register (CSR), located at internal address 00, indicates the status of the channel and its current operation. Figure 10.7 shows how the bits in the CSR are assigned. All status bits are active high. COC (channel operation complete) indicates the current state of the channel operation (completed when high). BTC (block transfer complete) indicates the status of the current block transfer operation. NDT (normal device termination), when high, indicates that the device has been terminated with $\overline{\text{DONE}}$. ERR (error) indicates that an error has occurred (and the CER must be read). ACT (channel active) indicates when the channel is active. RLD (reload) indicates when a reload has occurred. PCT (PCL transition) is high when a high-to-low transition occurred on PCL. PCS (PCL state) indicates the current state of the PCL line.

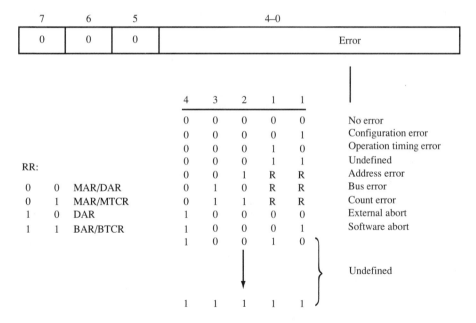

FIGURE 10.6 Types of 68440 channel errors indicated by the channel error register

7	6	5	4	3	2	1	0
COC	BTC	NDT	ERR	ACT	RLD	PCT	PCS

FIGURE 10.7 68440 channel status register

Four additional registers are used to control DMA in a 68440 channel. Each register holds 8 bits and controls a different aspect of the channel operation. The entire channel is controlled by the channel control register (CCR), located at internal address 07. The sequence control register (SCR), located at internal address 06, controls the sequencing of the MAR and DAR. The sequencing indicates how the register is adjusted during a DMA operation. The register may remain unchanged or be programmed to increment after each transfer is performed. Bit assignments for the CCR and SCR are listed in Figure 10.8.

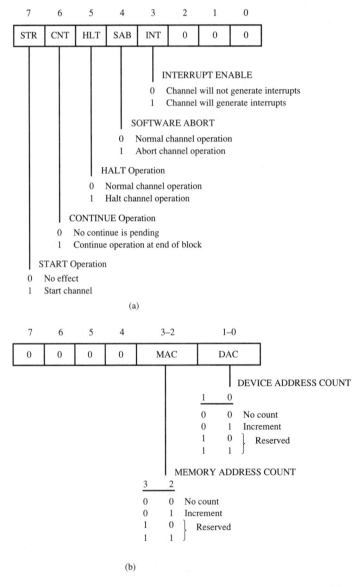

FIGURE 10.8   Bit assignments for the (a) channel control register and (b) sequence control register

The device control register (DCR), located at internal address 04, selects the operation of the device used in the channel operation. The DCR defines the size of the data transfer (byte or word), the type of DMA operation implemented (burst or cycle steal), and the type of device used (6800- or 68000-based peripheral). Burst operation means that DMA is performed in small bursts, with a number of bytes or words transferred during each burst. Cycle stealing involves accessing the system bus when the processor is not using it. The processor is not using the bus during times when it is performing internal instruction execution (register-to-register transfers, for example). Cycle-stealing DMA is much slower than burst DMA. Even so, it has its advantages. Burst DMA requires control of the system bus from time to time, and this involves placing the processor into a wait state for the duration of the burst cycle. This technique slows down program execution when large blocks of data must be transferred. Figure 10.9 shows the bit assignments for the DCR. In device-type selection we see choices between explicit and implicit addressing. Explicit addressing requires that the 68440 output the address of the device to select it. Implicit addressing utilizes the five control lines ($\overline{\text{REQ}}$, $\overline{\text{ACK}}$, PCL, $\overline{\text{DTC}}$, and $\overline{\text{DONE}}$) to control the peripheral. The function of the PCL line is programmed by bits 0 to 2 in the DCR.

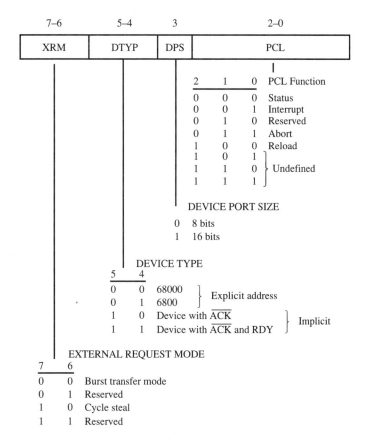

FIGURE 10.9   68440 device control register

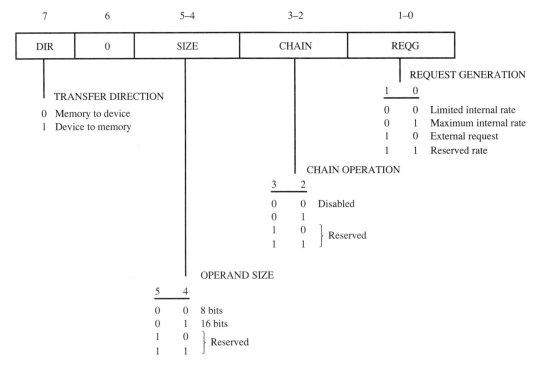

FIGURE 10.10   Bit assignments for the operation control register

The last control register is the operation control register (OCR), located at internal address 05. The OCR determines the direction of the data transfer (memory to device, or device to memory), the data size (byte or word), and the way in which transfer requests are detected. Chaining is not allowed in the 68440, although it is supported in other DMA controllers such as the 68450. Figure 10.10 shows the bit assignments for the operation control register.

So far we have been exposed to the registers used to control DMA in a 68440 channel. Example 10.1 will show how we can use the 68440 to transfer a 32KB block of RAM from one location to another.

Example 10.1:   Consider the following segment of code, which is used to copy a 32KB block of RAM from BLOCKA to BLOCKB:

```
BLOCKA      EQU       $10000              ;starting address of first block
BLOCKB      EQU       $20000              ;starting address of second block
SIZE        EQU       32768               ;size of block
  .
  .
  .
            MOVEA.L   #BLOCKA,A0          ;init source pointer
            MOVEA.L   #BLOCKB,A1          ;init destination pointer
            MOVE.W    #SIZE-1,D0          ;init loop counter
TRANSFER    MOVE.B    (A0)+,(A1)+         ;transfer a byte
            DBRA      D0,TRANSFER
```

From a previous discussion of instruction execution times, we recall that the MOVE.B and DBRA instructions will take a certain number of clock cycles to execute. Since they are in a loop, this number will be multiplied by 32,768. The minimum number of clock cycles required to execute MOVE.B and DBRA comes to 22. When this number is multiplied by 32,768, we get a total of 720,896 clock cycles required. Ignoring overhead, and assuming an 8-MHz clock, the time required for the processor to transfer the 32KB block is over 90 ms! Compare this with a 68440, transferring the same block at a 5 million-byte-per-second rate. Only 6.5 ms are needed to complete the transfer. This simple example shows how the 68440 greatly increases system performance by dedicating it to the transfer of large blocks of data.

A routine to implement the transfer of the same block of data is listed here. The code assumes that the 68440 has a base address of 3000, and the channel used for the DMA operation is channel 0.

```
MAR        EQU        $300C
DAR        EQU        $3014
MTCR       EQU        $300A
MFCR       EQU        $3029
DFCR       EQU        $3031
NIVR       EQU        $3025
EIVR       EQU        $3027
GCR        EQU        $30FF
SCR        EQU        $3006
OCR        EQU        $3005
DCR        EQU        $3004
CCR        EQU        $3007
           .
           .
           .
DMAMOVE    MOVEA.L    #BLOCKA,MAR    ;load MAR with source address
           MOVEA.L    #BLOCKB,DAR    ;load DAR with destination address
           MOVE.W     #SIZE,MTCR     ;load MTCR with size of block
           MOVEQ.B    #1,D0
           MOVE.B     D0,MFCR        ;load 'user data' function code into MFCR
           MOVE.B     D0,DFCR        ;load 'user data' function code into DFCR
           MOVE.B     #$90,NIVR      ;load NIVR with vector number
           MOVE.B     #$91,EIVR      ;load EIVR with vector number
           MOVE.B     #$06,GCR       ;select 32-cycle burst, 12.5% bandwidth
           MOVE.B     #$05,SCR       ;let MAR and DAR increment
           MOVE.B     #1,OCR         ;select memory-to-device byte transfer
           MOVE.B     #0,DCR         ;select explicit 68000 burst, byte device
           MOVE.B     #$88,CCR       ;start DMA and enable interrupt
           RTS
```

DMAMOVE initializes all control registers and provides for generation of a vector $90 interrupt when the DMA operation completes.

10.3   THE 68451 MEMORY MANAGEMENT UNIT

A multiuser operating system may consist of many users running programs on a single processor, or even more users running on multiple processors. Either way, the system's memory, at any point in time, may contain code and data for several users. Figure 10.11

shows two examples of how memory might be allocated for a number of users on two different occasions. Notice that memory in both cases has not been allocated sequentially to each user. This is due to the asynchronous nature of the users. User 5 may have submitted his or her job at a later time than users 1 and 2. Also, the location of each user's block of code in memory depends on what memory is available when the user's job is submitted. We can see from Figure 10.11 that the users, at a later point in time, have been allocated memory differently. This brings up an important point: User programs must be able to run *anywhere* in memory. If this were not possible, the operating system would always have to wait for a specific block of memory to become available before assigning it to a specific user. The only other option would be to reassemble or recompile the user's program each time it was loaded into memory at a new address. This would be time consuming and very inefficient. Also, although the 68000 is capable of relative addressing, there are times when absolute memory addresses must be used.

Consider a program that has been assembled into machine code from a source file containing an ORG $2000 statement. The program must be loaded into memory beginning at address 2000 for proper operation. Examine Figure 10.12, which shows the same program loaded into memory at two different addresses (and at two different times). Initially the program is loaded at the address it was assembled to run at, address 2000. The next time the program is submitted for execution, the operating system places it in memory beginning with address 9800, most likely due to the lack of free memory from 2000 to 3FFF. Two methods are commonly used to enable correct program execution in this case. The first method requires that all absolute addresses in the program be modified before it is loaded into memory. The absolute addresses within the program are changed by adding or subtracting an offset, the difference between the desired load address and the available load address. Figure 10.12 shows that the difference between these two load addresses is 7800. Each absolute address within the program must be modified by adding 7800 to it. For example, an instruction such as JMP   2B14 would become JMP   A314, where A314 equals 2B14 plus 7800. This address translation ensures that the JMP instruction goes to the proper location, no matter where the program is loaded.

FIGURE 10.11 Memory allocation for multiple users

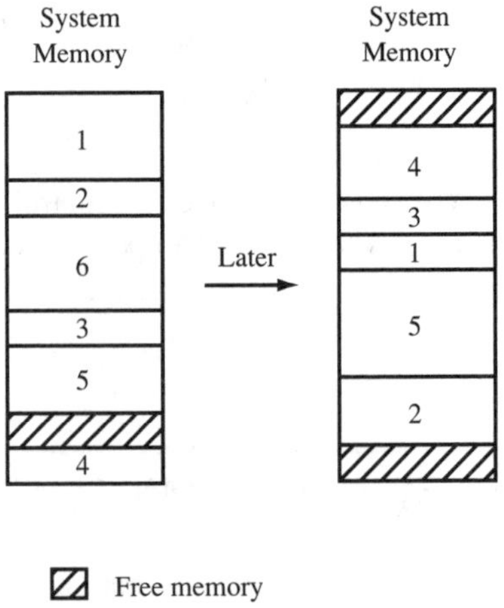

Free memory

FIGURE 10.12 A program encounters two different load addresses

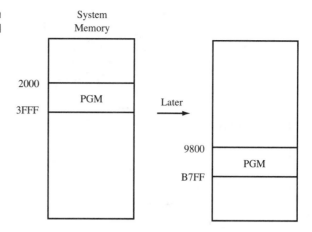

To save time, a second technique called **dynamic address translation** is used to accomplish the same goal. The difference is that the addresses are modified as they are used, during execution. The JMP   2B14 instruction is still loaded as JMP   2B14. During execution, the 2B14 is changed into A314. This process saves time, since the entire program does not have to be altered before loading. A device capable of performing this dynamic address translation is the 68451 memory management unit.

The 68451 will perform translation over the entire 16MB address space of the 68000. The 68451 breaks this address space into 32 **segments.** The size of each segment is variable but limited to block size multiples of 2, starting with 256 bytes. Thus 256, 1K, 16K, 64K, 128K, and 1,024KB are all valid segment sizes. The 68451 also allows segments to be combined to form different lengths as well. For example, if a user program requires 36K of memory, two segments of size 32K and 8K can be combined to obtain the required memory. If this were not possible, a 64K segment would have to be used, with 28K of the 64K segment wasted. The operating system defines segments by writing data to special registers called **descriptors.**

Another feature provided by the 68451 is protection. When many user programs reside in memory simultaneously, it must be possible to restrict memory references between programs. Otherwise, one user's program might be able to write over another's code, with certainly disastrous results. The 68451 is capable of providing this protection and interrupting the processor when an out-of-bounds address is generated from within any segment.

The 68451 performs memory management with a mixture of hardware and software. Details are presented in the next two sections.

Interfacing the 68451

Figure 10.13 shows a simplified diagram of a 68000 system utilizing the 68451. The 68451 is inserted between the processor and memory. This gives the 68451 control over all addresses presented to the memory section. Notice that only processor address lines A_1

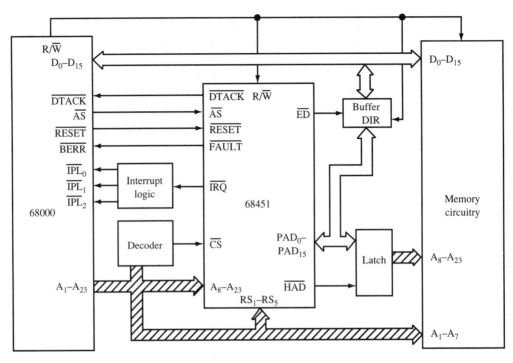

FIGURE 10.13   Interfacing the 68451 to the 68000

through A_7 go directly to the memory circuitry. Since the size of any segment must be at least 256 bytes, these seven address lines do not require translation. The remaining address lines, A_8 through A_{23}, must be translated and thus become inputs to the 68451. An address decoder is used to enable the 68451 when it is being programmed (more on this in the next section). For the 68451 to recognize processor memory requests, it must have access to the signals that govern such requests. The signals used for this purpose are $\overline{\text{DTACK}}$, $\overline{\text{AS}}$, and R/$\overline{\text{W}}$. $\overline{\text{RESET}}$ is also connected to ensure that the 68451 comes up in a valid state, and will respond to commands from the processor when it is initialized.

Programming the 68451 is accomplished by loading its internal registers with the necessary data to maintain the various segments required by the operating system. These registers are addressed via the five register-select lines, RS_1 through RS_5. All data enters and leaves the 68451 through the multiplexed address/data bus on lines PAD_0 through PAD_{15}. When these 16 lines are used for bidirectional data transfer between the 68451 and the processor, the $\overline{\text{ED}}$ (enable data) output will be low. This output is used to enable and disable a bidirectional data bus buffer. The buffer is used to gate data on the 68000's data bus onto the 68451's data bus. When $\overline{\text{ED}}$ is high, the 68451 is indicating that PAD_0 through PAD_{15} represent translated address information, and should be interpreted as address lines A_8 through A_{23}. Address information present on these lines is latched externally, whenever $\overline{\text{HAD}}$ (hold address) is low. Two additional outputs, $\overline{\text{WIN}}$ (write inhibit) and $\overline{\text{MAS}}$ (mapped address strobe), are not shown on the simplified schematic in Figure 10.13 but are required for correct operation of the memory section. $\overline{\text{MAS}}$ indicates to external circuitry that a valid **physical** address exists on PAD_0 through PAD_{15} (similar to what $\overline{\text{AS}}$ does for the 68000). $\overline{\text{WIN}}$ is used to inhibit writes to protected sections of memory.

$\overline{\text{FAULT}}$ is used to generate a bus error exception in the processor, whenever an attempt is made to write to protected memory. $\overline{\text{FAULT}}$ is also active when the processor tries to access a segment that has not yet been defined.

$\overline{\text{IRQ}}$ (interrupt request) is used to interrupt the processor when an access is made to a segment whose descriptor indicates that interrupts are enabled. This means that the 68000 can detect any read or write requests to a block of memory, something we may wish to avoid in a multiuser system. Remember that no user should be able to alter another user's code. The 68451 supplies the processor with an exception vector number during the interrupt acknowledge cycle (initiated by a low on the 68451's $\overline{\text{IACK}}$ input).

Three signals are included to allow for multiple 68451s to be used in a single system. They are $\overline{\text{GO}}$ (global operation), $\overline{\text{ANY}}$, and ALL. These three signals are all bidirectional and are used by the 68451s to communicate with each other. It becomes necessary, at times, for one 68451 to become a **master** and the other 68451s **slaves.** This hierarchical approach is used to prevent memory addressing errors in multiple 68451 systems.

One last signal deserves mention before we examine the software operation of the 68451. MODE, an input, is used to tell the 68451 which mode of operation it should be in. MODE is used to indicate one of three modes in the following way: If MODE is left open, mode A is selected. If MODE is tied to ground, mode S_2 is enabled. If MODE is tied high, mode S_1 is selected. Modes A and S_1 are used for asynchronous operation. Mode S_2 is used for synchronous operation. The design of the memory circuitry determines which mode should be used for proper operation.

Programming the 68451

The 68451 performs address translation by mapping **logical** addresses into **physical** addresses. A logical address is any address issued by the processor (or by a DMA device). A physical address is an actual address in the memory section. Logical to physical translation is performed by mapping logical addresses into physical addresses. This is accomplished inside the 68451 by special registers called **descriptors.** A descriptor is a set of six registers (9 bytes total) containing information about a single segment. The descriptor is loaded with data indicating the base logical and physical addresses of the segment, the segment size, the address space number (each segment occupies a particular address space), and some status information. There are a total of 32 descriptors available, and all are defined as shown in Figure 10.14.

The logical base address register (LBA) contains the upper 16 bits representing the starting address of the logical segment. The logical address mask register (LAM) is used to specify the length of the segment. This is done by writing 1s into certain bit positions within the LAM. These 1s are used to compare bits in the LBA with the incoming logical address. When a match occurs, the 68451 knows a segment is being addressed. The physical base address register (PBA) contains the starting address of the physical memory segment. Address bits in the incoming logical address that do not require translation will pass straight through to the physical address outputs. These bits are indicated by 0s in the LAM. Logical

FIGURE 10.14 Descriptor definition

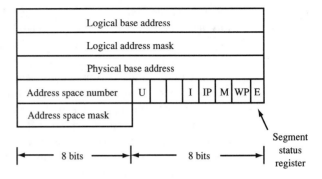

address bits that require translation are replaced by the corresponding bits from the PBA. Figure 10.15 shows an example of how a logical address is translated into a physical address. The upper 16 bits of the logical address yield C5C8. The LBA register contains C5C0, and the LAM register indicates that logical address bits 23 through 14 should be used to find an address match. Since these 10 bits in both the logical address and the LBA register match, we know that an address translation should be performed. The 0s in the LAM register indicate two things. First, logical address bits 13 through 8 should not be used to find an address match, and thus pass through to the physical address, without translation. Second, they indicate that the segment size is 16,384 bytes. Since 6 bits in the LAM are 0, we are requesting 64 (2 raised to the 6th power) 256-byte blocks for this segment.

The 1s in the LAM indicate that logical address bits 23 through 14 should be translated. Translation is accomplished by using the upper 10 bits of the PBA register to complete the physical address, which becomes 3A88xx.

If no match occurs between the logical address and the LBA register, this process is repeated for every descriptor that has been defined. If no match occurs in any descriptor, the 68451 will indicate an undefined segment error by activating $\overline{\text{FAULT}}$. When more

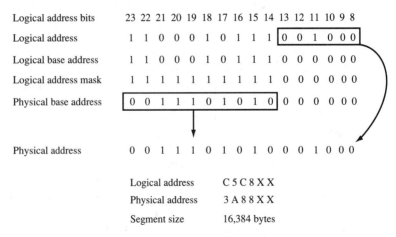

FIGURE 10.15 Mapping a logical address into a physical address

than one 68451 is used in a system, they must all fail to match the address, before $\overline{\text{FAULT}}$ is activated.

The address space number (ASN) is an 8-bit value used to assign a number to the segment. Segments can be grouped by giving them the same ASN. A previous example required the use of a 32K segment and an 8K segment so a 36K program could be loaded into memory. In this case, each segment would be given the same ASN. The ASN is used in conjunction with a table, called the address space table, to determine if the memory request should be allowed to proceed. This makes it possible for us to protect certain segments, since the address space table contains a list of the allowable address space numbers. The address space mask (ASM) is used to determine which bits in the ASN are used in comparisons.

The segment status register (SSR) is an 8-bit register containing the following status bits:

U: (U)sed. This bit is set if the segment has been accessed since it was defined.

I: (I)nterrupt control. If set, an interrupt is generated when the segment is accessed.

IP: (I)nterrupt (P)ending. This bit is set if the I bit is set, and the segment is accessed.

M: (M)odified. This bit is set if the segment has been written to since it was defined.

WP: (W)rite (P)rotect. Writes to the segment are not allowed if this bit is set.

E: (E)nable. If set, the segment is enabled and can be used in logical address comparisons.

An operating system will make good use of these status bits. To illustrate, suppose that a multiuser system has filled its entire memory with user programs, and another program is submitted. It becomes necessary to determine which user program should be swapped out, to make room for the new program. When the program to be replaced is determined, the operating system must decide if it should save the memory used by the program or simply overwrite it with the new program. If the old program did not change its memory image (by writing data into its segment), it can be overwritten by the new code. If it has modified its contents, it must be saved on disk before the new program can be loaded. The operating system will make use of the M status bit when deciding if it should back up the old program on disk.

Each of the 32 descriptors is written to by first loading the data into the 68451's accumulator. The accumulator consists of nine 8-bit registers that are defined in the following way:

Logical base address (MSB):	Address 20
Logical base address (LSB):	Address 21
Logical address mask (MSB):	Address 22
Logical address mask (LSB):	Address 23
Physical base address (MSB):	Address 24
Physical base address (LSB):	Address 25

Address space number (ASN):	Address 26
Segment status register (SSR):	Address 27
Address space mask (ASM):	Address 28

The internal 68451 addresses (20 through 28) are selected by register select lines RS_1 through RS_5.

The address space table (AST) is a collection of 16 8-bit registers that begin at address 00 and occupy even addresses up to 1E. The AST is loaded with address space numbers by the operating system. The AST is then accessed, depending on the state of the processor's function codes. For example, when the function code outputs indicate a request to user program memory, the 68451 will examine the third number in the AST. If this number does not match the address space number in the segment being addressed, an error will result.

Descriptors are loaded by writing all required information for the segment into the accumulator, and then loading the number of the descriptor (0 through 31) into the 68451's descriptor pointer register (DP). Once this has been done, the descriptor can be saved using a load descriptor command. The load descriptor command is issued by an access to address 3F.

A routine to load descriptor 7 from the 9 bytes stored in memory (beginning at NEWSEG) follows. The 68451's address decoder responds to a base address of 8000 in this example.

```
NEWSEG    DS.B      9      ;A 9-byte area containing the data to be placed
                          ;into the descriptor
LBAHI     EQU       $8020
LOAD      EQU       $803F
.

.

.
LOADES    MOVE.W    #8,D0                    ;init loop counter
          MOVEA.L   #NEWSEG,A0               ;init pointer to descriptor data
          MOVEA.L   #LBAHI,A1                ;init pointer to 68451 accumulator
FILLA     MOVE.B    (A0)+,(A1)+              ;load accumulator with data
          DBRA      D0,FILLA
          MOVE.B    LOAD,D1                  ;begin load descriptor operation
          CMPI.B    #0,D1                    ;was load successful?
          BNE       FAILED
          RTS
FAILED    JMP       LOADFAULT
```

The load descriptor operation begins by reading address 3F. If the load is successful, the 68451 will return 00. If the load fails because a previously defined descriptor is mapped to the same logical address (called a collision), the 68451 will return FF.

When the 68451 detects a protection violation, or if a logical address does not match any of the descriptor LBAs, it will issue an interrupt request to the processor. When the processor enters into its interrupt acknowledge cycle, the 68451 will place an 8-bit vector number on data lines D_0 through D_7. The vector number used by the 68451 comes from its interrupt vector register (IVR), located at address 2B. The operating system must have code capable of handling the interrupt, and this code must be accessed by the exception generated by the IVR vector number.

The following section of code can be used to load a new vector number into the IVR. This may not be necessary, since the IVR is loaded with vector number 0F after a RESET. Once again, the 68451 has a base address of 8000.

```
VECTOR   DS.B     1       ;place vector number here
IVR      EQU      $802B
.
.
.
LOADVEC  MOVEA.L  #IVR,A0              ;init pointer to IVR
         MOVE.B   VECTOR,D0           ;load vector number into D0
         MOVE.B   D0,(A0)             ;load IVR with new vector number
         RTS
```

The 68451 is a very complicated device. Even so, use of the 68451 in a new operating system will provide the designers with a reliable method of managing memory, with a minimum of hardware and software.

10.4   THE 68681 DUAL UART

As we saw in Chapter 9, a UART is used to convert parallel data into serial data, and vice versa, at different baud rates, and in a variety of formats. A computer system might use serial communication with its modem, mouse, or printer. Thus, there is often a need for more than one serial channel in a small system. The 68681 contains two separate UARTs, both of which are independently programmable. As Figure 10.16 shows, a small 68000-based system uses the 68681 to communicate at two different speeds. A low-speed, 2,400-baud channel connects to the user's ASCII terminal (which itself may utilize a dual UART) and a 9,600-baud high-speed channel connects to a central computer. Simple keystrokes on the user terminal may result in multiple high-speed transfers between the small system and the central computer.

Let us see how the 68681 is used.

FIGURE 10.16   The 68681 Dual UART in a small system

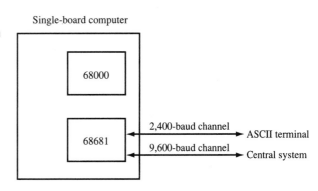

Interfacing the 68681

Figure 10.17 shows a simplified block diagram of the 68681 connected to the 68000. Asynchronous data transfers between the 68681 and the 68000 are implemented via $\overline{\text{DTACK}}$, R/$\overline{\text{W}}$, $\overline{\text{LDS}}$, $\overline{\text{AS}}$, and the lower half of the processor's data bus. An address decoder is used to place the 68681 at the desired location within the address space of the system. The lower four address lines (A_1 through A_4) are used to select the internal registers that configure the UART. The 68681 may request interrupt processing via $\overline{\text{IRQ}}$. Timing for the 68681 can be generated in two ways. A crystal may be directly connected to the 68681 (on X_1 and X_2). An internal oscillator will then generate the required timing signals. If a crystal is not used, a TTL clock may be connected to X_1 (and its complement to X_2).

Two separate serial data channels are implemented. Both channels are capable of simultaneous transmission and reception. Six parallel inputs and eight parallel outputs are also provided. These pins can be programmed for standard parallel I/O, or they can be used to implement handshaking required by some serial systems. An internal clock/timer circuit is also available and can be programmed to generate square waves on OP_3. Figure 10.18 shows each function available on the parallel input and output lines. The six parallel inputs (IP_0 through IP_5) can be programmed as clear-to-send inputs for each channel, or as transmitter/receiver clock inputs (for use with custom baud rates). The eight parallel outputs can be programmed as request-to-send outputs, as transmitter/receiver clock outputs, or as status outputs indicating when characters have been transmitted or received.

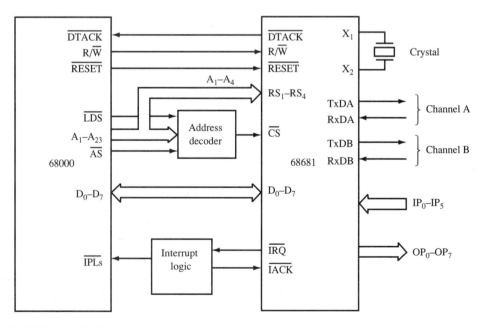

FIGURE 10.17   Using the 68681 in a 68000-based system

FIGURE 10.18 Alternate function pins on the 68681

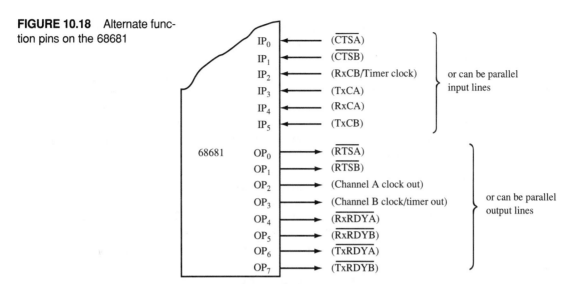

Programming the 68681

Due to the identical nature of the dual serial channels, we will only discuss how channel A may be programmed. Channel A's mode register 1, located at internal address 00, is selected after a RESET operation. Figure 10.19 shows the bit assignments for this register. Mode register 1 allows selection of the number of data bits used for each char-

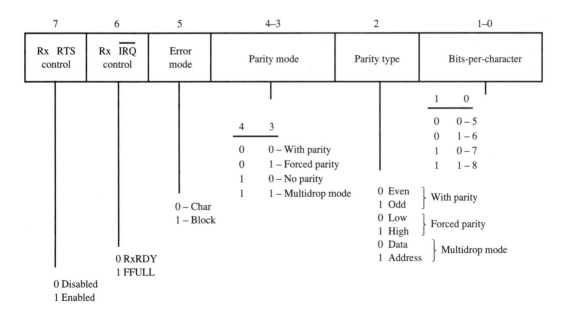

FIGURE 10.19   Mode register 1 bit assignments

acter, the type of parity (if any), and what type of interrupt is generated. An RxRDY (receiver ready) interrupt is generated when a new character has been received. A FFULL (FIFO Full) interrupt indicates that the three-character FIFO and receiver buffer are full (four characters have been received but not read from the UART yet and a fifth one is now arriving).

Channel A may be programmed to operate in *multidrop* mode. This mode of operation is used to connect many receivers, with each one receiving the same data. The receivers will ignore the received data until they detect their individual address characters, at which point they begin capturing received data until a block of characters has been received.

A second register is used to further specify operational characteristics for channel A. Channel A mode register 2 is located at the same address as mode register 1, and will be accessed only after mode register 1 has been written to. Mode register 1 will not be accessed again, unless a RESET is performed by the CPU. The bit assignments for mode register 2 are shown in Figure 10.20. The lower 4 bits of mode register 2 select the number of stop bits. Usually this number will be 1, 1.5, or 2, but it is possible to select a portion of a stop bit between these values. For example, if bits 0 through 3 are all low, 0.563 stop bits are used in a 6–8 bits/character transmission, and 1.063 stop bits are used in a 5 bits/character transmission. We will not consider any of these fractional stop bit selections in this discussion.

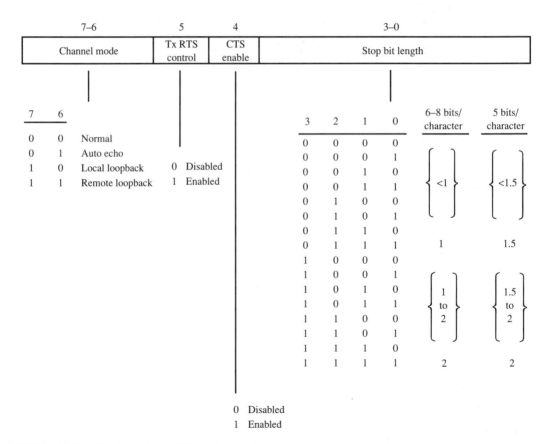

FIGURE 10.20   Mode register 2 bit assignments

A channel may operate in one of four modes: normal, automatic echo, local loopback, and remote loopback. Normal mode means that the transmitter and receiver operate independently of each other. In automatic echo mode, the channel transmits data as it is received, on a bit-by-bit basis. No data may be sent to the transmitter by the CPU in this mode. Local loopback mode internally connects the output of the transmitter to the input of the receiver. Thus, as a character is transmitted, it is also received. This mode is useful for testing the channel via software. The CPU may continue to use the transmitter and receiver sections. The last mode, remote loopback, is used to test a remote device. The channel automatically transmits received data on a bit-by-bit basis. The CPU may not access the receiver. Both local and remote loopback modes are useful for channel testing through software.

Baud rates for the transmitter and receiver are programmed through the clock select register. Figure 10.21 shows the bit assignments for this register, which is located at internal address 02. The baud rates in parentheses are selected when bit 7 in the auxiliary control register is set. All other rates are selected when it is cleared. We will assume that bit 7 is cleared for this discussion. The auxiliary control register is used to select counter/timer clock sources, and will not be discussed here.

One last register is used to control channel A. The channel A command register, located internally at address 04, is used to enable/disable the transmitter and receiver sections and to reset various functions in the channel. Figure 10.22 shows the bit assignments for this register. The lower 4 bits are used to enable/disable the transmitter and receiver. Once these two sections have been enabled, writing 0 to the lower 4 bits will have no effect on their operation. In this way, the upper bits can be changed to issue new commands to the channel without disabling the transmitter or receiver.

7–4					3–0				
Receiver clock set					Transmitter clock select				
7	6	5	4	Baud rate	3	2	1	0	Baud rate
0	0	0	0	50 (75)	0	0	0	0	50 (75)
0	0	0	1	110	0	0	0	1	110
0	0	1	0	134.5	0	0	1	0	134.5
0	0	1	1	200 (150)	0	0	1	1	200 (150)
0	1	0	0	300	0	1	0	0	300
0	1	0	1	600	0	1	0	1	600
0	1	1	0	1200	0	1	1	0	1200
0	1	1	1	1050 (2000)	0	1	1	1	1050 (2000)
1	0	0	0	2400	1	0	0	0	2400
1	0	0	1	4800	1	0	0	1	4800
1	0	1	0	7200 (1800)	1	0	1	0	7200 (1800)
1	0	1	1	9600	1	0	1	1	9600
1	1	0	0	38.4K (19.2K)	1	1	0	0	38.4K (19.2K)
1	1	0	1	Timer	1	1	0	1	Timer
1	1	1	0	IP_4–16X	1	1	1	0	IP_3–16X
1	1	1	1	IP_4–1X	1	1	1	1	IP_3–1X

FIGURE 10.21 Bit assignments in the clock select register

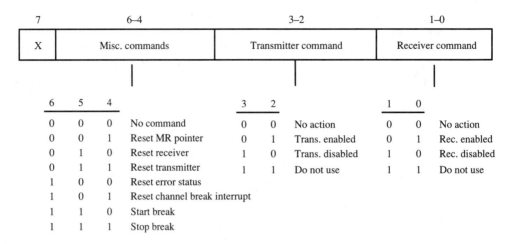

FIGURE 10.22   Bit assignments for the channel A command register

When interrupts are not used to indicate the completion of a transmission or the reception of a new character, the CPU must be able to examine the status of the channel to determine these conditions. The channel A status register, located at internal address 02, can be used for this purpose. The channel A status register indicates the states of the transmitter and receiver sections, and their associated FIFOs, and also any error conditions that may have arisen in the channel, such as parity, overrun, and framing. Figure 10.23 shows the bit assignments for the status register.

Any routine that desires access to the transmitter or receiver sections must first check the appropriate status bit, to see if the section is busy. When it is acceptable to read the received data, or send new data to the transmitter, the channel A receiver/transmitter buffer, located at internal address 06, must be accessed. Reading address 06 places a received character onto the processor's data bus. This character will either come directly from the receiver or from the receiver's FIFO (if more than one character has arrived since the last read). To transmit a character, data must be written to address 06.

The two examples that follow show how channel A in the 68681 can be initialized, and how a character-input routine can be written to implement serial data reception.

7	6	5	4	3	2	1	0
Received break	Framing error	Parity error	Overrun error	TxEMT	TxRDY	FFULL	RxRDY

Note: All bits are active when high.

FIGURE 10.23   Channel A status register

Example 10.2:   The following code is used to program channel A for normal mode operation, 1,200 baud in both transmitter and receiver, 7 data bits/character, even parity, and 2 stop bits. A base 68681 address of 6,000 is used. Odd external addresses are used to generate the correct internal addresses.

```
MODER1     EQU      $6000                      ;mode registers 1 and 2
CLKSEL     EQU      $6002                      ;clock select register
COMMAND    EQU      $6004                      ;command register
.
.
.
INITA      MOVE.B   #2,MODER1                  ;select even parity, 7 data bits
           MOVE.B   #$0F,MODER1                ;select normal mode, 2 stop bits
           MOVE.B   #$66,CLKSEL                ;select 1200 BAUD
           MOVE.B   #5,COMMAND                 ;enable transmitter/receiver
           RTS
```

Once the channel has been initialized, a routine similar to CHARIN, presented in the next example, can be used to read received characters.

Example 10.3:

```
STATUS     EQU      $6002                  ;status register
RECDATA    EQU      $6006                  ;receiver buffer
.
.
.
CHARIN     MOVE.B   STATUS,D0              ;get channel status
           ANDI.B   #$01,D0               ;mask off all other bits
           BEQ      CHARIN                ;loop until character received
           MOVE.B   RECDATA,D1            ;read received character
           ANDI.B   #$7F,D1               ;clear MSB for true ASCII
           RTS
```

Note that CHARIN will not return until a character has been received.

10.5   THE 68230 PI/T (PARALLEL INTERFACE/TIMER)

Parallel input and output differs from serial I/O in a number of ways. Serial data transfer occurs 1 bit at a time, and thus requires multiple bit times (based on the baud rate) to transfer an entire chunk of data. With parallel I/O, all bits get transferred at the same time, resulting in a much faster data transfer rate. As an example, a serial output line, running at 19,200 baud, takes almost 417 µs to transmit 8 bits. At 417 µs per byte, the serial output would take over 0.4 second to transmit 1,024 bytes. A parallel output port, capable of sending 8 bits at once, would take a much smaller amount of time. If the parallel port could output a new byte every 10 µs (under CPU control), only 0.01 seconds is needed to transfer all 1,024 bytes. This is almost 42 times faster than the serial output rate. When high-speed data transfers are needed (as is the case during DMA operations), parallel I/O is usually used.

Although parallel I/O is faster than serial I/O, it still contains a slight disadvantage. Parallel I/O often requires the use of handshaking signals to ensure that the data gets transferred properly. Serial I/O, due to its asynchronous nature, does not. We will see, however, that parallel I/O devices come equipped with the necessary handshaking logic already implemented. The handshaking signals are usually represented by strobes, or by clock pulses. Figure 10.24 shows the action of handshaking signals on parallel input and output ports. Figure 10.24 indicates that the external circuitry sending or receiving the parallel data will make use of the handshaking signal during the transfer.

One device capable of performing parallel I/O is the 68230 parallel interface/timer. The 68230 contains three parallel ports, with each port consisting of eight parallel lines. The three ports are called PA, PB, and PC. Ports A and B can be programmed as output or

FIGURE 10.24 Handshaking signals used in parallel I/O: (a) input port and (b) output port

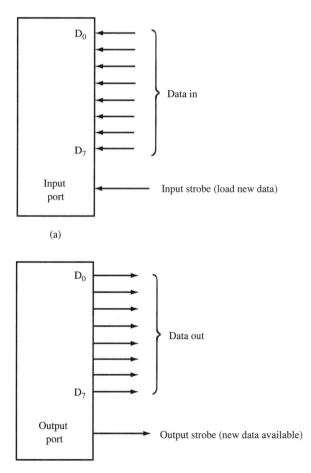

input ports (or even both at the same time). The direction of individual bits can also be programmed, and a special mode of operation allows for combining ports A and B into a single 16-bit port. The 68230 can be programmed to interrupt the processor when any port receives new data. This is very helpful and eliminates the need for the CPU to constantly poll the port as it looks for new data.

The 68230 also contains a 24-bit counter. The counter is loaded by the programmer and counts down each time it is clocked. The 68230 can be programmed to interrupt the processor when the counter reaches a count of 000000. Other programming modes exist that can turn the output of the counter into a square-wave generator whose frequency is controlled by the initial value of the counter.

In the next two sections, we will see how the 68230 is connected to the 68000 and how it is programmed.

Interfacing the 68230

Figure 10.25 shows a simplified diagram detailing the required connections between the 68000 and the 68230. Since the 68230 is designed for use with the 68000, most connections go directly from one chip to the other. The exception is the $\overline{\text{CS}}$ (chip-select) pin, which requires an external address decoder. The address decoder will place the 68230 at the desired location within the address space of the processor. The 68230 is then programmed and used by reading and writing data to the correct memory-mapped locations.

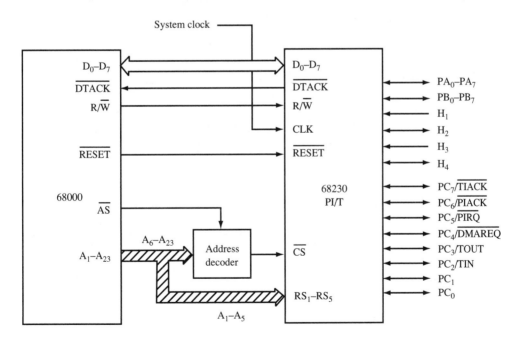

FIGURE 10.25   Interfacing the 68230 PI/T

These locations are selected by the state of the five register-select inputs (RS_1 through RS_5). The register-select inputs address the 23 internal registers of the 68230. H_1 and H_2 are the handshaking signals associated with port A (PA_0 through PA_7). H_3 and H_4 are the handshaking signals for port B (PB_0 through PB_7). A number of signals on port C (PC_0 through PC_7) can be programmed to provide special functions. $PC_5/\overline{PIRQ}$ can be programmed as a parallel I/O pin, or as an interrupt request output. PC_6 and PC_7 may be programmed as parallel I/O, timer (PC_7), or port (PC_6) interrupt acknowledge inputs. PC_4 has an alternate function as $\overline{DMAREQ}$ (DMA request). Finally, PC_3 and PC_2 can be programmed to serve as the timer output and input signals.

Programming the 68230

The 68230 is programmed after a CPU RESET by writing the appropriate control codes to its internal registers. Ports A and B are capable of operating in one of four modes. These modes offer the following functions:

Mode 0:   Unidirectional 8-bit transfers on A and B.

Mode 1:   Unidirectional 16-bit transfers. Port A is the MSB, port B the LSB.

Mode 2:   Bidirectional 8-bit transfers on B, bit I/O on A.

Mode 3:   Bidirectional 16-bit transfers. Port A is the MSB, port B the LSB.

Each mode contains an additional submode that defines the operation of the four handshaking signals (H_1 through H_4). The mode is programmed by writing a control word to the 68230's port general control register (PGCR). Figure 10.26 shows how the bits in the PGCR are defined. Bits 6 and 7 are used to select the mode of operation for the ports. Bits 4 and 5 are used to enable/disable the handshaking pins for each port. The remaining bits control the sense of each handshaking line. For example, if bit 0 (H_1 sense) is zero, H_1 will become activated by a low-level (logic 0) voltage at its input. If bit 0 is high, H_1 will be activated by a high-level (logic 1) voltage. H_2, H_3, and H_4 operate the same way.

When programming the mode bits, the 68230 requires that the H_{12} and H_{34} enable bits be 0. Only after the mode has been programmed can these two bits be changed. The

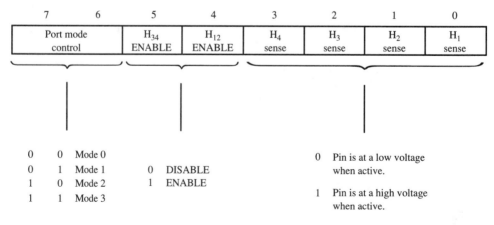

FIGURE 10.26   PGCR bit assignments

following section of code shows how the 68230 can be programmed for mode 1, H_{12} and H_{34} enabled, and active high handshaking:

```
MOVEA.L     #PGCR,A0          ;point to PGCR address
MOVE.B      #$40,(A0)         ;select port mode 1
MOVE.B      #$7F,(A0)         ;enable H12 and H34, active high handshaking
```

The PGCR address is a function of the $\overline{CS}$ circuitry and the 68230's internal address for the PGCR register (0). Thus, if the address decoder has been designed to respond to addresses 4000 through 403F, the PGCR can be programmed by writing data to location 4001.

Interrupts and DMA are controlled by the port service request register (PSRR). The PSRR defines the operation of multifunction pins PC_4 through PC_7. Interrupt priority levels are also programmed by the PSRR. Figure 10.27 shows how the PSRR is defined. Bit 7 is unused and thus thought of as always low.

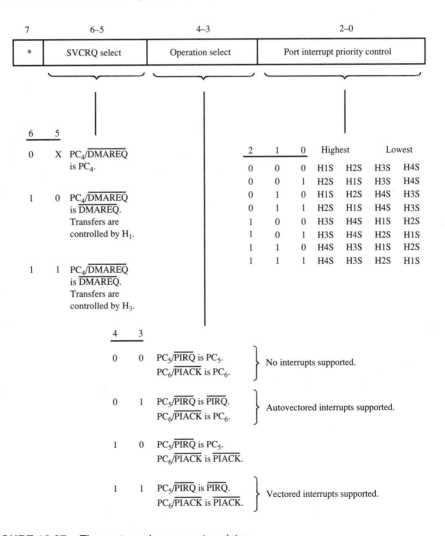

FIGURE 10.27   The port service request register

Bits 5 and 6 control the operation of the $PC_4/\overline{DMAREQ}$ pin. Two modes of DMA are possible, one controlled by H_1 and the other by H_3. If no DMA is required, PC_4 is available for parallel I/O.

Bits 3 and 4 control the operation of $PC_5/\overline{PIRQ}$ and $PC_6/\overline{PIACK}$. When port interrupts are required, these pins should be programmed as $\overline{PIRQ}$ and $\overline{PIACK}$. Otherwise, PC_5 and PC_6 may be used as parallel I/O lines.

The 3 lower bits in the PSRR are used to select the port interrupt priority. Eight priority schemes are available, with each one ranking the four handshake lines from highest to lowest.

The PSRR has an internal 68230 address of 02. To continue with the address decoding example previously used, the PSRR is accessed by writing to location 4003. The following section of codes shows how this can be done. In this example the 68230 is being programmed for I/O on PC_4 (no DMA), vectored interrupts, and an interrupt priority of HS2 . . . H1S . . . H4S . . . H3S.

```
PSRR      EQU       $4003
.
.
.
          MOVE.B    #$1B,D0          ;bit pattern 0 00 11 011 for PSRR
          MOVE.B    D0,PSRR          ;write pattern to the PSRR
```

When vectored interrupts have been selected, the vector number should be written into the 68230's port interrupt vector register (PIVR), located internally at address 0A. Only the upper 6 bits of the PIVR can be written to by the programmer. The lower 2 bits generated depend on which handshaking line produced the interrupt request, as shown in Figure 10.28. The programmer sets the base vector number. H_1 through H_4 are then used to determine the actual vector number when the interrupt is acknowledged.

A base vector number of 50 can be written to the 68230 by the following code:

```
PIVR      EQU       $400B            ;example 68230 PIVR address
.
.
.
          MOVE.B    #$50,D2
          MOVE.B    D2,PIVR
```

FIGURE 10.28 The port interrupt service register

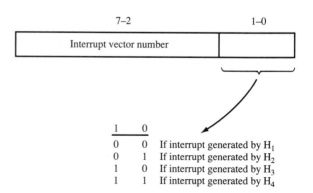

The four interrupt vectors possible after execution of this code sequence are 50, 51, 52, and 53. When using the timer to generate interrupts, the timer interrupt vector register (TIVR), located within the 68230 at address 22, should be loaded with an 8-bit vector number. Port and timer interrupts are serviced separately.

Each of the three parallel ports has a data direction register associated with it. The data direction register (DDR) is used to program the direction of I/O on each bit in the port. Each bit in the DDR controls the direction of the corresponding pin in the I/O port. For example, bit 0 in port A's DDR controls the direction of PA_0. To program a pin for output, its corresponding bit in the DDR should be a 1. A 0 in any bit of a DDR programs the pin for input.

The three data direction registers and their internal 68230 addresses are:

Port A DDR (PADDR):   Address 04

Port B DDR (PBDDR):   Address 06

Port C DDR (PCDDR):   Address 08

The following section of code programs all bits on port A as output, all bits on port B as input, and half of port C (the lower 4 bits) as input and the other half as output:

```
PADDR     EQU        $4005                   ;example DDR address
.
.
.
          MOVEA.L    #PADDR,A0               ;init pointer to DDR
          ST         D0                      ;set lower byte to all 1s
          MOVE.B     D0,(A0)                 ;program port A for all output
          CLR.B      D0                      ;lower byte is all 0s now
          MOVE.B     D0,2(A0)                ;program port B for all input
          MOVE.B     #$0F,D0                 ;pattern is 0000 1111
          MOVE.B     D0,4(A0)                ;program port C half and half
```

Once the ports have been programmed, they are accessed via their respective port data register. If port A has been programmed for output, writing data into the port A data register (PADR) will send the data to the output lines. Reading port input data is done by reading its port data register. Each port has its own data register and internal 68230 address, as seen here:

Port A data register (PADR):   Address 10

Port B data register (PBDR):   Address 12

Port C data register (PCDR):   Address 18

Assuming that port B has been programmed for ouput, the following code will output an 8-bit binary count to the port:

```
PBDR      EQU        $4013                   ;example 68230 PBDR address
.
.
.
COUNT     MOVE.B     D0,PBDR                 ;output count to port
          ADDQ.B     #1,D0
          BRA        COUNT
```

A more exotic parallel application involves the use of a digital-to-analog converter (DAC). In Figure 10.29(a), a 1408 8-bit DAC is connected to port B of the 68230. A 741 operational amplifier is used to convert the 1408's output current into a proportional voltage. The table within Figure 10.29(a) indicates that the analog output voltage will be –2.5 volts when a 0 byte is output to port B. Voltage steps of 19.6 millivolts are possible by adjusting the LSB of port B.

With an 8-bit DAC connected to the output of the 68230, we can generate some useful and interesting waveforms (see Figure 10.29(b)). First is the *square wave:*

```
SQRWAVE    MOVE.B    #0,PBDR      ;output low level
           JSR       DELAY        ;and delay
           MOVE.B    #$FF,PBDR    ;output high level
           JSR       DELAY        ;and delay
           BRA       SQRWAVE      ;repeat cycle
```

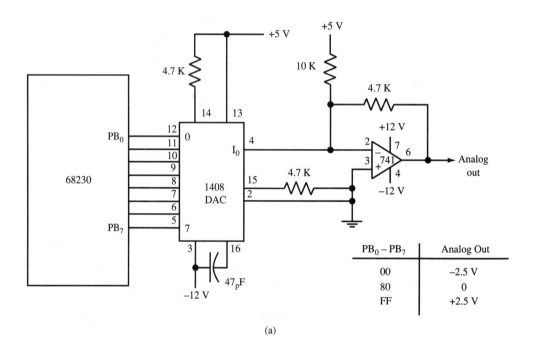

(a)

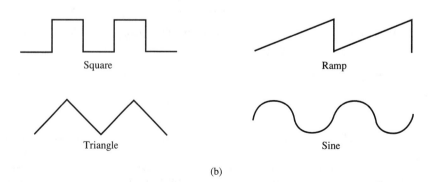

(b)

FIGURE 10.29   Eight-bit digital-to-analog converter circuit: (a) schematic and (b) waveforms generated

The use of a DELAY between outputs allows the user to control the frequency of the waveform that is generated.

The next waveform is the *ramp:*

```
RAMP       CLR.B      D0              ;init output register
NEXTR      MOVE.B     D0,PBDR         ;output pattern
           JSR        DELAY           ;and delay
           ADDQ.B     #1,D0           ;increment output register
           BRA        NEXTR           ;and repeat
```

The ramp waveform drops sharply at the end of every cycle when the output register wraps around from FF to 00.

A *triangle* waveform is generated by repeatedly ramping up and down:

```
TRIANGLE   CLR.B      D0              ;ramp up until 00 is seen
TRIONE     MOVE.B     D0,PBDR
           JSR        DELAY
           ADDQ.B     #1,D0
           CMPI.B     #0,D0
           BNE        TRIONE          ;exit TRIONE loop if 00
           SUBQ.B     #1,D0
TRITWO     MOVE.B     D0,PBDR         ;ramp down until 00 is seen
           JSR        DELAY
           SUBQ.B     #1,D0
           CMPI.B     #0,D0
           BNE        TRITWO          ;start over if 00
           BRA        TRIONE
```

Other interesting waveforms can be generated (such as sine waves and synthesized speech) with the use of a data table. These are left for you to implement on your own.

So far we have only examined the operation of the parallel ports on the 68230. In addition to these 24 I/O lines, the 68230 contains an internal 24-bit counter. The counter circuitry can be programmed for a variety of functions to generate square waves and various kinds of interrupts.

The counter is loaded by writing the 24-bit value into three 8-bit registers. These three registers are called the counter preload registers, with each one holding 8 counter bits. The addresses of the three counter registers are:

> CPRH (upper 8 bits of counter value):   Address 26
>
> CPRM (middle 8 bits of counter value):   Address 28
>
> CPRL (lower 8 bits of counter value):   Address 2A

To load the counter, write individual bytes into each preload register.

```
CPRH       EQU        $4027           ;example 68230 CPRH address
  .
  .
  .
           MOVEA.L    #CPRH,A0        ;init pointer to CPRH
           MOVE.B     #$3C,(A0)       ;fill CPRH
           MOVE.B     #$05,2(A0)      ;fill CPRM
           MOVE.B     #$2F,4(A0)      ;fill CPRL
```

Once the counter has been loaded, its mode of operation is selected by writing a control word to the 68230's timer control register (TCR), located at internal address 50. Figure 10.30 shows how the bits in the TCR are defined. Bits 5–7 select one of six modes of operation. While doing so, they configure multifunction pins PC$_3$/TOUT and PC$_7$/$\overline{\text{TIACK}}$ on port C. The timer is capable of supporting vectored interrupts, autovectored interrupts, or no interrupts at all.

Bit 4 is the zero detect control bit, which decides what the counter will do when it has been decremented to 000000. When 0, it allows for automatic reloading of the counter

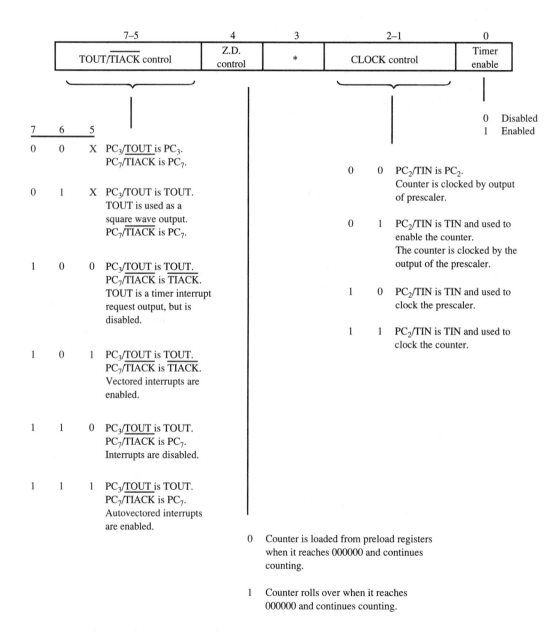

FIGURE 10.30   The timer control register

FIGURE 10.31 Four ways the counter can be clocked: (a) by CLK and prescaler; (b) by CLK and prescaler (TIN enables counting); (c) TIN clocks the prescaler; (d) TIN clocks the counter

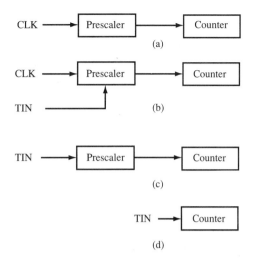

(from the preload registers). When high, the counter rolls over (to FFFFFF). Counting resumes in both cases.

Bits 2 and 1 control the method used to clock the counter. The 68230 contains an internal 5-bit prescaler (a second counter) that may be programmed to clock the 24-bit counter. Figure 10.31 shows the four possible ways the 24-bit counter can be clocked. The prescaler divides its input clock by 32. Thus, if bits 2 and 1 are both low and the CLK frequency is 2 MHz, the counter's clock rate is 62.5 kHz.

Bit 0 is the timer enable bit. When low, it prevents the counter from counting even if clock pulses are supplied. The counter will count only when bit 0 is high.

Some sample control words and their resulting operations are:

101 0 0 00 1:   Periodic vectored interrupts are generated each time the counter reaches 000000.

010 0 0 00 1:   A square wave is generated at TOUT. Its period is twice the time needed for the counter to count down to 000000.

111 1 0 00 1:   Generate a single autovectored interrupt when the counter reaches 000000.

000 1 0 11 1:   Clock pulses arriving at TIN are counted. Can be used for keeping time or measuring frequency.

Now that most of the basics have been covered, we can examine two applications for the 68230. The first application uses the 68230 to scan 16 push buttons (arranged as a hexadecimal keypad). The second application uses the 68230 to multiplex four 7-segment displays.

Example 10.4:   Figure 10.32 shows the schematic of a 68230 interfaced to 16 push buttons, arranged in a 4-by-4 matrix. The push buttons are all normally open, momentary

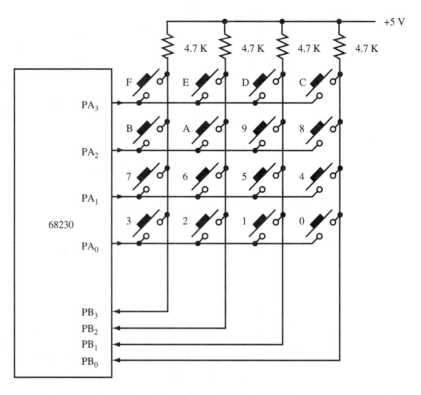

FIGURE 10.32   Sixteen-key keypad scanner using the 68230

contact. The four rows of buttons are scanned by port A bits 0–3. The four columns are sensed by port B bits 0–3. Initialization software for the 68230 requires that port A be programmed for output and port B for input. The technique used to scan the keyboard matrix assumes that only *one* button is pushed at any instant.

From the schematic, you will see that whenever a button is pushed it will short out its connecting row and column lines. For example, if button 5 is pushed, the column connected to PB_1 and the row connected to PA_1 are shorted together. To detect any button pushed in a single row, we must output a 0 to the row line. Any button connected to that row will place a 0 on its associated port B pin. So, to scan the second row of buttons—buttons 7, 6, 5, and 4—we output a 0 on PA_1 (and 1s on PA_0, PA_2, and PA_3), and look for 0s on PB_0 through PB_3. Table 10.1 shows how all 16 buttons may be scanned with this technique.

The subroutine READKB, which follows, scans the keypad matrix using the technique just described. If a single button is pushed, the number of the button (0 through F) will be returned in the lower byte of D0. If no buttons are pushed, D0 will contain $80 in its lower byte. The column bits are checked by rotating them into the carry flag and checking for a 0.

TABLE 10.1 Keyboard scanning codes

	Parallel outputs				
	PA_3	PA_2	PA_1	PA_0	Buttons scanned
	1	1	1	0	3, 2, 1, 0
	1	1	0	1	7, 6, 5, 4
	1	0	1	1	B, A, 9, 8
	0	1	1	1	F, E, D, C

```
PADR      EQU      $4011               ;sample 68230 port A address
PBDR      EQU      $4013               ;sample 68230 port B address
  .
  .
  .
READKB    CLR.B    D0                  ;init D0 to button 0
          MOVEQ    #3,D1               ;init row counter
          MOVE.B   #$FE,D2             ;init row-pattern generator
NEXTROW   MOVEQ    #3,D3               ;init column counter
          MOVE.B   D2,PADR             ;select a row to scan
          MOVE.B   PBDR,D4             ;read button information
NEXTCOL   ROR.B    #1,D4               ;move column bit into carry
          BCC      GOTKEY
          ADDQ.B   #1,D0               ;next button number
          DBF      D3,NEXTCOL          ;check other columns
          ROL.B    #1,D2               ;generate next row pattern
          DBF      D1,NEXTROW          ;check other rows
          MOVE.B   #$80,D0             ;no buttons pushed
GOTKEY    RTS                          ;return with button code in D0
```

Since mechanical contact bounce is always a problem, we will need to read the keypad a number of times, hoping that each read produces the same button information. Usually a short delay of 10 ms between reads is used to eliminate contact bounce. The following subroutine, GETKEY, reads the keypad twice, returning only when it gets the same button code each time. DELAY is a subroutine that produces 10 ms of delay.

```
GETKEY    BSR      READKB              ;get a button code
          CMPI.B   #$80,D0             ;was a button pushed?
          BEQ      GETKEY              ;no, keep reading
          MOVE.B   D0,KEY              ;save button code
          BSR      DELAY               ;wait 10 ms
          BSR      READKB              ;check keypad again
          CMP.B    KEY,D0              ;do we have a match?
          BNE      GETKEY
          RTS
```

Example 10.5: Figure 10.33 shows the 68230 connections required to implement a multiplexed display. A multiplexed display consists of a number of displays connected in parallel, so that each display receives the same information. Only one display will actually become energized, however, because we only enable one display at a time. When the data on the segment lines of the displays is changed rapidly and they are enabled in a round-robin fashion, our eyes perceive the illusion of all displays being on at the same time. If the displays are not multiplexed quickly enough, they will flicker (as the segments dim during disabled periods). To avoid flicker, the displays should be enabled at least 60 times every second.

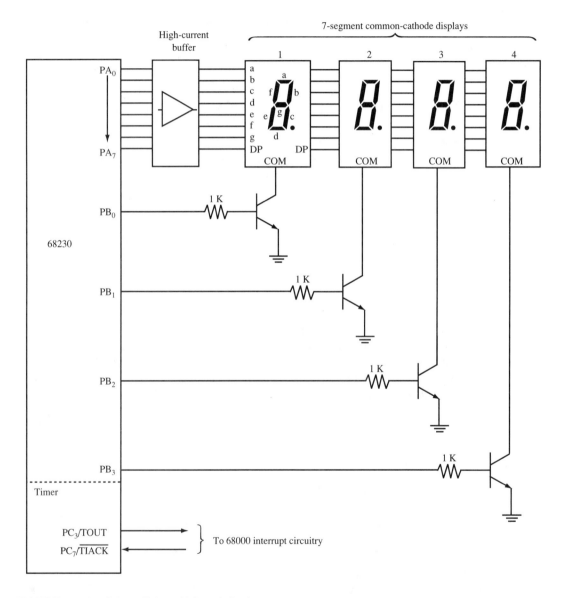

FIGURE 10.33   A four-digit multiplexed display

Figure 10.33 shows that port A is used to drive the segments of each display, and port B controls the common pin. We need to output a 1 on one of the PB_0 through PB_3 lines to enable any display. The other three port B outputs should be low to keep their displays off. For example, to turn display 3 on, the pattern output to port B must be 0100. The subroutine TIMERINT is used to multiplex the displays. The 68230's timer will be initialized so that it generates a vectored interrupt once every 4 ms. TIMERINT will take care of enabling the next display in sequence, with a scan order of 1 . . . 2 . . . 3 . . . 4. TIMERINT will automatically wrap around to display 1, after display 4 has been enabled for its time period. Initialization software is required to program ports A and B for output, to load the 24-bit timer counter with the proper value needed to generate 4-ms vectored interrupts, and to load the timer interrupt vector register with the interrupt vector number that will cause TIMERINT to be serviced.

```
DISPLAY     DS.B       1              ;storage for display enable pattern
SEGMENTS    DS.B       4              ;buffer area for segment information
SEGPTR      DS.B       4              ;segment pointer address buffer
PADR        EQU        $4011          ;example 68230 port A address
PABR        EQU        $4013          ;example 68230 port B address
.
.
.
TIMERINT    MOVE.B     DISPLAY,D0     ;get enabled display pattern
            CLR.B      PADR           ;turn off segments in current display
            ROR.B      #1,D0          ;generate next display enable pattern
            BCC        NEXTDISP       ;scanned all four yet?
            MOVE.B     #$08,D0        ;load initial enable pattern
            MOVEA.L    #SEGMENTS,A0   ;load initial segment data pointer
            MOVE.L     A0,SEGPTR      ;save it
NEXTDISP    MOVE.B     D0,PBDR        ;turn on next display
            MOVEA.L    SEGPTR,A0      ;get address of next segment data
            MOVE.B     (A0)+,D1       ;get segment data
            MOVE.B     D1,PADR        ;display new data
            MOVE.L     A0,SEGPTR      ;save segment pointer
            MOVE.B     D0,DISPLAY     ;save current enable pattern
            RTS
```

10.6   THE 68901 MULTIFUNCTION PERIPHERAL

The previous two sections dealt with peripherals capable of supporting serial and parallel I/O. The usefulness of programmable I/O is clear. Many different applications can be performed with the same hardware, with the hardware configured differently by the software, for each application. The peripheral covered in this section brings parallel and serial I/O together in one chip. The 68901 multifunction peripheral contains a programmable 8-bit port and a single-channel serial transmitter/receiver. In addition, the 68901 contains four internal timers, which can be used to generate square waveforms, perform pulse width measurements, and generate periodic interrupts. Since we have covered parallel and serial I/O in some detail in the previous sections, we will not spend a great deal of time on them here. Instead, we will concentrate on the other features of the 68901. These features include a daisy-chained prioritized interrupt structure and 16 types of maskable interrupts.

Interfacing the 68901

Figure 10.34 shows a simplified schematic of how the 68901 is connected to a 68000-based system. The 68901 communicates with the processor via an 8-bit bidirectional data bus. This data bus is connected to the lower half (D_0 through D_7) of the processor's data bus when the 68000, 68008, and 68010 are used. The 68020 requires that the 68901 be connected to D_{24} through D_{31} for proper operation.

Five register-select inputs (RS_1 through RS_5) are used to select the 24 internal registers used by the 68901. Interrupts are generated and acknowledged by the $\overline{IRQ}$ and $\overline{IACK}$ pins. Prioritized (daisy-chained) interrupts are made possible by the $\overline{IEI}$ (interrupt enable in) and $\overline{IEO}$ (interrupt enable out) pins. Multiple 68901s can be cascaded together by connecting the $\overline{IEO}$ of one 68901 to the $\overline{IEI}$ of another (as shown in Figure 10.35). The $\overline{IEI}$ pin of the highest priority 68901 must be grounded to make interrupts possible. The first 68901 in the chain will disable all others, via $\overline{IEO}$, should it request an interrupt. By connecting the 68901s in this fashion, we end up with a prioritized interrupt scheme. The first 68901 will have the highest priority. The last 68901 in the chain will have the lowest interrupting priority.

Returning to Figure 10.34, we see four separate functional sections. The first section is composed of eight programmable I/O lines, I_0 through I_7. The direction of each one of these bits is programmable through an internal data direction register. Furthermore, each line is capable of generating a separate vectored interrupt when activated.

The second section is composed of signals connected to the four internal timers. TAI and TBI are clock inputs to timers A and B. TAO, TBO, TCO, and TDO are the four timer

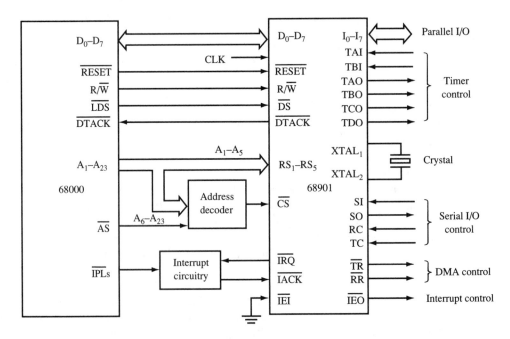

FIGURE 10.34   Interfacing the 68901 to the 68000

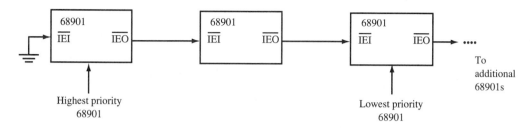

FIGURE 10.35 Daisy-chaining multiple 68901s

outputs. The timers are also capable of generating vectored interrupts. Timing signals for the timers are derived from a crystal connected to $XTAL_1$ and $XTAL_2$.

The third section is composed of signals associated with the serial transmitter and receiver. SI (serial in) and SO (serial out) are the data lines connected to the receiver and transmitter. RC and TC are receiver and transmitter clock inputs, used to control the bit rate of each section. Alternatively, the transmitter and receivers may be clocked by a timer output.

The last section is used to control DMA operations and consists of two pins, $\overline{TR}$ (transmitter ready) and $\overline{RR}$ (receiver ready).

Programming the 68901

The 68901 is configured for various modes of operation through the use of 24 internal registers. The registers, and their external addresses, are listed here:

Address	Register
01	GPDR, general purpose I/O data register
03	AER, active edge register
05	DDR, data direction register
07	IERA, interrupt enable register A
09	IERB, interrupt enable register B
0B	IPRA, interrupt pending register A
0D	IPRB, interrupt pending register B
0F	ISRA, interrupt in-service register A
11	ISRB, interrupt in-service register B
13	IMRA, interrupt mask register A
15	IMRB, interrupt mask register B
17	VR, vector register
19	TACR, timer A control register
1B	TBCR, timer B control register
1D	TCDCR, timers C and D control register

1F	TADR, timer A data register
21	TBDR, timer B data register
23	TCDR, timer C data register
25	TDDR, timer D data register
27	SCR, synchronous character register
29	UCR, USART control register
2B	RSR, receiver status register
2D	TSR, transmitter status register
2F	UDR, USART data register

The GPDR is used to send and receive data from the parallel port (pins I_0 through I_7). The direction of data on these lines is controlled by the DDR. Each bit in the DDR specifies the direction of the corresponding bit in the I/O port. A 0 is used to program a pin for input, and a 1 to program the pin for output. The AER is used to program the transition required on a parallel input to generate an interrupt. High-to-low and low-to-high transitions are programmed by writing 0s and 1s to the desired bits in the AER.

The four timers are controlled by seven registers: TACR, TBCR, TCDCR, TADR, TBDR, TCDR, and TDDR. Timers A and B are programmed via the TACR and TBCR. Each control register consists of five user-programmed bits. Bit 4 is used to reset the timer. Bits 0 through 3 select the timer mode. Sixteen modes are possible with timers A and B, and are as follows:

0:	Timer stopped
1:	Delay mode, /4 prescaler
2:	Delay mode, /10 prescaler
3:	Delay mode, /16 prescaler
4:	Delay mode, /50 prescaler
5:	Delay mode, /64 prescaler
6:	Delay mode, /100 prescaler
7:	Delay mode, /200 prescaler
8:	Event count mode
9:	Pulse width mode, /4 prescaler
A:	Pulse width mode, /10 prescaler
B:	Pulse width mode, /16 prescaler
C:	Pulse width mode, /50 prescaler
D:	Pulse width mode, /64 prescaler
E:	Pulse width mode, /100 prescaler
F:	Pulse width mode, /200 prescaler

The delay mode of operation causes the timer to be decremented at a rate determined by the prescaler output. When the timer reaches a count of 01, it is reloaded from its data register. Additionally, the associated timer output will toggle and an interrupt may be generated if enabled by the interrupt mask register.

Timers C and D are limited in operation, and each timer's mode is specified via two groups of 3 bits in the TCDCR. Bits 0 through 2 control timer D, and bits 4 through 6 control timer C. Only timer modes 0 through 7 are available with timers C and D.

The USART (universal synchronous/asynchronous receiver/transmitter) section is controlled by a number of registers. The UCR selects the clock rate (1X or 16X), the character length (5 to 8 bits), the number of stop bits, and the parity. Two status registers are included (RSR and TSR) to provide status concerning the receiver and transmitter (buffers empty or full, parity and framing errors, and so on). The interrupt mask registers can be programmed to allow USART interrupts when a character is received or transmitted.

Vectored interrupts are generated by 16 different conditions. The vector number generated depends on the condition and the number stored in the VR. The upper 4 bits of the VR form the base vector number. The lower 4 bits are generated by the type of interrupt requested. All possible interrupts are listed here:

0:	General purpose interrupt 7 (I_7)
1:	General purpose interrupt 6 (I_6)
2:	Timer A
3:	Receiver buffer full
4:	Receive error
5:	Transmitter buffer error
6:	Transmit error
7:	Timer B
8:	General purpose interrupt 5 (I_5)
9:	General purpose interrupt 4 (I_4)
A:	Timer C
B:	Timer D
C:	General purpose interrupt 3 (I_3)
D:	General purpose interrupt 2 (I_2)
E:	General purpose interrupt 1 (I_1)
F:	General purpose interrupt 0 (I_0)

As shown, all major sections of the 68901 are capable of generating interrupts. The actual vector number generated is a combination of the upper 4 bits in the VR (that are supplied by the programmer) and the lower 4 bits (the interrupt type from the previous list). For example, if the upper bits have been set to 1000 (8) and timer C generates an interrupt, the vector number produced is 8A.

When a small system requires parallel I/O, serial I/O, and timing functions, the 68901 is an ideal way to provide all functions with a minimum of hardware.

10.7   THE 68153 BUS INTERRUPT MODULE

In the previous sections, we have seen that many peripheral interrupts exist and need to be serviced. Designing hardware to service a large number of interrupts can easily become a complicated task. Additionally, the hardware will be designed to produce certain fixed results. Future changes are thus difficult to implement without changing the hardware.

A peripheral designed to handle independent interrupt requests for up to four devices and be configured through software is the 68153 bus interrupt module. Asynchronous interrupts coming from the four sources will cause the 68153 to interrupt the processor and supply a preprogrammed vector number. If the system requirements change sometime in the future, no hardware redesign is needed: The 68153 must simply be initialized differently to handle the new requirements.

Interfacing the 68153

Figure 10.36 shows the interfacing required to connect the 68153 to the 68000. As usual, the asynchronous bus signals are used to control the transfer data between the CPU and the peripheral, with the address decoder setting the base address for the device. Address lines A_1 through A_3 serve two purposes. First, they are used to select one of eight internal registers used by the 68153 to control generation of vectored interrupts. Second, they supply the interrupt level dur-

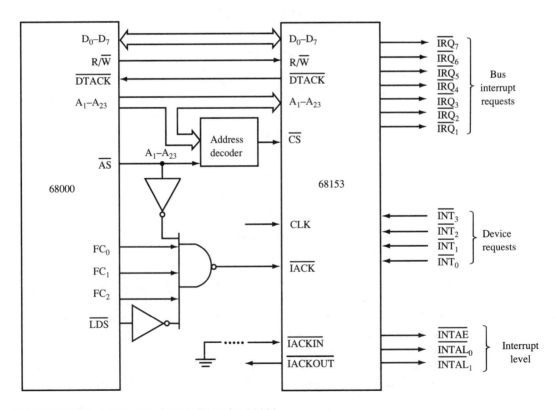

FIGURE 10.36   Interfacing the 68153 to the 68000

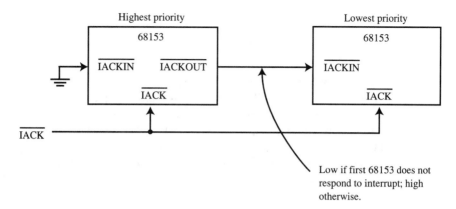

FIGURE 10.37 Daisy-chaining 68153s

ing interrupt acknowledge cycles. $\overline{\text{IACK}}$ (interrupt acknowledge) is driven low during an interrupt acknowledge cycle to inform the 68153 that it must supply an interrupt vector. $\overline{\text{IACKIN}}$ and $\overline{\text{IACKOUT}}$ are included to implement daisy-chaining of the 68153 with other devices (for example, a second 68153). When $\overline{\text{IACKIN}}$ is high, the 68153 will not respond to an interrupt acknowledge cycle. When low, the 68153 will check its internal registers (during an interrupt acknowledge cycle) to see if it should respond. If the interrupt levels stored in the internal registers match the level supplied by A_1 through A_3, the 68153 will respond with a vector number and output a high on $\overline{\text{IACKOUT}}$ (to disable other devices in the chain). If no match is found, $\overline{\text{IACKOUT}}$ will be driven low, enabling other devices in the chain and allowing them to respond instead. Figure 10.37 shows two 68153s daisy-chained together.

Getting back to Figure 10.36, we see seven bus interrupt request outputs, $\overline{\text{IRQ}}_1$ through $\overline{\text{IRQ}}_7$. The level of the interrupt being serviced determines which of these outputs goes low. Since $\overline{\text{IRQ}}_1$ through $\overline{\text{IRQ}}_7$ are open-collector outputs, a pull-up resistor is required for proper operation.

Device interrupt request signals $\overline{\text{INT}}_0$ through $\overline{\text{INT}}_3$ are used by up to four external devices (peripherals) to request a bus interrupt. The 68153 will output the encoded device number (0 through 3) it is responding to on outputs $\overline{\text{INTAL}}_0$ and $\overline{\text{INTAL}}_1$. These two outputs are valid when $\overline{\text{INTAE}}$ (interrupt acknowledge enable) is active. These outputs are provided to allow the external device the option of supplying its own vector number.

Programming the 68153

The 68153 contains eight internal registers that must be programmed to provide interrupt processing. These registers are assigned as follows:

0:	Control register 0 (for $\overline{\text{INT}}_0$)
1:	Control register 1 (for $\overline{\text{INT}}_1$)
2:	Control register 2 (for $\overline{\text{INT}}_2$)
3:	Control register 3 (for $\overline{\text{INT}}_3$)
4:	Vector register 0

5:      Vector register 1

6:      Vector register 2

7:      Vector register 3

Each control register is paired with a vector register. The vector register is loaded with an 8-bit vector number that will be supplied when the appropriate interrupt is requested. The control register, depicted in Figure 10.38, is used to store the interrupt level and other control information. The $\overline{\text{IRQ}}$ output that will be active when the interrupt is acknowledged is coded in the lower 3 bits. Device interrupts are enabled/disabled through bit 4, IRE. Bit 3, IRAC, is used to automatically clear the IRE bit when an interrupt is acknowledged. This will require the interrupt service routine to set IRE again, if further interrupts are desired.

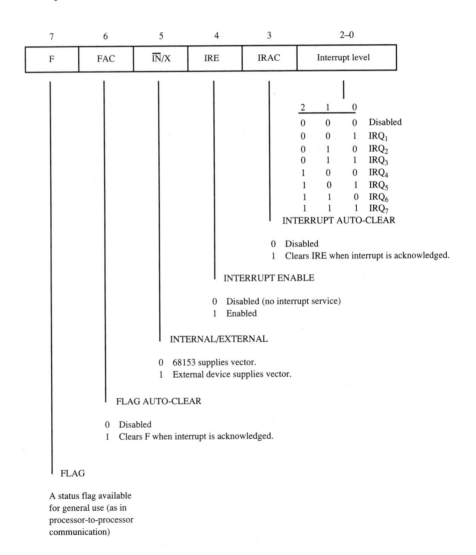

FIGURE 10.38   68153 control register

If the external device is going to supply the vector number instead of the 68153, the $\overline{\text{IN}}$/X bit should be set. When cleared, the 68153 supplies the vector number from the appropriate vector register.

The 68153 is an ideal device for use with older peripherals or non-68000-based devices that are incapable of generating interrupt vector numbers by themselves.

Example 10.6: The subroutine INTSET, listed here, is used to set up all four interrupt vector registers and their respective control registers for a 68153 located at base address B000.

```
CONTROL0    EQU         $B000                   ;base 68153 register address
   .
   .
   .
INTSET      MOVEA.L     #CONTROL0,A0            ;init register pointer
            MOVE.B      #$11,1(A0)              ;INT0 enabled, level IRQ1
            MOVE.B      #$1C,3(A0)              ;INT1 enabled, level IRQ4
            MOVE.B      #$B7,5(A0)              ;INT2 enabled, level IRQ7
            MOVE.B      #$22,7(A0)              ;INT3 disabled, level IRQ2
            MOVE.B      #$94,9(A0)              ;vector number for INT0
            MOVE.B      #$95,11(A0)             ;vector number for INT1
            MOVE.B      #$96,13(A0)             ;vector number for INT2
            MOVE.B      #$97,15(A0)             ;vector number for INT3
            RTS
```

10.8   THE 68881 FLOATING-POINT COPROCESSOR

Coprocessors have been mentioned numerous times throughout this text. A coprocessor is used to enhance the power of its host processor, which, by design, does not include an extensive set of instructions capable of handling floating-point numbers. A floating-point number (defined by the *IEEE Standard for Floating-Point Arithmetic,* and implemented by the 68881) consists of up to 64 bits of mantissa, a sign bit, and a 15-bit exponent. This 80-bit number does not fit into any of the processor's registers. For this reason, the 68881 contains a set of internal floating-point registers, each 80 bits long, that are accessed and manipulated by the coprocessor's set of floating-point instructions. These instructions are mixed in with the instructions of the host processor, and when encountered, cause the coprocessor to perform the specified function. As far as the programmer is concerned, the combination of processor and coprocessor (any 68000 family CPU connected to the 68881) operate as a single processor, capable of executing the additional floating-point instructions. The 68881 is most efficiently operated when interfaced to the 68020, but may also be connected to the 68000, 68008, 68010, and 68030. The only difference is the amount of hardware required in the interface.

The 68881 has a large set of floating-point instructions (a total of 40), composed of move, arithmetic, branch, and other miscellaneous instructions. Seven data formats are supported. Three of these formats—byte, word, and longword—are already known to us.

The other four are single-, double-, and extended-precision, and packed decimal. A floating-point instruction extension is used to specify the data format.

Many of the 68881's floating-point instructions require less than 100 clock cycles to execute. Others require many hundreds of clock cycles to complete. Even so, the co-processor offers a great speed advantage not available when performing similar operations with an ordinary processor.

Interfacing the 68881

Figure 10.39 shows the required connections used to interface the 68881 to the 68000. The size difference between the processor's 16-bit data bus and the coprocessor's 32-bit data bus requires that the upper and lower halves of the 68881's data bus be wired in parallel. A_0 and $\overline{\text{SIZE}}$ are used to configure the 68881 for the size of the processor's data bus. $\overline{\text{DS}}$ (data strobe) is used to indicate that valid information exists on the data bus. The AND gate is used to assert $\overline{\text{DS}}$ whenever $\overline{\text{UDS}}$ or $\overline{\text{LDS}}$ goes low. $\overline{\text{DSACK}}_1$ (data transfer and size acknowledge) is used to signal the completion of an asynchronous bus transfer. When the host processor is the 68020, $\overline{\text{DSACK}}_1$ and a second signal, $\overline{\text{DSACK}}_0$, are used to indicate the size of the data transfer between the processor and the 68881.

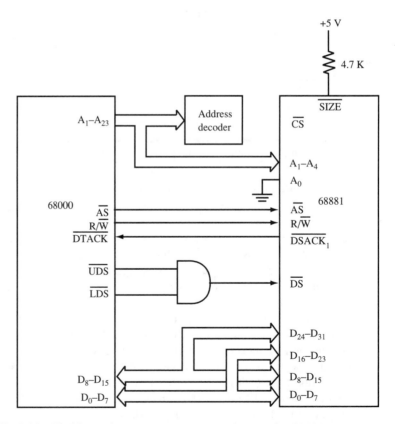

FIGURE 10.39   Floating-point coprocessor connections to the 68000

Both the host processor and the coprocessor are capable of running at different speeds. It is therefore not necessary to provide both with an identical clock signal.

Programming the 68881

The 68881 contains a number of internal registers that can be accessed by the programmer. Eight floating-point registers (FP0 through FP7) are used to store 80-bit numbers for use in 68881 calculations. A 32-bit control register is used to program the types of exceptions the 68881 will generate (that is, divide-by-zero, operand error, inexact decimal input), and also selects the type of rounding to be used in calculations (toward zero, toward positive or negative infinity). A 32-bit status register contains the floating-point condition codes, quotient bits, and exception status data, all accessible by the programmer. Another 32-bit register is used to save the host processor's address of the last floating-point instruction that was executed.

The floating-point instructions available with the 68881 are separated into four groups: monadic (one operand); dyadic (two operands); branch, set, and trap-on-condition; and miscellaneous. All instructions are listed according to group as follows:

| *Monadic* | *Monadic (cont.)* |
| | *Dyadic* |

Monadic

FABS, absolute value
FACOS, arc cosine
FASIN, arc sine
FATAN, arc tangent
FATANH, hyperbolic arc tangent
FCOS, cosine
FCOSH, hyperbolic cosine
FETOX, e to the xth power
FETOXM1, e to the $(x - 1)$th power
FGETEXP, get exponent
FGETMAN, get mantissa
FINT, integer part
FINTRZ, integer part (truncated)
FLOG10, log base 10
FLOG2, log base 2
FLOGN, log base e
FLOGNPI, log base e of $(x + 1)$
FNEG, negate
FSIN, sine
FSINCOS, simultaneous sine/
   cosine
FSINH, hyperbolic sine
FSQRT, square root

Monadic (cont.)

FTAN, tangent
FTANH, hyperbolic tangent
FTEXTOX, 10 to the xth power
FTST, test
FTWOTOX, 2 to the xth power

Dyadic

FADD, add
FCMP, compare
FDIV, divide
FMOD, modulo remainder
FMUL, multiply
FREM, IEEE remainder
FSCALE, scale exponent
FSGLDIV, single-precision divide
FSGLMUL, single-precision multiply
FSUB, subtract

Branch, set, and trap-on-condition

FBcc, branch-on-condition
FDBcc, decrement and branch-on-condition
FScc, set byte-on-condition
FTRAPcc, trap-on-condition

Miscellaneous

FMOVE, move to/from 68881 register

FMOVEM, move multiple registers

FSAVE, save virtual machine state

FRESTORE, restore virtual machine state

FNOP, no operation

The 68881 also contains 22 internal constants. These constants may be directly loaded into a floating-point register, without the need for an external access. The 22 constants are as follows:

pi	log10(2)	e
log2(e)	log10(e)	0.0
1n(2)	1n(10)	1
10	100	10^4
10^8	10^{16}	10^{32}
10^{64}	10^{128}	10^{256}
10^{512}	10^{1024}	10^{2048}
10^{4096}		

Seven data types are supported by the 68881. The first three—byte, word, and long-word—are also supported by the host processor. These three data types allow binary integers to be specified in the following ranges:

8 bits: −128 to +127

16 bits: −32768 to +32767

32 bits: −2147483648 to +2147483647

Note that these ranges are for signed (2's complement) numbers.

Single-precision real numbers are specified by a 23-bit fraction, an 8-bit exponent, and a sign bit. Since all 68881 numbers are internally normalized, they each contain an additional bit of precision. For example, the fraction part of a single-precision number may look like this: 1.11010110001011010100101 (1. followed by 23 data bits). The fractional part of a normalized number always begins with 1; therefore, we do not need to specify the first bit of any number. We can use the 23 bits available for the fractional part only.

The 8 exponent bits specify an unsigned exponent (in the range 0 to 255). An internal exponent bias of 127 is subtracted from the specified exponent to determine the true, signed exponent. For example, if the 8 exponent bits represent an exponent of 100, the actual exponent of the single-precision number is 100 – 127, or –27. The signed exponent is a base 2 exponent.

Double-precision real numbers are specified by a 52-bit fraction, an 11-bit exponent, and a sign bit. The exponent bias for these numbers is 1,023.

Extended-precision real numbers, the largest possible in the 68881, are specified by a 64-bit fraction, a 15-bit exponent (with a bias of 16383), and a sign bit. The 68881 performs all calculations involving real numbers, in extended-precision format. Single- and double-precision numbers are converted into this format before the calculations begin and reconverted back when finished. This ensures the smallest loss of accuracy.

The seventh data type supported is packed decimal real. BCD numbers are specified with this data type. These numbers may contain up to 17 digits in the mantissa and a three-digit exponent. The mantissa and exponent signs are specified with separate bits.

The data type (or format) of an operand is specified in the instruction, by use of a single-character extension. Each data format has its own extension character as follows:

Byte	.B
Word	.W
Longword	.L
Single	.S
Double	.D
Extended	.X
Packed Decimal	.P

The best place for floating-point numbers to be saved is in the internal floating-point registers (FP0 through FP7), although the 68881 is also capable of accessing external memory as well. The 68881 supports all addressing modes available on 68000-based systems. Thus, instructions such as those listed here are possible:

FSUB.B	#3,FP0
FADD.L	D4,FP2
FADD.L	NUMBER,FP6
FADD.S	#2.7182818,FP1
FADD.D	(A2)+,FP5
FSUB.E	−(A1),FP4

The 68881 is too complicated to cover in detail in this discussion. Even so, two simple examples will serve to show how powerful this coprocessor is, and, hopefully, how much easier it is to use. Remember that floating-point instructions would need to be implemented in software, through complicated time-consuming subroutines, if the 68881 were not available.

Example 10.7:   The floating-point instructions in this example are used to calculate the standard deviation of a group of samples. There are *n* samples in the group, and they are saved in memory beginning at SAMPLES. The following equation will be used to compute the standard deviation (assuming that n is larger than 10):

$$\text{S.D.} = \sqrt{\frac{1}{n} \sum_{i=1}^{n} (X_i - \overline{X})^2}$$

where n = the number of samples
$\overline{X}$ = the average of the samples

First the average ($\overline{X}$) will be computed, since this is needed during the calculations. The final result (the standard deviation) is found in FP2.

```
N           EQU       ?                 ;the number of samples goes here
SAMPLES     DS.B      ?                 ;the samples go here
   .
   .
   .
            MOVEA.L   #SAMPLES,A0       ;init pointer to samples
            MOVE.W    #N-1,D0           ;init loop counter
            FSUB.D    FP1,FP1           ;clear FP1
DPADD       FADD.D    (A0)+,FP1         ;total samples
            DBRA      D0,DPADD
            FDIV.D    #N,FP1            ;compute average
            FSUB.D    FP2,FP2           ;clear FP2
            MOVE.W    #N-1,D0           ;init loop counter
            MOVEA.L   #SAMPLES,A0       ;init sample pointer
SQRSUM      FMOVE.D   (A0)+,FP3         ;get sample into FP3
            FSUB.D    FP1,FP3           ;subtract sample average
            FMUL.D    FP3,FP3           ;square result
            FADD.D    FP3,FP2           ;total result
            DBRA      D0,SQRSUM
            FDIV.D    #N,FP2            ;divide by N
            FSQRT.D   FP2               ;use square root to get standard deviation
```

Note: It is possible to eliminate the second MOVEA instruction by using predecrement addressing during the second loop.

Example 10.8:   A resistor-capacitor circuit is shown in Figure 10.40(a). It consists of a resistor, a capacitor, and a DC voltage source, E. At time t equal to 0, the switch is closed and the capacitor begins to charge. The time constant for the circuit, which determines how fast the capacitor charges, is found by the product of R and C (and is 1 ms in this example). Figure 10.40(b) shows a sketch of the charging curve for the capacitor. Notice that V_C reaches the applied voltage E in five time constants.

The equation that governs the capacitor's charging rate is also shown in Figure 10.40(b). This is the equation implemented by CHARGE, the 68881 routine listed here, which is used to compute the capacitor's voltage at any time T. The result is returned in FP1.

```
R           EQU       1000              ;resistance value
C           EQU       1E-6              ;capacitance value
E           EQU       10                ;voltage source value
T           DS.B      ?                 ;storage for time value
```

.
.
.

```
CHARGE    FMOVE.D    #R,FP0       ;load FP0 with resistance value
          FMUL.D     #C,FP0       ;multiply FP0 by capacitance
          FMOVE.D    #T,FP1       ;load time into FP1
          FDIV.D     FP0,FP1      ;compute exponent for e (time/time-constant)
          FNEG.D     FP1          ;make exponent sign negative
          FETOX.D    FP1          ;compute e to xth power
          FNEG.D     FP1          ;change sign
          FADD.D     #1.0,FP1     ;compute difference
          FMUL.D     #E,FP1       ;calculate capacitor's voltage
          RTS
```

FIGURE 10.40 Exponential charging in a resistor-capacitor network: (a) a resistor-capacitor circuit and (b) a capacitor charging curve

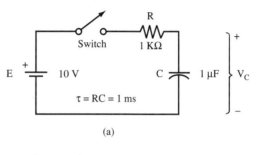

(a)

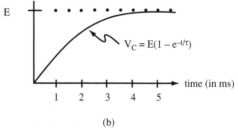

(b)

10.9 INTERFACING WITH NON-MOTOROLA PERIPHERALS

A common frustration encountered in the hardware and software fields is the lack of compatibility between equipment produced by different manufacturers. For example, Motorola peripherals are designed to work with Motorola processors. Intel peripherals are designed for use with Intel processors. It is difficult but not impossible to get one manufacturer's devices to work on another's system. In this section we will see how a non-Motorola peripheral, the 8279 programmable keyboard/display interface, can be interfaced to a 68000-based system. Figure 10.41 shows one way this non-Motorola device can be connected. As usual, an address decoder is included to chip-select the device when a specific address is encountered on the address bus. Notice that address line A_1 does not go

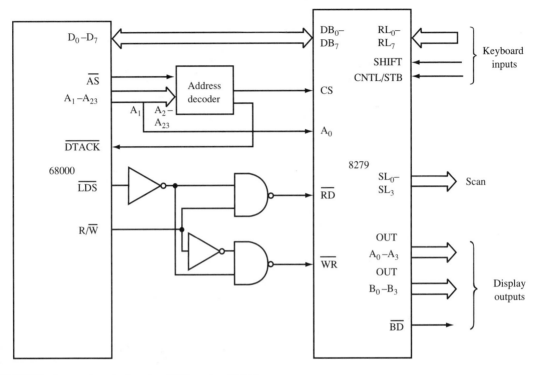

FIGURE 10.41   Interfacing the 8279 to the 68000

to the decoder but instead connects to the A_0 input on the 8279. Both address lines (A_1 on the 68000 and A_0 on the 8279) represent the least significant address bit for our purposes, and are thus connected together. A_0 is used on the 8279 to select control/data operations within the device.

One common hardware feature of Intel peripherals is the use of separate $\overline{RD}$ (read) and $\overline{WR}$ (write) signals. Since the 68000 outputs a single signal, R/$\overline{W}$, we must artificially create $\overline{RD}$ and $\overline{WR}$. This is accomplished with the two NAND gates. Only one of the NAND gates is allowed to output a 0 at any time. Since $\overline{LDS}$ is used to control this circuit, we must transfer data bytes to/from the 8279 over the lower half of the data bus.

Software interfacing is not a problem, since it only requires the transfer of data between the device and the processor. The 8279 is thus treated like any other memory-mapped device.

Being able to interface non-Motorola devices with the 68000 increases the flexibility of the system during the design stage, since a non-Motorola part may be used, if necessary, to perform some desired function. We will see, however, that this possibility is remote, given the wide pool of devices available for 68000 systems. These devices are summarized in the next section.

10.10   A SUMMARY OF OTHER 68000-BASED PERIPHERALS

This section summarizes a number of 68000-based peripherals. The intention is to let you know what kinds of devices are available. As you read this section you will see that, through its peripherals, the 68000 is capable of performing many complex functions in many different environments. Ideally, you should obtain a copy of a peripheral handbook and look up the data sheets for each peripheral, or search the Web for additional information.

The 68652 Multiprotocol Communications Controller

This device is used to provide several communication formats. Bit- and byte-oriented protocols are available, as is synchronous serial data transmission and reception. The 68652 will also generate and check 16-bit CRC codes.

The 68184 Broadband Interface Controller

When used with the 68824 token bus controller, the 68184 implements a broadband IEEE 802.4 token bus node. Bit rates up to 10 million bits per second are available, as are scrambling/descrambling capability and posterror correction.

The 68824 Token-Passing Bus Controller

This device implements the media access control portion of the IEEE 802.4 standard and the receiver portion of the IEEE 802.2 logical link control. The 68824 is the interface between the CPU, its memory, and the token bus. This device implements high-speed serial data communication in a local area network.

The 68452 Bus Arbitration Module

The 68452 allows up to eight users access to a common system bus in a 68000-based system. The 68452 is suitable for use in a multiprocessor environment, controlling each CPU's access to the main system address and data bus.

The 68590 LAN Controller for Ethernet

The device implements the required communication protocols for Ethernet LAN data transmission and reception. Ethernet is a widely used communication network that can easily link thousands of users together.

The 68465 Floppy Disk Controller

This device is used to interface two single- or double-density floppy disk drives to the 68000. The 68465 performs all hardware functions required by the floppy disk drives. The 68000 must supply the software required to implement a disk operating system.

The 68486/68487 Raster Memory System

These two peripherals are used to implement raster scan graphics on a 68000-based system. Color graphics, collision detection, and light-pen input are provided. The 68486/68487 uses a block of system RAM as a video memory.

10.11   TROUBLESHOOTING TECHNIQUES

The peripherals in this chapter are suitably advanced. Although they cover a wide variety of topics, there are many other, even more specialized, peripherals. When you come across a new device and are faced with the challenging task of getting it to work, keep these suggestions in mind:

* Look over the data manual on the new peripheral. If you do not have a data manual, try searching the Web. Motorola has a very useful site, offering downloads of manuals (680x0 series plus many others) in PDF format. Many other educational institutions post important information on the Web also, as part of class projects.

    Skim the figures and captions, look at the register and bit assignments, read the tables. Read about the hardware signals. Study the timing diagrams. Look at any sample interface designs provided by the manufacturer. Be sure you understand why the signals are used the way they are.

    Read about the software architecture of the peripheral. How is it controlled? How do you send data to it, or read data from it? How many different functions does it perform?
* Get the hardware interface working properly. This requires your skill in designing I/O hardware. Some software may be required to fully test the interface.
* If the peripheral has many modes of operation, begin with the simplest. Program the peripheral to operate in this mode to be sure you have control over it. Expand to other modes of operation as you learn more about the device.

Even if all you are doing is modifying someone else's code, written long ago, for a peripheral that is already operational, it is still good to learn as much about the device as possible. This will help you avoid typical problems, such as forgetting to issue the master reset command, even though power was just applied.

SUMMARY

In this chapter we have looked at a number of peripherals in some detail. Each peripheral performs a different, necessary function. Parallel and serial I/O, memory management, DMA, interrupt handling, and floating-point operations were all covered. Any 68000-based system may include any or all of these functions. Each peripheral requires some hardware interfacing, consisting of an address decoder. In some cases, other hardware was needed to get the peripheral to communicate with the processor or the external circuitry connected to the peripheral. Examples of software routines, designed to program the devices for specific functions, were included to complete the interface.

An example showing how a non-Motorola peripheral could be connected to the 68000 was also covered. In addition, a brief summary of other peripherals in the 68000 family was given, along with a set of troubleshooting tips.

STUDY QUESTIONS

1. Modify DMAMOVE so that it uses channel 1 instead of channel 0. All channel 1 registers are offset by 40H.
2. What instructions must be included in DMAMOVE to ensure that channel 0 is the highest priority channel?
3. What is the GCR code for a 64-cycle burst rate and a bus bandwidth of 25.00 percent?
4. Explain the differences between burst DMA and cycle-steal DMA. Is one technique better for transferring large blocks of data?
5. The processor can read the channel status register at any time to determine the channel status. Write the required code to detect a complete channel operation.
6. How can the 68440 be used to load a 1KB block of RAM from a device, using implicit addressing?
7. What is the largest block of data that can be transferred with the 68440 with a single-channel operation? Explain how larger blocks could be transferred.
8. What are the possible segment sizes available with the 68451?
9. What must the contents of a LAM register be to define a segment size of 256KB?
10. How can any number of segments be combined to make a 94KB block with a minimum of wasted space?
11. What is the physical address mapped by the following descriptor?

     LBA: A000

     LAM: F000

     PBA: E000

12. What is the segment size in Question 11?
13. Show the contents of the LBA, LAM, and PBA to allow translation from a logical address of 2C0000 to a physical address of C90000 in a 64KB segment.
14. Suppose that the SSR of a descriptor has been copied into the lower byte of D3. Write a section of code that will JMP to NOTUSED if the U bit is cleared, and to MODIFIED if the M bit is set. Jump to CONTINUE if neither condition holds.
15. What channel A mode register 1 byte is needed to program channel A for 8 data bits, no parity, and FFULL interrupts?
16. What channel A mode register 2 byte is needed to program channel A for 1 stop bit and automatic echo mode?
17. How can channel A be programmed for 9,600 baud in the receiver and 600 baud in the transmitter?

18. What channel A command code is needed to reset the error status?
19. Modify the CHARIN subroutine so that the RxRDY status bit is checked by a BTST instruction.
20. Write a subroutine called CHAROUT that will transmit the ASCII character located in the lower byte of D3.
21. Write a routine called ERRORDET that will examine the contents of channel A's status register, and enter exception processing with a TRAP   #5 instruction if any error bit is set. The error bits consist of parity, overrun, and framing. If no errors are present, simply return from ERRORDET.
22. Explain how two different functions are possible with the multifunction pins on port C of a 68230.
23. Generate a data table of sine wave values that contains 32 bytes. The 32 bytes should be spread evenly through a complete cycle (e.g., evaluate the sine function every 11 degrees).
24. What are the port A pin directions for each of the following data direction register byte values: 11111111, 00000000, 11110000, 10101010, and 10000001?
25. How might two-key rollover be implemented in the keypad scanner?
26. Write the initialization code for the keypad scanner. The 68230 must be configured for port A out and port B in.
27. What hardware and software is required to detect a pushed button at any time? Show how the processor can be interrupted by action on the keypad.
28. Rewrite KEYBD so that it supports scanning of a 64-key matrix.
29. Rewrite TIMERINT so that the display scanning order is 4 . . . 3 . . . 2 . . . 1.
30. Write the initialization code required for TIMERINT. The 68230 must be configured for ports A and B out and vectored interrupts from the timer. Assume that loading the counter with $1400 will generate the proper timing, if the counter is clocked by the output of the prescaler. The TIVR should be loaded with vector number $86 to service TIMERINT.
31. Modify TIMERINT to support eight multiplexed displays.
32. Write a routine that will flash HELP on the four displays (described in Figure 10.33).
33. Explain how the general purpose I/O lines on the 68901 can function as inputs or outputs. How is the direction programmed?
34. What timer modes must be selected so that timers A, B, and C generate frequencies that differ by a factor of 4 (i.e., A equals 100 Hz, B equals 400 Hz, and C equals 1600 Hz)?
35. What is the advantage of interrupting the processor when a character is received by the USART? If the VR contains 1011xxxx, what vector is generated by USART when a character is received?
36. Explain how the 68901 can be used to service interrupt requests from eight different devices.
37. How many parallel I/O lines are available when four 68901s are daisy-chained? How many interrupts? How many serial channels?
38. How many vectored interrupts are available with five daisy-chained 68153s?
39. Which control registers in INTSET have the FLAG bit set?
40. Write a set of instructions that will enable interrupts in control register 2 without affecting any other bits. Assume CR2 is located at address B002.
41. Which control registers in INTSET are set up to allow an external device to supply the vector number?

42. Write a 68881 routine to convert degrees into radians. The number of degrees is passed to your routine in DEGREES. Return the number of radians in FP4.
43. Write a floating-point routine to compute the volume of a sphere with radius R.
44. Write a floating-point routine to convert a polar number (magnitude and phase angle) into a rectangular number (X- and Y-axis values).
45. Write a floating-point routine to normalize a set of data values. The data set is normalized by dividing each number in the set by the largest number. Assume that the data is sorted in ascending order and begins at POINTS.
46. Write a floating-point routine to compute the final velocity of an object that starts out with an initial velocity, V_i, and accelerates due to gravity, for a period of time T.
47. What changes need to be made to the 8279 interface if data is to be transferred over the upper half of the processor's data bus?
48. What peripherals would you choose to use in a new system that required four levels of interrupts, two serial channels, a timer, and 16 bits of parallel I/O?
49. What kind of initialization would be needed in the system of Question 48?
50. In a system employing multiple peripheral devices, is there any design advantage in putting all peripheral I/O addresses close to each other?
51. Explain how the circuits of Figures 10.32 and 10.33 are ideal for use in the calculator project.

CHAPTER 11

Building a Working 68000 System

OBJECTIVES

In this chapter you will learn about

- The main parts of a single-board computer
- The design of custom circuitry for the major sections of the microcomputer system
- How to generate and answer the necessary questions for the design or modification of a single-board computer
- The operation of a software monitor program
- How the single-board computer can interface with a PC

11.1   INTRODUCTION

This chapter deals exclusively with the design of a custom 68000-based microcomputer system. The system is an ideal project for students wishing to get some hands-on experience, and is also a very educational way of utilizing all of the concepts we have studied so far.

Ideally, we wish to design a system that is easy to build, has a minimal cost, and yet gives the most for the money. The very least we expect the system to do is execute programs written in 68000 code. It is therefore necessary to have some kind of software monitor that will provide us with the ability to enter 68000 code into memory, execute programs, and even aid in debugging. This chapter then will consist of two parts. The first deals with the design of the minimal system, and the second with the design of a software monitor and the use of its commands.

Pay close attention to the trade-offs that we will be making during the design process. A difficult hardware task can often be performed by cleverly written machine code, and the same goes for the reverse. Do not forget our main goal: to design a *minimal* 68000-based system suitable for custom programming.

Section 11.2 covers the minimal requirements of the system we will design. Section 11.3 describes the design of the system hardware. Section 11.4 contains the parts list for the system. Section 11.5 gives hints on how the system may be constructed. Section 11.6 shows how a PAL might be used to generate several system signals. Section 11.7 deals with the design of the software monitor program for the system. Section 11.8 explains how additional commands may be added to the monitor program. Section 11.9 gives examples of some simple programs that can be used to test the hardware and software. Finally, troubleshooting techniques are presented in Section 11.10.

11.2 MINIMAL SYSTEM REQUIREMENTS

The requirements of our minimal system are the same as those of any computer system, and consist of four main sections: timing, CPU, memory, and input/output. Since we are the designers, building this system for our personal use, it is up to us to answer the following questions:

1. How fast should the CPU clock speed be?
2. How much EPROM memory is needed?
3. How much RAM memory is needed?
4. Should we use static or dynamic RAMs?
5. What kind of I/O should be used—parallel, serial, or both?
6. Do we want interrupt capability?
7. Will future expansion (of memory, I/O, and so on) be required?
8. What kind of software is required?

It should be clear that we have a big task ahead of us. During the design, all of these questions will be answered and the reasons for choosing one answer over another explained. Make sure you understand each step before proceeding to the next one. In this fashion, you should be able to design your *own* computer system, from scratch, without any outside help.

11.3 DESIGNING THE HARDWARE

In this section, the four main functional sections of the system will be designed. In each case, there will be questions to answer regarding specific choices that must be made. You may want to make a list of all important questions as you go.

The Timing Section

The timing section has the main responsibility of providing the CPU with a nice, stable clock. Any type of digital oscillator will work. It is then necessary to decide on an operating frequency for the oscillator. Many times this frequency is the operating frequency of the CPU being used. Microprocessors are commonly available in different clock speeds, two examples being the 8 and 12 MHz versions of the 68000.

One important factor limiting the clock speed is the speed of the memories being used in the system. A 12 MHz CPU might require RAMs or EPROMs with access times less than 100 ns! In our design, we will use a 4 MHz clock. This is fast enough to provide very quick execution of programs, while at the same time allowing for the use of less expensive RAMs with longer access times (250 ns).

The circuit of Figure 11.1 shows a 4 MHz oscillator package driving two inverters. This is done so that any loading from external sources will not affect its operation. One of the outputs, CPU-CLK, is the master CPU clock signal. Since many other circuits in our system might also require the use of this master clock, we make the CLK signal available, too. The CPU therefore gets its own clock signal. It is desirable to separate the clocks in this fashion, to aid in any digital troubleshooting that may need to be done. By making multiple clocks available, it is easier to trace the cause of a missing clock, should that problem occur.

In addition to the clock, we will need a circuit to provide the CPU with a reset pulse upon the application of power. It is very important to properly reset the CPU at power-on, to ensure that it begins executing its main program correctly. There are two requirements that must be met to correctly reset the 68000 upon power-on. One is to pull both the $\overline{\text{HALT}}$ and $\overline{\text{RESET}}$ pins on the CPU low at the same time. The second is to keep them low for at least 100 ms. The circuit of Figure 11.2 satisfies both of these needs. Figure 11.2 shows a 555 timer connected as a monostable multivibrator. The R-C combination on pins 6 and 7 sets the output pulse width to at least 100 ms. The instant power is applied, pin 3 of the 555 timer will go high. This causes the output of the inverter to go low, which causes the outputs of the open-collector buffers connected to $\overline{\text{HALT}}$ and $\overline{\text{RESET}}$ to also go low. These open-collector gates are needed here, because $\overline{\text{HALT}}$ and $\overline{\text{RESET}}$ are **bidirectional** signal pins.

The instant the 68000 recognizes the reset request, it will force the $\overline{\text{RESET}}$ pin low, which causes the $\overline{\text{RESET}}$ pin to act like an output! If a regular TTL gate is used instead of an open-collector gate, fireworks may result when this happens.

When the 555 times out, pin 3 will go low. This will cause the inverter to send a logic 1 to the open-collector buffers, which (via their 4.7 K ohm pull-up resistors) will pull both $\overline{\text{HALT}}$ and $\overline{\text{RESET}}$ high, ending the reset request. Since we may want to reset the processor at a random point in time (maybe during the execution of a runaway program),

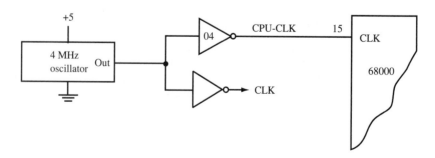

FIGURE 11.1   4 MHz system timing clock

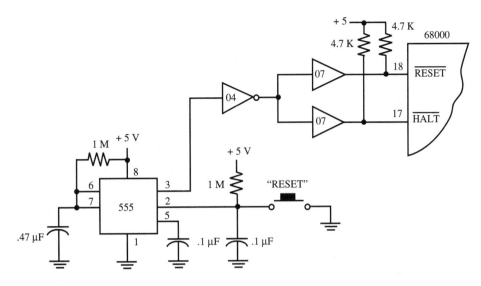

FIGURE 11.2 Power-on reset circuitry

we can get the 555 timer to repeat this process by momentarily grounding pin 2. The normally open push button connected to pin 2 will accomplish this for us, and provide a panic button for those hard-to-fix programs.

If we want a visual indication that the processor has entered the HALT state, we can also add the circuitry of Figure 11.3. A buffer is used here to drive a light-emitting diode. A zero on the $\overline{\text{HALT}}$ line will cause the LED to glow. The LED will also light when we reset the processor, since $\overline{\text{HALT}}$ is pulled low by the open-collector gate. If the HALT LED always stays lit, no matter what we do, we have an indication that something is wrong with the system hardware or software.

FIGURE 11.3 HALT indicator (optional)

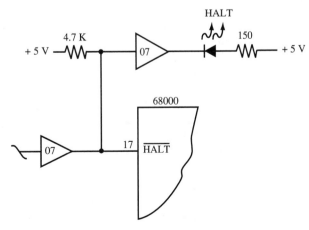

The buffer is used to provide forward-bias current for the LED. Directly connecting the LED to the $\overline{\text{HALT}}$ pin may draw too much current and possibly result in erratic operation of the CPU. We must remember to always protect the CPU from this kind of problem by buffering the signal lines whenever possible.

The CPU Section

Now that we have a working timing section, we must design the CPU portion of our system. It is during the design of this section that we answer our question about interrupts, and pose a few more important questions. For instance, do we need to buffer the address and data lines? Do we want to give bus-grant capability to an external device? How will the CPU respond to a bus error? Take a good look at Figure 11.4 before continuing with the reading.

The figure details the connections we must make to the CPU for it to function in our minimal system. On the right side of the CPU we see the data and address lines. These signal lines are used in both the memory and I/O sections. Since we are working with a minimal system, there is no need to buffer them with high-current drivers, which are required for large memory systems. Our memory section will contain four EPROMs and RAMs, and the 68000 will have no difficulty driving the handful of loads these devices will present.

The top left side of the CPU shows the $\overline{\text{BR}}$ and $\overline{\text{BGACK}}$ signals pulled up to 5 volts. The logic 1s on these inputs will disable the CPU from ever granting its buses to another system master. Again we choose this condition for simplicity. If it were important to provide bus-grant capability, so that an external peripheral can perform DMA, extra logic would be needed here to provide the correct signals to $\overline{\text{BR}}$ and $\overline{\text{BGACK}}$.

The three interrupt lines ($\overline{\text{IPL}}_0$ through $\overline{\text{IPL}}_2$) are also pulled high to prevent an external interrupt request. For systems that may need only one interrupt, the three inputs may be tied together to form a single interrupt input ($\overline{\text{INT}}$). If this input line is left floating, all three $\overline{\text{IPL}}$ lines will be high. The 68000 recognizes this state as a level-0 priority and thus takes no action. But pulling the $\overline{\text{INT}}$ line to ground makes all the $\overline{\text{IPL}}$ inputs zero, which creates a level-7 request. We must remember to provide the necessary software to handle a level-7 interrupt when we write the monitor program.

The last bit of circuitry in our CPU section is a counting circuit that helps to detect errors on the 68000's address bus. Remember, the $\overline{\text{BERR}}$ input informs the 68000 that a problem exists with the current bus cycle. Maybe a problem has occurred in the memory section and the $\overline{\text{DTACK}}$ signal has not been properly generated. For whatever reason, once the bus error has been detected, we inform the CPU of the error by pulling $\overline{\text{BERR}}$ low. The circuitry connected to the $\overline{\text{BERR}}$ input is designed to do this. Here is how it works: The 68000 begins a bus cycle with $\overline{\text{AS}}$ high. This causes the clear input on the 74LS393 binary counter to be high, which forces its four outputs low. The $\overline{\text{BERR}}$ input of the 68000 will then be high, via the inverter and the 74LS393's C output. When $\overline{\text{AS}}$ goes low, to indicate a stable memory address, the 74LS393 will be allowed to count at a rate determined by the E clock of the 68000. If the 74LS393 is allowed to reach the binary count 0100, the C output will go high. This causes $\overline{\text{BERR}}$ to go low and a bus error to be recognized by the CPU. During normal bus operation, the $\overline{\text{AS}}$ signal will remain low only for short periods of time, and the 74LS393 will be cleared before reaching the 0100 state. Circuits like these are commonly called **watchdog monitors,** and are usually utilized to reset the processor when things begin to go wrong.

You may have noticed that we are not using the function code outputs (FC_0, FC_1, and FC_2). Since they are primarily used to indicate user/supervisor states, we have no need for them in our minimal system. Even the interrupt acknowledge function code is not needed,

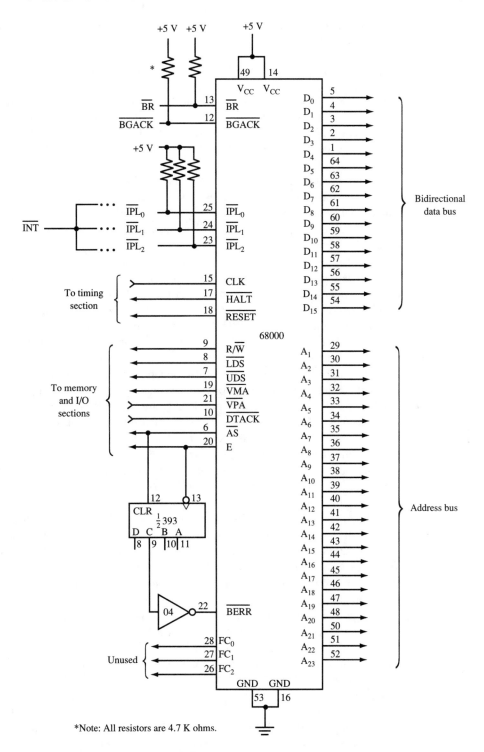

*Note: All resistors are 4.7 K ohms.

FIGURE 11.4 CPU section circuitry

since we have no elaborate interrupt-generating circuitry. Thus we have designed the simplest CPU section for our needs, and may now move on to the design of our memory section.

The Memory Section

There are a number of questions that must be answered before we get involved in the design of our memory section. For instance, how much EPROM memory is needed? How much RAM? Should we use static or dynamic RAM? Should we use full- or partial-addressing decoding? Will we allow DMA operations?

The answer to each of these questions will help to specify the required hardware for the memory section. If we first consider what *applications* we will be using our computer for, the previous questions will almost answer themselves. Our application at this time is educational. We desire a 68000-based system that will run short machine language programs. Keeping this point in mind, we will now proceed to come up with answers to our design questions.

A programmer, through experience, can estimate the required amount of machine code needed to perform a desired task. The software monitor that we will need to control our system will have to be placed in the EPROMs of our memory section. Two standard 2716 EPROMs will provide us with 2,048 words of programmable memory. This is more than enough EPROM to implement our software monitor. We will still have space left over in the EPROMs in case we want to add more functions to the monitor in the future.

The amount of RAM required also depends on our application. Since we will be using our system only to test short, educational programs, we can get by with a few hundred words or so. Since dynamic RAMs are generally used in very large memory systems (64K, 256K, and more), we will not use them because most of the memory would go to waste. Other reasons exist for not choosing dynamic RAMs at this time. They require complex timing and refresh logic, and will also need to be wired very carefully to prevent messy noise problems from occurring. Even if we use a DRAM controller, we will need some external logic to support the controller, which itself could be a very costly item.

For these reasons, we decide to use static RAM. Even though a few hundred words will cover our needs, we will use two 6116 static RAMs, thus making our RAM memory 2,048 words long also. The 6116s are low-power static RAMs, with pinouts almost exactly identical to the 2716s we are using for our EPROM memory. So, by adding only four more integrated circuits (plus a few for control), our memory needs are taken care of.

Figure 11.5 shows how we use a two- to four-line decoder to perform partial-address decoding for us. Since we are not concerned with future expansion on a large scale, partial-address decoding becomes the cheapest way to generate our addressing signals. Address lines A_{15} and A_{16} are used because they map the 68000's memory into convenient ranges. With both A_{15} and A_{16} low, the 74LS139 decoder will output a 0 on the $\overline{PROM}$ line; thus, $\overline{PROM}$ will be low whenever the 68000 addresses memory in the 00000 to 07FFF range. This allows us to use EPROMs as large as 16KB each for the monitor program. If A_{15} is high while A_{16} is low, the $\overline{RAM}$ signal will go low. This corresponds to an address range from 08000 to 0FFFF. If either $\overline{PROM}$ or $\overline{RAM}$ goes low, (as one will during a valid memory reference), the output of the AND gate will go low

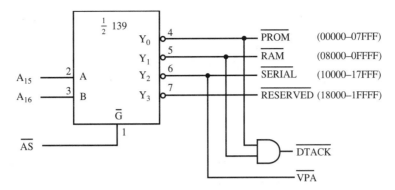

FIGURE 11.5 Partial address decoder and $\overline{\text{DTACK}}$ circuit

issuing $\overline{\text{DTACK}}$. The $\overline{\text{SERIAL}}$ signal is low whenever the 68000 addresses memory from 10000 to 17FFF, and this signal is used in the serial section and also to issue $\overline{\text{VPA}}$. That leaves the fourth ouput of the 74LS139 free for use in the future. If we need to put more EPROM, RAM, or memory-based I/O into our system, we will be able to stick it into the 18000 to 1FFFF address range. Note in Figure 11.5 that the enable input of the 74LS139 ($\overline{\text{G}}$) is connected to $\overline{\text{AS}}$. This ensures that the 74LS139 will work only during valid bus cycles (when $\overline{\text{AS}}$ is low). If we were allowing DMA operations, we might not want the 74LS139 to operate the same way. The $\overline{\text{G}}$ input gives us a way to disable the 74LS139 during a DMA operation, so that the DMA hardware will have full control of the memory section.

Figure 11.6 shows how we generate the required memory read and write signals. We need separate read and write signals for both lower and upper addresses, to allow for byte

FIGURE 11.6 Memory read/ write logic

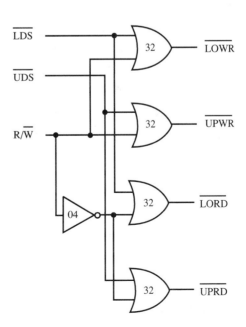

operations in memory. By using only four OR gates and an inverter, we are able to generate these signals rather easily.

And now only the actual EPROM and RAM chips need to be connected. Figure 11.7 shows how we use the four memories to finish our memory section. Notice that the EPROMs and RAMs are separately labeled *upper* and *lower*. The lower memories have

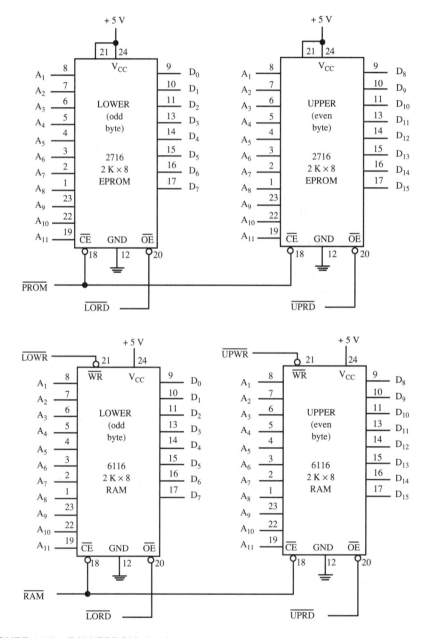

FIGURE 11.7   RAM/EPROM circuitry

their data bus pins connected to D_0 through D_7, and the upper memories have their pins connected to D_8 through D_{15}. To perform a byte read from memory, only one of the four memories will be fully enabled by the signals present on $\overline{\text{LORD}}$, $\overline{\text{UPRD}}$, $\overline{\text{PROM}}$, and $\overline{\text{RAM}}$. Address bits A_1 through A_{11} are used to point to 1 of 2,048 locations within the memories. The $\overline{\text{LOWR}}$ and $\overline{\text{UPWR}}$ signals are only connected to the RAMs because we cannot write into the EPROMs once they are programmed.

Before we continue, let us examine how the whole memory system works. For instance, what happens when the 68000 tries to read a word of data from location 81F4? The decoder (in Figure 11.5) will pull the $\overline{\text{RAM}}$ signal low, due to A_{15} being high and A_{16} being low in the address 81F4. This low on $\overline{\text{RAM}}$ will chip-enable both 6116s, and also activate the $\overline{\text{DTACK}}$ signal. Since $\overline{\text{LDS}}$ and $\overline{\text{UDS}}$ will be low, both $\overline{\text{LORD}}$ and $\overline{\text{UPRD}}$ will go low to enable the output circuitry of the 6116s. This will place 16 bits of data from RAM location 1F4 onto the data bus. When the memory cycle finishes and $\overline{\text{AS}}$ goes high again, $\overline{\text{RAM}}$ and $\overline{\text{DTACK}}$ will revert to their inactive, high states.

Since the address and data lines in our memory section are connected only to four chips, we need not buffer the address or data lines from the CPU, since the 68000 is capable of driving this many loads by itself.

The Serial Section

The serial section of our computer will contain all hardware required to communicate with the outside world (via an EIA-compatible data terminal). The only question that must be answered concerns the baud rate we will be transmitting and receiving at. A very acceptable speed is 2,400 baud. Speeds higher than this will be too fast to read on the screen, and slower speeds will take too long to read.

The circuit that will take care of transmitting and receiving the serial data is the 6850 ACIA chip. We only have to supply external logic to enable the chip and generate its required baud rate clocks. The $\overline{\text{SERIAL}}$ output of our memory address decoder (see Figure 11.5) must be connected to the processor's $\overline{\text{VPA}}$ input. Asserting $\overline{\text{VPA}}$ indicates to the 68000 that it is addressing a slow Motorola peripheral and must therefore synchronize its bus cycle with that of the slower E clock.

Figure 11.8 shows how the baud rate clocks are generated. We will use software to select an X16 clock *inside* the ACIA. This means we must supply $2,400 \times 16$, or 38.4 kHz to the ACIA to get our desired speed of 2,400 baud. The crystal oscillator uses a special baud rate crystal (2.4576 MHz), whose frequency is an even multiple of all usable baud rates. The two D flip-flops and the other half of the 74LS393 binary counter are used to divide the 2.4576 MHz frequency by 64, which results in the 38.4 kHz that we need for the ACIA. The separate oscillator is needed, since we have no way of evenly dividing the master system clock (4 MHz) down to usable baud rates.

In Figure 11.9 we see the remainder of the circuitry that makes up the serial section. Here we can see that the upper 8 bits of the data bus carry information to and from the ACIA. Address bit A_1 is used to select the internal registers of the ACIA (status/control and data). The ACIA's timing is synchronized with the E clock from the 68000, which is one-tenth the speed of the master CPU clock. The ACIA is chip-enabled when $\overline{\text{CS}}_2$ is low and CS_0 and CS_1 are high. $\overline{\text{UDS}}$ is used to provide the logic 0 on $\overline{\text{CS}}_2$, since we are already using the upper data bus bits for data transfer. The 68000's $\overline{\text{VMA}}$ output is used to complete the enabling of the ACIA. Remember that $\overline{\text{VMA}}$ will

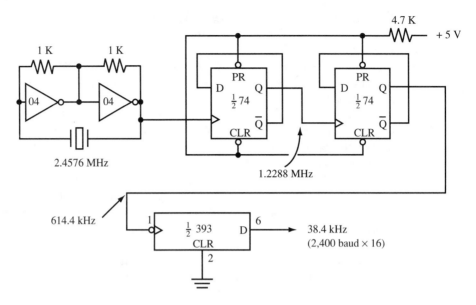

FIGURE 11.8 Serial section timing circuitry

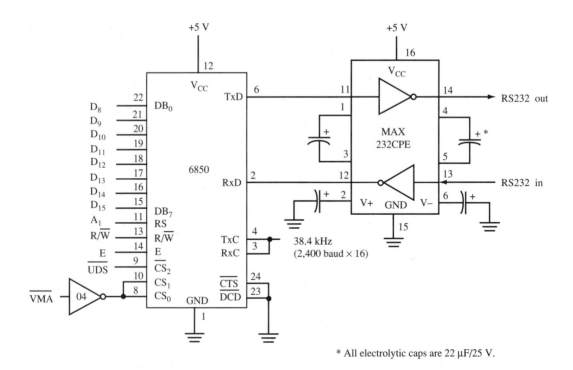

FIGURE 11.9 Serial transmitter/receiver circuitry

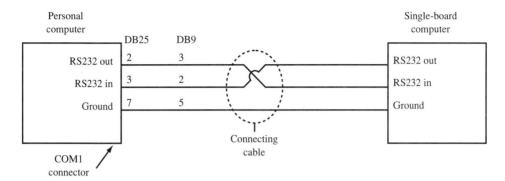

FIGURE 11.10   Serial connection to a personal computer

go low after the CPU receives a low on $\overline{VPA}$ (which has already been generated by the circuit of Figure 11.5).

For simplicity the $\overline{CTS}$ and $\overline{DCD}$ pins on the ACIA are grounded. If handshaking is required, these pins may be wired accordingly, depending on the I/O device used. Both TxC and RxC (the transmitter and receiver clocks) are tied together and driven by the same baud rate clock. That leaves just the serial-in and serial-out pins. Here we use a MAX232 chip to convert the TTL serial information into RS232C levels that swing plus and minus 10 volts. The MAX232 has an internal oscillator circuit that generates these voltages from a single 5-volt source. This lets us eliminate the need for a negative power supply, whose only job would be to drive the RS232C outputs. The serial-in and serial-out signals from the MAX232 can be wired to a DB25 connector or other suitable connector.

Figure 11.10 shows one way the single-board computer may be connected to the personal computer. The transmit and receive lines (RS232 out and RS232 in) on the single-board are connected to the receive and transmit lines on the PC's COM1 connector. The connecting cable is sometimes referred to as a **null modem,** since it cross-couples the transmit and receive lines for full-duplex communication.

If more than one serial channel is needed, a second one may be added very easily by duplicating the circuitry of Figure 11.9, and making some small wiring changes. Connect the lower data bus bits to the ACIA's data lines, and use $\overline{LDS}$ to enable the ACIA.

11.4   THE MINIMAL SYSTEM'S PARTS LIST

Now that we have finished designing the minimal system, we can look back on all of the figures and decide how many ICs we will need to build it. It might not have been clear to you at the time why we used halves of the 74LS393 binary counter, one in the $\overline{BERR}$ circuit and the other in the baud rate circuit. We could have used two 74LS93 binary counters instead, but the 74LS393 lets us do both jobs with one IC, not two! The whole idea behind the design was to build a working system with a minimum of parts. The following list summarizes all ICs that are needed. Pull-up resistors, discrete components, and sockets are not included.

one 68000 CPU

two 2716 EPROMS

two 6116 2K-by-8 RAMs

one 6850 ACIA

one 555 timer

two 74LS04 hex inverters

one 74LS07 open-collector buffer

one 74LS393 dual binary counter

one 74LS32 quad two-input OR gate

one 74LS08 quad two-input AND gate

one 74LS139 dual two- to four-line decoder

one 74LS74 dual D flip-flop

one MAX232CPE TTL to RS232 converter

one 4 MHz crystal (or digital) oscillator

one 2.4576 MHz crystal (or digital) oscillator

11.5   CONSTRUCTION TIPS

The easiest way to build the minimal system is to wire-wrap it. A printed circuit board may be used, but would be very complex and most likely double-sided!

The minimal system will work the first time power is applied, if the following points are kept in mind:

1. Keep all wires as short as possible. Long wires pick up noise.
2. Connect 0.1-μF bypass capacitors across +5 volts and GND on all ICs.
3. Trim all excess component leads to avoid short circuits.
4. Connect power and ground to all ICs before wiring anything else.
5. Pull all unused TTL inputs to +5 volts with 4.7K ohm resistors.
6. Mark off connections on a copy of the schematic, as they are made.
7. Make sure no ICs are plugged in backward before applying power.
8. Use an ohmmeter to check each connection as it is made.
9. Check that each IC has proper power before beginning any major troubleshooting.
10. Plug in only the clock ICs first. If the clock does not work, neither will the rest of the system.
11. Check the $\overline{\text{RESET}}/\overline{\text{HALT}}$ circuit for proper operation.

Experience, of course, is the best teacher, but these hints should be enough to get you started. There is nothing like the feeling of building a circuit that works the first time! If it fails to operate properly, do not get discouraged. With your knowledge of TTL, you should be able to track down the source of the problem in no time. You might be surprised that most problems will be due to wrong wiring. Always check your wiring very carefully!

11.6   USING PROGRAMMABLE LOGIC IN THE MINIMAL SYSTEM

Although the parts requirements for the minimal system are few, it is possible to further reduce the number of components needed, by adding programmable logic arrays (PLAs, PALs). These devices usually consist of any number of internal gates, such as AND, OR, exclusive OR, and flip-flops, whose inputs are connected by blowing selected fuses in an internal grid. The designer must write Boolean expressions that define the operation of all of the PAL outputs. The expressions are evaluated by software in a device called a PAL programmer (similar to an EPROM burner), which decides exactly how the internal PAL connections should be made.

It is not uncommon that a single PAL can be programmed to replace the logic operation of five TTL ICs! The following example shows how a PALs may be used to generate some of the required memory section signals, thus replacing the circuitry of Figures 11.5 and 11.6.

Figure 11.11 shows the connections to a 10L8 PAL, a device that contains 10 inputs and eight programmed outputs. The inputs to the PAL are on pins 1 through 9 and 11. The four NC inputs are no-connects, inputs that are not needed for this application. The eight programmed outputs are on pins 12 through 19. The PAL program listed in Figure 11.12 defines the input/output signal names and the required equations needed to implement the desired functions. The use of this single PAL eliminates the need for two ICs, the 74LS32 quad OR gate, and the 74LS139 decoder. Even though we have only reduced our chip count by one so far, we have gained a tremendous advantage in the area of flexibility. Changing the addresses of all main sections (EPROM, RAM, SERIAL) is now as easy as reprogramming and replacing this one PAL!

FIGURE 11.11   PAL connections for 10L8 memory controller

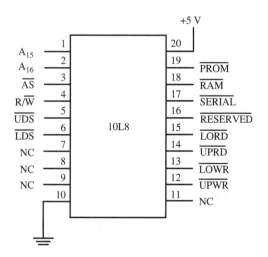

```
PAL10L8                                          PAL DESIGN SPECIFICATION
68K.MEM                                          JAMES L. ANTONAKOS
68000 MEMORY CONTROLLER
BROOME COMMUNITY COLLEGE
A15 A16 /AS RW /UDS /LDS NC NC NC GND
NC /LOWR /UPWR /LORD /UPRD /RES /SER /RAM /PROM VCC

IF (VCC) PROM = /A16 * /A15 * AS               ;PROM --- 000000-007FFF
IF (VCC) RAM = /A16 * A15 * AS                 ;RAM ---- 008000-00FFFF
IF (VCC) SER = A16 * /A15 * AS                 ;SERIAL - 010000-017FFF
IF (VCC) RES = A16 * A15 * AS                  ;RESVD -- 018000-01FFFF

IF (VCC) LOWR = /RW * LDS                       ;LOWER WRITE
IF (VCC) UPWR = /RW * UDS                       ;UPPER WRITE
IF (VCC) LORD = RW * LDS                        ;LOWER READ
IF (VCC) UPRD = RW * UDS                        ;UPPER READ

FUNCTION TABLE
/AS A16 A15 RW /UDS /LDS
/PROM /RAM /SER /RES /LOWR /UPWR /LORD /UPRD
```

;/AS	A16	A15	RW	/UDS	/LDS	/PROM	/RAM	/SER	/RES	/LOWR	/UPWR	/LORD	/UPRD
H	X	X	X	X	X	H	H	H	H	X	X	X	X
L	L	L	X	X	X	L	H	H	H	X	X	X	X
L	L	L	X	X	X	L	H	H	H	X	X	X	X
L	H	L	X	X	X	H	H	L	H	X	X	X	X
L	H	H	X	X	X	H	H	H	L	X	X	X	X
X	X	X	L	L	L	X	X	X	X	L	L	H	H
X	X	X	L	L	H	X	X	X	X	H	L	H	H
X	X	X	L	H	L	X	X	X	X	L	H	H	H
X	X	X	H	L	L	X	X	X	X	H	H	L	L
X	X	X	H	L	H	X	X	X	X	H	H	H	L
X	X	X	H	H	L	X	X	X	X	H	H	L	H

FIGURE 11.12   PAL program for 68000 memory control

11.7   WRITING THE SOFTWARE MONITOR

Now that we have the system hardware designed, we must tackle the job of writing the system software. Since our goal is to use the system for the testing of custom 68000 programs, the monitor must be capable of performing every step that is needed for us to get the new program into memory, edit it if needed, display it in hexadecimal format, and also execute it. This will require the creation of three monitor commands: EDIT (place hexadecimal data into memory), DUMP (display memory contents in hexadecimal), and GO (execute a program). Two more commands, while not completely necessary, might come in handy. They are MOVE (move data in memory) and

TABLE 11.1 Auxiliary subroutines

Name	Function
INIT_ACIA	Initialize 6850 ACIA device
INSTAT	Check keyboard status for character received
CHAR_IN	Read ASCII character from keyboard
CHAR_OUT	Send ASCII character to display
CRLF	Send ASCII CR, LF to display
BLANK	Send ASCII space to display
PRINT_MSG	Send ASCII text string to display
SIGN_ON	Send monitor start-up message to display
UPPER_CASE	Convert into uppercase if possible
PRINT_BYTE	Output hex pair to display
PRINT_WORD	Output two hex pairs to display
PRINT_LONG	Output four hex pairs to display
VALDIG	Check input data for correct hexadecimal range
ERROR	Send error message to display and restart monitor
TO_HEX	Convert ASCII character into 4-bit hex
GETLINE	Read a line of user input text
GET_BYTE	Read a hex pair from the keyboard
GET_ADDR	Read an address from the keyboard
PANIC	Check keyboard for Control-C
FREEZE	Check keyboard for Control-S

HELP. The MOVE command is very useful if large blocks of memory need to be cleared, or filled with the same value. The HELP command is included to provide a quick reference on command syntax.

To implement the required monitor commands, we have to write machine language subroutines that EDIT, DUMP, GO, MOVE, and HELP. In addition to these routines, we will need a host of smaller ones to do some necessary chores for us, like sending data to the terminal, reading the keyboard, and doing simple conversions from hex to ASCII. These routines will be referred to as **auxiliary** subroutines, and are summarized in Table 11.1. We will study the operation of each main routine (DUMP, GO, and so on) and all auxiliary routines, and then see how they are combined to become the software monitor.

Keep in mind that the monitor has been written using simple instructions and addressing modes. It is entirely possible that many of the routines presented can be simplified. You are encouraged to do so, after seeing all the routines first.

The Auxiliary Subroutines

The auxiliary subroutines are used by many of the monitor command subroutines. For example, all of the command subroutines must call the GET_ADDR subroutine to get their respective starting memory addresses. On the other hand, only DUMP uses the FREEZE subroutine. The use of auxiliary subroutines greatly simplifies the overall task of writing the monitor.

INIT_ACIA   This routine is used to initialize the 6850. On power-on, the ACIA may be in some random state, with selected options (internal clock divide, number of bits/character, parity) possibly very different from the options we desire. For this reason, we give the ACIA a master reset command, followed by a control byte that indicates the options we need for proper operation (X16 clock, 8 data bits, no parity). The following code is used to initialize the ACIA.

```
INIT_ACIA    MOVEA.L    #$10000,A0        ;point to ACIA control register
             MOVE.B     #3,(A0)           ;ACIA master reset
             MOVE.B     #$15,(A0)         ;select options
             RTS
```

INSTAT   This routine is used to check the status of the keyboard. It is often necessary to check the ACIA receiver status, to determine if a character is waiting (or has been received). It is not important at this time to actually read the character. The input status is checked by the following code. If a character is waiting in the receiver, the zero flag will be cleared.

```
INSTAT    MOVEA.L    #$10000,A0        ;point to ACIA status register
          MOVE.B     (A0),D0           ;get ACIA status
          ANDI.B     #1,D0             ;test RRDY bit
          RTS
```

CHAR_IN   This routine is used to read a character from the keyboard. INSTAT is called to see if a character is waiting, and the character is read if one has been received. If the receiver is empty, the routine waits for a character to arrive, by continually calling INSTAT. The received character is in the lower byte of register D_1 upon return.

```
CHAR_IN    BSR        INSTAT           ;check receiver status
           BEQ        CHAR_IN          ;loop if no character waiting
           MOVEA.L    #$10002,A0       ;point to ACIA data register
           MOVE.B     (A0),D1          ;get the ASCII character
           ANDI.B     #$7F,D1          ;strip off MSB
           RTS
```

CHAR_OUT   This routine is used to send an ASCII character to the display. The ACIA status register is checked to see if the transmitter is empty. If so, the ASCII character is transmitted. If the transmitter is already busy transmitting, the routine will wait until it is empty, and then send the new character. The ASCII character to transmit must be in the lower byte of register D_1 before calling CHAR_OUT.

```
CHAR_OUT     MOVEA.L    #$10000,A0        ;point to ACIA status register
CHAR_OUT2    MOVE.B     (A0),D0           ;read the ACIA status
             ANDI.B     #2,D0             ;check the TRDY bit
             BEQ        CHAR_OUT2         ;loop if not ready
             MOVEA.L    #$10002,A0        ;point to ACIA data register
             MOVE.B     D1,(A0)           ;send the character
             RTS
```

CRLF   This routine is used to send carriage return and line feed characters to the display.

```
CRLF    MOVE.B    #$0D,D1        ;send ASCII CR
        BSR       CHAR_OUT
        MOVE.B    #$0A,D1        ;send ASCII LF
        BSR       CHAR_OUT
        RTS
```

BLANK   This routine is used to output an ASCII space character. The space is used to separate hex characters as they are DUMPed.

```
BLANK      MOVE.B     #$20,D1           ;send ASCII SP
           BSR        CHAR_OUT
           RTS
```

PRINT_MSG   This routine is used to send a message to the display. The ASCII message characters are read from memory, and sent to the display, until a 0 is encountered. Upon entry, register A3 must point to the first message character in memory.

```
PRINT_MSG    MOVE.B     (A3),D1          ;get a character
             CMPI.B     #0,D1            ;end of message?
             BEQ        PRINT_MSG2
             BSR        CHAR_OUT         ;send character to display
             ADDQ.L     #1,A3            ;point to next character
             BRA        PRINT_MSG
PRINT_MSG2   RTS
```

SIGN_ON   This routine is used to greet the user when power is first applied to the system (or after a RESET has occurred). It is a nice way of seeing that the system is up and running.

```
SIGN_ON    MOVEA.L    #HELLO,A3         ;get starting message address
           BSR        PRINT_MSG         ;send the message
           RTS
HELLO      DC.B       $0D,$0A           ;newline
           DC.B       '68000 Monitor, Version 7.5, C1995 JLACD'
           DC.B       $0D,$0A,0         ;newline and end characters
```

It is very important that an even number of bytes are used in the HELLO message. An odd number of bytes will cause an assembly error resulting from the odd memory address generated by the odd length of the message data.

UPPER_CASE   This routine checks the lower byte of D_1 to see if it represents a lowercase letter. If so, it converts the character code into uppercase format by clearing bit 5.

```
UPPER_CASE    CMPI.B  #'a',D1           ;check for lowercase
              BMI     NO_CHG
              CMPI.B  #'{',D1           ;first code after 'z'
              BPL     NO_CHG
              ANDI.B  #$DF,D1           ;switch to uppercase
NO_CHG        RTS
```

PRINT_BYTE   This routine is used to convert a byte of binary data into two ASCII characters that represent the hexadecimal value contained in the byte. Suppose the byte's value is 3F. PRINT_BYTE will send the two ASCII characters '3' ($30) and 'F' ($46) to the display. PRINT_BYTE contains a subroutine that converts the lower 4 bits of register D_1 into the correct ASCII code that represents the hex value of the 4 bits. To get the upper 4 bits of the lower byte in D_1, a rotate command is used. The data to be converted must be in register D_2 before calling.

```
PRINT_BYTE    MOVE.L  D2,D1             ;init conversion register
              ROL.B   #4,D1             ;look at upper nybble first
              BSR     TO_ASCII          ;print ASCII equivalent
              MOVE.L  D2,D1             ;repeat for lower nybble
TO_ASCII      ANDI.B  #$0F,D1           ;strip off upper nybble
```

```
            ADDI.B      #$30,D1             ;add ASCII bias
            CMPI.B      #$3A,D1             ;test for alpha conversion
            BMI         NO_ADD
            ADDI.B      #7,D1               ;correct to $41-$47 (A-F)
NO_ADD      BSR         CHAR_OUT            ;send character
            RTS
```

PRINT_WORD   This routine is used to print four ASCII characters that represent the binary data in the lower 16 bits of register D_2. An 8-bit rotate is used so that the upper 8 bits get converted first.

```
PRINT_WORD  ROL.W       #8,D2               ;get upper 8 bits
            BSR         PRINT_BYTE          ;output first 2 characters
            ROL.W       #8,D2               ;now do the lower 8 bits
            BSR         PRINT_BYTE
            RTS
```

PRINT_LONG   This routine is used to output 8 ASCII characters representing the entire 32-bit contents of register D_2. The SWAP command is used to exchange the upper and lower 16-bit halves of D_2, so that the upper word gets converted first.

```
PRINT_LONG  SWAP        D2                  ;get upper 16 bits
            BSR         PRINT_WORD          ;do 4-character conversion
            SWAP        D2                  ;and repeat for lower word
            BSR         PRINT_WORD
            RTS
```

VALDIG   This routine is used to check the ASCII value in register D1 for the correct hexadecimal range of values $30–$39 or $41–$47. Any value not in this range is interpreted as an incorrect hex value, and VALDIG jumps to the ERROR routine to indicate this. Otherwise, good values cause a return to the caller.

```
VALDIG      CMPI.B      #'G',D1             ;greater than F?
            BPL         ERROR
            CMPI.B      #'0',D1             ;less than 0?
            BMI         ERROR
            CMPI.B      #':',D1             ;is it now A-F?
            BPL         NEXT_TST
            RTS                             ;range is 0-9
NEXT_TST    CMPI.B      #'A',D1             ;less than A?
            BMI         ERROR
            RTS                             ;range is A-F
```

ERROR   This routine is used to print an error message that indicates the user has entered an incorrect command or used an incorrect character when entering a hexadecimal value. A jump to the beginning of the monitor is performed after ERROR prints the error message. A jump is used to avoid having to clean up the stack, which may have temporary values saved on it.

```
ERROR       MOVEA.L     #WHAT,A3            ;get message pointer
            BSR         PRINT_MSG
            BRA         GETCMD              ;restart monitor program
WHAT        DC.B        $0D,$0A             ;newline
            DC.B        'What?'
            DC.B        $0D,$0A,0           ;newline and end characters
```

Once again you will notice that an even number of bytes (10) were used in the error message (including the CR, LF, and 00). This ensures that the assembler's next instruction address is even.

TO_HEX   This routine is used to convert the ASCII value in register D_1 into a 4-bit hex value. For example, if the lower byte of D_1 contains $37 (an ASCII 7), D_1 will be converted into $07. If D_1 contains $43 (an ASCII C), it will be converted into $0C.

```
TO_HEX      SUBI.B      #$30,D1         ;remove ASCII bias
            CMPI.B      #$0A,D1         ;0-9?
            BMI         FIN_CONV
            SUBI.B      #7,D1           ;remove alpha bias
FIN_CONV    RTS
```

GETLINE   This routine is used to get a line of input text from the user. All characters entered by the user, including the final carriage return, are stored in memory pointed to by A_6. The number of characters entered is returned in D_6.

Two simple editing features are included in GETLINE. The first allows the use of Backspace (or Delete) to edit text already typed in, one character at a time. The second feature deletes the entire line by restarting GETLINE if a Control-U is entered.

```
GETLINE       MOVEA.L A6,A5           ;copy pointer
GETLINE2      MOVEA.L A5,A6           ;reset pointer
              CLR.B   D6             ;reset character counter
GET_CHARS     BSR     CHAR_IN        ;get character
              CMPI.B  #$0D,D1        ;check for carriage return
              BEQ     EXIT
              CMPI.B  #$08,D1        ;check for backspace key
              BEQ     BKSPS
CHECK_DELETE  CMPI.B  #$7F,D1        ;check for delete key
              BEQ     BKSPS
              CMPI.B  #$15,D1        ;check for ^U and scrap the
              BEQ     LINE_REDO      ;line if encountered
              MOVE.B  D1,(A6)+       ;otherwise place character
              ADDQ.B  #1,D6          ;in buffer, update counter and
              BSR     CHAR_OUT       ;echo it to the screen
              BRA     GET_CHARS      ;get another character
EXIT          MOVE.B  D1,(A6)        ;on exit place $0D in buffer
              ADDQ.B  #1,D6          ;update character counter
              BSR     CRLF           ;fix display
              RTS
LINE_REDO     BSR     CRLF           ;clean up the screen
              BRA     GETLINE2       ;start over
BKSPS         CMPI.B  #0,D6          ;anything to delete?
              BEQ     GET_CHARS
              SUBQ.B  #1,D6          ;decrease character counter
              SUBA.L  #1,A6          ;move pointer backwards
              MOVE.B  #8,D1          ;send backspace code to display
              BSR     CHAR_OUT
              MOVE.B  #' ',D1        ;erase last character
              BSR     CHAR_OUT
              MOVE.B  #8,D1          ;send another backspace
              BSR     CHAR_OUT
              BRA     GET_CHARS      ;go get new character
```

GETLINE requires that A_6 be initialized by the calling code. Since the monitor routines make use of GETLINE, a default address of 8F00 is used for all user input. This is the beginning of the last 256 bytes of user RAM, and is reserved for use by the monitor when it calls GETLINE.

GET_BYTE   This routine is used to get a byte of hexadecimal data from the user. Incoming characters are stripped of their ASCII bias and placed in a temporary register. The final 8-bit value is returned in the lower 8 bits of register D_1. Any illegal characters cause immediate restarting of the monitor.

If a carriage return is entered in place of the first digit, the routine exits with an ASCII '*' in the lower byte of D_2. This character is used to indicate that no data has been entered by the user. If two valid hex characters are entered, D_2 will contain an ASCII '0'.

Note that GET_BYTE reads data directly from the input buffer pointed to by A_6.

```
GET_BYTE        MOVE.B    (A6)+,D1        ;skip blanks if necessary
                CMPI.B    #$20,D1
                BEQ       GET_BYTE
                BSR       UPPER_CASE      ;adjust first digit
                CMPI.B    #$0D,D1         ;test for CR
                BEQ       NO_CHAN
                BSR       VALDIG          ;check for valid digit
                BSR       TO_HEX          ;convert into hex
                ROL.B     #4,D1           ;move first digit
                MOVE.B    D1,D2           ;save first digit
                MOVE.B    (A6)+,D1        ;get second digit
                BSR       UPPER_CASE
                BSR       VALDIG          ;check for valid digit
                BSR       TO_HEX          ;convert into hex
                ADD.B     D2,D1           ;form final result
                MOVE.B    #'0',D2         ;change entered
                RTS
NO_CHAN         MOVE.B    #'*',D2         ;no change character
                RTS
```

GET_ADDR   This routine is used to get an address from the user. Up to eight hexadecimal values may be specified. The right-justified value is returned in register D_2. An ASCII carriage return or space terminates the address conversion. All other illegal characters will cause the monitor to restart.

```
GET_ADDR        MOVE.B    (A6)+,D1        ;skip over leading blanks
                CMPI.B    #$20,D1
                BEQ       GET_ADDR
                CLR.L     D1              ;init temp register
                CLR.L     D2              ;init result register
NEXT_CHAR       MOVE.B    (A6)+,D1        ;get a character
                BSR       UPPER_CASE
                CMPI.B    #$0D,D1         ;exit if CR
                BEQ       EXIT_ADR
                CMPI.B    #$20,D1         ;exit if space
                BEQ       EXIT_ADR
                BSR       VALDIG          ;test for valid digit
                BSR       TO_HEX          ;convert digit into hex
                ROL.L     #4,D2           ;prepare D2 for new digit
                ANDI.B    #$F0,D2
                ADD.B     D1,D2           ;insert new digit
                BRA       NEXT_CHAR       ;and continue
EXIT_ADR        RTS
```

PANIC   This routine is used to check for incoming keyboard data. If no character has been received, or if the character received is not a Control-C, the routine will simply return to the caller. If the character is a Control-C, the monitor will restart.

```
PANIC           BSR       INSTAT          ;check for key
                BEQ       EXIT_BRK        ;return if none hit
TEST_KEY        BSR       CHAR_IN         ;get key
```

```
TEST_KEY2      CMPI.B    #3,D1              ;Control-C?
               BEQ       GETCMD             ;if yes, restart monitor
EXIT_BRK       RTS
```

FREEZE   This routine is used to freeze the screen during a memory dump. It checks for a Control-S from the keyboard. If no character has been entered, the routine returns to the caller. If a character has been received that is not a Control-S, the routine jumps into the PANIC routine to check for a Control-C. If a Control-S is received, the routine will wait for a second character to be received before jumping to PANIC to check for a Control-C.

```
FREEZE         BSR       INSTAT             ;check for key
               BEQ       EXIT_FREZ          ;return if none hit
               BSR       CHAR_IN            ;get key
               CMPI.B    #$13,D1            ;Control-S?
               BEQ       HOLD_IT
               BRA       TEST_KEY2          ;Control-C?
EXIT_FREZ      RTS
HOLD_IT        BSR       INSTAT             ;wait for another character
               BEQ       HOLD_IT
               BRA       TEST_KEY           ;let PANIC check for Control-C
```

The use of the auxiliary subroutines will greatly aid in the creation of the monitor commands. By breaking down the larger tasks required by the commands into smaller ones, most of the command software may be written by selectively calling groups of auxiliary routines, and moving data between registers. The next section will show just how the monitor command routines are written.

The Monitor Commands

The monitor commands are really the heart of the software monitor. Through the use of the monitor commands, the job of creating new and useful software becomes much easier. We will now examine just how these commands are implemented. Study the methods used to perform I/O with the user, and how decisions are made within the routines. You should be able to gain a very good understanding of the structure of the command routine, and be able to use that knowledge to write your own routine to perform a job that the basic monitor cannot perform.

The Command Recognizer

In order for us to use a command routine, we must be able to get to its starting address in memory. It is much simpler and more convenient to enter DUMP or GO, as opposed to a multidigit starting address. The purpose of the command recognizer is to determine which of the monitor commands the user has entered, and jump to the routine for execution. The routine accepts either full command names (DUMP, EDIT, GO, MOVE, or HELP), or single-letter abbreviations (D, E, G, M, or H).

Once the command is recognized, the address of that routine is read from a table, and the routine is branched to. For this reason, the command routines must branch back to the start of the monitor, and not return as a subroutine would. The command recognizer is written so that additional commands may be added to the monitor with a minimum of change in its code.

```
GETCMD     MOVEA.L    #STACK,A7       ;init stack
           BSR        CRLF
           MOVE.B     #'*',D1         ;output command prompt
           BSR        CHAR_OUT
           CLR.L      D2              ;init command-text buffer
           MOVEA.L    #$8F00,A6       ;load buffer location
           BSR        GETLINE         ;fill buffer with text
           MOVEA.L    #$8F00,A6       ;reset buffer pointer
SKIPBL     MOVE.B     (A6),D1         ;skip leading blanks
           CMPI.B     #$20,D1
           ADDA.L     #1,A6
           BEQ        SKIPBL
INCOM      MOVE.B     (A6)+,D1        ;read command letter
           BSR        UPPER_CASE
           CMPI.B     #$0D,D1         ;test for CR
           BEQ        SEARCH
           CMPI.B     #$20,D1         ;test for SP
           BEQ        SEARCH
           ROL.L      #8,D2           ;prepare D2 for new letter
           MOVE.B     D1,D2           ;insert it into D2
           BRA        INCOM           ;get next letter
SEARCH     MOVE.L     #5,D5           ;number of commands to check
           MOVEA.L    #COMMANDS,A2    ;init command text pointer
           MOVEA.L    #COM_ADRS,A1    ;init command address pointer
TEST_NEXT  CMP.L      (A2),D2         ;compare command text
           BEQ        DO_JUMP         ;branch if match
           ADDA.L     #4,A2           ;point to abbreviated command
           CMP.L      (A2),D2         ;test again
           BEQ        DO_JUMP
           ADDA.L     #4,A2           ;point to next command
           ADDA.L     #4,A1           ;point to next address
           SUBQ.L     #1,D5           ;all commands checked yet?
           BNE        TEST_NEXT
           BRA        ERROR           ;illegal command entered
DO_JUMP    MOVE.L     (A1),A1         ;get command address
           JMP        (A1)            ;and go execute command
COMMANDS   DC.B       'DUMP'          ;full command name
           DC.B       0,0,0,'D'       ;abbreviated name
           DC.B       'GO  '
           DC.B       0,0,0,'G'
           DC.B       'EDIT'
           DC.B       0,0,0,'E'
           DC.B       'MOVE'
           DC.B       0,0,0,'M'
           DC.B       'HELP'
           DC.B       0,0,0,'H'
COM_ADRS   DC.W       0,DUMP          ;DUMP execution address
           DC.W       0,GO
           DC.W       0,EDIT
           DC.W       0,MOOV
           DC.W       0,HELP
```

Notice how the routine brings in a command. Register D_2 is initially cleared. When a valid command letter is received, D_2 is rotated 8 bits to the left, and the new ASCII letter value is placed in the lower 8 bits. If four characters are received, the first letter has been rotated into the uppermost byte position of D_2. If only one letter is entered (as is the case for abbreviated command recognition), it will appear in the lower 8 bits of D_2, with all other bits equal to 0. The COMMANDS table contains the values that will appear in D_2 for any of the acceptable command entries.

Additional commands may be added very easily by increasing the initial value put into D_5, and extending the data table sizes of COMMANDS and COM_ADRS.

The DUMP Routine

This routine is used to display a range of memory locations and their associated ASCII values. The format of the output is as follows: 8 hex characters representing the address, followed by 16 hex pairs, which indicate the data contained in the memory locations following the printed address. These 16 hex pairs are followed by 16 ASCII characters that are the printable ASCII equivalents of the 16 bytes. The ASCII output is very useful in determining when the code displayed is simply hex data, and not machine code. A sample output line might look like this:

```
00008100   4E  75  E1  5A  61  D8  E1  5A  61  D4  4E  75  48  42  61  F2   Nu.Za..Za.NuHBa.
```

The user must specify a starting and ending address in the DUMP command line. The command to DUMP the previous data would be DUMP 8100 810F. For starting addresses that do not start on an even 16-byte address (such as 8136), the DUMP routine will convert the starting address, by making the lower digit of the address 0 (thus making 8136 into 8130). For ending addresses that do not end on an even 16-byte address, the routine will show the remaining locations also. Thus, an ending address of 82F7 will cause DUMP to continue to 82FF (and so end on the even address 8300).

To freeze the display during a DUMP, just press Control-S. Any key other than Control-C may then be used to restart the DUMP.

Study the routine carefully. You may already have ideas on how to improve the output format.

```
DUMP        BSR     GET_ADDR        ;get the starting address
            ANDI.B  #$F0,D2         ;make lower nybble zero
            MOVEA.L D2,A4           ;A4 is memory read register
            BSR     GET_ADDR        ;get the ending address
            MOVEA.L D2,A5
ADR_OUT     BSR     CRLF            ;new line please
            MOVE.L  A4,D2           ;print address
            BSR     PRINT_LONG
            BSR     BLANK           ;and some blanks
            BSR     BLANK
BYTE_OUT    MOVE.B  (A4)+,D2        ;get a byte and increment A4
            BSR     PRINT_BYTE      ;print the byte
            BSR     BLANK
            MOVE.L  A4,D1           ;done 16 yet?
            ANDI.L  #$0F,D1
            BNE     BYTE_OUT
            SUBA.L  #16,A4          ;back up 16 bytes
            BSR     BLANK
ASCII_OUT   MOVE.B  (A4)+,D1        ;get a byte
            ANDI.B  #$7F,D1
            CMPI.B  #$20,D1         ;is it printable?
            BMI     UN_PRINT
            CMPI.B  #$7D,D1
            BMI     SEND_IT
UN_PRINT    MOVE.B  #$2E,D1         ;use period for unprintables
SEND_IT     BSR     CHAR_OUT        ;print the ASCII equivalent
            MOVE.L  A4,D2           ;done 16 yet?
```

```
        ANDI.L    #$0F,D2
        BNE       ASCII_OUT
        BSR       FREEZE            ;hold display?
        CMPA.L    A4,A5             ;done with dump?
        BMI       GETCMD
        BRA       ADR_OUT
```

The DUMP routine could be modified to output words (two hex pairs) for each even address location, or just hex data with no ASCII equivalent. The format of the output is designed specifically to help the user understand what information presently resides in a section of memory.

The GO Routine

This routine is used to execute a user program stored in memory. The starting address of the user routine must be entered in the command. For example, GO 8500 will cause execution of whatever code starts in location 8500. The routine makes use of an instruction that permits an absolute jump to a specific address.

```
GO      BSR       GET_ADDR          ;get execution address
        MOVEA.L   D2,A1
        JMP       (A1)
```

It is possible to modify the GO routine to load any or all of the CPU registers (data or address) before performing the jump. This would give a nice way of executing your custom program with starting data.

The MOVE Routine

This routine is used to move data around in memory. Suppose a section of code needs to be placed in a higher memory area. If the starting and ending addresses of the data block are 8000 and 80FF, and the new starting address should be 8650, MOVE 8000 80FF 8650 will make the required copy of the data.

```
MOOV    BSR       GET_ADDR          ;get starting address
        MOVEA.L   D2,A1
        BSR       GET_ADDR          ;get ending address
        MOVEA.L   D2,A2
        ADDA.L    #1,A2             ;include last location
        BSR       GET_ADDR          ;get destination address
        MOVEA.L   D2,A3
MOOVEM  MOVE.B    (A1)+,(A3)+       ;move and increment pointers
        CMPA.L    A1,A2             ;at ending address yet?
        BNE       MOOVEM
        BRA       GETCMD
```

The routine name is not MOVE, since this is a reserved assembler name.

The EDIT Routine

This routine is used to place user program data into memory. The starting address must be specified in the command. The EDIT routine will output an address and the byte of data contained there. After the data byte is printed comes a question mark. This indicates to the user that the data at the present location may be changed. To change the data, the user simply enters a hex pair. Any illegal characters will cause the monitor to restart. The only exception is the Carriage Return key. Entering this immediately following the question mark causes EDIT to skip over the current address and not alter the data contained there. A sample EDIT command might look like this:

```
EDIT   8200
00008200   3E?   16
00008201   27?   00
00008202   FE?   <CR>
00008203   9A?   20
00008204   2B?   <Ctrl-C>
```

Notice that the user changed the first 2 bytes, kept the third one the same, changed the fourth, and then exited when the fifth byte was printed. The following code implements the last of the monitor commands.

```
EDIT        BSR       GET_ADDR          ;get starting address
            MOVEA.L   D2,A2             ;A2 is the memory pointer
NEW_DATA    BSR       CRLF              ;new line please
            MOVE.L    A2,D2             ;print data address
            BSR       PRINT_LONG
            BSR       BLANK
            MOVE.B    (A2),D2           ;get the data
            BSR       PRINT_BYTE        ;and show it
            MOVE.B    #'?',D1           ;output change prompt
            BSR       CHAR_OUT
            MOVEA.L   #$8F00,A6
            BSR       GETLINE
            MOVEA.L   #$8F00,A6
            BSR       GET_BYTE          ;get new data
            CMPI.B    #'*',D2           ;no change requested?
            BNE       ENTER_IT          ;jump if new data entered
            MOVE.B    (A2),D1           ;get old byte back
ENTER_IT    MOVE.B    D1,(A2)+          ;save data and increment pointer
            BSR       NEW_DATA
```

The only exit out of this routine is to enter an illegal character after the question mark, which GET_BYTE will detect.

The Body of the Monitor

At this point we have covered the creation of all the routines (auxiliary and command) that we need inside our monitor. The last step we need to perform is to collect all the routines and organize them into a source file. The source file must provide an exception vector table, which must (at the very least) contain initial SSP and PC addresses for the RESET exception. Since these exceptions are at addresses 0 and 4, the exception vector table must be ORGed at 0. The 256 possible exception vectors will occupy the first 1,024 locations (0–3FF), making address 400 the first usable address for the machine code of the monitor. The following software details the creation of the exception vector table and startup monitor code.

```
            ORG       0
            DC.W      0,STACK           ;RESET: initial SSP
            DC.W      0,START           ;RESET: initial PC
            DC.W      0,BUS_ERROR
            DC.W      0,ADRS_ERROR
            DC.W      0,ILLEGAL_INST
            DC.W      0,DIV_ZERO
```

```
              DC.W      0,START                  ;CHK not implemented
              DC.W      0,TRAP_V
              ORG       $7C                      ;skip reserved vectors
              DC.W      0,LEVEL_7
              DC.W      0,TO_CHAR_IN             ;TRAP vector 0
              DC.W      0,TO_CHAR_OUT            ;TRAP vector 1
              DC.W      0,TO_CRLF                ;TRAP vector 2
              DC.W      0,TO_PRINT_MSG           ;TRAP vector 3
              DC.W      0,TO_PRINT_BYTE          ;TRAP vector 4
              DC.W      0,TO_PRINT_WORD          ;TRAP vector 5
              DC.W      0,TO_PRINT_LONG          ;TRAP vector 6
              DC.W      0,TO_GET_BYTE            ;TRAP vector 7
              DC.W      0,TO_GET_ADDR            ;TRAP vector 8
              DC.W      0,TO_GETCMD              ;TRAP vector 9
              ORG       $400                     ;start of monitor
STACK         EQU       $9000
START         BSR       INIT_ACIA                ;init serial chip
              BSR       SIGN_ON                  ;greet user
```

Notice that we have provided vectors for major system exceptions such as bus and address errors, and also for the first 10 TRAP vectors. Since the vectors represent usable auxiliary routine addresses, we have provided an easy way for the user to use system software that already exists. To use CHAR_IN, we just have to include TRAP #0 in our code. To call PRINT_LONG, we use TRAP #6.

An example of the code that lets us use the TRAPs is as follows:

```
TO_CHAR_IN    BSR       CHAR_IN
              RTE
```

Since the TRAP instruction requires exception processing, we cannot simply use the name CHAR_IN in the vector. The RTS at the end of CHAR_IN would not do the same job as the required RTE needed for correct exception recovery. If we instead vector to a routine that *calls* CHAR_IN, we will both use the desired subroutine, and also return correctly to the main program.

The source statements that follow the BSR   SIGN_ON instruction must begin with the code for the command recognizer. After that, the routines may be in any order. All auxiliary and command routines should be included. The only remaining source code to write is the code for the exception handlers, which is listed here:

```
BUS_ERROR     MOVEA.L   #MSG_1,A3
              BRA       REPORT
ADRS_ERROR    MOVEA.L   #MSG_2,A3
              BRA       REPORT
ILLEGAL_INST  MOVEA.L   #MSG_3,A3
              BRA       REPORT
DIV_ZERO      MOVEA.L   #MSG_4,A3
              BRA       REPORT
TRAP_V        MOVEA.L   #MSG_5,A3
              BRA       REPORT
LEVEL_7       MOVEA.L   #MSG_6,A3
              BRA       REPORT
HELP          MOVEA.L   #H_MSG,A3
REPORT        BSR       CRLF                     ;new line, thank you
              BSR       PRINT_MSG                ;print message pointed to by A3
              BSR       CRLF
              BRA       GETCMD
MSG_1         DC.B      'Bus Error',0
MSG_2         DC.B      'Address Error',0
```

```
MSG_3          DC.B          'Illegal Instruction Error',0
MSG_4          DC.B          'Divide-by-Zero Error',0
MSG_5          DC.B          'TRAPV Overflow',0
MSG_6          DC.B          'Level 7 Interrupt',0
H_MSG          DC.B          'The HELP message goes here!!!',0
```

Notice that the HELP command routine is implemented here. Since we need only output a help message (possibly containing the syntax of all commands), we include HELP in the exception processing code.

The next section will give ideas on how to improve the basic monitor by adding more commands, and making changes to existing ones. *But if you plan to build the minimal system, make sure the basic monitor works before you begin changing it!*

11.8   ADDING COMMANDS TO THE MONITOR

This section will describe some possible ways the power of the monitor may be increased. A good understanding of the 68000's instruction set is required before any attempt is made to change the operation of the monitor. Barring this slight warning, let us proceed with some details.

One command that might be useful to add could be called XREG. This would be the examine-register command. Suppose that all eight data registers and address registers are saved every time the monitor is restarted. If the monitor is restarted as the result of a user program exception, it would be very desirable to see what data was in the data registers prior to the exception. The XREG command would display the contents of the registers. A sample format might be:

D0:	05687786
D1:	FFFFFFFE
D2:	01040345
D3:	4E4060FA
etc.	

A third command, LREG, could be used to load the eight data and address registers (either singly via LREG D2 or LREG A6, or all at once) with program data prior to using the GO command. LREG would provide a means for getting data into a user program. A change to the GO routine would be needed to load the registers before JMPing to the user program. Using XREG and LREG, we can test a custom program for proper operation very quickly.

Since large programs (greater than 25 words) will take a long time to enter with the EDIT command, it might be useful to include a command capable of **downloading** a program into memory. Consider a command called SREC that downloads and converts Motorola S-record files (called a **hex** file). Hex files are created by ASM68K. An example of a hex file is as follows:

```
S004000020DB
S1258100267C000081974E434E400C01005366F84E400C01003067F00C01003966000018343C6C
S125812200074E4051CAFFFC267C000081B84E4360007ECC0C01003166CC420661000090570274
S12581441E0261000088E14A42853A026100007E8A0228456100007618C2530766F66100006CD2
S12581660C0600FF679C267C0000817A4E43240C4E4660FE0D0A436865636B73756D20657272DC
```

```
S12581886F72206174206164647265737320000D0A57616974696E6720666F7220532D726563AA
S12581AA6F72642066696C652E2E2E0D0A000D0A446F776E6C6F616420636F6D706C65746564DD
S12581CC2E0D0A0061000010E9091401610000088401DC024E754E40040100300C01000A650002
S10981EE00045F014E7560
S9030000FC
```

All lines begin with an S. Embedded within the line of text is a record type (0, 1, or 9), a length byte (04 in the first line, 25 in the next seven lines, then 09 and 03 in the last two lines), a four-digit load address (8100 in the second line, 81EE in the second-to-last line), up to 34 data bytes, and finally a checksum byte (DB in the first line, FC in the last line). The SREC command must read in each line of a hex file and convert its contents into the actual bytes that get stored in memory. This is a big job and makes the SREC routine rather lengthy.

In addition to SREC, even more advanced commands that would provide breakpoints, and possibly assembly or disassembly, could be added. All that is required is cleverly written code. The more functions available in the monitor, the easier the job of writing new software becomes.

11.9   EXAMPLE PROGRAMS

The following routines are included to test the operation of the monitor. They make use of the TRAP instructions and exercise the auxiliary routines.

TV Typewriter

This routine simply echoes all received characters.

```
        ORG     $8000
TOP     TRAP    #0              ;get a character
        TRAP    #1              ;send it to display
        BRA     TOP
```

Improved TV Typewriter

This routine echoes all received characters also. When CR is received, both CR and LF are sent to the display. This causes the screen to scroll.

```
        ORG     $8000
TOP     TRAP    #0              ;get a character
        CMPI.B  #$0D,D1         ;test for CR
        BNE     SKIP
        TRAP    #2              ;send CR and LF to display
        BRA     TOP
SKIP    TRAP    #1              ;echo character
        BRA     TOP
```

Counter with Message

This routine outputs a constant message saying "This is message number", with a different number at the end of each message.

```
        ORG     $8000
        MOVE.L  #0,D2           ;init counter
TOP     MOVEA.L #MESS,A3        ;get message pointer
```

```
            TRAP       #3                                      ;print message
            TRAP       #6                                      ;output count
            ADDQ.L     #1,D2                                   ;increment counter
            TRAP       #2           .                         ;get a new line
            BRA        TOP
MESS        DC.B       'This is message number',0
```

Hexadecimal Adder

This routine asks for 2 hexadecimal bytes, adds them together, and displays the result.

```
            ORG        $8000
TOP         MOVEA.L    #NUMA,A3                                ;display first message
            TRAP       #3
            TRAP       #7                                      ;get first number
            MOVE.B     D1,D6                                   ;save it here
            MOVEA.L    #NUMB,A3                                ;display second message
            TRAP       #3
            TRAP       #7                                      ;get second number
            ADD.B      D6,D1                                   ;add numbers together
            MOVE.W     D1,-(A7)                                ;save D1 on stack
            MOVEA.L    #ANSWER,A3                              ;display answer message
            TRAP       #3
            MOVE.W     (A7)+,D1                                ;pop D1 off stack
            TRAP       #4                                      ;display answer
            BRA        TOP
NUMA        DC.B       'Enter first number: ',0
NUMB        DC.B       'Enter second number: ',0
ANSWER      DC.B       'The sum is:  ',0
```

Memory Search

This routine asks the user for a byte value, and then proceeds to find all memory locations that contain the same value. When a location is found, its address is displayed.

```
            ORG        $8000
            MOVEA.L    #ASKIT,A3                               ;ask for data byte
            TRAP       #3
            TRAP       #7                                      ;get data byte
            MOVE.B     D1,D6                                   ;save data byte here
            MOVEA.L    #0,A0                                   ;start comparing at address 0
CHECK       CMP.B      (A0),D6                                 ;compare
            BNE        NEXT                                    ;skip if not equal
            MOVE.L     A0,D2                                   ;display address
            TRAP       #6
NEXT        ADDA.L     #1,A0                                   ;point to next location
            BRA        CHECK
ASKIT       DC.B       'Enter search byte:   ',0
```

Other examples, such as multiplying binary numbers, entering data for sorting, hi-lo games, interactive programs, and more may be written efficiently with good use of the TRAP auxiliary routines. Can you think of still more auxiliary routines that may be added (and called with a TRAP), to further improve the monitor? How about one for converting binary data to decimal (much like PRINT_LONG, except the output would be decimal and not hex), or others for sorting, comparing, or doing math?

If you are planning on using a personal computer as the communication console for the single-board computer, it will be necessary to wire both computers together (as previously shown in Figure 11.10), and then execute a *terminal emulation* program on the PC.

If no terminal emulator is available, the SBCIO program included on the companion disk will do all that is necessary to provide serial I/O between both machines. SBCIO continuously checks COM1's UART status, as well as the status of the PC's keyboard. Whenever a key is pressed on the keyboard, it is transmitted to the single-board computer as soon as the transmitter in COM1 is ready. Likewise, if a character is received by COM1's UART, it is immediately output to the display screen. So, full duplex communication is implemented through the SBCIO program. Please note that COM1's baud rate and other parameters must be set up before executing SBCIO.

11.10   TROUBLESHOOTING TECHNIQUES

Here are some things we can do if the single-board computer does not work when we turn it on:

- Feel around the board for hot components. A chip that is incorrectly wired or placed backwards in its socket can get very hot. You may even smell the hot component.
- Make sure all the ICs have power by measuring with a DMM or an oscilloscope. Put the probe right on the pin of the IC, not on the socket lead.
- Use an oscilloscope to examine the CLK input of the 68000. Push the RESET button to verify that the $\overline{\text{RESET}}$ and $\overline{\text{HALT}}$ signals are being generated properly.
- Look at the address, data, and control lines with an oscilloscope or logic analyzer. Activity is a good sign; there may be something as simple as a missing address or data line, or crossed lines. No activity means the processor is not receiving the correct information. By examining the logic analyzer traces, you should be able to determine if the memory and I/O address decoders are working properly.
- Verify that the EPROMs were burned correctly, and are in the correct sockets and not switched with the RAM. Check that the even and odd EPROMs are in the even and odd sockets. You should be able to connect a logic analyzer to verify that the processor fetches the first instruction from address $000400. You should also be able to see the first instruction word come out of the EPROMs as well.
- Examine the TxD output of the 6850. Activity at power-on or RESET is a good sign, since the monitor program is designed to output a short greeting to let us know it is alive. If TxD wiggles around, but the serial output of the MAX232 does not, there could be a wiring problem there. Putting the capacitors in backwards is bad for the MAX chip.
- Check every connection again from a fresh schematic. Many times a missing connection is found, even when every attempt was made to be careful during construction.
- Change all the chips, one by one. Look for bent or missing pins when you remove them.
- When all else fails, tell someone else everything you've done and see if they can suggest anything else.
- You could also set the project aside for a while to get your mind off it. The problem may present itself to you when you least expect it. You may suddenly remember that

you commented out an important I/O routine in the monitor because you were having trouble with the assembler. Silly things like that are really fun to find.

SUMMARY

This chapter dealt with the design of a single-board computer, equipped with 2K words of EPROM and RAM, and a 2,400 baud serial I/O section. Many digital design ideas were suggested, and reasons for choosing one design over another were given.

The software design proceeded in the same fashion. Minimum system requirements were established, and the software commands were implemented in groups of code that made good use of a number of auxiliary subroutines.

Example programs were included to show the ease and power of using the working system, through the use of special system calls and monitor commands. There is room for improvement in the monitor program, which should be done after the basic monitor is up and running.

STUDY QUESTIONS

1. Explain how the $\overline{\text{RESET}}$ circuitry shown in Figure 11.13 works.
2. How can the $\overline{\text{BERR}}$ circuit of Figure 11.4 be disabled when an external DMA signal is low?

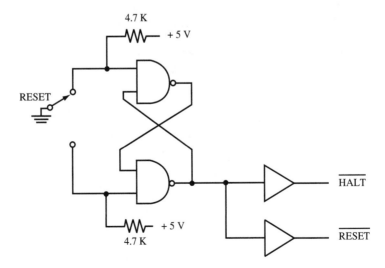

FIGURE 11.13 For Question 11.1

FIGURE 11.14   For Question 11.3

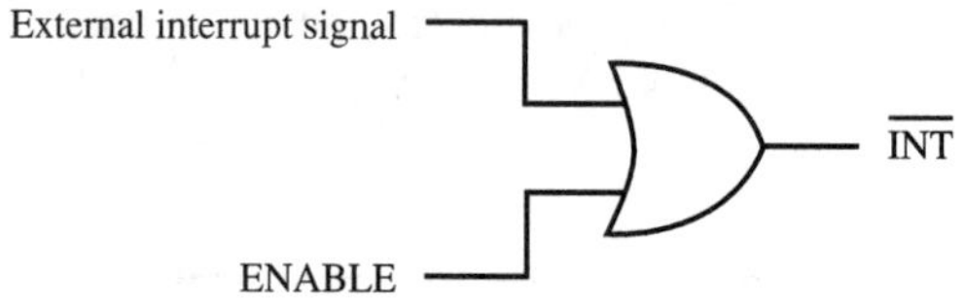

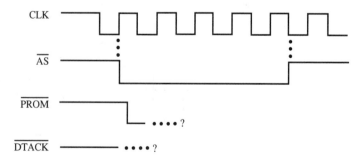

3. Explain how interrupts are enabled/disabled by the ENABLE signal in the circuit shown in Figure 11.14.
4. Show how octal buffers (74LS244) may be added to the 68000's address lines to buffer them. Assume the buffers are always enabled.
5. Modify the circuitry of Question 4 so that the buffers are enabled only when an external DMA signal is high.
6. What type of circuitry is needed to buffer the bidirectional data bus? Can the processor's R/$\overline{W}$ signal be used to control the direction?
7. Design a circuit to fully decode the processor's address bus, before enabling the 74LS139 decoder of Figure 11.5.
8. Modify the memory decoder of Figure 11.5 so that it only works when an external DMA signal is high.
9. Redesign the memory R/$\overline{W}$ signal generator of Figure 11.6, and use only NAND circuitry.
10. Would it be possible to build a minimal system without RAM?
11. What circuitry is needed to add another 2K words of RAM to the minimal system?
12. How must the memory section be changed to allow the use of 2732 EPROMs? How about 2764s?
13. Since the 68000 always reads words during instruction fetches, why is there a need for byte addressing in the EPROM memory?
14. How must the $\overline{DTACK}$ circuit be modified so that its output is always high if an external DMA signal is low?
15. Finish the timing diagram in Figure 11.15 detailing the operation of the $\overline{DTACK}$ signal.

FIGURE 11.15   For Question 11.15

16. What type of gate should replace the two-input AND gate that issues $\overline{\text{DTACK}}$, if two additional signals—$\overline{\text{DRAM}}$ and $\overline{\text{IOCHAN}}$—need to be added?
17. The circuit shown in Figure 11.16 may be used as a 1-bit output port. How does it work?
18. Modify the serial circuit so that it communicates with the lower half of the data bus. What signals need to be changed? How does this affect the software that controls the ACIA?
19. Design a circuit that will control the timing circuitry in the serial section, so that two baud rates (2,400 or 300) may be selected by the flip of a switch. The circuit must send either 38.4 kHz or 4,800 Hz to the ACIA.
20. Show the circuitry required to add a second serial port to the minimal system.
21. Write a subroutine called SKIP_BLANK that skips over any leading blanks in memory pointed to by A6.
22. Explain how GETCMD is modified to allow the SREC command.
23. Complete the following table, showing the use of registers in the auxiliary routines. Which registers pass data to the routine, which are destroyed, and which are the result registers?

Routine	Input Data	Output Data	Registers Used
PRINT_MSG	A3	—	D0,D1,A0

24. Write a routine that divides a data register by 0, to check the DIV_ZERO exception.
25. How can EDIT and MOVE be used to clear a block of RAM?

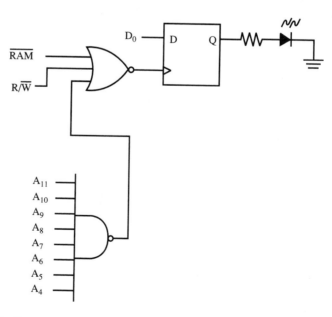

FIGURE 11.16 For Question 11.17

26. What does the following program do?

```
TRAP        #7
MOVE.B      D1,D5
TRAP        #2
TRAP        #7
ADD.B       D1,D5
TRAP        #2
MOVE.B      D5,D2
TRAP        #4
TRAP        #9
```

27. Write a routine called INSERT that allows data to be entered in one long string of bytes, as in I8000 2E 3C 00 08 1C AF, and so on.
28. Assume that the contents of all eight data registers are stored in memory beginning at address DATAREG, with D0 stored in the first 4 bytes and D7 stored in the last 4 bytes of the 32-byte area. Write a subroutine that uses PRINT_LONG to display the contents of each data register.
29. Modify the EXECUTE routine, so that only execution addresses in the 8000-to-877F range are allowed.
30. How could PRINT_MSG be modified so that the end-of-message code is '$' instead of 00?
31. Design a circuit that allows the 16-bit instruction word being fetched to be displayed on a set of 16 LEDs.

PART 5

Advanced Topics

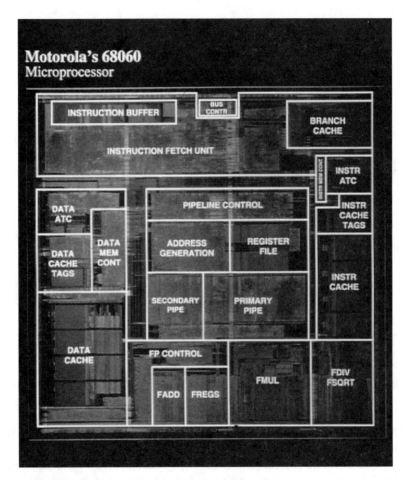

Architectural layout of the 68060 microprocessor

CHAPTER 12

An Introduction to the Advanced 680x0 Series Microprocessors

OBJECTIVES

In this chapter you will learn about:

- New addressing modes and instructions
- The mechanics of virtual memory
- The operation of cache memory
- Demand paging
- Bus snooping
- Superscalar architecture
- Microcontrollers

12.1   INTRODUCTION

Many new microprocessors have been introduced by Motorola since the 68000 first came out. These 680x0 series processors offer additional addressing modes, new instructions, faster clock speeds, and many other enhancements. In this final chapter we will take a short look at each new 680x0 series processor, from both hardware and software viewpoints.

Section 12.2 provides a brief overview of the 68008, a 16-bit machine with an 8-bit data bus. Sections 12.3 through 12.6 discuss the features of the 68010, 68020, 68030, and 68040, respectively. Section 12.7 covers the moset advanced 680x0 microprocessor, the 68060. Troubleshooting techniques complete the chapter in Section 12.8.

12.2   THE 68008: AN 8-BIT SPINOFF

Before we begin with the advanced processors, let us take a quick look at the 68008, a machine equivalent to the 16-bit 68000, except for its 8-bit external connections. Shortly after the 68000 was introduced, it became evident that this new processor offered computing power far beyond that of an 8-bit machine. For users of 8-bit machines this created a problem: how to get the power and speed of the 68000 without starting over from scratch. After all, it would be a terrible waste to get rid of existing 8-bit systems merely because a new processor required 16-bit interfacing. For this reason, and to satisfy designers who were comfortable working with 8-bit buses, Motorola came up with the 68008. Externally, the 68008 is an 8-bit machine, but internally it is a full 16-bit processor, identical to the 68000. As Figure 12.1 shows, some hardware changes were made to accommodate the 8-bit external data bus. Since only 8 bits of data may be exchanged over the data bus, the 68000's $\overline{\text{UDS}}$ and $\overline{\text{LDS}}$ signals were replaced by address line A_0 and a signal called $\overline{\text{DS}}$ (data strobe). The $\overline{\text{DS}}$ signal is used to control the memory or I/O address decoders, indicating when data may be placed onto the data bus, or read from it. The overall addressing capability of the 68008 is 1MB of memory.

Another hardware modification is in the use of external hardware interrupts. To reduce the number of pins on the 68008's package to 48 (the 68000 comes in a 64-pin

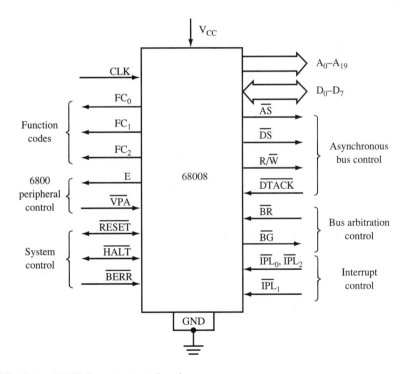

FIGURE 12.1   68008 input/output signals

TABLE 12.1 External
interrupts on the 68008

$\overline{IPL_2}/\overline{IPL_0}$	$\overline{IPL_1}$	Interrupt priority level
\multicolumn{2}{c}{Request}	
0	0	7
0	1	5
1	0	2
1	1	0

package), the designers decided to eliminate an external signal by combining two signals into one. $\overline{IPL_0}$ and $\overline{IPL_2}$ have been tied together internally, resulting in only four levels of hardware interrupts, which are indicated by Table 12.1. Another omission is that of a signal called $\overline{VMA}$ (valid memory address). This signal is used to control 6800-type peripherals, and must be implemented by external hardware when these peripherals are used in a 68008 system.

From a software standpoint, the 68008 is identical to the 68000. All address and data registers have been included, and any program written for the 68000 will run on the 68008, although at a slower speed. The decrease in speed results from the extra memory accesses that must take place in order for the 68008 to fetch the same amount of information as the 68000.

The 68008 can be retrofitted to replace the 68000 CPU in the single-board computer of Chapter 11, with a small number of changes. First, eliminate one EPROM and one RAM, rewire the UART, and change the address decoder inputs to reflect the difference in the address lines. Reassemble the monitor source file after changing all memory-mapped addresses as necessary. Burn a new EPROM and start testing.

12.3   THE 68010: A VIRTUAL-MEMORY MICROPROCESSOR

This processor uses the same input and output signals as the 68000, but contains a number of functional differences. These differences help support the use of **virtual** memory. Virtual memory is a term used to describe a technique that gives the user of a computer the impression that the entire memory space is available for use, no matter how much actual memory exists in the hardware. We often use the term **physical** memory to refer to the actual amount of storage implemented in hardware, and **logical** memory to refer to the entire address space of the microprocessor. Thus, even though a particular system may only have 512KB of physical memory, if it is a 68000-based system, its logical memory space is over 16 million bytes!

The 68010 virtual memory processor was designed to bring virtual memory capability to the microprocessor level. Virtual memory had previously been available only on larger mainframe computers. The 68010's software model is compatible with the 68000's. All address and data registers are included, and the instruction sets are identical, although the 68010 has some additional instructions to support its virtual-memory environment.

This environment is largely implemented via some additional registers and software functions. The VBR (vector base register) and the SFC and DFC—two additional function code registers that are only 3 bits in length—have been included to support virtual memory in the 68010. The VBR is used to relocate the processor's exception vector table, and the function code registers are used to control external hardware. The VBR is a 32-bit register that sets the base address of the exception vector table. Recall that exception vectors are located in memory between addresses 000000 and 0003FF, a 1KB block of storage. The 68010 uses the VBR and the 8-bit vector number generated during an exception to access the vector table, as shown in Figure 12.2. Thus, the exception vector table can begin at any address that is a multiple of 1,024. Suppose the VBR contains the address 002AF300. This is the address of vector 0. The last vector has an address of 002AF300 plus 3FC, or 002AF6FC. The VBR is automatically loaded with address 00000000 when the 68010 is reset. The programmer may use the new privileged instruction MOVEC (Move Control register) to load the VBR from any of the eight data registers, as in:

```
MOVE.L   #$2AF300,D0
MOVEC    D0,VBR
```

So, it is possible that multiple copies of the exception vector table may exist in memory simultaneously, with software controlling the way the 68010 will handle exceptions.

In the operation of a virtual machine, a small portion of a program will reside in the system's physical memory, the rest being stored on disk. During execution of the program, a reference to a memory location outside of the space allocated to the program creates a **page fault.** In the 68010, a page fault is indicated by a bus error ($\overline{\text{BERR}}$ is active), and causes the processor to enter exception processing. The exception handler will take care of loading the next portion of the program from disk into physical memory and resuming execution. Motorola calls this resumption of execution **instruction continuation.** In order for instruction continuation to work, the processor must save important system information on the stack during exception processing caused by a page fault. Data—such as the address that caused the page fault, the contents of internal CPU data registers, and system status information—is saved at the beginning of exception processing, to be reloaded later when program execution resumes. This information makes up the 68010's **stack frame,** and is organized as indicated in Figure 12.3. Let us see how the contents of the stack frame help support virtual memory.

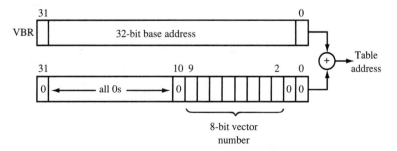

FIGURE 12.2 Generating the exception vector table address in the 68010

FIGURE 12.3 68010 exception stack frame for bus error

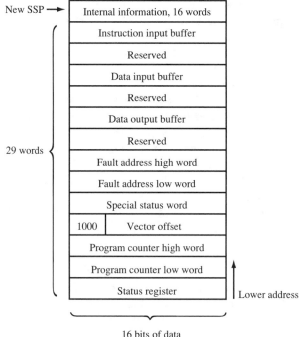

New SSP →

Internal information, 16 words
Instruction input buffer
Reserved
Data input buffer
Reserved
Data output buffer
Reserved
Fault address high word
Fault address low word
Special status word
1000 · Vector offset
Program counter high word
Program counter low word
Status register

29 words

Lower address

16 bits of data

Suppose that 1MB of physical memory is available in a 68010-based system. The memory is divided into 16 blocks of 64KB, each of which has its own base address (via a programmable address decoder). One 64KB block has been loaded with a program. A MOVE.W instruction located within this block attempts to read a word from address 9B8000, an uninitialized area of memory. None of the programmable address decoders responds to the address, which causes a bus error. The 68010 writes the bus error stack frame into supervisor stack memory and then starts the bus error exception handler. The handler reads the fault address out of the stack frame, and uses it to determine how to load one of the available 64KB blocks. Next, one of the address decoders is programmed with the base address of the new 64KB block. Finally, instruction continuation (via RTE and the stack frame) is used to resume execution of the MOVE.W instruction. The 68451 Memory Management Unit would be a big help in the necessary memory allocation of this virtual memory system and eliminate the need for the programmable address decoders.

Particularly important to the bus error handler are the contents of the special status word, whose bit assignments are shown in Figure 12.4. This status word indicates what type of bus cycle was being performed when the bus error occurred, and is critical to the proper resumption of program execution.

Another enhancement found in the 68010 is in the use of **loop mode** for special types of loops written around the DB*cc* (Decrement-and-Branch) instructions. Consider the following loop:

```
TOP    MOVE.W    D1,(A0)+
       DBRA      D0,TOP
```

FIGURE 12.4 Special status word details

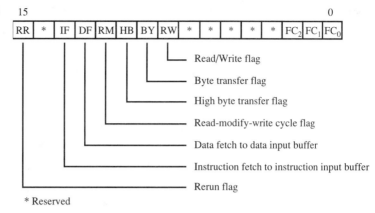

* Reserved

When a DB*cc* instruction is used to branch back to a single loop instruction, the 68010 will store the loop instruction (MOVE.W) and the loop displacement in special internal buffers, and only perform operand accesses to memory. This eliminates the need to constantly fetch and decode the MOVE.W and DBRA instructions, resulting in fewer clock cycles for each loop. The addressing modes that the loop instruction may use are limited to three address register indirect forms: (A*n*), –(A*n*), and (A*n*)+. This is sufficient, since we often access sequential memory locations during execution of a loop.

Due to the similarities between the 68000 and the 68010, it should be possible to upgrade a 68000-based system to a 68010 virtual-memory system with a minimum of change to the software and hardware.

12.4   THE 68020: A TRUE 32-BIT MACHINE

This processor is the first Motorola 680x0 series CPU that can really be called a 32-bit processor. Motorola improved on the virtual-memory capabilities of the 68010 by expanding its address and data buses into full 32-bit lengths. The 68020's 32-bit address bus may now address almost 4.3 *billion* bytes of information. Together with the ability to grab 32 bits of data at once, we see an increase in computing power and speed.

Figure 12.5 shows the various input and output signals of the 68020. Let us take a brief look at their meaning and operation.

Memory transfers may take place in 8-, 16-, or 32-bit quantities. Processor signals SIZ_0 and SIZ_1 are used by the 68020 to inform external hardware about the remaining number of bytes to be transferred in a memory access. Two other signals, $\overline{DSACK_0}$ and $\overline{DSACK_1}$, are used by the external memory system to tell the 68020 how many bytes may be transferred during the current operation. The use of these signals makes it possible for the 68020 to interface with bytewide memories (seen on earlier systems) and with the larger 16- or 32-bit-wide memories as well.

The $\overline{ECS}$ (External Cycle Start) and $\overline{OCS}$ (Operand Cycle Start) outputs are used to indicate to external hardware that a bus cycle is beginning. $\overline{RMC}$ (Read-Modify-write

FIGURE 12.5   68020 input/output signals

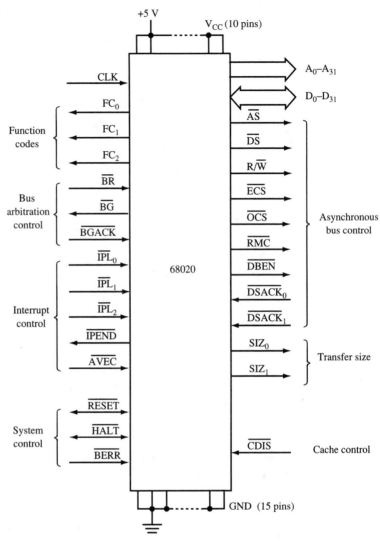

Cycle) goes low to indicate that a particular bus cycle should not be interrupted, so that data can be correctly read and rewritten. $\overline{\text{DBEN}}$ (Data Buffer Enable) is used to control an external bidirectional buffer and operates similar to the R/$\overline{\text{W}}$ signal.

$\overline{\text{IPEND}}$ (Interrupt Pending) is used to indicate that the requested interrupt level (on $\overline{\text{IPL}}_0$ through $\overline{\text{IPL}}_2$) is higher than the interrupt mask stored in the status register. $\overline{\text{AVEC}}$ (Autovector) is used to force the processor to generate an autovector during exception processing.

The last signal, $\overline{\text{CDIS}}$ (Cache Disable), is used to disable the on-chip, 256-byte **instruction cache.** A cache is a special type of high-speed memory used to decrease execution time. To be most effective, a cache is loaded with the data of the most frequently accessed locations, and their addresses. Whenever the CPU tries to access a memory location that matches one stored in the cache, a **hit** is made and the cache returns the data to the CPU much faster than the external memory system can. If no matches are found, we have a **miss.** A miss usually causes the least frequently used data in the cache to be replaced by the new data that caused the miss. The idea is to fill up the cache so that the hits far outnumber the misses. The 68020 stores as many instructions as it can in its 256 bytes of internal cache memory (organized as 64 longwords). Programs that employ any kind of looping will run faster, because the instructions of the loop will be stored in the cache. Thus, as the loop repeats itself over and over, the CPU continually fetches the repeated instructions from the cache, resulting in a program that runs much faster.

The 68020 has two registers that are used to control the cache. They are the CACR (cache control register) and the CAAR (cache address register). The CACR is used to clear, modify, freeze, and enable the cache. The CAAR points to the cache entry to be cleared. Figure 12.6 shows a block diagram of the on-chip cache, which is referred to as a **direct-mapped** cache. Address lines A_2 through A_7 point to one of 64 **tags** within the cache. Each tag is used to store the contents of address lines A_8 through A_{31} for a particular instruction, and its associated FC_2 bit (user/supervisor access). During an instruction fetch, address lines A_8 through A_{31} are compared with the contents of the tag selected by address lines A_2 through A_7. If there is a match, and the tag's **valid** bit is also set, the cache indicates a hit, and address line A_1 selects one instruction word for output to the instruction stream.

If the tag does not match, or its valid bit is clear, a miss is indicated. The actual instruction is then fetched from memory and written into the cache, so that the next access will not miss. This entire process can be disabled by the $\overline{CDIS}$ input.

The 68020 also contains support for the 68881 floating-point coprocessor. As Table 12.2 shows, exception vectors 13 and 48 through 54 are assigned to various coprocessor

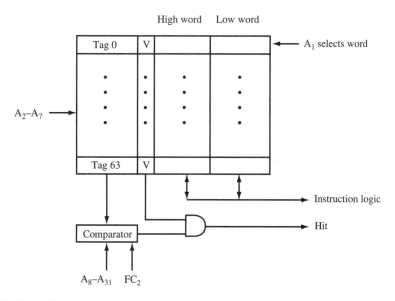

FIGURE 12.6   Block diagram of instruction cache

TABLE 12.2 68881 coprocessor exceptions on the 68020

Vector number	68881 event
13	Protocol violation
48	Branch/set on unordered condition
49	Inexact result
50	Divide-by-zero
51	Underflow
52	Operand error
53	Overflow
54	Signaling NAN

results. The 68881 was designed to interface directly with the 68020, so that its instructions constitute an extension of the 68020's instruction set.

Virtual memory is available on the 68020, with the VBR and bus error exception performing the same roles as they did in the 68010. A new stack frame format (called the **Module Call stack frame**) is utilized by the 68020 to support the implementation of high-level language procedures that require parameter passing. Two new instructions, CALLM (Call Module) and RTM (Return from Module), use the module call stack frame to properly manipulate the stack.

The 68020 contains many other new instructions. These instructions are summarized in Table 12.3. One notable addition concerns the instructions that manipulate **bit fields.** A bit field is a range of bits (up to 32) that reside anywhere in memory. The bit field instructions are suitable for use in systems that employ a **GUI (Graphic User Interface)**. One example of such a system is the Macintosh II, which contains a 68020 running at 16 MHz. Like other Macintosh computers, the operating system interfaces with the user via graphic screen information. Since the screen image is merely a copy of each bit in a memory-mapped video block of RAM, the bit field instructions provide an efficient way for the operating system to update the display.

Lastly, the 68020 supports a number of new addressing modes. These are summarized in Table 12.4. Most significant is the addition of the **memory indirect** addressing mode, which is represented by square brackets [] in the assembler syntax. Thus, (A0) and ([A0]) refer to totally different effective addresses. ([A0]) causes the 68020 to read the longword from the location pointed to by A0, and then use the longword as the address of the operand, whereas (A0) simply reads the operand from the location pointed to by A0.

Altogether, the architectural enhancements of the 68020 provide a significant improvement over the 68010.

12.5   THE 68030: AN ENHANCED 32-BIT MACHINE

Motorola created the 68030 by adding a paged memory management unit, a 256-byte data cache, and an enhanced bus controller to the 68020. Let us take a short look at the new processor signals and then address the new features of the 68030.

TABLE 12.3   New
68020 instructions

Instruction	Meaning
BFCHG	Test bit field and change
BFCLR	Test bit field and clear
BFEXTS	Extract bit field signed
BFEXTU	Extract bit field unsigned
BFFFO	Find first one in bit field
BFINS	Insert bit field
BFSET	Set bit field
BFTST	Test bit field
BKPT	Breakpoint
CALLM	Call module
CAS, CAS2	Compare and swap with operand
CHK2, CMP2	Compare register against bounds
DIVSL	Signed 64-bit divide
DIVUL	Unsigned 64-bit divide
EXTB	Extend byte to longword
MULS.L	Signed 32-bit multiply
MULU.L	Unsigned 32-bit multiply
PACK	Pack BCD
RTM	Return from module
TRAPcc	TRAP on condition
UNPACK	Unpack BCD

TABLE 12.4   New
68020 addressing modes

Addressing mode	Assembler syntax/example
Address register indirect with index (8-bit displacement)	(d8,An,Xn.SIZE*SCALE) (6,A2,D3.L*4)
Address register indirect with index (base displacement)	(bd,An,Xn.SIZE*SCALE) ($2E00,A6,A2.W*8)
Memory indirect postindexed	([bd,An],Xn.SIZE*SCALE,od) ([100,A0],D5.W,$3C)
Memory indirect preindexed	([bd,An,Xn.SIZE*SCALE],od) ([A5,D2.W])
Program counter indirect with index (8-bit displacement)	(d8,PC,Xn.SIZE*SCALE) ($80,PC,A4.L*2)
Program counter indirect with index (base displacement)	(bd,PC,Xn.SIZE*SCALE) ($7E204800,PC,D0.W)
Program counter memory indirect postindexed	([bd,PC],Xn.SIZE*SCALE,od) ([143,PC],A2.L)
Program counter memory indirect preindexed	([bd,PC,Xn.SIZE*SCALE],od) ([PC,A6.L*4],10)

Notes: [ ]—Memory indirect operators
d8—Optional signed 8-bit displacement
bd—Optional signed 16- or 32-bit base displacement
od—Optional signed 32-bit outer displacement
Xn—Any address or data register
SIZE—.W or .L
SCALE—1, 2, 4, or 8

Figure 12.7 shows the input and output signals of the 68030. You will notice that many of the signals are the same as those for the 68020. Four new signals are used for cache control. They are $\overline{\text{CIIN}}$ (Cache Inhibit In), $\overline{\text{CIOUT}}$ (Cache Inhibit Output), $\overline{\text{CBREQ}}$ (Cache Burst Request), and $\overline{\text{CBACK}}$ (Cache Burst Acknowledge).

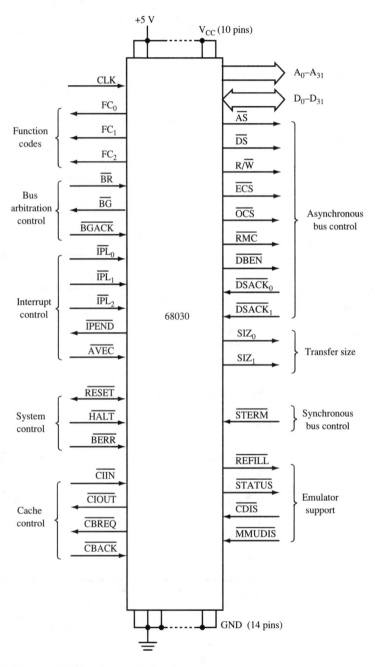

FIGURE 12.7   68030 input/output signals

The instruction and data caches are not loaded with new information when $\overline{\text{CIIN}}$ is active. If an external cache is used with the 68030, $\overline{\text{CIOUT}}$ may be used to inhibit the external cache when active. Both $\overline{\text{CBREQ}}$ and $\overline{\text{CBACK}}$ are used to control the cache's burst mode of operation, which allows the instruction or data cache to be loaded very quickly.

The $\overline{\text{HALT}}$ signal is now only used as an input. On all previous CPUs the $\overline{\text{HALT}}$ signal was bidirectional. $\overline{\text{HALT}}$ may be asserted with $\overline{\text{BERR}}$ to cause the 68030 to retry the current bus cycle.

$\overline{\text{MMUDIS}}$ (MMU Disable) prevents the internal memory management unit from performing address translations. Recall from the discussion of the 68451 memory management unit in Chapter 10 that address translation involves the conversion of a logical address into a physical address. This is the essence of a virtual-memory machine.

Both the 68020 and the 68030 are equipped with a three-stage instruction **pipeline,** which is capable of decoding portions of multiple instructions simultaneously. When the 68030 begins to fill the pipeline, $\overline{\text{REFILL}}$ (Pipeline Refill) will go low. The pipeline works in conjunction with an internal **microsequencer** that controls the operation of many processor tasks, such as instruction execution and effective address generation. The $\overline{\text{STATUS}}$ signal reflects the state of the microsequencer, and is low for different numbers of CLK cycles depending on what the microsequencer is busy doing.

The last new 68030 signal is $\overline{\text{STERM}}$ (Synchronous Termination). This signal is used to inform the processor that a 32-bit transfer is to be performed synchronously.

As mentioned before, the 68030 contains two internal 256-byte cache memories, one for instructions and the other for data. Considerable research has been done in the area of both internal and external cache memory systems, and there is general agreement that using separate caches for instruction and data results in better performance than using a single cache for both. Each cache is accessed by logical addresses generated by the processor, and has its own set of address and data buses. This allows faster exchange of data within the processor by avoiding address/data conflicts that are present when a single internal bus is used. Each cache is direct-mapped, with 16 entries of 16 bytes each. Tags are once again compared against address lines A_8 through A_{31}, as they were for the 68020. Address lines A_2 through A_7 are used slightly differently, however, as shown in Figure 12.8. Since each cache entry stores 16 bytes (four longwords), address lines A_2 and A_3 are used to select a longword/valid-bit pair. Address lines A_4 through A_7 are used to select one of 16 entries for tag comparison.

Each cache is capable of operating in **burst-fill** mode (when enabled via the cache control register). A burst-fill loads an entire 16-byte cache entry when a miss occurs or when all 4 valid bits are clear (which indicates that none of the longwords stored in the entry are valid). This is accomplished with help from the internal bus controller, which also controls instruction prefetching, data bus alignment (reading/writing words or longwords at odd addresses), and bus cycle timing.

The memory management unit of the 68030 uses a technique called **demand paging** to manage pages of memory. In a demand paging system, a large program is not loaded entirely into memory. Instead, only a portion of the program is initially loaded, with future portions loaded by demand. Each portion of the program is referred to as a **page.** Physical memory is divided into equally sized page **frames;** thus, on demand, a page frame is loaded with a page of an executing program. Performing memory allocation in this fashion allows many programs to reside in memory simultaneously, since each program may only require a few page frames at a time to execute properly. As

FIGURE 12.8   68030 cache organization

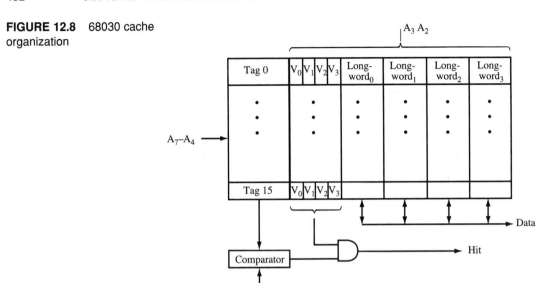

shown in Figure 12.9, a 2MB memory is divided into 64 page frames of 32KB each. Portions of six different programs are loaded into various page frames. Program 6 is the largest, with 11 frames allocated. Notice that the allocated page frames are all scattered about through the 2MB physical addressing space. This is what actually happens when many programs compete for page frames. When a program terminates, all page frames allocated to it are freed up for other uses. Since page frames are requested at different rates by executing programs, eventually the pages for a single program will become scattered. This does not affect the program's execution. As a matter of fact, as far as the

FIGURE 12.9   A demand-paging system

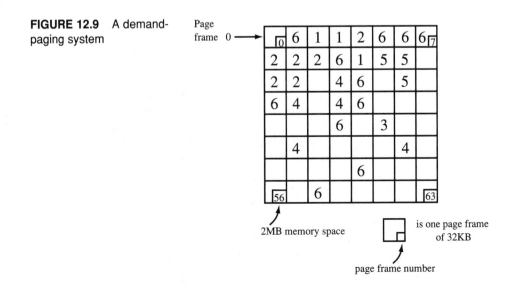

program is concerned, it is stored in a continuous logical memory. This is accomplished by translating the logical addresses generated within a program's page into the respective physical addresses of the program's page frame. This is accomplished through the use of **translation tables,** where the upper bits of the logical address are replaced by the upper bits of the physical address for the allocated page frame. The translation table must be maintained by the operating system.

The 68030's memory management unit translates 32-bit logical addresses into 32-bit physical addresses, supports page sizes from 256 bytes up to 32KB (in multiples of 2), and is capable of accessing physical memory in only two CLK cycles. Two new control registers control how translation is performed. These are the Translation Control and Transparent Translation registers. A third register provides status information about the memory management unit. Five new instructions are used to control the operation of the memory management unit. They are PFLUSHA, PFLUSH, PLOAD, PMOVE, and PTEST, and are mostly used to manipulate the address translation mechanism.

One last point: With so many of the 68030's signals identical to those used in the 68020, it is possible to upgrade a 68020 system by replacing the 68020 CPU with a 68030 logic board. The logic board is required, since the 68020 and 68030 have different pin assignments, and because the new 68030 signals need to be connected. Some changes to system software will then allow the older 68020 programs to take advantage of the memory management unit and dual cache.

12.6 AN OVERVIEW OF THE 68040

One look at Figure 12.10 shows that the 68040 is a radical departure from the previous Motorola CPUs. Almost every signal is different. Even the address bus is different, being bidirectional on the 68040! These architectural changes help support a technique called **bus snooping.** This technique is designed for use when the 68040 is connected to other 68040s in a multiprocessor system. At any time in this type of system, one CPU exists as the master, while the rest are slaves. All CPUs share common buses so that they may all access system RAM and EPROM. The bidirectional nature of the 68040's signals allows the processor (when acting as a slave) to examine the current bus cycle to determine if it should intervene. Bus snooping allows each 68040 in a multiprocessor system to examine each other's data cache, and force cache updates if necessary. This helps solve a nasty problem in multiprocessor systems called **cache coherency,** which requires that data stored for location X in one cache is the same data stored for location X in another cache. This is the only way to guarantee that all processors agree on the contents of shared memory.

The 68040's instruction cache and data cache are larger than those used in the 68030, and have been reorganized as **four-way, set-associative** caches, with their own memory management hardware. This is almost like having four direct-mapped caches running in parallel. Each of the four sets contains 64 entries, with each entry storing four longwords. This gives a total cache size of 1,024 longwords (or 4,096 bytes). In the case of cache memory, a larger cache has advantages and generally leads to a performance improvement.

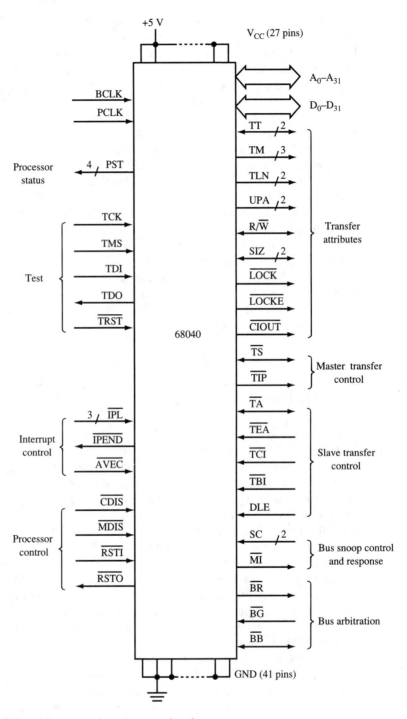

FIGURE 12.10   68040 input/output signals

The 68040 runs the same instruction set as all earlier machines, but now also contains a built-in coprocessor. Many of the floating-point instructions associated with the 68881 floating-point coprocessor are now included in the 68040's instruction set. Since the coprocessor itself contains eight 80-bit floating-point registers, the internal register set of the 68040 is much larger than any other 680x0 machine.

12.7   THE 68060: A SUPERSCALAR MICROPROCESSOR

The latest microprocessor in the 680x0 family is the 68060, which follows the 68040 (there is no 68050 processor). The 68060 has many technical advances and differences over the 68040, such as superscalar execution (two instructions complete in the same clock cycle), internal memory management and floating-point units, instruction pipelining, instruction and data cache, and a high bandwidth bus architecture. Figure 12.11 shows the input and output signals of the 68060. Although many of the signals look the same as those found on the 68040, the internal workings of the 68060 are much different. To see why, let us take a look at the development of the RISC movement.

The last decade has delivered a great deal of change in the area of computer architecture, due to advances in microelectronic manufacturing technology. As more and more logic was crammed into the small space of the silicon wafer, computer architects were able to implement processors with increasingly complex instructions and addressing modes. In general, these processors fall into a category called **CISC,** for Complex Instruction Set Computer. Unfortunately, cramming so much logic into a single package leads to a performance bottleneck, since many CISC instructions require multiple clock cycles to execute.

One solution to the performance bottleneck was the emergence of a new design philosophy called **RISC,** for Reduced Instruction Set Computer. RISC designers chose to make the instruction sets *smaller,* using fewer instructions and simpler addressing modes. This reduced set of operations was easier to implement on silicon, resulting in faster performance. Coupled with other architectural differences and improvements, RISC gained popularity, acceptance, and commercial interest. Video games and laser printers now boast of their internal RISC processors.

In general, a RISC machine is designed with the following goals in mind:

- Reduce accesses to main memory.
- Keep instructions and addressing modes simple.
- Make good use of registers.
- Pipeline everything.
- Utilize the compiler extensively.

Let us briefly examine each goal with the 68060 in mind.

Reduce Accesses to Main Memory

Even with the improvements made in memory technology, there remains the problem of the processor speed being much faster than the speed of the memory. For example, a

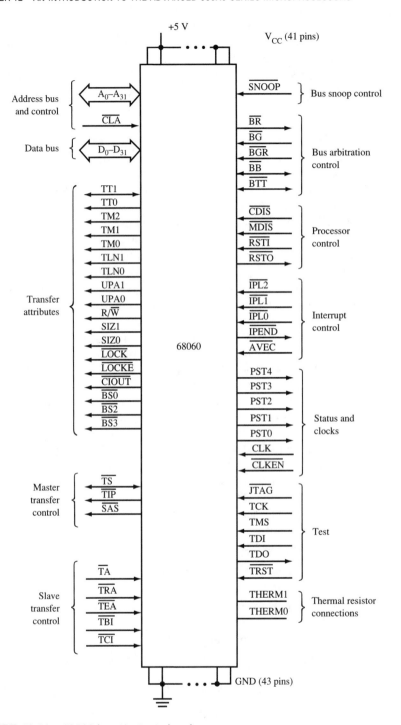

FIGURE 12.11   68060 input/output signals

processor clocked at 100 MHz would like to access memory in 10 nanoseconds, the period of its 100 MHz clock. Unfortunately, the memory interfaced to the processor might require 60 nanoseconds for an access. So, the processor ends up waiting during each memory access, wasting execution cycles.

To reduce the number of accesses to main memory, RISC designers added instruction and data **cache** to the design. Cache is commonly 10 times faster than main memory, so you can see the advantage of getting data in 10 nanoseconds instead of 60. Only when we *miss* and do not find the required data in the cache, does it take the full access time of 60 nanoseconds. But this can only happen once, since a copy of the new data is written into the cache after a miss. The data will be there next time we need it.

Instruction cache is used to store frequently used instructions, such as those in a short loop. Data cache is used to store frequently used data. Each cache is initially empty and fills as a program executes. The 68060 microprocessor utilizes an 8KB instruction cache and an 8KB data cache. The instruction cache does not allow self-modifying code.

Keep Instructions and Addressing Modes Simple

Computer scientists learned, after studying the operation of many different types of programs, that programmers use only a small subset of the instructions available on the processor they are using. The same is true for the processor's addressing modes.

Implementing fewer instructions and addressing modes on silicon will reduce the complexity of the instruction decoder, the addressing logic, and the execution unit. This allows the machine to be clocked at a faster speed, since less work needs to be done each clock period.

Unfortunately, the 68060 processor cannot meet this design goal. In order to remain compatible with the installed base of software in the entire 680x0 family, the 68060 designers had to keep each and every instruction and addressing mode of the previous machine, the 68040. The 68060 looks like a CISC machine from this point of view.

Make Good Use of Registers

RISC machines typically have large sets of registers. The number of registers available in a processor can affect performance in the same way a memory access does. A complex calculation may require the use of several data values. If the data values all reside in memory during the calculations, many memory accesses must be used to utilize them. If the data values are stored instead in the internal registers of the processor, their access during calculations will be much faster. It is good, then, to have lots of internal registers.

The 8-bit processors of the late 1970s had few registers (two on the 6800 and seven on the 8085) for general purpose use. These registers were also only 8 bits wide. The 68060 has eight general purpose registers (D_0 through D_7), all of which are 32 bits wide. The 68060 also contains eight 32-bit address registers, and eight **floating-point registers,** each of which is *80 bit*s wide, which are used by the floating-point unit. So, the 68060 has a fairly large set of registers to work with.

Pipeline Everything

Pipelining is a technique used to enable one instruction to complete with each clock cycle. Compare the two three-instruction sequences in Figure 12.12. On a nonpipelined machine, nine clock cycles are needed for the individual fetch, decode, and execute cycles. On a pipelined machine, where fetch, decode, and execute operations are performed *in parallel,* only five cycles are needed to execute the same three instructions. The first instruction requires three cycles to complete. Additional instructions then complete at a rate of one per cycle. As Figure 12.12 indicates, during clock cycle 5 we have I_3 completing, I_4 being decoded, and I_5 being fetched. A long sequence of instructions, say 1,000 of them, might require 3,000 clock cycles on a nonpipelined machine, and only 1,002 clock cycles when pipelined (three cycles for the first instruction, and then one cycle for the remaining 999). So, pipelining results in a tremendous performance gain.

The 68060 employs dual four-stage instruction pipelines that are capable of independent operation. This allows two instructions to complete each clock cycle in many cases (one in each pipeline). This type of architecture is called **superscalar** architecture.

Furthermore, the 68060 employs a technique called **branch folding** that helps identify possible interruptions to the normal flow of instructions through the instruction pipelines. By predicting which instructions might branch and change program flow, it is possible to keep a steady stream of instructions flowing into the pipelines. Once again, an attempt is made to keep instructions completing at a rate of one per clock cycle in each pipeline. The 68060 is very like a RISC machine in this respect.

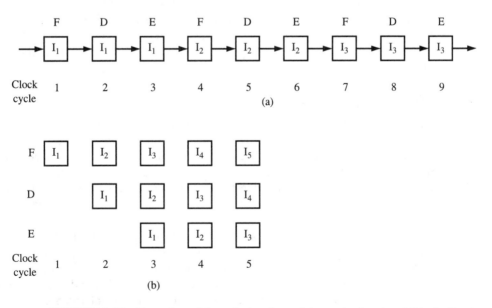

FIGURE 12.12   Execution of three instructions: (a) nonpipelined and (b) pipelined

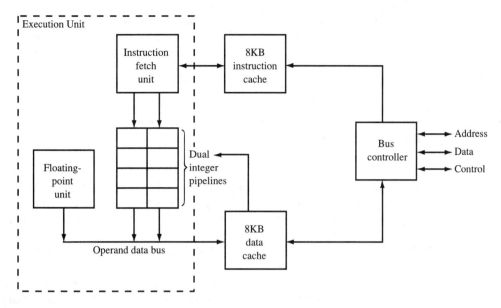

FIGURE 12.13   68060 block diagram

Figure 12.13 shows a simplified block diagram of the 68060. Note the separate instruction and data cache, dual integer pipelines, and internal floating-point unit. The internal hardware is designed to communicate over an internal bus so that all of the technical features just discussed have their full impact. One last goal still needs to be met, however.

Utilize the Compiler Extensively

When a high-level language program (such as a C program) is compiled, the individual statements within the program's source file are converted into assembly language instructions or groups of instructions. A 68060 compiler, if written properly, can perform many optimizations on the assembly language code to take advantage of the 68060's architectural advances. For instance, the compiler may reorder certain pairs of instructions to allow them to execute in parallel in the floating-point unit or dual integer pipelines. Instructions may also be rearranged to take advantage of the 68060's branch folding strategy. Other optimizations may involve use of the instruction/data cache, or algorithms to allocate the minimum number of processor registers during parsing of an arithmetic statement. Sometimes it is possible to substitute an instruction (such as MOVE.L   #0,D7) with an equivalent instruction (such as SUB.L   D7,D7) to reduce the number of clock cycles or the number of bytes of machine code.

Overall, the compiler plays an important role in helping a CISC or RISC machine achieve high performance. As a matter of fact, programs written for earlier 680x0 machines, even the 68040, can still be improved by recompiling their sources with a 68060 compiler.

What we have seen is that the 68060 contains both CISC *and* RISC characteristics. This is due in part to Motorola's commitment to support for the world of 680x0 users. Through pipelining, branch folding, instruction and data cache, and clever compilation, the 68060 delivers impressive performance.

12.8   TROUBLESHOOTING TECHNIQUES

With all the attention paid to microprocessors, it is easy to overlook another great source of computing power: the microcontroller. Microcontrollers are actually complete systems on a single chip, configured with EPROM, RAM, A/D converters, timers, counters, I/O peripherals, and other goodies. Table 12.5 shows a list of a few of Motorola's 683xx microcontrollers. These machines contain a processing core called CPU32 that is object code compatible with the 68000 and contains many 68010 and 68020 improvements.

Deciding between a microcontroller-based system or a microprocessor-based system depends on many factors. These may include design and prototyping time, troubleshooting time, cost, functionality, and performance. A good designer should be familiar with the other types of computing resources available, such as microcontrollers, in order to play an active role in the design and troubleshooting processes.

SUMMARY

This chapter provided a brief look at the other microprocessors in the 680x0 series. Each new machine is backward-compatible with all earlier models, guaranteeing that a program

TABLE 12.5 Motorola 683xx microcontrollers

Part	Description
68302	Integrated multiprotocol processor
68306	Integrated EC000 processor
68307	Integrated multiple bus processor
68322	Bandit-integrated printer processor
68330	Integrated CPU32 processor
68340	Integrated processor with DMA
68341	Integrated CD-I engine
68349	Dragon 1 high-performance integrated processor
68356	Signal processing communication engine
68360	Quad integrated communication controller

written on any machine will always execute on a more advanced model. We saw architectural enhancements with each new model, beginning with full 32-bit buses on the 68020 to the data and instruction cache, memory management unit, pipelines, and coprocessor of the 68060. And it all began with the 68000 microprocessor we covered in detail in this textbook.

STUDY QUESTIONS

1. Why will a 68008 system run slower than a 68000 system even if they have the same clock speeds?
2. Is it possible to generate a level-4 interrupt on the 68008?
3. What is virtual memory?
4. What is a page fault?
5. What is the difference between a logical address and a physical address?
6. What does the vector base register do on the 68010?
7. What is required to support instruction continuation?
8. What are the requirements for loop mode in a DB*cc* instruction on the 68010?
9. What are the main differences between the 68010 and the 68020?
10. Why would anyone need 4.3 billion memory locations? What applications might require this much memory?
11. What is meant by direct-mapping?
12. What is required to get a hit in the 68020's instruction cache?
13. How does an instruction cache speed up loop execution?
14. What is memory indirect addressing?
15. What are the main differences between the 68020 and the 68030?
16. Why split cache memory into instruction cache and data cache?
17. What is demand paging? How is it implemented?
18. How does the 68030 perform address translation?
19. What are the main differences between the 68030 and the 68040?
20. Draw a diagram of the architectural changes in the 680x0 series, beginning with the 68000.
21. What is bus snooping? What is it used for in a multiprocessor system?
22. What is meant by cache coherency?
23. How does a set-associative cache differ from a direct-mapped cache?
24. Why add a coprocessor to the 68040? Wouldn't it be simpler to just add a 68881 to the 68040's logic board in a new system?
25. Explain in detail why a 500-byte 68000 program that contains loops will run much faster on the 68040 than on the 68020 or 68030.
26. What does the term *superscalar* mean? Why is the 68060 a superscalar machine?
27. What are the design goals of a RISC processor?
28. How is a microcontroller different from a processor?

APPENDIX A

68000 Data Sheets

MC68000

Technical Summary
16-/32-Bit Microprocessor

This document contains both a summary of the MC68000 as well as a detailed set of parametrics. The purpose is twofold - to provide an introduction to the MC68000 and support for the sophisticated user. For detailed information on the MC68000, refer to the *MC68000 16- 32-Bit Microprossor Advance Information Data Sheet.*

The MC68000 is the first implementation of the M68000 16 32 microprocessor architecture. The MC68000 has a 16-bit data bus and 24-bit address bus while the full architecture provides for 32-bit address and data buses. It is completely code-compatible with the MC68008 8-bit data bus implementation of the M68000 and is upward code compatible to the MC68010 MC68012 virtual extensions and the MC68020 32-bit implementation of the architecture. Any user-mode programs written using the MC68000 instruction set will run unchanged on the MC68008, MC68010, MC68020. This is possible because the user programming model is identical for all five processors and the instruction sets are proper sub-sets of the complete architecture. Resources available to the MC68000 user consists of the following:

- 17 32-Bit Data and Address Registers
- 16 Megabyte Direct Addressing Range
- 56 Powerful Instruction Types
- Operations on Five Main Data Types
- Memory Mapped I O
- 14 Addressing Modes

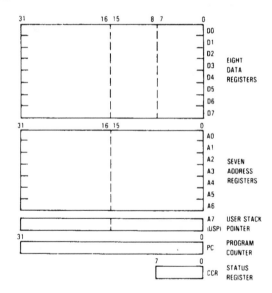

Figure 1. User Programming Model

This document contains information on a new product. Specifications and information herein are subject to change without notice.

INTRODUCTION

As shown in the user programming model (Figure 1), the MC68000 offers 16 32-bit registers and a 32-bit program counter. The first eight registers (D0-D7) are used as data registers for byte (8-bit), word (16-bit), and long word (32-bit) operations. The second set of seven registers (A0-A6) and the user stack pointer (USP) may be used as software stack pointers and base address registers. In addition, the registers may be used for word and long word operations. All of the 16 registers may be used as index registers.

In supervisor mode, the upper byte of the status register and the supervisor stack pointer (SSP) are also available to the programmer. These registers are shown in Figure 2.

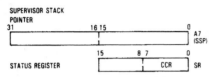

Figure 2. Supervisor Programming Model Supplement

The status register (Figure 3) contains the interrupt mask (eight levels available) as well as the condition codes: extend (X), negative (N), zero (Z), overflow (V), and carry (C). Additional status bits indicate that the processor is in a trace (T) mode and in a supervisor (S) or user state.

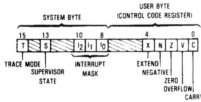

Figure 3. Status Register

DATA TYPES AND ADDRESSING MODES

Five basic data types are supported. These data types are:

- Bits
- BCD Digits (4-Bits)
- Bytes (8 Bits)
- Words (16 Bits)
- Long Words (32 Bits)

In addition, operations on other data types such as memory addresses, status word data, etc. are provided in the instruction set.

The 14 addressing modes, shown in Table 1, include six basic types:

- Register Direct
- Register Indirect
- Absolute

- Program Counter Relative
- Immediate
- Implied

Included in the register indirect addressing modes is a capability to do postincrementing, predecrementing, offsetting, and indexing. The program counter relative mode can also be modified via indexing and offsetting.

Table 1. Addressing Modes

Addressing Modes	Syntax
Register Direct Addressing	
Data Register Direct	Dn
Address Register Direct	An
Absolute Data Addressing	
Absolute Short	xxx.W
Absolute Long	xxx.L
Program Counter Relative Addressing	
Relative with Offset	$d_{16}(PC)$
Relative with Index Offset	$d_8(PC,Xn)$
Register Indirect Addressing	
Register Indirect	(An)
Postincrement Register Indirect	(An)
Predecrement Register Indirect	(An)
Register Indirect with Offset	$d_{16}(An)$
Indexed Register Indirect with Offset	$d_8(An,Xn)$
Immediate Data Addressing	
Immediate	#xxx
Quick Immediate	#1-#8
Implied Addressing	
Implied Register	SR USP SP PC

NOTES:

Dn	Data Register
An	Address Register
Xn	Address of Data Register used as Index Register
SR	Status Register
PC	Program Counter
SP	Stack Pointer
USP	User Stack Pointer
( )	Effective Address
d_8	8-Bit Offset (Displacement)
D_{16}	16-Bit Offset (Displacement)
#xxx	Immediate Data

INSTRUCTION SET OVERVIEW

The MC68000 instruction set is shown in Table 2. Some additional instructions are variations, or sub-sets, of these and they appear in Table 3. Special emphasis has been given to the instruction set's support of structured high-level languages to facilitate ease of programming. Each instruction, with few exceptions, operates on bytes, words, and long words and most instructions can use any of the 14 addressing modes. Combining instruction types, data types, and addressing modes, over 1000 useful instructions are provided. These instructions include signed and unsigned, multiply and divide, "quick" arithmetic operations, BCD arithmetic, and expanded operations (through traps).

Table 2. Instruction Set Summary

Mnemonic	Description
ABCD	Add Decimal With Extend
ADD	Add
AND	Logical AND
ASL	Arithmetic Shift Left
ASR	Arithmetic Shift Right
Bcc	Branch Conditionally
BCHG	Bit Test and Change
BCLR	Bit Test and Clear
BRA	Branch Always
BSET	Bit Test and Set
BSR	Branch to Subroutine
BTST	Bit Test
CHK	Check Register Against Bounds
CLR	Clear Operand
CMP	Compare
DBcc	Test Condition, Decrement and Branch
DIVS	Signed Divide
DIVU	Unsigned Divide
EOR	Exclusive OR
EXG	Exchange Registers
EXT	Sign Extend
JMP	Jump
JSR	Jump to Subroutine
LEA	Lead Effective Address
LINK	Link Stack
LSL	Logical Shift Left
LSR	Logical Shift Right
MOVE	Move
MULS	Signed Multiply
MULU	Unsigned Multiply
NBCD	Negate Decimal with Extend
NEG	Negate
NOP	No Operation
NOT	One's Complement
OR	Logical OR
PEA	Push Effective Address
RESET	Reset External Devices
ROL	Rotate Left without Extend
ROR	Rotate Right without Extend
ROXL	Rotate Left with Extend
ROXR	Rotate Right with Extend
RTE	Return from Exception
RTR	Return and Restore
RTS	Return from Subroutine
SBCD	Subtract Decimal with Extend
Scc	Set Conditional
STOP	Stop
SUB	Subtract
SWAP	Swap Data Register Halves
TAS	Test and Set Operand
TRAP	Trap
TRAPV	Trap on Overflow
TST	Test
UNLK	Unlink

Table 3. Variations of Instruction Types

Instruction Type	Variation	Description
ADD	ADD	Add
	ADDA	Add Address
	ADDQ	Add Quick
	ADDI	Add Immediate
	ADDX	Add with Extend
AND	AND	Logical AND
	ANDI	AND Immediate
	ANDI to CCR	AND Immediate to Condition Codes
	ANDI to SR	AND Immediate to Status Register
CMP	CMP	Compare
	CMPA	Compare Address
	CMPM	Compare Memory
	CMPI	Compare Immediate
EOR	EOR	Exclusive OR
	EORI	Exclusive OR Immediate
	EORI to CCR	Exclusive OR Immediate to Condition Codes
	EORI to SR	Exclusive OR Immediate to Status Register
MOVE	MOVE	Move
	MOVEA	Move Address
	MOVEM	Move Multiple Registers
	MOVEP	Move Peripheral Data
	MOVEQ	Move Quick
	MOVE from SR	Move from Status Register
	MOVE to SR	Move to Status Register
	MOVE to CCR	Move to Condition Codes
	MOVE USP	Move User Stack Pointer
NEG	NEG	Negate
	NEGX	Negate with Extend
OR	OR	Logical OR
	ORI	OR Immediate
	ORI to CCR	OR Immediate to Condition Codes
	ORI to SR	OR Immediate to Status Register
SUB	SUB	Subtract
	SUBA	Subtract Address
	SUBI	Subtract Immediate
	SUBQ	Subtract Quick
	SUBX	Subtract with Extend

446

Reprinted with permission of Motorola Inc.

SIGNAL DESCRIPTION

The input and output signals are illustrated functionally in Figure 4 and are described in the following paragraphs.

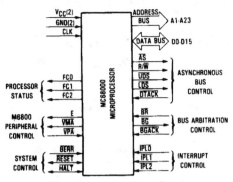

Figure 4. Input and Output Signals

ADDRESS BUS (A1 THROUGH A23)

This 32-bit, unidirectional, three-state bus is capable of addressing 16 megabytes of data. It provides the address for bus operation during all cycles except interrupt cycles. During interrupt cycles, address lines A1, A2, and A3 provide information about what level interrupt is being serviced while address lines A4 through A23 are set to a logic high.

DATA BUS (D0 THROUGH D15)

This 16-bit, bidirectional, three-state bus is the general purpose data path. It can transfer and accept data in either word or byte lenght. During an interrupt acknowledge cycle, the external device supplies the vector number on data lines D0-D7.

ASYNCHRONOUS BUS CONTROL

Asynchronous data transfers are handled using the following control signals: address strobe, read/write, upper and lower data strobes, and data transfer acknowledge. These signals are explained in the following paragraphs.

Address Strobe ($\overline{AS}$)

This signal indicates that there is a valid address on the address bus.

Read/Write (R/$\overline{W}$)

This signal defines the data bus transfer as a read or write cycle. The R/$\overline{W}$ signal also works in conjunction with the data strobes as explained in the following paragraph.

Upper and Lower Data Strobe ($\overline{UDS}$, $\overline{LDS}$)

These signals control the flow of data on the data bus, as shown in Table 4. When the R/$\overline{W}$ line is high, the processor will read from the data bus as indicated. When the R/$\overline{W}$ line is low, the processor will write to the data bus as shown.

Data Transfer Acknowledge ($\overline{DTACK}$)

This input indicates that the data transfer is completed. When the processor recognizes $\overline{DTACK}$ during a read cycle, data is latched and the bus cycle terminated. When $\overline{DTACK}$ is recognized during a write cycle, the bus cycle is terminated.

BUS ARBITRATION CONTROL

The three signals, bus request, bus grant, and bus grant acknowledge, form a bus arbitration circuit to determine which device will be the bus master device.

Bus Request ($\overline{BR}$)

This input is wire-ORed with all other devices that could be bus masters. This input indicates to the processor that some other device desires to become the bus master.

Bus Grant ($\overline{BG}$)

This output indicates to all other potential bus master devices that the processor will release bus control at the end of the current bus cycle.

Bus Grant Acknoweldge ($\overline{BGACK}$)

This input indicates that some other device has become the bus master. This signal should not be asserted until the following four conditions are met:

1. a bus grant has been received,
2. address strobe is inactive which indicates that the microprocessor is not using the bus,
3. data transfer acknowledge is inactive which indicates that neither memory nor peripherals are using the bus, and
4. bus grant acknowledge is inactive which indicates that no other device is still claiming bus mastership.

Table 4. Data Strobe Control of Data Bus

$\overline{UDS}$	$\overline{LDS}$	R/$\overline{W}$	D8-D15	D0-D7
High	High	—	No Valid Data	No Valid Data
Low	Low	High	Valid Data Bits 8-15	Valid Data Bits 0-7
High	Low	High	No Valid Data	Valid Data Bits 0-7
Low	High	High	Valid Data Bits 8-15	No Valid Data
Low	Low	Low	Valid Data Bits 8-15	Valid Data Bits 0-7
High	Low	Low	Valid Data Bits 0-7*	Valid Data Bits 0-7
Low	High	Low	Valid Data Bits 8-15	Valid Data Bits 8-15*

*These conditions are a result of current implementation and may not appear on future devices.

INTERRUPT CONTROL (IPL0, IPL1, IPL2)

These input pins indiacte the encoded priority level of the device requesting an interrupt. Level seven is the highest priority while level zero indicates that no interrupts are requested. Level seven cannot be masked. The least significant bit is given in IPL0 and the most significant bit is contained in IPL2. These lines must remain stable until the processor signals interrupt acknowledge (FC0-FC2 are all high) to insure that the interrupt is recognized.

SYSTEM CONTROL

The system control inputs are used to either reset or halt the processor and to indicate to the processor that bus errors have occurred. The three system control inputs are explained in the following paragraphs.

Bus Error (BERR)

This input informs the processor that there is a problem with the cycle currently being executed. Problems may be a result of:

1. nonresponding devices,
2. interrupt vector number acquisition failure,
3. illegal access request as determined by a memory management unit, or
4. other application dependent errors.

The bus error signal interacts with the halt signal to determine if the current bus cycle should be re-executed or if exception processing should be performed.

Reset (RESET)

This bidirectional signal line acts to reset (start a system initialization sequence) the processor in response to an external reset signal. An internally generated reset (result of a RESET instruction) causes all external devices to be reset and the internal state of the processor is not affected. A total system reset (processor and external devices) is the result of external HALT and RESET signals applied at the same time.

Halt (HALT)

When this bidirectional line is driven by an external device, it will cause the processor to stop at the completion of the current bus cycle. When the processor has been halted using this input, all control signals are inactive and all three-state lines are put in their high-impedance state.

When the processor has stopped executing instructions, such as in a double bus fault condition, the HALT line is driven by the processor to indicate to external devices that the processor has stopped.

M6800 PERIPHERAL CONTROL

These control signals are used to allow the interfacing of synchronous M6800 peripheral devices with the asynchronous MC68000. These signals are explained in the following paragraphs.

Enable (E)

This signal is the standard enable signal common to all M6800 type peripheral devices. The period for this output is ten MC68000 clock periods (six clocks low, four clocks high). Enable is generated by an internal ring counter which may come up in any state (i.e., at power on, it is impossible to guarantee phase relationship of E to CLK). E is a free-running clock and runs regardless of the state of the bus on the MPU.

Valid Peripheral Address (VPA)

This input indicates that the device or region addressed is an M68000 Family device or region addressed is an M68000 Family device and that data transfer should be synchronized with the enable (E) signal. This input also indicates that the processor should use automatic vectoring for an interrupt during an IACK cycle.

Valid Memory Address (VMA)

This output is used to indicate to M68000 peripheral devices that there is a valid address on the address bus and the processor is synchronized to enable. This signal only responds to a valid peripheral address (VPA) input which indicates that the peripheral is an M68000 Family device.

PROCESSOR STATUS (FC0, FC1, FC2)

These function code outputs indicate the state (user or supervisor) and the cycle type currently being executed, as shown in Table 5. The information indicated by the function code outputs is valid whenever address strobe (AS) is active.

Table 5. Function Code Outputs

Function Code Output			Cycle Time
FC2	FC1	FC0	
Low	Low	Low	(Undefined, Reserved)
Low	Low	High	User Data
Low	High	Low	User Program
Low	High	High	(Undefined, Reserved)
High	Low	Low	(Undefined, Reserved)
High	Low	High	Supervisor Data
High	High	Low	Supervisor Program
High	High	High	Interrupt Acknowledge

CLOCK (CLK)

The clock input is a TTL-compatible signal that is internally buffered for development of the internal clocks needed by the processor. The clock input should not be gated off at any time and the clock signal must conform to minimum and maximum pulse width times.

DATA TRANSFER OPERATIONS

Transfer of data between devices involves the following leads:

1. address bus A1 through A23,
2. data bus D0 through D15, and
3. control signals.

The address and data buses are separate parallel buses used to transfer data using an asynchronous bus structure. In all cycles, the bus master assumes responsibility for deskewing all signals it issues at both the start and end of a cycle. In addition, the bus master is responsible for deskewing the acknowledge and data signals from the slave device.

The following paragraphs explain the read, write, and read-modify-write cycles. The indivisible read-modify-write cycle is the method used by the MC68000 for interlocked multiprocessor communications.

READ CYCLE

During a read cycle, the processor receives data from the memory or a peripheral device. The processor reads bytes of data in all cases. If the instruction specifies a word (or double word) operation, the processor reads both upper and lower bytes simultaneously by asserting both upper and lower data strobes. When the instruction specifies byte operation, the processor uses an internal A0 bit to determine which byte to read and then issues the data strobe required for that byte. For byte operations, when the A0 bit equals zero, the upper data strobe is issued. When the A0 bit equals one, the lower data strobe is issued. When the data is received, the processor correctly positions it internally.

WRITE CYCLE

During a write cycle, the processor sends data to either the memory or a peripheral device. The processor writes bytes of data in all cases. If the instruction specifies a word operation, the processor writes both bytes. When the instruction specifies a byte operation, the processor uses an internal A0 bit to determine which byte to write and then issues the data strobe required for that byte. For byte operations, when the A0 bit equals zero, the upper data strobe is issued. When the A0 bit equals one, the lower data strobe is issued.

READ-MODIFY-WRITE CYCLE

The read-modify-write cycle performs a read, modifies the data in the arithmetic-logic unit, and writes the data back to the same address. In the MC68000, this cycle is indivisible in that the address strobe is asserted throughout the entire cycle. The test and set (TAS) instruction uses this cycle to provide meaningful communication between processors in a multiple processor environment.

This instruction is the only instruction that uses the read-modify-write cycles and since the test and set instruction only operates on bytes, all read-modify-write cycles are byte operations.

PROCESSING STATES

The MC68000 is always in one of three processing states: normal, exception, or halted.

NORMAL PROCESSING

The normal processing state is that associated with instruction execution; the memory references are to fetch instructions and operands, and to store results. A special case of normal state is the stopped state which the processor enters when a stop instruction is executed. In this state, no further references are made.

EXCEPTION PROCESSING

The exception processing state is associated with interrupts, trap instructions, tracing, and other exception conditions. The exception may be internally generated by an instruction or by an unusual condition arising during the execution of an instruction. Externally, exception processing can be forced by an interrupt, by a bus error, or by a reset. Exception processing is designed to provide an efficient context switch so that the processor may handle unusual conditions.

HALTED PROCESSING

The halted processing state is an indication of catastrophic hardware failure. For example, if during the exception processing of a bus error another bus error occurs, the processor assumes that the system is unusable and halts. Only an external reset can restart a halted processor. Note that a processor in the stopped state is not in the halted state, nor vice versa.

INTERFACE WITH M6800 PERIPHERALS

Motorola's extensive line of M6800 peripherals are directly compatible with the MC68000. Some of these devices that are particularly useful are:

MC6821   Peripheral Interface Adapter
MC6840   Programmable Timer Module
MC6843   Floppy Disk Controller
MC6845   CRT Controller
MC6850   Asynchronous Communications Interface Adapter
MC6854   Advanced Data Link Controller

To interface the synchronous M6800 peripherals with the asynchronous MC68000, the processor modifies its bus cycle to meet the M6800 cycle requirements whenever an M6800 device address is detected. This is possible since both the processors use memory mapped I O.

ELECTRICAL SPECIFICATIONS

MAXIMUM RATINGS

Rating	Symbol	Value	Unit
Supply Voltage	V_{CC}	-0.3 to $+7.0$	V
Input Voltage	V_{in}	-0.3 to $+7.0$	V
Operating Temperature Range MC68000 MC68000C	T_A	T_L to T_H 0 to 70 -40 to 85	°C
Storage Temperature	T_{stg}	-55 to 150	°C

The device contains circuitry to protect the inputs against damage due to high static voltages or electric fields; however, normal precautions should be taken to avoid application of voltages higher than maximum-rated voltages to these high-impedance circuits. Tying unused inputs to the appropriate logic voltage level (e.g., either GND or V_{CC}) enhances reliability of operation.

THERMAL CHARACTERISTICS

Characteristic	Symbol	Value	Symbol	Value	Rating
Thermal Resistance (Still Air) Ceramic, Type L/LC Ceramic, Type R/RC Plastic, Type P Plastic, Type FN	θ_{JA}	30 33 30 45	θ_{JC}	15* 15 15* 25*	°C/W

*Estimated

DC ELECTRICAL CHARACTERISTICS ($V_{CC} = 5.0$ Vdc $\pm 5\%$; GND $= 0$ Vdc; $T_A = T_L$ to T_H)

Characteristic		Symbol	Min	Max	Unit
Input High Voltage		V_{IH}	2.0	V_{CC}	V
Input Low Voltage		V_{IL}	GND -0.3	0.8	V
Input Leakage Current @ 5.25 V	BERR, BGACK, BR, DTACK, CLK, IPL0-IPL2, VPA HALT, RESET	I_{IN}	— —	2.5 20	μA
Three-State (Off State) Input Current @ 2.4 V/0.4 V	AS, A1-A23, D0-D15, FC0-FC2, LDS, R/W, UDS, VMA	I_{TSI}	—	20	μA
Output High Voltage ($I_{OH} = -400$ μA) ($I_{OH} = -400$ μA)	E* E, AS, A1-A23, BG, D0-D15, FC0-FC2, LDS, R/W, UDS, VMA	V_{OH}	$V_{CC} - 0.75$ 2.4	— 2.4	V
Output Low Voltage ($I_{OL} = 1.6$ mA) ($I_{OL} = 3.2$ mA) ($I_{OL} = 5.0$ mA) ($I_{OL} = 5.3$ mA)	HALT A1-A23, BG, FC0-FC2 RESET E, AS, D0-D15, LDS, R/W, UDS, VMA	V_{OL}	— — — —	0.5 0.5 0.5 0.5	V
Power Dissipation (see **POWER CONSIDERATIONS**)		P_D***	—	—	W
Capacitance ($V_{in} = 0$ V, $T_A = 25$°C, Frequency = 1 MHz)**		C_{in}	—	20.0	pF
Load Capacitance	HALT All Others	C_L	— —	70 130	pF

*With external pullup resistor of 1.1 Ω.
**Capacitance is periodically sampled rather than 100% tested.
***During normal operation instantaneous V_{CC} current requirements may be as high as 1.5 A.

Reprinted with permission of Motorola Inc.

POWER CONSIDERATIONS

The average die-junction temperature, T_J, in °C can be obtained from:

$$T_J = T_A + (P_D \cdot \theta_{JA}) \qquad (1)$$

where:

T_A = Ambient Temperature, °C
θ_{JA} = Package Thermal Resistance, Junction-to-Ambient, °C/W
P_D = $P_{INT} + P_{I/O}$
P_{INT} = $I_{CC} \times V_{CC}$, Watts — Chip Internal Power
$P_{I/O}$ = Power Dissipation on Input and Output Pins — User Determined

For most applications $P_{I/O} < P_{INT}$ and can be neglected.

An appropriate relationship between P_D and T_J (if $P_{I/O}$ is neglected) is:

$$P_D = K \div (T_J + 273 \text{ °C}) \qquad (2)$$

Solving equations (1) and (2) for K gives:

$$K = P_D \cdot (T_A + 273°C) + \theta_{JA} \cdot P_D^2 \qquad (3)$$

where K is a constant pertaining to the particular part. K can be determined from equation (3) by measuring P_D (at thermal equilibrium) for a known T_A. Using this value of K, the values of P_D and T_J can be obtained by solving equations (1) and (2) iteratively for any value of T_A.

The curve shown in Figure 11-1 gives the graphic solution to the above equations for the specified power dissipation of 1.5 watts over the ambient temperature range of -55 °C to 125 °C using a maximum θ_{JA} of 45

°C/W. Ambient temperature is that of the still air surrounding the device. Lower values of θ_{JA} cause the curve to shift downward slightly; for instance, for θ_{JA} of 40 °/W, the curve is just below 1.4 watts at 25 °C.

The total thermal resistance of a package (θ_{JA}) can be separated into two components, θ_{JC} and θ_{CA}, representing the barrier to heat flow from the semiconductor junction to the package (case) surface (θ_{JC}) and from the case to the outside ambient air (θ_{CA}). These terms are related by the equation:

$$\theta_{JA} = \theta_{JC} + \theta_{CA} \qquad (4)$$

θ_{JC} is device related and cannot be influenced by the user. However, θ_{CA} is user dependent and can be minimized by such thermal management techniques as heat sinks, ambient air cooling, and thermal convection. Thus, good thermal management on the part of the user can significantly reduce θ_{CA} so that θ_{JA} approximately equals θ_{JC}. Substitution of θ_{JC} for θ_{JA} in equation 1 results in a lower semiconductor junction temperature.

Table 6 summarizes maximum power dissipation and average junction temperature for the curve drawn in Figure 5, using the minimum and maximum values of ambient temperature for different packages and substituting θ_{JC} for θ_{JA} (assuming good thermal management). Table 7 provides the maximum power dissipation and average junction temperature assuming that no thermal management is applied (i.e., still air).

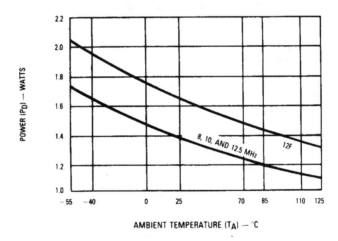

Figure 5. MC68000 Power Dissipation (P_D) vs Ambient Temperature (T_A)

Table 6. Power Dissipation and Junction Temperature vs Temperature ($\theta_{JC} = \theta_{JA}$)

Package	T_A Range	θ_{JC} (°C/W)	P_D (W) @ T_A Min.	T_J (°C) @ T_A Min.	P_D (W) @ T_A Max.	T_J (°C) @ T_A Max.
L/LC	0°C to 70°C	15	1.5	23	1.2	88
	−40°C to 85°C	15	1.7	−14	1.2	103
	0°C to 85°C	15	1.5	23	1.2	103
P	0°C to 70°C	15	1.5	23	1.2	88
R/RC	0°C to 70°C	15	1.5	23	1.2	88
	−40°C to 85°C	15	1.7	−14	1.2	103
	0°C to 85°C	15	1.5	23	1.2	103
FN	0°C to 70°C	25	1.5	38	1.2	101

NOTE: Table does not include values for the MC68000 12F.

Table 7. Power Dissipation and Junction Temperature vs Temperature ($\theta_{JC} \neq \theta_{JC}$)

Package	T_A Range	θ_{JA} (°C/W)	P_D (W) @ T_A Min.	T_J (°C) @ T_A Min.	P_D (W) @ T_A Max.	T_J (°C) @ T_A Max.
L/LC	0°C to 70°C	30	1.5	23	1.2	88
	−40°C to 85°C	30	1.7	−14	1.2	103
	0°C to 85°C	30	1.5	23	1.2	103
P	0°C to 70°C	30	1.5	23	1.2	88
R/RC	0°C to 70°C	33	1.5	23	1.2	88
	−40°C to 85°C	33	1.7	−14	1.2	103
	0°C to 85°C	33	1.5	23	1.2	103
FN	0°C to 70°C	40	1.5	38	1.2	101

NOTE: Table does not include values for the MC68000 12F.

AC ELECTRICAL SPECIFICATIONS — CLOCK TIMING (see Figure 6)

Num.	Characteristic	Symbol	8 MHz* Min	8 MHz* Max	10 MHz* Min	10 MHz* Max	12.5 MHz* Min	12.5 MHz* Max	16.67 MHz '12F' Min	16.67 MHz '12F' Max	Unit
	Frequency of Operation	f	4.0	8.0	4.0	10.0	4.0	12.5	8.0	16.7	MHz
1	Cycle Time	t_{cyc}	125	250	100	250	80	250	60	125	ns
2,3	Clock Pulse Width (Measured from 1.5 V to 1.5 V for 12F)	t_{CL} t_{CH}	55 55	125 125	45 45	125 125	35 35	125 125	27 27	62.5 62.5	ns
4,5	Clock Rise and Fall Times	t_{Cr} t_{Cf}	— —	10 10	— —	10 10	— —	5 5	— —	5 5	ns

*These specifications represent an improvement over previously published specifications for the 8-, 10-, and 12.5-MHz MC68000 and are valid only for product bearing date codes of 8827 and later.

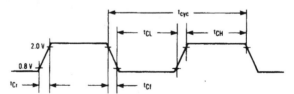

NOTE: Timing measurements are referenced to and from a low voltage of 0.8 volt and high a voltage of 2.0 volts, unless otherwise noted. The voltage swing through this range should start outside and pass through the range such that the rise or fall will be linear between 0.8 volt and 2.0 volts.

Figure 6. Clock Input Timing

APPENDIX B

68000 Instruction Set

APPENDIX A
CONDITION CODE COMPUTATION

A.1 INTRODUCTION

This appendix provides a discussion of the development of condition codes, the meaning of each bit, the computing of each bit, and the representation of the condition code in the instruction set details.

Two criteria were used in developing the condition codes:
- Consistency – across instructions, uses, and instances
- Meaningful Results – no change unless it provides useful information

Consistency across instructions means instructions that are special cases of more general instructions affect the condition codes in the same way. Consistency across uses means that conditional instructions test the condition codes similarly and provide the same results whether the condition codes are set by a compare, test, or move instruction. Consistency across instances means that all instances of an instruction affect the condition codes in the same way. The tests used for the conditional instructions and the code computations are listed in Table A-2.

A.2 CONDITION CODE REGISTER

The condition code register portion of the status register contains the following five bits:

X - Extend
N - Negative
Z - Zero
V - Overflow
C - Carry

The X bit is an operand for multiprecision computations. The next four bits are true condition code bits in that they reflect the condition of the result of a processor operation. The carry bit (C) and the multiprecision operand extend bit (X) are separate in the M68000 Family to simplify the programming model.

A.3 CONDITION CODE REGISTER NOTATION

In **APPENDIX B INSTRUCTION SET DETAILS**, the effect of the instruction on the condition codes is shown in the following form:

Condition Codes:

X	N	Z	V	C

where
X (extend)   Transparent to data movement. When affected by arithmetic operations, it is set the same as the C bit.

454

N (negative) Set if the most significant bit of the result is set. Cleared otherwise.
Z (zero) Set if the result equals zero. Cleared otherwise.
V (overflow) Set if there was arithmetic overflow. This implies that the result is not representable in the operand size. Cleared otherwise.
C (carry) Set if a carry is generated out of the most significant bit of the operands for an addition. Also, set if a borrow is generated in a subtraction. Cleared otherwise.

The notation that is used in the condition code register representation is as follows:
* set according to the result of the operation
— not affected by the operation
0 cleared
1 set
U undefined after the operation

A.4 CONDITION CODE COMPUTATION

Most operations compute a result from a source operand and a destination operand and store the result in the destination location. Unary operations compute a result from a destination operand and store the result in the destination location. Table A-1 lists condition code computations used by all instructions.

A.5 CONDITIONAL TESTS

Table A-2 lists the condition names, encodings, and tests for the condition branch and set instructions. The test associated with each condition is a logical formula based on the current state of the condition codes. If this formula evaluates to one, the condition succeeds, or is true. If the formula evaluates to zero, the condition is unsuccessful, or false. For example, the true (T) condition always succeeds, and the equal (EQ) condition succeeds only if the Z condition code bit is currently set.

Table A-1. Condition Code Computations

Operations	X	N	Z	V	C	Special Definition
ABCD	*	U	?	U	?	$C = $ Decimal Carry $Z = Z \wedge \overline{Rm} \wedge \ldots \wedge \overline{R0}$
ADD, ADDI, ADDQ	*	*	*	?	?	$V = Sm \wedge Dm \wedge \overline{Rm} \vee \overline{Sm} \wedge \overline{Dm} \wedge Rm$ $C = Sm \wedge Dm \vee \overline{Rm} \wedge Dm \vee Sm \wedge \overline{Rm}$
ADDX	*	*	?	?	?	$V = Sm \wedge Dm \wedge \overline{Rm} \vee \overline{Sm} \wedge \overline{Dm} \wedge Rm$ $C = Sm \wedge Dm \vee \overline{Rm} \wedge Dm \vee Sm \wedge \overline{Rm}$ $Z = Z \wedge \overline{Rm} \wedge \ldots \wedge \overline{R0}$
AND, ANDI, EOR, EORI, MOVEQ, MOVE, OR, ORI, CLR, EXT, NOT, TAS, TST	—	*	*	0	0	
CHK	—	*	U	U	U	
CHK2, CMP2	—	U	?	U	?	$Z = (R = LB) \vee (R = UB)$ $C = (LB < = UB) \wedge (IR < LB) \vee (R > UB)) \vee (UB < LB)$ $\wedge (R > UB) \wedge (R < LB)$
SUB, SUBI, SUBQ	*	*	*	?	?	$V = \overline{Sm} \wedge Dm \wedge \overline{Rm} \vee Sm \wedge \overline{Dm} \wedge Rm$ $C = Sm \wedge \overline{Dm} \vee Rm \wedge \overline{Dm} \vee Sm \wedge Rm$
SUBX	*	*	?	?	?	$V = \overline{Sm} \wedge Dm \wedge \overline{Rm} \vee Sm \wedge \overline{Dm} \wedge Rm$ $C = Sm \wedge \overline{Dm} \vee Rm \wedge \overline{Dm} \vee Sm \wedge Rm$ $Z = Z \wedge \overline{Rm} \wedge \ldots \wedge \overline{R0}$

A

455

Operations	X	N	Z	V	C	Special Definition
CAS, CAS2, CMP, CAMPI, CMPM	—	*	*	?	?	$V = \overline{Sm} \wedge Dm \wedge \overline{Rm} \vee Sm \wedge \overline{Dm} \wedge Rm$ $C = Sm \wedge \overline{Dm} \vee Rm \wedge \overline{Dm} \vee Sm \wedge Rm$
DIVS, DUVI	—	*	*	?	0	V = Division Overflow
MULS, MULU	—	*	*	0	0	V = Multiplication Overflow
SBCD, NBCD	*	U	?	U	?	C = Decimal Borrow $Z = Z \wedge \overline{Rm} \wedge \ldots \wedge \overline{Ro}$
NEG	*	*	*	?	?	$V = Dm \wedge Rm$ $C = Dm \vee Rm$
NEGX	*	*	?	?	?	$V = Dm \wedge Rm$ $C = Dm \vee Rm$ $Z = Z \wedge \overline{Rm} \wedge \ldots \wedge \overline{R0}$
BTST, BCHG, BSET, BCLR	—	—	?	—	—	$Z = \overline{Dn}$
BFTST, BFCHG, BFSET, BFCLR	—	?	?	0	0	$N = Dm$ $Z = \overline{Dm} \wedge \overline{DM-1} \wedge \ldots \wedge \overline{D0}$
BFEXTS, BFEXTU, BFFFO	—	?	?	0	0	$N = Sm$ $Z = \overline{Sm} \wedge \overline{Sm-1} \wedge \ldots \wedge \overline{S0}$
BFINS	—	?	?	0	0	$N = Dm$ $Z = \overline{Dm} \wedge \overline{DM-1} \wedge \ldots \wedge \overline{D0}$
ASL	*	*	*	?	?	$V = Dm \wedge (\overline{Dm-1} \vee \ldots \vee \overline{Dm-r}) \vee \overline{Dm} \wedge$ $(DM-1 \vee \ldots \vee Dm-r)\ C = \overline{Dm-r+1}$
ASL (R=0)	—	*	*	0	0	
LSL, ROXL	*	*	*	0	?	$C = Dm-r+1$
LSR (r=0)	—	*	*	0	0	
ROXL (r=0)	—	*	*	0	?	$C = X$
ROL	—	*	*	0	?	$C = Dm-r+1$
ROL (r=0)	—	*	*	0	0	
ASR, LSR, ROXR	*	*	*	0	?	$C = Dr-1$
ASR, LSR (r=0)	—	*	*	0	0	
ROXR (r=0)	—	*	*	0	?	$C = X$
ROR	—	*	*	0	?	$C = Dr-1$
ROR (r=0)	—	*	*	0	0	

— = Not Affected
U = Undefined, Result Meaningless
? = Other — See Special Definition
* = General Case
    X = C
    N = Rm
    Z = $\overline{Rm} \wedge \ldots \wedge \overline{R0}$
Sm = Source Operand — Most Significant Bit
Dm = Destination Operand — Most Significant Bit

Rm = Result Operand — Most Significant Bit
R = Register Tested
n = Bit Number
r = Shift Count
LB = Lower bound
UB = Upper Bound
$\wedge$ = Boolean AND
$\vee$ = Boolean OR
$\overline{Rm}$ = NOT Rm

A

Table A-2. Conditional Tests

Mnemonic	Condition	Encoding	Test
T*	True	0000	1
F*	False	0001	0
HI	High	0010	$\overline{C} \cdot \overline{Z}$
LS	Low or Same	0011	$C + Z$
CC(HS)	Carry Clear	0100	$\overline{C}$
CS(LO)	Carry Set	0101	C
NE	Not Equal	0110	$\overline{Z}$
EQ	Equal	0111	Z
VC	Overflow Clear	1000	$\overline{V}$
VS	Overflow Set	1001	V
PL	Plus	1010	$\overline{N}$
MI	Minus	1011	N
GE	Greater or Equal	1100	$N \cdot V + \overline{N} \cdot \overline{V}$
LT	Less Than	1101	$N \cdot \overline{V} + \overline{N} \cdot V$
GT	Greater Than	1110	$N \cdot V \cdot \overline{Z} + \overline{N} \cdot \overline{V} \cdot \overline{Z}$
LE	Less or Equal	1111	$Z + N \cdot \overline{V} + \overline{N} \cdot V$

$\cdot$ = Boolean AND
$+$ = Boolean OR
$\overline{N}$ = Boolean NOT N

*Not available for the Bcc instruction.

A

457

INSTRUCTION SET DETAILS

This appendix contains detailed information about each instruction in the M68000 instruction set. Instruction descriptions are arranged in alphabetical order with the mnemonic heading set in large bold type for easy reference.

B.1 ADDRESSING CATEGORIES

Effective address modes can be categorized by the ways in which they are used. The following classifications are used in the instruction definitions.

Data   If an effective address mode is used to refer to data operands, it is considered a data addressing effective address mode.

Memory   If an effective address mode is used to refer to memory operands, it is considered a memory addressing effective address mode.

Alterable   If an effective address mode is used to refer to alterable (writeable) operands, it is considered an alterable addressing effective address mode.

Control   If an effective address mode is used to refer to memory operands without associated sizes, it is considered a control addressing effective address mode.

Table B-1 shows the categories of each of the effective address modes.

Table B-1. Effective Address Mode Categories

Address Modes	Mode	Register	Data	Memory	Control	Alterable	Assembler Syntax
Data Register Direct	000	reg. no.	X	—	—	X	Dn
Address Register Direct	001	reg. no.	—	—	—	X	An
Address Register Indirect	010	reg. no.	X	X	X	X	(An)
Address Register Indirect with Postincrement	011	reg. no.	X	X	—	X	(An) +
Address Register Indirect with Predecrement	100	reg. no.	X	X	—	X	– (An)
Address Register Indirect with Displacement	101	reg. no.	X	X	X	X	(d_{16},An) or $d_{16}(An)$
Address Register Indirect with Index	110	reg. no.	X	X	X	X	(d_8,An,Xn) or $d_8(An,Xn)$
Absolute Short	111	000	X	X	X	X	(xxx).W
Absolute Long	111	001	X	X	X	X	(xxx).L
Program Counter Indirect with Displacement	111	101	X	X	X	—	(d_{16},PC) or $d_{16}(PC)$
Program Counter Indirect with Index	111	011	X	X	X	—	(d_8,PC,Xn) or $d_8(PC,Xn)$
Immediate	111	100	X	X	—	—	#(data)

These categories can be combined to define additional, more restrictive classifications. For example, the instruction descriptions use such classifications as memory alterable and data alterable. Memory alterable memory refers to addressing modes that are both alterable and memory addresses. Data alterable refers to addressing modes that are both data and alterable.

B.2 INSTRUCTION DESCRIPTION

The instruction descriptions in this section contain detailed information about the instructions. The format of these descriptions is shown in Figure B-1.

B.3 OPERATION DESCRIPTION DEFINITIONS

The following notation is used in the instruction descriptions.

OPERANDS

An	—Address register
Dn	—Data register
Rn	—Any data or address register
PC	—Program counter
SR	—Status register
CCR	—Condition codes (low-order byte of status)
SSP	—Supervisor stack pointer
USP	—User stack pointer
SP	—Active stack pointer (equivalent to A7)
X	—Extend operand condition code
N	—Negative condition code
Z	—Zero condition code
V	—Overflow condition code
C	—Carry condition code
Immediate data	—Immediate data from the instruction
d	—Address displacement
Source	—Source contents
Destination	—Destination contents
Vector	—Location of exception vector
ea	—Any valid effective address

SUBFIELDS AND QUALIFIERS

⟨bit⟩ of ⟨operand⟩	Selects a single bit of the operand
(⟨operand⟩)	The contents of the referenced location
⟨operand⟩$_{10}$	The operand is binary coded decimal; operations are to be performed in decimal
(⟨address register⟩) −(⟨address register⟩) (⟨address register⟩)+	The register indirect operator, which indicates that the operand register points to the memory location of the instruction operand
#xxx or #⟨data⟩	Immediate data operand from the instruction

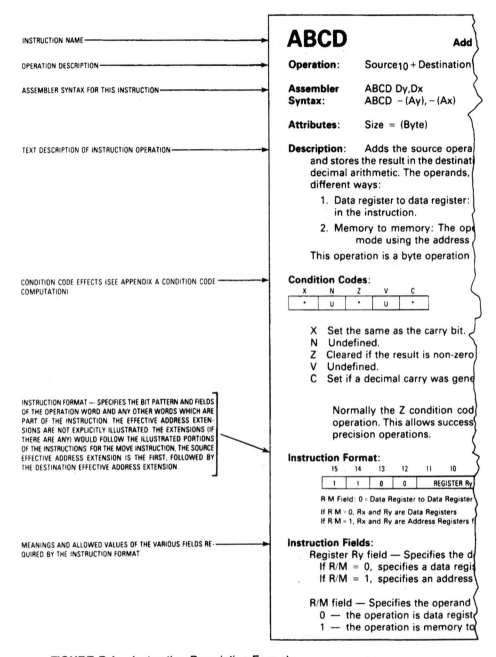

INSTRUCTION NAME

OPERATION DESCRIPTION

ASSEMBLER SYNTAX FOR THIS INSTRUCTION

TEXT DESCRIPTION OF INSTRUCTION OPERATION

CONDITION CODE EFFECTS (SEE APPENDIX A CONDITION CODE COMPUTATION)

INSTRUCTION FORMAT — SPECIFIES THE BIT PATTERN AND FIELDS OF THE OPERATION WORD AND ANY OTHER WORDS WHICH ARE PART OF THE INSTRUCTION. THE EFFECTIVE ADDRESS EXTENSIONS ARE NOT EXPLICITLY ILLUSTRATED. THE EXTENSIONS (IF THERE ARE ANY) WOULD FOLLOW THE ILLUSTRATED PORTIONS OF THE INSTRUCTIONS FOR THE MOVE INSTRUCTION, THE SOURCE EFFECTIVE ADDRESS EXTENSION IS THE FIRST, FOLLOWED BY THE DESTINATION EFFECTIVE ADDRESS EXTENSION.

MEANINGS AND ALLOWED VALUES OF THE VARIOUS FIELDS REQUIRED BY THE INSTRUCTION FORMAT.

ABCD   **Add**

Operation: Source$_{10}$ + Destination

Assembler   ABCD Dy,Dx
Syntax: ABCD – (Ay), – (Ax)

Attributes: Size = (Byte)

Description: Adds the source opera
and stores the result in the destinati
decimal arithmetic. The operands,
different ways:

1. Data register to data register:
   in the instruction.

2. Memory to memory: The op
   mode using the address

This operation is a byte operation

Condition Codes:

X	N	Z	V	C
*	U	*	U	*

X   Set the same as the carry bit.
N   Undefined.
Z   Cleared if the result is non-zero
V   Undefined.
C   Set if a decimal carry was gene

Normally the Z condition cod
operation. This allows success
precision operations.

Instruction Format:

15	14	13	12	11	10
1	1	0	0	REGISTER Ry	

R/M Field: 0 = Data Register to Data Register
If R/M = 0, Rx and Ry are Data Registers
If R/M = 1, Rx and Ry are Address Registers f

Instruction Fields:
Register Ry field — Specifies the d
If R/M = 0, specifies a data regis
If R/M = 1, specifies an address

R/M field — Specifies the operand
0 — the operation is data regis
1 — the operation is memory to

FIGURE B.1   Instruction Description Format

460

BINARY OPERATIONS

These operations are written ⟨operand⟩ ⟨op⟩ ⟨operand⟩, where ⟨op⟩ is one of the following:

⬧	The left operand is moved to the right operand
⬥	The two operands are exchanged
+	The operands are added
−	The right operand is subtracted from the left operand
*	The operands are multiplied
/	The first operand is divided by the second operand
Λ	The operands are logically ANDed
V	The operands are logically ORed
⟨	Relational test, true if the left operand is less than the right operand
⟩-	Relational test, true if the left operand is greater than the right operand
shifted by	The left operand is shifted or rotated by the number of positions specified
rotated by	by the right operand

UNARY OPERATIONS

~⟨operand⟩	The operand is logically complemented
⟨operand⟩sign extended	The operand is sign extended; bits equal to the high-order bit of the operand are inserted to extend the operand to the left
⟨operand⟩tested	The operand is compared to zero; the condition codes are set to the result.

OTHER OPERATIONS

TRAP
: Equivalent to: SSP - 2 ⬧ SSP; format/offset word ⬧ (SSP); SSP - 4 ⬧ SSP; PC ⬧ (SSP); SSP - 2 ⬧ ; SR ⬧ (SSP); (vector) ⬧ PC

STOP
: Enter the stopped state, waiting for interrupts

if ⟨condition⟩ then ⟨operations⟩ else ⟨operations⟩;

The condition is tested. If true, the operations after "then" are performed. If the condition is false and the optional "else" clause is present, the "else" clause operations are performed. If the condition is false and the "else" clause is omitted, the instruction performs no operation.

The semicolon is used to separate operations and terminate the if/then/else operation.

ABCD

ABCD

Add Decimal with Extend

Operation: Source$_{10}$ + Destination$_{10}$ + X ♦ Destination

Assembler Syntax: ABCD Dy,Dx
ABCD – (Ay),– (Ax)

Attributes: Size = (Byte)

Description: Adds the source operand to the destination operand along with the extend bit, and stores the result in the destination location. The addition is performed using binary coded decimal arithmetic. The operands, which are packed BCD numbers, can be addressed in two different ways:

1. Data register to data register: The operands are contained in the data registers specified in the instruction.

2. Memory to memory: The operands are addressed with the predecrement addressing mode using the address registers specified in the instruction.

This operation is a byte operation only.

Condition Codes:

X	N	Z	V	C
*	U	*	U	*

X Set the same as the carry bit.
N Undefined
Z Cleared if the result is non-zero. Unchanged otherwise.
V Undefined
C Set if a decimal carry was generated. Cleared otherwise.

NOTE

Normally the Z condition code bit is set via programming before the start of an operation. This allows successful tests for zero results upon completion of multiple-precision operations.

Instruction Format:

15	14	13	12	11	10	9	8	7	6	5	4	3	2	1	0
1	1	0	0	REGISTER Rx			1	0	0	0	0	R/M	REGISTER Ry		

Instruction Fields:

Register Rx field — Specifies the destination register:
If R/M = 0, specifies a data register
If R/M = 1, specifies an address register for the predecrement addressing mode
R/M field — Specifies the operand addressing mode:
0 — the operation is data register to data register
1 — the operation is memory to memory
Register Ry field — Specifies the source register:
If R/M = 0, specifies a data register
If R/M = 1, specifies an address register for the predecrement addressing mode

---

ADD

ADD

Add

Operation: Source + Destination ♦ Destination

Assembler Syntax: ADD (ea),Dn
ADD Dn,(ea)

Attributes: Size = (Byte, Word, Long)

Description: Adds the source operand to the destination operand using binary addition, and stores the result in the destination location. The size of the operation may be specified as byte, word, or long. The mode of the instruction indicates which operand is the source and which is the destination as well as the operand size.

Condition Codes:

X	N	Z	V	C
*	*	*	*	*

X Set the same as the carry bit.
N Set if the result is negative. Cleared otherwise.
Z Set if the result is zero. Cleared otherwise.
V Set if an overflow is generated. Cleared otherwise.
C Set if a carry is generated. Cleared otherwise.

Instruction Format:

15	14	13	12	11	10	9	8	7	6	5	4	3	2	1	0
1	1	0	1	REGISTER			OP-MODE			EFFECTIVE ADDRESS					
										MODE			REGISTER		

Instruction Fields:

Register field — Specifies any of the eight data registers.
Op-Mode field —

Byte	Word	Long	Operation
000	001	010	(ea) + (Dn) ♦ (n)
100	101	110	(Dn) + (ea) ♦ (ea)

ADD   Add   ADD

Effective Address Field — Determines addressing mode:

a. If the location specified is a source operand, all addressing modes are allowed as shown:

Addressing Mode	Mode	Register
Dn	000	reg. number:Dn
An*	001	reg. number:An
(An)	010	reg. number:An
(An)+	011	reg. number:An
-(An)	100	reg. number:An
(d16,An)	101	reg. number:An
(d8,An,Xn)	110	reg. number:An

Addressing Mode	Mode	Register
(xxx).W	111	000
(xxx).L	111	001
#(data)	111	100
(d16,PC)	111	010
(d8,PC,Xn)	111	011

*Word and Long only.

b. If the location specified is a destination operand, only memory alterable addressing modes are allowed as shown:

Addressing Mode	Mode	Register
Dn	—	—
An	—	—
(An)	010	reg. number:An
(An)+	011	reg. number:An
-(An)	100	reg. number:An
(d16,An)	101	reg. number:An
(d8,An,Xn)	110	reg. number:An

Addressing Mode	Mode	Register
(xxx).W	111	000
(xxx).L	111	001
#(data)	—	—
(d16,PC)	—	—
(d8,PC,Xn)	—	—

Notes:

1. The Dn mode is used when the destination is a data register; the destination (ea) mode is invalid for a data register.

2. ADDA is used when the destination is an address register. ADDI and ADDQ are used when the source is immediate data. Most assemblers automatically make this distinction.

ADDA   Add Address   ADDA

Operation: Source + Destination ♦ Destination

Assembler Syntax: ADDA (ea), An

Attributes: Size = (Word, Long)

Description: Adds the source operand to the destination address register, and stores the result in the address register. The size of the operation may be specified as word or long. The entire destination address register is used regardless of the operation size.

Condition Codes:
Not affected

Instruction Format:

15	14	13	12	11	10	9	8	7	6	5	4	3	2	1	0
1	1	0	1	REGISTER			OP-MODE			EFFECTIVE ADDRESS					
										MODE			REGISTER		

Op-Mode Field:		Operation
Word	Long	
011	111	((ea)) + ((An)) ♦ (An)

Instruction Fields:

Register field — Specifies any of the eight address registers. This is always the destination.

Op-Mode field — Specifies the size of the operation:
011 — Word operation. The source operand is sign-extended to a long operand and the operation is performed on the address register using all 32 bits.
111 — Long operation.

Effective Address field — Specifies the source operand. All addressing modes are allowed as shown:

Addressing Mode	Mode	Register
Dn	000	reg. number:Dn
An	001	reg. number:An
(An)	010	reg. number:An
(An)+	011	reg. number:An
-(An)	100	reg. number:An
(d16,An)	101	reg. number:An
(d8,An,Xn)	110	reg. number:An

Addressing Mode	Mode	Register
(xxx).W	111	000
(xxx).L	111	001
#(data)	111	100
(d16,PC)	111	010
(d8,PC,Xn)	111	011

ADDI ADDI ADDI

Add Immediate

Operation: Immediate Data + Destination → Destination

Assembler Syntax: ADDI #<data>,<ea>

Attributes: Size = (Byte, Word, Long)

Description: Adds the immediate data to the destination operand, and stores the result in the destination location. The size of the operation may be specified as byte, word, or long. The size of the immediate data matches the operation size.

Condition Codes:

X	N	Z	V	C
*	*	*	*	*

X — Set the same as the carry bit.
N — Set if the result is negative. Cleared otherwise.
Z — Set if the result is zero. Cleared otherwise.
V — Set if an overflow is generated. Cleared otherwise.
C — Set if a carry is generated. Cleared otherwise.

Instruction Format:

15	14	13	12	11	10	9	8	7	6	5	4	3	2	1	0
0	0	0	0	0	1	1	0	SIZE		EFFECTIVE ADDRESS MODE			REGISTER		
WORD DATA (16 BITS)								BYTE DATA (8 BITS)							
LONG DATA (32 BITS)															

Instruction Fields:
Size field — Specifies the size of the operation:
00 — Byte operation
01 — Word operation
10 — Long operation

ADDI ADDI

Add Immediate

Effective Address field — Specifies the destination operand.
Only data alterable addressing modes are allowed as shown:

Addressing Mode	Mode	Register
Dn	000	reg. number:Dn
An	—	—
(An)	010	reg. number:An
(An)+	011	reg. number:An
-(An)	100	reg. number:An
(d16,An)	101	reg. number:An
(d8,An,Xn)	110	reg. number:An

Addressing Mode	Mode	Register
(xxx).W	111	000
(xxx).L	111	001
#<data>	—	—
(d16,PC)	—	—
(d8,PC,Xn)	—	—

Immediate field — (Data immediately following the instruction):
If size = 00, the data is the low order byte of the immediate word
If size = 01, the data is the entire immediate word
If size = 10, the data is the next two immediate words

ADDQ   ADDQ   ADDQ

Add Quick

Operation: Immediate Data + Destination → Destination

Assembler Syntax: ADDQ #<data>,<ea>

Attributes: Size = (Byte, Word, Long)

Description: Adds an immediate value of 1 to 8 to the operand at the destination location. The size of the operation may be specified as byte, word, or long. Word and long operations are also allowed on the address registers. When adding to address registers, the condition codes are not altered, and the entire destination address register is used regardless of the operation size.

Condition Codes:

X	N	Z	V	C
*	*	*	*	*

X — Set the same as the carry bit.
N — Set if the result is negative. Cleared otherwise.
Z — Set if the result is zero. Cleared otherwise.
V — Set if an overflow occurs. Cleared otherwise.
C — Set if a carry occurs. Cleared otherwise.

The condition codes are not affected when the destination is an address register.

Instruction Format:

15	14	13	12	11	10	9	8	7	6	5	4	3	2	1	0
0	1	0	1		DATA		0		SIZE			EFFECTIVE ADDRESS			
												MODE		REGISTER	

Instruction Fields:

Data field — Three bits of immediate data, 0-7 (with the immediate value 0 representing a value of 8).
Size field — Specifies the size of the operation:
00 — Byte operation
01 — Word operation
10 — Long operation

Effective Address field — Specifies the destination location. Only alterable addressing modes are allowed as shown:

Addressing Mode	Mode	Register
Dn	000	reg. number:Dn
An*	001	reg. number:An
(An)	010	reg. number:An
(An)+	011	reg. number:An
-(An)	100	reg. number:An
(d16,An)	101	reg. number:An
(d8,An,Xn)	110	reg. number:An

Addressing Mode	Mode	Register
(xxx).W	111	000
(xxx).L	111	001
#<data>	—	—
(d16,PC)	—	—
(d8,PC,Xn)	—	—

*Word and Long only.

ADDX   Add Extended   ADDX

Operation: Source + Destination + X $\rightarrow$ Destination

Assembler Syntax:
ADDX Dy,Dx
ADDX -(Ay),-(Ax)

Attributes: Size = (Byte, Word, Long)

Description: Adds the source operand to the destination operand along with the extend bit and stores the result in the destination location. The operands can be addressed in two different ways:

1. Data register to data register: The data registers specified in the instruction contain the operands.
2. Memory to memory: The address registers specified in the instruction address the operands using the predecrement addressing mode.

The size of the operation can be specified as byte, word, or long.

Condition Codes:

X	N	Z	V	C
*	*	*	*	*

X Set the same as the carry bit.
N Set if the result is negative. Cleared otherwise.
Z Cleared if the result is non-zero. Unchanged otherwise.
V Set if an overflow occurs. Cleared otherwise.
C Set if a carry is generated. Cleared otherwise.

NOTE

Normally the Z condition code bit is set via programming before the start of an operation. This allows successful tests for zero results upon completion of multiple-precision operations.

ADDX   Add Extended   ADDX

Instruction Fields:

Register Rx field — Specifies the destination register:
If R/M = 0, specifies a data register
If R/M = 1, specifies an address register for the predecrement addressing mode

Size field — Specifies the size of the operation:
00 — Byte operation
01 — Word operation
10 — Long operation

R/M field — Specifies the operand address mode:
0 — The operation is data register to data register
1 — The operation is memory to memory

Register Ry field — Specifies the source register:
If R/M = 0, specifies a data register
If R/M = 1, specifies an address register for the predecrement addressing mode

Instruction Format:

15	14	13	12	11	10	9	8	7	6	5	4	3	2	1	0
1	1	0	1		REGISTER Rx		1		SIZE	0	0	R/M		REGISTER Ry	

AND

And Logical

Operation: Source∧Destination → Destination

Assembler
Syntax: AND ⟨ea⟩,Dn
AND Dn,⟨ea⟩

Attributes: Size = (Byte, Word, Long)

Description: Performs an AND operation of the source operand with the destination operand and stores the result in the destination location. The size of the operation can be specified as byte, word, or long. The contents of an address register may not be used as an operand.

Condition Codes:

X	N	Z	V	C
—	*	*	0	0

X Not affected
N Set if the most-significant bit of the result is set. Cleared otherwise.
Z Set if the result is zero. Cleared otherwise.
V Always cleared
C Always cleared

Instruction Format:

15	14	13	12	11	10	9	8	7	6	5	4	3	2	1	0
1	1	0	0	REGISTER			OP-MODE			EFFECTIVE ADDRESS					
										MODE			REGISTER		

Instruction Fields:

Register field — Specifies any of the eight data registers

Op-Mode field —

Byte	Word	Long	Operation
000	001	010	(⟨ea⟩)∧(⟨Dn⟩) → Dn
100	101	110	(⟨Dn⟩)∧(⟨ea⟩) → ea

AND

And Logical

Effective Address field — Determines addressing mode:
If the location specified is a source operand only data addressing modes are allowed as shown:

Addressing Mode	Mode	Register	Addressing Mode	Mode	Register
Dn	000	reg. number:Dn	(xxx).W	111	000
An	—	—	(xxx).L	111	001
(An)	010	reg. number:An	#⟨data⟩	111	100
(An)+	011	reg. number:An			
−(An)	100	reg. number:An			
(d16,An)	101	reg. number:An	(d16,PC)	111	010
(d8,An,Xn)	110	reg. number:An	(d8,PC,Xn)	111	011

If the location specified is a destination operand only memory alterable addressing modes are allowed as shown:

Addressing Mode	Mode	Register	Addressing Mode	Mode	Register
Dn	—	—	(xxx).W	111	000
An	—	—	(xxx).L	111	001
(An)	010	reg. number:An	#⟨data⟩	—	—
(An)+	011	reg. number:An			
−(An)	100	reg. number:An			
(d16,An)	101	reg. number:An	(d16,PC)	—	—
(d8,An,Xn)	110	reg. number:An	(d8,PC,Xn)	—	—

Notes:
1. The Dn mode is used when the destination is a data register; the destination ⟨ea⟩ mode is invalid for a data register.
2. Most assemblers use ANDI when the source is immediate data.

467

ANDI

AND Immediate

Operation: Immediate Data∧Destination ♦ Destination

Assembler
Syntax: ANDI #(data),(ea)

Attributes: Size = (Byte, Word, Long)

Description: Performs an AND operation of the immediate data with the destination operand and stores the result in the destination location. The size of the operation can be specified as byte, word, or long. The size of the immediate data matches the operation size.

Condition Codes:

X	N	Z	V	C
—	*	*	0	0

X Not affected
N Set if the most-significant bit of the result is set. Cleared otherwise.
Z Set if the result is zero. Cleared otherwise.
V Always cleared
C Always cleared

Instruction Format:

15	14	13	12	11	10	9	8	7	6	5	4	3	2	1	0
0	0	0	0	0	0	1	0		SIZE		MODE			REGISTER	

EFFECTIVE ADDRESS

| WORD DATA (16 BITS) | BYTE DATA (8 BITS) |
| LONG DATA (32 BITS) | |

Instruction Fields:
Size field — Specifies the size of the operation:
00 — Byte operation
01 — Word operation
10 — Long operation

ANDI

AND Immediate

Effective Address field — Specifies the destination operand
Only data alterable addressing modes are allowed as shown:

Addressing Mode	Mode	Register
Dn	000	reg. number:Dn
An	—	—
(An)	010	reg. number:An
(An)+	011	reg. number:An
–(An)	100	reg. number:An
(d16,An)	101	reg. number:An
(d8,An,Xn)	110	reg. number:An

Addressing Mode	Mode	Register
(xxx).W	111	000
(xxx).L	111	001
#(data)	—	—
(d16,PC)	—	—
(d8,PC,Xn)	—	—

Immediate field — (Data immediately following the instruction):
If size = 00, the data is the low order byte of the immediate word
If size = 01, the data is the entire immediate word
If size = 10, the data is the next two immediate words

ANDI to CCR

AND Immediate to Condition Codes

Operation: Source∧CCR ♦ CCR

Assembler Syntax: ANDI #(data),CCR

Attributes: Size = (Byte)

Description: Performs an AND operation of the immediate operand with the condition codes and stores the result in the low-order byte of the status register.

Condition Codes:

X	N	Z	V	C
*	*	*	*	*

X Cleared if bit 4 of immediate operand is zero. Unchanged otherwise.
N Cleared if bit 3 of immediate operand is zero. Unchanged otherwise.
Z Cleared if bit 2 of immediate operand is zero. Unchanged otherwise.
V Cleared if bit 1 of immediate operand is zero. Unchanged otherwise.
C Cleared if bit 0 of immediate operand is zero. Unchanged otherwise.

Instruction Format:

15	14	13	12	11	10	9	8	7	6	5	4	3	2	1	0
0	0	0	0	0	0	1	0	0	0	1	1	1	1	0	0
0	0	0	0	0	0	0	0				BYTE DATA (8 BITS)			0	0

ANDI to SR

**AND Immediate to the Status Register
(Privileged Instruction)**

Operation: If supervisor state
then Source∧SR ♦ SR
else TRAP

Assembler Syntax: ANDI #(data),SR

Attributes: Size = (Word)

Description: Performs an AND operation of the immediate operand with the contents of the status register and stores the result in the status register. All implemented bits of the status register are affected.

Condition Codes:

X	N	Z	V	C
*	*	*	*	*

X Cleared if bit 4 of immediate operand is zero. Unchanged otherwise.
N Cleared if bit 3 of immediate operand is zero. Unchanged otherwise.
Z Cleared if bit 2 of immediate operand is zero. Unchanged otherwise.
V Cleared if bit 1 of immediate operand is zero. Unchanged otherwise.
C Cleared if bit 0 of immediate operand is zero. Unchanged otherwise.

Instruction Format:

15	14	13	12	11	10	9	8	7	6	5	4	3	2	1	0
0	0	0	0	0	0	1	0	0	1	1	1	1	1	0	0
							WORD DATA (16 BITS)								

469

Arithmetic Shift

Operation: Destination Shifted by ⟨count⟩ ▶ Destination

Assembler
Syntax:
ASd Dx,Dy
ASd #⟨data⟩,Dy
ASd ⟨ea⟩
where d is direction, L or R

Attributes: Size = (Byte, Word, Long)

Description: Arithmetically shifts the bits of the operand in the direction (L or R) specified. The carry bit receives the last bit shifted out of the operand. The shift count for the shifting of a register may be specified in two different ways:

1. Immediate — The shift count is specified in the instruction (shift range, 1-8).
2. Register — The shift count is the value in the data register specified in instruction modulo 64.

The size of the operation can be specified as byte, word, or long. An operand in memory can be shifted one bit only, and the operand size is restricted to a word.

For ASL, the operand is shifted left; the number of positions shifted is the shift count. Bits shifted out of the high-order bit go to both the carry and the extend bits; zeros are shifted into the low-order bit. The overflow bit indicates if any sign changes occur during the shift.

ASL

For ASR, the operand is shifted right; the number of positions shifted is the shift count. Bits shifted out of the low-order bit go to both the carry and the extend bits; the sign-bit (MSB) is shifted into the high-order bit.

ASR

Arithmetic Shift

Condition Codes:

X	N	Z	V	C
*	*	*	*	*

X Set according to the last bit shifted out of the operand. Unaffected for a shift count of zero.
N Set if the most-significant bit of the result is set. Cleared otherwise.
Z Set if the result is zero. Cleared otherwise.
V Set if the most significant bit is changed at any time during the shift operation. Cleared otherwise.
C Set according to the last bit shifted out of the operand. Cleared for a shift count of zero.

Instruction Format (Register Shifts):

15	14	13	12	11	10	9	8	7	6	5	4	3	2	1	0
1	1	1	0	COUNT/REGISTER			dr	SIZE		i/r	0	0	REGISTER		

Instruction Fields (Register Shifts):

Count/Register field — Specifies shift count or register that contains the shift count:
If i/r = 0, this field contains the shift count. The values 1-7 represent counts of 1-7; value of zero represents a count of 8.
If i/r = 1, this field specifies the data register that contains the shift count (modulo 64).

dr field — Specifies the direction of the shift:
0 — Shift right
1 — Shift left

Size field — Specifies the size of the operation:
00 — Byte operation
01 — Word operation
10 — Long operation

i/r field:
If i/r = 0, specifies immediate shift count
If i/r = 1, specifies register shift count

Register field — Specifies a data register to be shifted

Instruction Format (Memory Shifts):

15	14	13	12	11	10	9	8	7	6	5	4	3	2	1	0
1	1	1	0	0	0	0	dr	1	1	MODE			REGISTER		
										EFFECTIVE ADDRESS					

ASL, ASR — Arithmetic Shift

Instruction Fields (Memory Shifts):

dr field — Specifies the direction of the shift:

0 — Shift right
1 — Shift left

Effective Address field — Specifies the operand to be shifted

Only memory alterable addressing modes are allowed as shown:

Addressing Mode	Mode	Register
Dn	—	—
An	—	—
(An)	010	reg. number:An
(An)+	011	reg. number:An
−(An)	100	reg. number:An
(d₁₆,An)	101	reg. number:An
(d₈,An,Xn)	110	reg. number:An

Addressing Mode	Mode	Register
(xxx).W	111	000
(xxx).L	111	001
#(data)	—	—
(d₁₆,PC)	—	—
(d₈,PC,Xn)	—	—

Bcc — Branch Conditionally

Operation: If (condition true) then PC + d ➧ PC

Assembler Syntax: Bcc (label)

Attributes: Size = (Byte, Word)

Description: If the specified condition is true, program execution continues at location (PC) + displacement. The PC contains the address of the instruction word of the Bcc instruction plus two. The displacement is a twos complement integer that represents the relative distance in bytes from the current PC to the destination PC. If the 8-bit displacement field in the instruction word is zero, a 16-bit displacement (the word immediately following the instruction) is used. Condition code cc specifies one of the following conditions:

CC	carry clear	0100	C̄	LS	low or same	0011	C + Z
CS	carry set	0101	C	LT	less than	1101	N·V̄ + N̄·V
EQ	equal	0111	Z	MI	minus	1011	N
GE	greater or equal	1100	N·V + N̄·V̄	NE	not equal	0110	Z̄
GT	greater than	1110	N·V·Z̄ + N̄·V̄·Z̄	PL	plus	1010	N̄
HI	high	0010	C̄·Z̄	VC	overflow clear	1000	V̄
LE	less or equal	1111	Z + N·V̄ + N̄·V	VS	overflow set	1001	V

Condition Codes:
Not affected

Instruction Format:

15	14	13	12	11	10	9	8	7	6	5	4	3	2	1	0
0	1	1	0		CONDITION						8-BIT DISPLACEMENT				
				16-BIT DISPLACEMENT IF 8-BIT DISPLACEMENT = $00											

Instruction Fields:
Condition field — The binary code for one of the conditions listed in the table.

8-Bit Displacement field — Twos complement integer specifying the number of bytes between the branch instruction and the next instruction to be executed if the condition is met.

16-Bit Displacement field — Used for the displacement when the 8-bit displacement field contains $00.

NOTE

A branch to the immediately following instruction automatically uses the 16-bit displacement format because the 8-bit displacement field contains $00 (zero offset).

Test a Bit and Change

Operation: ~((number) of Destination) → Z;
~((number) of Destination) → (bit number) of Destination

Assembler
Syntax: BCHG Dn,⟨ea⟩
BCHG #⟨data⟩,⟨ea⟩

Attributes: Size = (Byte, Long)

Description: Tests a bit in the destination operand and sets the Z condition code appropriately, then inverts the specified bit in the destination. When the destination is a data register, any of the 32 bits can be specified by the modulo 32-bit number. When the destination is a memory location, the operation is a byte operation, and the bit number is modulo 8. In all cases, bit zero refers to the least-significant bit. The bit number for this operation may be specified in either of two ways:
1. Immediate — The bit number is specified in a second word of the instruction.
2. Register — The specified data register contains the bit number.

Condition Codes:

X	N	Z	V	C
—	—	*	—	—

X Not affected
N Not affected
Z Set if the bit tested is zero. Cleared otherwise.
V Not affected
C Not affected

Instruction Format (Bit Number Dynamic, specified in a register):

15	14	13	12	11	10	9	8	7	6	5	4	3	2	1	0
0	0	0	0		REGISTER		1	0	1		MODE			REGISTER	
										EFFECTIVE ADDRESS					

Instruction Fields (Bit Number Dynamic):
Register field — Specifies the data register that contains the bit number
Effective Address field — Specifies the destination location. Only data alterable addressing modes are allowed as shown:

Addressing Mode	Mode	Register		Addressing Mode	Mode	Register
Dn*	000	reg. number:Dn		(xxx).W	111	000
An	—	—		(xxx).L	111	001
(An)	010	reg. number:An		#⟨data⟩	—	—
(An)+	011	reg. number:An				
-(An)	100	reg. number:An		(d16,PC)	—	—
(d16,An)	101	reg. number:An		(d8,PC,Xn)	—	—
(d8,An,Xn)	110	reg. number:An				

*Long only; all others are byte only.

BCHG

Test a Bit and Change

Instruction Format (Bit Number Static, specified as immediate data):

15	14	13	12	11	10	9	8	7	6	5	4	3	2	1	0
0	0	0	0	1	0	0	0	0	1		MODE			REGISTER	
0	0	0	0	0	0	0	0				BIT NUMBER				
										EFFECTIVE ADDRESS					

Instruction Fields (Bit Number Static):
Effective Address field — Specifies the destination location
Only data alterable addressing modes are allowed as shown:

Addressing Mode	Mode	Register		Addressing Mode	Mode	Register
Dn*	000	reg. number:Dn		(xxx).W	111	000
An	—	—		(xxx).L	111	001
(An)	010	reg. number:An		#⟨data⟩	—	—
(An)+	011	reg. number:An				
-(An)	100	reg. number:An		(d16,PC)	—	—
(d16,An)	101	reg. number:An		(d8,PC,Xn)	—	—
(d8,An,Xn)	110	reg. number:An				

*Long only; all others are byte only.

Bit Number field — Specifies the bit number

BCLR

Test a Bit and Clear

BCLR

Operation: ~((bit number) of Destination) → Z;
0 → (bit number) of Destination

Assembler BCLR Dn,(ea)
Syntax: BCLR #(data),(ea)

Attributes: Size = (Byte, Long)

Description: Tests a bit in the destination operand and sets the Z condition code appropriately, then clears the specified bit in the destination. When a data register is the destination, any of the 32 bits can be specified by a modulo 32-bit number. When a memory location is the destination, the operation is a byte operation, and the bit number is modulo 8. In all cases, bit zero refers to the least-significant bit. The bit number for this operation can be specified in either of two ways:

1. Immediate — The bit number is specified in a second word of the instruction.
2. Register — The specified data register contains the bit number.

Condition Codes:

X	N	Z	V	C
—	—	*	—	—

X Not affected
N Not affected
Z Set if the bit tested is zero. Cleared otherwise.
V Not affected
C Not affected

Instruction Format (Bit Number Dynamic, specified in a register):

15	14	13	12	11	10	9	8	7	6	5	4	3	2	1	0
0	0	0	0		REGISTER			1	1	0		MODE	EFFECTIVE ADDRESS		REGISTER

Instruction Fields (Bit Number Dynamic):
Register field — Specifies the data register that contains the bit number

---

BCLR

Test a Bit and Clear

BCLR

Effective Address field — Specifies the destination location
Only data alterable addressing modes are allowed as shown:

Addressing Mode	Mode	Register		Addressing Mode	Mode	Register
Dn*	000	reg. number:Dn		(xxx).W	111	000
An	—	—		(xxx).L	111	001
(An)	010	reg. number:An		#(data)	—	—
(An)+	011	reg. number:An				
-(An)	100	reg. number:An		(d16,PC)	—	—
(d16,An)	101	reg. number:An		(d8,PC,Xn)	—	—
(d8,An,Xn)	110	reg. number:An				

*Long only; all others are byte only.

Instruction Format (Bit Number Static, specified as immediate data):

15	14	13	12	11	10	9	8	7	6	5	4	3	2	1	0
0	0	0	0	1	0	0	0	1	0		MODE	EFFECTIVE ADDRESS			REGISTER
0	0	0	0	0	0	0	0					BIT NUMBER			

Instruction Fields (Bit Number Static):
Effective Address field — Specifies the destination location
Only data alterable addressing modes are allowed as shown:

Addressing Mode	Mode	Register		Addressing Mode	Mode	Register
Dn*	000	reg. number:Dn		(xxx).W	111	000
An	—	—		(xxx).L	111	001
(An)	010	reg. number:An		#(data)	—	—
(An)+	011	reg. number:An				
-(An)	100	reg. number:An		(d16,PC)	—	—
(d16,An)	101	reg. number:An		(d8,PC,Xn)	—	—
(d8,An,Xn)	110	reg. number:An				

*Long only; all others are byte only.

Bit Number field — Specifies the bit number

473

BKPT

BKPT   **Breakpoint**

Operation: Execute breakpoint acknowledge bus cycle;
                 Trap as illegal instruction

**Assembler
Syntax:**   BKPT #⟨data⟩

Attributes: Unsized

Description: This instruction is used to support the program breakpoint function for
debug monitors and real-time hardware emulators, and the operation will be dependent on the implementation. Execution of this instruction will cause the MC68010
to run a breakpoint acknowledge bus cycle (all function codes driven high) and zeros
on all address lines.

Whether the breakpoint acknowledge bus cycle is terminated with DTACK, BERR, or
VPA, the processor always takes an illegal instruction exception. During exception
processing, a debug monitor can distinguish eight different software breakpoints by
decoding the field in the BKPT instruction.

For the MC68000, MC68HC000, and MC68008, this instruction causes an illegal instruction exception but does not run the breakpoint acknowledge bus cycle.

Condition Codes:
Not affected

Instruction Format:

15	14	13	12	11	10	9	8	7	6	5	4	3	2	1	0
0	1	0	0	1	0	0	0	0	1	0	0	1	VECTOR		

Instruction Fields:
Vector field — Contains the immediate data, a value in the range of 0-7. This is the
breakpoint number.

BRA

BRA   **Branch Always**

Operation: PC + d ➧ PC

**Assembler
Syntax:**   BRA ⟨label⟩

Attributes: Size = (Byte, Word)

Description: Program execution continues at location (PC) + displacement. The PC contains the address of the instruction word of the BRA instruction plus two. The displacement is a twos complement integer that represents the relative distance in bytes
from the current PC to the destination PC. If the 8-bit displacement field in the instruction word is zero, a 16-bit displacement (the word immediately following the instruction) is used.

Condition Codes:
Not affected

Instruction Format:

15	14	13	12	11	10	9	8	7	6	5	4	3	2	1	0
0	1	1	0	0	0	0	0			8-BIT DISPLACEMENT					
16-BIT DISPLACEMENT IF 8-BIT DISPLACEMENT = $00															

Instruction Fields:
8-Bit Displacement field — Twos complement integer specifying the number of bytes
between the branch instruction and the next instruction to be executed.
16-Bit Displacement field — Used for a larger displacement when the 8-bit displacement is equal to $00.

NOTE

A branch to the immediately following instruction automatically uses the 16-
bit displacement format because the 8-bit displacement field contains $00
(zero offset).

BSET   BSET

Test a Bit and Set

Operation: ~((bit number) of Destination) → Z;
1 → (bit number) of Destination

Assembler Syntax: BSET Dn,(ea)
BSET #(data),(ea)

Attributes: Size = (Byte, Long)

Description: Tests a bit in the destination operand and sets the Z condition code appropriately. Then sets the specified bit in the destination operand. When a data register is the destination, any of the 32 bits can be specified by a modulo 32-bit number. When a memory location is the destination, the operation is a byte operation, and the bit number is modulo 8. In all cases, bit zero refers to the least-significant bit. The bit number for this operation can be specified in either of two ways:

1. Immediate — The bit number is specified in the second word of the instruction.
2. Register — The specified data register contains the bit number.

Condition Codes:

X	N	Z	V	C
—	—	*	—	—

X Not affected
N Not affected
Z Set if the bit tested is zero. Cleared otherwise.
V Not affected
C Not affected

Instruction Format (Bit Number Dynamic, specified in a register):

15	14	13	12	11	10	9	8	7	6	5	4	3	2	1	0
0	0	0	0		REGISTER		1	1	1		MODE			REGISTER	
											EFFECTIVE ADDRESS				

BSET   BSET

Test a Bit and Set

Instruction Fields (Bit Number Dynamic):
Register field — Specifies the data register that contains the bit number
Effective Address field — Specifies the destination location. Only data alterable addressing modes are allowed as shown:

Addressing Mode	Mode	Register	Addressing Mode	Mode	Register
Dn*	000	reg. number:Dn	(xxx).W	111	000
An	—	—	(xxx).L	111	001
(An)	010	reg. number:An	#(data)	—	—
(An)+	011	reg. number:An			
-(An)	100	reg. number:An			
(d16,An)	101	reg. number:An	(d16,PC)	—	—
(d8,An,Xn)	110	reg. number:An	(d8,PC,Xn)	—	—

*Long only; all others are byte only.

Instruction Format (Bit Number Static, specified as immediate data):

15	14	13	12	11	10	9	8	7	6	5	4	3	2	1	0
0	0	0	0	1	0	0	0	1	1		MODE			REGISTER	
0	0	0	0	0	0	0	0				BIT NUMBER				

*Long only; all others are byte only.

Instruction Fields (Bit Number Static):
Effective Address field — Specifies the destination location. Only data alterable addressing modes are allowed as shown:

Addressing Mode	Mode	Register	Addressing Mode	Mode	Register
Dn*	000	reg. number:Dn	(xxx).W	111	000
An	—	—	(xxx).L	111	001
(An)	010	reg. number:An	#(data)	—	—
(An)+	011	reg. number:An			
-(An)	100	reg. number:An			
(d16,An)	101	reg. number:An	(d16,PC)	—	—
(d8,An,Xn)	110	reg. number:An	(d8,PC,Xn)	—	—

*Long only; all others are byte only.

Bit Number field — Specifies the bit number

BSR

BSR

Branch to Subroutine

Operation: SP − 4 ⇨ SP; PC ⇨ (SP); PC + d ⇨ PC

Assembler
Syntax: BSR (label)

Attributes: Size = (Byte, Word)

Description: Pushes the long word address of the instruction immediately following the BSR instruction onto the system stack. The PC contains the address of the instruction word plus two. Program execution then continues at location (PC) + displacement. The displacement is a twos complement integer that represents the relative distance in bytes from the current PC to the destination PC. If the 8-bit displacement field in the instruction word is zero, a 16-bit displacement (the word immediately following the instruction) is used.

Condition Codes:
Not affected

Instruction Format:

15	14	13	12	11	10	9	8	7	6	5	4	3	2	1	0
0	1	1	0	0	0	0	1				8-BIT DISPLACEMENT				

16-BIT DISPLACEMENT IF 8-BIT DISPLACEMENT = $00

Instruction Fields:

8-Bit Displacement field — Twos complement integer specifying the number of bytes between the branch instruction and the next instruction to be executed.

16-Bit Displacement field — Used for a larger displacement when the 8-bit displacement is equal to $00.

NOTE

A branch to the immediately following instruction automatically uses the 16-bit displacement format because the 8-bit displacement field contains $00 (zero offset).

BTST

BTST

Test a Bit

Operation: −((bit number) of Destination) ⇨ Z;

Assembler
Syntax: BTST Dn,(ea)
BTST #(data),(ea)

Attributes: Size = (Byte, Long)

Description: Tests a bit in the destination operand and sets the Z condition code appropriately. When a data register is the destination, any of the 32 bits can be specified by a modulo 32 bit number. When a memory location is the destination, the operation is a byte operation, and the bit number is modulo 8. In all cases, bit zero refers to the least significant bit. The bit number for this operation can be specified in either of two ways:
1. Immediate — The bit number is specified in a second word of the instruction.
2. Register — The specified data register contains the bit number.

Condition Codes:

X	N	Z	V	C
—	—	*	—	—

X Not affected
N Not affected
Z Set if the bit tested is zero. Cleared otherwise.
V Not affected
C Not affected

Instruction Format (Bit Number Dynamic, specified in a register):

15	14	13	12	11	10	9	8	7	6	5	4	3	2	1	0
0	0	0	0		REGISTER		1	0	0		EFFECTIVE ADDRESS				
										MODE			REGISTER		

BTST

Test a Bit

Instruction Fields (Bit Number Dynamic):
Register field — Specifies the data register that contains the bit number
Effective Address field — Specifies the destination location. Only data addressing modes are allowed as shown:

Addressing Mode	Mode	Register
Dn*	000	reg. number:Dn
An	—	—
(An)	010	reg. number:An
(An)+	011	reg. number:An
-(An)	100	reg. number:An
(d16,An)	101	reg. number:An
(d8,An,Xn)	110	reg. number:An

Addressing Mode	Mode	Register
(xxx).W	111	000
(xxx).L	111	001
#(data)	111	100
(d16,PC)	111	010
(d8,PC,Xn)	111	011

*Long only; all others are byte only.

Instruction Format (Bit Number Static, specified as immediate data):

15	14	13	12	11	10	9	8	7	6	5	4	3	2	1	0
0	0	0	0	1	0	0	0	0	0	— EFFECTIVE ADDRESS MODE			REGISTER		
0	0	0	0	0	0	0	0	BIT NUMBER							

Instruction Fields (Bit Number Static):
Effective Address field — Specifies the destination location. Only data addressing modes are allowed as shown:

Addressing Mode	Mode	Register
Dn	000	reg. number:Dn
An	—	—
(An)	010	reg. number:An
(An)+	011	reg. number:An
-(An)	100	reg. number:An
(d16,An)	101	reg. number:An
(d8,An,Xn)	110	reg. number:An

Addressing Mode	Mode	Register
(xxx).W	111	000
(xxx).L	111	001
#(data)	111	—
(d16,PC)	111	010
(d8,PC,Xn)	111	011

Bit Number field — Specifies the bit number

CHK

Check Register Against Bounds

Operation: If Dn < 0 or Dn > Source then TRAP

Assembler Syntax: CHK (ea),Dn

Attributes: Size = (Word)

Description: Compares the value in the data register specified in the instruction to zero and to the upper bound (effective address operand). The upper bound is a twos complement integer. If the register value is less than zero or greater than the upper bound, a CHK instruction exception, vector number 6, occurs.

Condition Codes:

X	N	Z	V	C
—	*	U	U	U

X Not affected
N Set if Dn < 0; cleared if Dn > effective address operand. Undefined otherwise.
Z Undefined
V Undefined
C Undefined

Instruction Format:

15	14	13	12	11	10	9	8	7	6	5	4	3	2	1	0
0	1	0	0	REGISTER			1	1	0	EFFECTIVE ADDRESS MODE			REGISTER		

Instruction Fields:
Register field — Specifies the data register that contains the value to be checked

Effective Address field — Specifies the upper bound operand. Only data addressing modes are allowed as shown:

Addressing Mode	Mode	Register
Dn	000	reg. number:Dn
An	—	—
(An)	010	reg. number:An
(An)+	011	reg. number:An
-(An)	100	reg. number:An
(d16,An)	101	reg. number:An
(d8,An,Xn)	110	reg. number:An

Addressing Mode	Mode	Register
(xxx).W	111	000
(xxx).L	111	001
#(data)	111	100
(d16,PC)	111	010
(d8,PC,Xn)	111	011

CLR

Clear an Operand

Operation: 0 ♦ Destination

Assembler Syntax: CLR ⟨ea⟩

Attributes: Size = (Byte, Word, Long)

Description: Clears the destination operand to zero. The size of the operation may be specified as byte, word, or long.

Condition Codes:

X	N	Z	V	C
–	0	1	0	0

X Not affected
N Always cleared
Z Always set
V Always cleared
C Always cleared

Instruction Format:

15	14	13	12	11	10	9	8	7	6	5	4	3	2	1	0
0	1	0	0	0	0	1	0	SIZE		EFFECTIVE ADDRESS					

| | | | | | | | | | | MODE | | | REGISTER | | |

Instruction Fields:

Size field — Specifies the size of the operation
00 — Byte operation
01 — Word operation
10 — Long operation

Effective Address field — Specifies the destination location. Only data alterable addressing modes are allowed as shown:

Addressing Mode	Mode	Register
Dn	000	reg. number:Dn
An	–	–
(An)	010	reg. number:An
(An)+	011	reg. number:An
-(An)	100	reg. number:An
(d16,An)	101	reg. number:An
(d8,An,Xn)	110	reg. number:An

Addressing Mode	Mode	Register
(xxx).W	111	000
(xxx).L	111	001
#⟨data⟩	–	–
(d16,PC)	–	–
(d8,PC,Xn)	–	–

NOTE

In the MC68000, MC68HC000, and MC68008 a memory destination is read before it is cleared.

CMP

Compare

Operation: Destination — Source ♦ cc

Assembler Syntax: CMP ⟨ea⟩, Dn

Attributes: Size = (Byte, Word, Long)

Description: Subtracts the source operand from the destination data register and sets the condition codes according to the result; the data register is not changed. The size of the operation can be byte, word, or long.

Condition Codes:

X	N	Z	V	C
–	*	*	*	*

X Not affected
N Set if the result is negative. Cleared otherwise.
Z Set if the result is zero. Cleared otherwise.
V Set if an overflow occurs. Cleared otherwise.
C Set if a borrow occurs. Cleared otherwise.

Instruction Format:

15	14	13	12	11	10	9	8	7	6	5	4	3	2	1	0
1	0	1	1	REGISTER			OP-MODE			EFFECTIVE ADDRESS					

| | | | | | | | | | | MODE | | | REGISTER | | |

Instruction Fields:

Register field — Specifies the destination data register

Op-Mode field —

Byte	Word	Long	Operation
000	001	010	((Dn)) – ((ea))

CMP — Compare

Effective Address field — Specifies the source operand. All addressing modes are allowed as shown:

Addressing Mode	Mode	Register
Dn	000	reg. number:Dn
An*	001	reg. number:An
(An)	010	reg. number:An
(An)+	011	reg. number:An
-(An)	100	reg. number:An
(d$_{16}$,An)	101	reg. number:An
(d$_8$,An,Xn)	110	reg. number:An

Addressing Mode	Mode	Register
(xxx).W	111	000
(xxx).L	111	001
#(data)	111	100
(d$_{16}$,PC)	111	010
(d$_8$,PC,Xn)	111	011

*Word and Long only.

NOTE

CMPA is used when the destination is an address register. CMPI is used when the source is immediate data. CMPM is used for memory to memory compares. Most assemblers automatically make the distinction.

CMPA — Compare Address

Operation: Destination − Source

Assembler Syntax: CMPA (ea), An

Attributes: Size = (Word, Long)

Description: Subtracts the source operand from the destination address register and sets the condition codes according to the result; the address register is not changed. The size of the operation can be specified as word or long. Word length source operands are sign extended to 32-bits for comparison.

Condition Codes:

X	N	Z	V	C
-	*	*	*	*

X Not affected
N Set if the result is negative. Cleared otherwise.
Z Set if the result is zero. Cleared otherwise.
V Set if an overflow is generated. Cleared otherwise.
C Set if a borrow is generated. Cleared otherwise.

Instruction Format:

15	14	13	12	11 10 9	8 7 6	5 4 3	2 1 0
1	0	1	1	REGISTER	OP-MODE	EFFECTIVE ADDRESS MODE	REGISTER

Op-Mode Field:
Word   Long   Operation

Instruction Fields:
Register field — Specifies the destination address register
Op-Mode field — Specifies the size of the operation:
011 — Word operation. The source operand is sign-extended to a long operand and the operation is performed on the address register using all 32 bits.
111 — Long operation

CMPA — Compare Address — CMPA

Effective Address field — Specifies the source operand. All addressing modes are allowed as shown:

Addressing Mode	Mode	Register
Dn	000	reg. number:Dn
An	001	reg. number:An
(An)	010	reg. number:An
(An)+	011	reg. number:An
-(An)	100	reg. number:An
(d16,An)	101	reg. number:An
(d8,An,Xn)	110	reg. number:An

Addressing Mode	Mode	Register
(xxx).W	111	000
(xxx).L	111	001
#(data)	111	100
(d16,PC)	111	010
(d8,PC,Xn)	111	011

CMPI — Compare Immediate — CMPI

Operation: Destination − Immediate Data

Assembler Syntax: CMPI #(data),(ea)

Attributes: Size = (Byte, Word, Long)

Description: Subtracts the immediate data from the destination operand and sets the condition codes according to the result; the destination location is not changed. The size of the operation may be specified as byte, word, or long. The size of the immediate data matches the operation size.

Condition Codes:

X	N	Z	V	C
–	*	*	*	*

X Not affected
N Set if the result is negative. Cleared otherwise.
Z Set if the result is zero. Cleared otherwise.
V Set if an overflow occurs. Cleared otherwise.
C Set if a borrow occurs. Cleared otherwise.

Instruction Format:

15	14	13	12	11	10	9	8	7	6	5	4	3	2	1	0
0	0	0	0	1	1	0	0	SIZE		EFFECTIVE ADDRESS					
										MODE			REGISTER		
WORD DATA (16 BITS)										BYTE DATA (8 BITS)					
LONG DATA (32 BITS)															

Instruction Fields:
Size field — Specifies the size of the operation:
00 — Byte operation
01 — Word operation
10 — Long operation

CMPI — Compare Immediate

Effective Address field — Specifies the destination operand. Only data addressing modes are allowed as shown:

Addressing Mode	Mode	Register		Addressing Mode	Mode	Register
Dn	000	reg. number:Dn		(xxx).W	111	000
An	—	—		(xxx).L	111	001
(An)	010	reg. number:An		#(data)	—	—
(An)+	011	reg. number:An				
-(An)	100	reg. number:An				
(d$_{16}$,An)	101	reg. number:An		(d$_{16}$,PC)	111	010
(d$_8$,An,Xn)	110	reg. number:An		(d$_8$,PC,Xn)	111	011

Immediate field — (Data immediately following the instruction):

If size = 00, the data is the low order byte of the immediate word

If size = 01, the data is the entire immediate word

If size = 10, the data is the next two immediate words

CMPM — Compare Memory

Operation: Destination — Source ♦ cc

Assembler Syntax: CMPM (Ay) + ,(Ax) +

Attributes: Size = (Byte, Word, Long)

Description: Subtracts the source operand from the destination operand and sets the condition codes according to the results; the destination location is not changed. The operands are always addressed with the postincrement addressing mode, using the address registers specified in the instruction. The size of the operation may be specified as byte, word, or long.

Condition Codes:

X	N	Z	V	C
—	*	*	*	*

X Not affected

N Set if the result is negative. Cleared otherwise.

Z Set if the result is zero. Cleared otherwise.

V Set if an overflow is generated. Cleared otherwise.

C Set if a borrow is generated. Cleared otherwise.

Instruction Format:

15	14	13	12	11	10	9	8	7	6	5	4	3	2	1	0
1	0	1	1		REGISTER Ax		1		SIZE	0	0	1		REGISTER Ay	

Instruction Fields:

Register Ax field — (always the destination) Specifies an address register in the postincrement addressing mode

Size field — Specifies the size of the operation:

00 — Byte operation

01 — Word operation

10 — Long operation

Register Ay field — (always the source) Specifies an address register in the postincrement addressing mode

DBcc

Test Condition, Decrement, and Branch DBcc

Operation: If condition false then (Dn − 1 → Dn;
If Dn ≠ −1 then PC + d → PC)

Assembler Syntax: DBcc Dn,⟨label⟩

Attributes: Size = (Word)

Description: Controls a loop of instructions. The parameters are: a condition code, a data register (counter), and a displacement value. The instruction first tests the condition (for termination); if it is true, no operation is performed. If the termination condition is not true, the low-order 16 bits of the counter data register are decremented by one. If the result is −1, execution continues with the next instruction. If the result is not equal to −1, execution continues at the location indicated by the current value of the PC plus the sign-extended 16-bit displacement. The value in the PC is the address of the instruction word of the DBcc instruction plus two. The displacement is a twos complement integer that represents the relative distance in bytes from the current PC to the destination PC.

Condition code cc specifies one of the following conditions:

CC	carry clear	0100	$\bar{C}$		LS	low or same	0011	$C+Z$
CS	carry set	0101	C		LT	less than	1101	$N\cdot\bar{V}+\bar{N}\cdot V$
EQ	equal	0111	Z		MI	minus	1011	N
F	never equal	0001	0		NE	not equal	0110	$\bar{Z}$
GE	greater or equal	1100	$N\cdot V+\bar{N}\cdot\bar{V}$		PL	plus	1010	$\bar{N}$
GT	greater than	1110	$N\cdot V\cdot\bar{Z}+\bar{N}\cdot\bar{V}\cdot\bar{Z}$		T	always true	0000	1
HI	high	0010	$\bar{C}\cdot\bar{Z}$		VC	overflow clear	1000	$\bar{V}$
LE	less or equal	1111	$Z+N\cdot\bar{V}+\bar{N}\cdot V$		VS	overflow set	1001	V

Condition Codes:
Not affected

Instruction Format:

15	14	13	12	11	10	9	8	7	6	5	4	3	2	1	0
0	1	0	1		CONDITION			1	1	0	0	1		REGISTER	
							DISPLACEMENT (16 BITS)								

DBcc

Test Condition, Decrement, and Branch DBcc

Instruction Fields:
Condition field — The binary code for one of the conditions listed in the table
Register field — Specifies the data register used as the counter
Displacement field — Specifies the number of bytes to branch

Notes:
1. The terminating condition is similar to the UNTIL loop clauses of high-level languages. For example: DBMI can be stated as "decrement and branch until minus".
2. Most assemblers accept DBRA for DBF for use when only a count terminates the loop (no condition is tested).
3. A program can enter a loop at the beginning or by branching to the trailing DBcc instruction. Entering the loop at the beginning is useful for indexed addressing modes and dynamically specified bit operations. In this case, the control index count must be one less than the desired number of loop executions. However, when entering a loop by branching directly to the trailing DBcc instruction, the control count should equal the loop execution count. In this case, if a zero count occurs, the DBcc instruction does not branch, and the main loop is not executed.

DIVS

Signed Divide

DIVS

Operation: Destination/Source ◆ Destination

Assembler Syntax: DIVS.W (ea),Dn   32/16 ◆ 16r:16q

Attributes: Size = (Word)

Description: Divides the signed destination operand by the signed source operand and stores the signed result in the destination. The instruction divides a long word by a word. The result is a quotient in the lower word (least-significant 16 bits) and the remainder is in the upper word (most-significant 16 bits) of the result. The sign of the remainder is the same as the sign of the dividend.

Two special conditions may arise during the operation:
1. Division by zero causes a trap
2. Overflow may be detected and set before the instruction completes. If the instruction detects an overflow, it sets the overflow condition code, and the operands are unaffected.

Condition Codes:

X	N	Z	V	C
—	*	*	*	0

X Not affected
N Set if the quotient is negative. Cleared otherwise. Undefined if overflow or divide by zero occurs.
Z Set if the quotient is zero. Cleared otherwise. Undefined if overflow or divide by zero occurs.
V Set if division overflow occurs; undefined if divide by zero occurs. Cleared otherwise.
C Always cleared

Instruction Format (word form):

15	14	13	12	11	10	9	8	7	6	5	4	3	2	1	0
1	0	0	0		REGISTER			1	1	1		EFFECTIVE ADDRESS			
												MODE		REGISTER	

DIVS

Signed Divide

DIVS

Instruction Fields:
Register field — Specifies any of the eight data registers. This field always specifies the destination operand.
Effective Address field — Specifies the source operand. Only data addressing modes are allowed as shown:

Addressing Mode	Mode	Register
Dn	000	reg. number:Dn
An	—	—
(An)	010	reg. number:An
(An)+	011	reg. number:An
-(An)	100	reg. number:An
(d₁₆,An)	101	reg. number:An
(d₈,An,Xn)	110	reg. number:An

Addressing Mode	Mode	Register
(xxx).W	111	000
(xxx).L	111	001
#(data)	111	100
(d₁₆,PC)	111	010
(d₈,PC,Xn)	111	011

NOTE

Overflow occurs if the quotient is larger than a 16-bit signed integer. The instruction checks for overflow at the start of execution. If the upper word of the dividend is greater than or equal to the divisor, the overflow bit is set in the condition codes, and the instruction terminates with the operands unchanged.

Operation: Destination/Source ♦ Destination

Assembler
Syntax: DIVU.W (ea),Dn   32/16 ♦ 16r:16q

Attributes: Size = (Word)

Description: Divides the unsigned destination operand by the unsigned source operand and stores the unsigned result in the destination. The instruction divides a long word by a word. The result is a quotient in the lower word (least-significant 16 bits) and the remainder is in the upper word (most significant 16 bits) of the result.

Two special conditions may arise during the operation:
1. Division by zero causes a trap
2. Overflow may be detected and set before the instruction completes. If the instruction detects an overflow, it sets the overflow condition code, and the operands are unaffected.

Condition Codes:

X	N	Z	V	C
—	*	*	*	0

X   Not affected
N   Set if the quotient is negative. Cleared otherwise. Undefined if overflow or divide by zero occurs.
Z   Set if the quotient is zero. Cleared otherwise. Undefined if overflow or divide by zero occurs.
V   Set if division overflow occurs; undefined if divide by zero occurs. Cleared otherwise.
C   Always cleared

Instruction Format (word form):

15	14	13	12	11	10	9	8	7	6	5	4	3	2	1	0
1	0	0	0			REGISTER		TYPE	1	1	1		EFFECTIVE ADDRESS		
													MODE		REGISTER

Instruction Fields:
Register field — Specifies any of the eight data registers. This field always specifies the destination operand.
Effective Address field — Specifies the source operand. Only data addressing modes are allowed as shown:

Addressing Mode	Mode	Register
Dn	000	reg. number:Dn
An	—	—
(An)	010	reg. number:An
(An)+	011	reg. number:An
−(An)	100	reg. number:An
(d16,An)	101	reg. number:An
(d8,An,Xn)	110	reg. number:An

Addressing Mode	Mode	Register
(xxx).W	111	000
(xxx).L	111	001
#(data)	111	100
(d16,PC)	111	010
(d8,PC,Xn)	111	011

NOTE

Overflow occurs if the quotient is larger than a 16-bit signed integer. The instruction checks for overflow at the start of execution. If the upper word of the dividend is greater than or equal to the divisor, the overflow bit is set in the condition codes, and the instruction terminates with the operands unchanged.

EOR

Exclusive OR Logical

EOR

Operation: Source $\oplus$ Destination $\rightarrow$ Destination

Assembler
Syntax: EOR Dn,(ea)

Attributes: Size = (Byte, Word, Long)

Description: Performs an exclusive OR operation on the destination operand using the source operand and stores the result in the destination location. The size of the operation may be specified to be byte, word, or long. The source operand must be a data register. The destination operand is specified in the effective address field.

Condition Codes:

X	N	Z	V	C
—	*	*	0	0

X Not affected
N Set if the most-significant bit of the result is set. Cleared otherwise.
Z Set if the result is zero. Cleared otherwise.
V Always cleared
C Always cleared

Instruction Format (word form):

15	14	13	12	11	10	9	8	7	6	5	4	3	2	1	0
1	0	1	1		REGISTER			OP-MODE			EFFECTIVE ADDRESS				
											MODE			REGISTER	

Instruction Fields:

Register field — Specifies any of the eight data registers
Op-Mode field —

	Byte	Word	Long	Operation
	100	101	110	$\langle\langle ea \rangle\rangle \oplus \langle\langle Dn \rangle\rangle \rightarrow \langle ea \rangle$

EOR

Exclusive OR Logical

EOR

Effective Address field — Specifies the destination operand. Only data alterable addressing modes are allowed as shown:

Addressing Mode	Mode	Register
Dn	000	reg. number:Dn
An	—	—
(An)	010	reg. number:An
(An)+	011	reg. number:An
−(An)	100	reg. number:An
(d$_{16}$,An)	101	reg. number:An
(d$_8$,An,Xn)	110	reg. number:An

Addressing Mode	Mode	Register
(xxx).W	111	000
(xxx).L	111	001
#(data)	—	—
(d$_{16}$,PC)	—	—
(d$_8$,PC,Xn)	—	—

NOTE

Memory to data register operations are not allowed. Most assemblers use EORI when the source is immediate data.

485

EORI

Exclusive OR Immediate

Operation: Immediate Data ⊕ Destination ♦ Destination

Assembler
Syntax: EORI #⟨data⟩,⟨ea⟩

Attributes: Size = (Byte, Word, Long)

Description: Performs an exclusive OR operation on the destination operand using the immediate data and the destination operand and stores the result in the destination location. The size of the operation may be specified as byte, word, or long. The size of the immediate data matches the operation size.

Condition Codes:

X	N	Z	V	C
—	*	*	0	0

X Not affected
N Set if the most significant bit of the result is set. Cleared otherwise.
Z Set if the result is zero. Cleared otherwise.
V Always cleared
C Always cleared

Instruction Format:

15	14	13	12	11	10	9	8	7	6	5	4	3	2	1	0
0	0	0	0	1	0	1	0	0	SIZE		EFFECTIVE ADDRESS				
											MODE		REGISTER		

WORD DATA (16 BITS)	BYTE DATA (8 BITS)
LONG DATA (32 BITS)	

Instruction Fields:
Size field — Specifies the size of the operation:
00 — Byte operation
01 — Word operation
10 — Long operation

EORI

Exclusive OR Immediate

Effective Address field — Specifies the destination operand. Only data alterable addressing modes are allowed as shown:

Addressing Mode	Mode	Register
Dn	000	reg. number:Dn
An	—	—
(An)	010	reg. number:An
(An)+	011	reg. number:An
−(An)	100	reg. number:An
(d16,An)	101	reg. number:An
(d8,An,Xn)	110	reg. number:An

Addressing Mode	Mode	Register
(xxx).W	111	000
(xxx).L	111	001
#⟨data⟩	—	—
(d16,PC)	—	—
(d8,PC,Xn)	—	—

Immediate field — (Data immediately following the instruction):
If size = 00, the data is the low-order byte of the immediate word
If size = 01, the data is the entire immediate word
If size = 10, the data is next two immediate words

EORI
to CCR

Exclusive OR Immediate to Condition Code

Operation: Source $\oplus$ CCR ▸ CCR

Assembler Syntax: EORI #(data),CCR

Attributes: Size = (Byte)

Description: Performs an exclusive OR operation on the condition code register using the immediate operand and stores the result in the condition code register (low-order byte of the status register). All implemented bits of the condition code register are affected.

Condition Codes:

X	N	Z	V	C
•	•	•	•	•

X Changed if bit 4 of immediate operand is one. Unchanged otherwise.
N Changed if bit 3 of immediate operand is one. Unchanged otherwise.
Z Changed if bit 2 of immediate operand is one. Unchanged otherwise.
V Changed if bit 1 of immediate operand is one. Unchanged otherwise.
C Changed if bit 0 of immediate operand is one. Unchanged otherwise.

Instruction Format:

15	14	13	12	11	10	9	8	7	6	5	4	3	2	1	0
0	0	0	0	1	0	1	0	0	0	1	1	1	1	0	0
0	0	0	0	0	0	0	0	BYTE DATA (8 BITS)						0	0

EORI
to SR

Exclusive OR Immediate to the Status Register
(Privileged Instruction)

Operation: If supervisor state
then Source $\oplus$ SR ▸ SR

else TRAP

Assembler Syntax: EORI #(data),SR

Attributes: Size = (Word)

Description: Performs an exclusive OR operation on the contents of the status register using the immediate operand and stores the result in the status register. All implemented bits of the status register are affected.

Condition Codes:

X	N	Z	V	C
•	•	•	•	•

X Changed if bit 4 of immediate operand is one. Unchanged otherwise.
N Changed if bit 3 of immediate operand is one. Unchanged otherwise.
Z Changed if bit 2 of immediate operand is one. Unchanged otherwise.
V Changed if bit 1 of immediate operand is one. Unchanged otherwise.
C Changed if bit 0 of immediate operand is one. Unchanged otherwise.

Instruction Format:

15	14	13	12	11	10	9	8	7	6	5	4	3	2	1	0
0	0	0	0	1	0	1	0	0	1	1	1	1	1	0	0
WORD DATA (16 BITS)															

487

EXG

Exchange Registers

EXG

Operation: Rx ↔ Ry

Assembler Syntax:
EXG Dx,Dy
EXG Ax,Ay
EXG Dx,Ay
EXG Ay,Dx

Attributes: Size = (Long)

Description: Exchanges the contents of two 32-bit registers. The instruction performs three types of exchanges:
1. Exchange data registers
2. Exchange address registers
3. Exchange a data register and an address register

Condition Codes:
Not affected

Instruction Format:

15	14	13	12	11	10	9	8	7	6	5	4	3	2	1	0
1	1	0	0	REGISTER Rx			1	OP-MODE					REGISTER Ry		

Instruction Fields:
Register Rx field — Specifies either a data register or an address register depending on the mode. If the exchange is between data and address registers, this field always specifies the data register.
Op-Mode field — Specifies the type of exchange:
01000 — Data registers
01001 — Address registers
10001 — Data register and address register
Register Ry field — Specifies either a data register or an address register depending on the mode. If the exchange is between data and address registers, this field always specifies the address register.

EXT

Sign Extend

EXT

Operation: Destination Sign-Extended → Destination

Assembler Syntax:
EXT.W Dn   Extend byte to word
EXT.L Dn   Extend word to long word

Attributes: Sizes (Word, Long)

Description: Extends a byte in a data register to a word or a word in a data register to a long word, by replicating the sign bit to the left. If the operation extends a byte to a word, bit [7] of the designated data register is copied to bits [15:8] of that data register. If the operation extends a word to a long word, bit [15] of the designated data register is copied to bits [31:16] of the data register.

Condition Codes:

X	N	Z	V	C
-	*	*	0	0

X Not affected
N Set if the result is negative. Cleared otherwise.
Z Set if the result is zero. Cleared otherwise.
V Always cleared
C Always cleared

Instruction Format:

15	14	13	12	11	10	9	8	7	6	5	4	3	2	1	0
0	1	0	0	1	0	0	0	OP-MODE			0	0	0	REGISTER	

Instruction Fields:
Op-Mode field — Specifies the size of the sign-extension operation:
010 — Sign-extend low-order byte of data register to word
011 — Sign-extend low-order word of data register to long
Register field — Specifies the data register is to be sign-extended

ILLEGAL

ILLEGAL   Take Illegal Instruction Trap   **ILLEGAL**

Operation: SSP − 2 ♦ SSP; Vector Offset ♦ (SSP);
SSP − 4 ♦ SSP; PC ♦ (SSP);
SSP − 2 ♦ SSP; SR ♦ (SSP);
Illegal Instruction Vector Address ♦ PC

Assembler
Syntax: ILLEGAL

Attributes: Unsized

Description: Forces an illegal instruction exception, vector number 4. All other illegal instruction bit patterns are reserved for future extension of the instruction set and should not be used to force an exception.

Only the MC68010 stores a four-word exception stack frame by first writing the exception vector offset and format code to the system stack. All processors write the PC, followed by the SR, to the system stack.

Condition Codes:
Not affected

Instruction Format:

15	14	13	12	11	10	9	8	7	6	5	4	3	2	1	0
0	1	0	0	1	0	1	0	1	1	1	1	1	1	0	0

JMP

JMP   Jump   **JMP**

Operation: Destination Address ♦ PC

Assembler
Syntax: JMP (ea)

Attributes: Unsized

Description: Program execution continues at the effective address specified by the instruction. The addressing mode for the effective address must be a control addressing mode.

Condition Codes:
Not affected

Instruction Format:

15	14	13	12	11	10	9	8	7	6	5	4	3	2	1	0
0	1	0	0	1	1	1	0	1	1		EFFECTIVE ADDRESS				
											MODE			REGISTER	

Instruction Fields:
Effective Address field — Specifies the address of the next instruction. Only control addressing modes are allowed as shown:

Addressing Mode	Mode	Register		Addressing Mode	Mode	Register
Dn	—	—		(xxx).W	111	000
An	—	—		(xxx).L	111	001
(An)	010	reg. number:An		#(data)	—	—
(An)+	—	—				
−(An)	—	—				
(d16,An)	101	reg. number:An		(d16,PC)	111	010
(d8,An,Xn)	110	reg. number:An		(d8,PC,Xn)	111	011

LEA

Load Effective Address

LEA

Operation: ⟨ea⟩ ⬥ An

Assembler Syntax: LEA ⟨ea⟩,An

Attributes: Size = (Long)

Description: Loads the effective address into the specified address register. All 32 bits of the address register are affected by this instruction.

Condition Codes: Not affected

Instruction Format:

15	14	13	12	11	10	9	8	7	6	5	4	3	2	1	0
0	1	0	0		REGISTER		1	1	1		EFFECTIVE ADDRESS MODE			REGISTER	

Instruction Fields:

Register field — Specifies the address register to be updated with the effective address

Effective Address field — Specifies the address to be loaded into the address register. Only control addressing modes are allowed as shown:

Addressing Mode	Mode	Register
Dn	—	—
An	—	—
(An)	010	reg. number:An
(An)+	—	—
-(An)	—	—
(d₁₆,An)	101	reg. number:An
(d₈,An,Xn)	110	reg. number:An

Addressing Mode	Mode	Register
(xxx).W	111	000
(xxx).L	111	001
#(data)	—	—
(d₁₆,PC)	111	010
(d₈,PC,Xn)	111	011

JSR

Jump to Subroutine

JSR

Operation: SP – 4 ⬥ Sp; PC ⬥ (SP)
Destination Address ⬥ PC

Assembler Syntax: JSR ⟨ea⟩

Attributes: Unsized

Description: Pushes the long word address of the instruction immediately following the JSR instruction onto the system stack. Program execution then continues at the address specified in the instruction.

Condition Codes: Not affected

Instruction Format:

15	14	13	12	11	10	9	8	7	6	5	4	3	2	1	0
0	1	0	0	1	1	1	0	1	0		EFFECTIVE ADDRESS MODE			REGISTER	

Instruction Fields:

Effective Address field — Specifies the address of the next instruction. Only control addressing modes are allowed as shown:

Addressing Mode	Mode	Register
Dn	—	—
An	—	—
(An)	010	reg. number:An
(An)+	—	—
-(An)	—	—
(d₁₆,An)	101	reg. number:An
(d₈,An,Xn)	110	reg. number:An

Addressing Mode	Mode	Register
(xxx).W	111	000
(xxx).L	111	001
#(data)	—	—
(d₁₆,PC)	111	010
(d₈,PC,Xn)	111	011

LINK

Link and Allocate

Operation: Sp − 4 ♦ Sp; An ♦ (SP);
SP ♦ An; SP + d ♦ SP

**Assembler
Syntax:** LINK An, #⟨displacement⟩

Attributes: Size = Unsized

Description: Pushes the contents of the specified address register onto the stack. Then loads the updated stack pointer into the address register. Finally, adds the 16-bit sign-extended displacement operand to the stack pointer. The address register occupies one long word on the stack. The user should specify a negative displacement in order to allocate stack area.

Condition Codes:
Not affected

Instruction Format:

15	14	13	12	11	10	9	8	7	6	5	4	3	2	1	0
0	1	0	0	1	1	1	0	0	1	0	1	0		REGISTER	

WORD DISPLACEMENT

Instruction Fields:
Register field — Specifies the address register for the link
Displacement field — Specifies the twos complement integer to be added to the stack pointer

NOTE

LINK and UNLK can be used to maintain a linked list of local data and parameter areas on the stack for nested subroutine calls.

LSL,LSR

Logical Shift

Operation: Destination Shifted by ⟨count⟩ ♦ Destination

**Assembler
Syntax:** LSd Dx,Dy
LSd #⟨data⟩,Dy
LSd ⟨ea⟩
where d is direction, L or R

Attributes: Size = (Byte, Word, Long)

Description: Shifts the bits of the operand in the direction specified (L or R). The carry bit receives the last bit shifted out of the operand. The shift count for the shifting of a register is specified in two different ways:
1. Immediate — The shift count (1-8) is specified in the instruction.
2. Register — The shift count is the value in the data register specified in the instruction modulo 64.

The size of the operation for register destinations may be specified as byte, word, or long. The contents of memory, ⟨ea⟩, can be shifted one bit only, and the operand size is restricted to a word.

The LSL instruction shifts the operand to the left the number of positions specified as the shift count. Bits shifted out of the high order bit go to both the carry and the extend bits; zeros are shifted into the low-order bit.

The LSR instruction shifts the operand to the right the number of positions specified as the shift count. Bits shifted out of the low order bit go to both the carry and the extend bits; zeros are shifted into the high order bit.

LSL,LSR   Logical Shift

LSL,LSR   Logical Shift

Condition Codes:

X	N	Z	V	C
*	*	*	0	*

X — Set according to the last bit shifted out of the operand. Unaffected for a shift count of zero.
N — Set if the result is negative. Cleared otherwise.
Z — Set if the result is zero. Cleared otherwise.
V — Always cleared
C — Set according to the last bit shifted out of the operand. Cleared for a shift count of zero.

Instruction Format (Register Shifts):

15	14	13	12	11	10	9	8	7	6	5	4	3	2	1	0
1	1	1	0	COUNT/REGISTER			dr	SIZE		i/r	0	1	REGISTER		

Instruction Field (Register Shifts):

Count/Register field:
If i/r = 0, this field contains the shift count. The values 1-7 represent shifts of 1-7; value of 0 specifies a shift count of 8.
If i/r = 1, the data register specified in this field contains the shift count (modulo 64).

dr field — Specifies the direction of the shift:
0 — Shift right
1 — Shift left

Size field — Specifies the size of the operation:
00 — Byte operation
01 — Word operation
10 — Long operation

i/r field:
If i/r = 0, specifies immediate shift count
If i/r = 1, specifies register shift count

Register field — Specifies a data register to be shifted

Instruction Format (Memory Shifts):

15	14	13	12	11	10	9	8	7	6	5	4	3	2	1	0
1	1	1	0	0	0	0	dr	1	1	MODE			REGISTER		
										EFFECTIVE ADDRESS					

Instruction Fields (Memory Shifts):

dr field — Specifies the direction of the shift:
0 — Shift right
1 — Shift left

Effective Address field — Specifies the operand to be shifted. Only memory alterable addressing modes are allowed as shown:

Addressing Mode	Mode	Register
Dn	—	—
An	—	—
(An)	010	reg. number:An
(An)+	011	reg. number:An
-(An)	100	reg. number:An
(d$_{16}$,An)	101	reg. number:An
(d$_8$,An,Xn)	110	reg. number:An

Addressing Mode	Mode	Register
(xxx).W	111	000
(xxx).L	111	001
#(data)	—	—
(d$_{16}$,PC)	—	—
(d$_8$,PC,Xn)	—	—

Operation: Source → Destination

Assembler Syntax: MOVE (ea),(ea)

Attributes: Size = (Byte, Word, Long)

Description: Moves the data at the source to the destination location, and sets the condition codes according to the data. The size of the operation may be specified as byte, word, or long.

Condition Codes:

X	N	Z	V	C
—	*	*	0	0

X Not affected
N Set if the result is negative. Cleared otherwise.
Z Set if the result is zero. Cleared otherwise.
V Always cleared
C Always cleared

Instruction Format:

15	14	13	12	11	10	9	8	7	6	5	4	3	2	1	0
0	0	SIZE		REGISTER (DESTINATION)			MODE (DESTINATION)			MODE (SOURCE)			REGISTER (SOURCE)		

Instruction Fields:
Size field — Specifies the size of the operand to be moved:
01 — Byte operation
11 — Word operation
10 — Long operation

Destination Effective Address field — Specifies the destination location. Only data alterable addressing modes are allowed as shown:

Addressing Mode	Mode	Register	Addressing Mode	Mode	Register
Dn	000	reg. number:Dn	(xxx).W	111	000
An	—	—	(xxx).L	111	001
(An)	010	reg. number:An	#(data)	—	—
(An)+	011	reg. number:An			
−(An)	100	reg. number:An			
(d16,An)	101	reg. number:An	(d16,PC)	—	—
(d8,An,Xn)	110	reg. number:An	(d8,PC,Xn)	—	—

Source Effective Address field — Specifies the source operand. All addressing modes are allowed as shown:

Addressing Mode	Mode	Register	Addressing Mode	Mode	Register
Dn	000	reg. number:Dn	(xxx).W	111	000
An*	001	reg. number:An	(xxx).L	111	001
(An)	010	reg. number:An	#(data)	111	100
(An)+	011	reg. number:An			
−(An)	100	reg. number:An			
(d16,An)	101	reg. number:An	(d16,PC)	111	010
(d8,An,Xn)	110	reg. number:An	(d8,PC,Xn)	111	011

*For byte size operation, address register direct is not allowed.

Notes:
1. Most assemblers use MOVEA when the destination is an address register.
2. MOVEQ can be used to move an immediate 8-bit value to a data register.

MOVEA

Move Address

Operation: Source ▶ Destination

Assembler Syntax: MOVEA (ea),An

Attributes: Size = (Word, Long)

Description: Moves the contents of the source to the destination address register. The size of the operation is specified as word or long. Word-size source operands are sign-extended to 32-bit quantities.

Condition Codes:
Not affected

Instruction Format:

15	14	13	12	11	10	9	8	7	6	5	4	3	2	1	0
0	0	SIZE		DESTINATION REGISTER			0	0	1	MODE			REGISTER		
										SOURCE					

Instruction Fields:

Size field — Specifies the size of the operand to be moved:
11 — Word operation. The source operand is sign-extended to a long operand and all 32 bits are loaded into the address register.
10 — Long operation

Destination Register field — Specifies the destination address register

Effective Address field — Specifies the location of the source operand. All addressing modes are allowed as shown:

Addressing Mode	Mode	Register
Dn	000	reg. number:Dn
An	001	reg. number:An
(An)	010	reg. number:An
(An)+	011	reg. number:An
-(An)	100	reg. number:An
(d16,An)	101	reg. number:An
(d8,An,Xn)	110	reg. number:An

Addressing Mode	Mode	Register
(xxx).W	111	000
(xxx).L	111	001
#(data)	111	100
(d16,PC)	111	010
(d8,PC,Xn)	111	011

MOVE from CCR

Move from the Condition Code Register

Operation: CCR ▶ Destination

Assembler Syntax: MOVE CCR,(ea)

Attributes: Size = (Word)

Description: Moves the condition code bits (zero extended to word size) to the destination location. The operand size is a word. Unimplemented bits are read as zeros.

Condition Codes:
Not affected

Instruction Format:

15	14	13	12	11	10	9	8	7	6	5	4	3	2	1	0
0	1	0	0	0	0	1	0	1	1	MODE			REGISTER		
										EFFECTIVE ADDRESS					

Instruction Fields:

Effective Address field — Specifies the destination location. Only data alterable addressing modes are allowed as shown:

Addressing Mode	Mode	Register
Dn	000	reg. number:Dn
An	—	—
(An)	010	reg. number:An
(An)+	011	reg. number:An
-(An)	100	reg. number:An
(d16,An)	101	reg. number:An
(d8,An,Xn)	110	reg. number:An

Addressing Mode	Mode	Register
(xxx).W	111	000
(xxx).L	111	001
#(data)	—	—
(d16,PC)	—	—
(d8,PC,Xn)	—	—

NOTE

MOVE from CCR is a word operation. ANDI, ORI, and EORI to CCR are byte operations.

MOVE
from SR

Move from the Status Register
(Privileged Instruction – MC68010 Only)

Operation: SR ♦ Destination
MC68010 only:
If Supervisor state
then SR ♦ Destination
else TRAP

Assembler
Syntax: MOVE SR,(ea)

Attributes: Size = (Word)

Description: Moves the data in the status register to the destination location. The destination is word length. Unimplemented bits are read as zeros.

Condition Codes:
Not affected

Instruction Format:

15	14	13	12	11	10	9	8	7	6	5	4	3	2	1	0
0	1	0	0	0	0	0	0	1	1		EFFECTIVE ADDRESS				
											MODE			REGISTER	

Instruction Fields:
Effective Address field — Specifies the destination location. Only data alterable addressing modes are allowed as shown:

Addressing Mode	Mode	Register
Dn	000	reg. number:Dn
An	–	–
(An)	010	reg. number:An
(An)+	011	reg. number:An
-(An)	100	reg. number:An
(d16,An)	101	reg. number:An
(d8,An,Xn)	110	reg. number:An

Addressing Mode	Mode	Register
(xxx).W	111	000
(xxx).L	111	001
#(data)	–	–
(d16,PC)	–	–
(d8,PC,Xn)	–	–

NOTE

Use the MOVE from CCR instruction to access only the condition codes. In the MC68000, MC68HC000, and MC68008, memory destination is read before it is written to.

MOVE
to CCR

Move to Condition Codes

Operation: Source ♦ CCR

Assembler
Syntax: MOVE (ea),CCR

Attributes: Size = (Word)

Description: Moves the low-order byte of the source operand to the condition code register. The upper byte of the source operand is ignored; the upper byte of the status register is not altered.

Condition Codes:

X	N	Z	V	C
*	*	*	*	*

X — Set to the value of bit 4 of the source operand
N — Set to the value of bit 3 of the source operand
Z — Set to the value of bit 2 of the source operand
V — Set to the value of bit 1 of the source operand
C — Set to the value of bit 0 of the source operand

Instruction Format:

15	14	13	12	11	10	9	8	7	6	5	4	3	2	1	0
0	1	0	0	0	1	0	0	1	1		EFFECTIVE ADDRESS				
											MODE			REGISTER	

Instruction Fields:
Effective Address field — Specifies the location of the source operand. Only data addressing modes are allowed as shown:

Addressing Mode	Mode	Register
Dn	000	reg. number:Dn
An	–	–
(An)	010	reg. number:An
(An)+	011	reg. number:An
-(An)	100	reg. number:An
(d16,An)	101	reg. number:An
(d8,An,Xn)	110	reg. number:An

Addressing Mode	Mode	Register
(xxx).W	111	000
(xxx).L	111	001
#(data)	111	100
(d16,PC)	111	010
(d8,PC,Xn)	111	011

NOTE

MOVE to CCR is a word operation. ANDI, ORI, and EORI to CCR are byte operations.

MOVE to SR

Move to the Status Register
(Priviledged Instruction)

Operation: If supervisor state
then Source ⊅ SR
else TRAP

Assembler Syntax: MOVE (ea),SR

Attributes: Size = (Word)

Description: Moves the data in the source operand to the status register. The source operand is a word and all implemented bits of the status register are affected.

Condition Codes:
Set according to the source operand

Instruction Format:

15	14	13	12	11	10	9	8	7	6	5	4	3	2	1	0
0	1	0	0	0	1	1	0	1	1	EFFECTIVE ADDRESS					

15	14	13	12	11	10	9	8	7	6	5	4	3	2	1	0
0	1	0	0	0	1	1	0	1	1	MODE			REGISTER		

Instruction Fields:
Effective Address field — Specifies the location of the source operand. Only data addressing modes are allowed as shown:

Addressing Mode	Mode	Register
Dn	000	reg. number: Dn
An	—	—
(An)	010	reg. number: An
(An) -	011	reg. number: An
(An)	100	reg. number: An
(d16,An)	101	reg. number: An
(d8,An,Xn)	110	reg. number: An

Addressing Mode	Mode	Register
(xxx).W	111	000
(xxx).L	111	001
#(data)	111	100
(d16,PC)	111	010
(d8,PC,Xn)	111	011

MOVE USP

Move User Stack Pointer
(Privileged Instruction)

Operation: If supervisor state
then USP ⊅ An or An ⊅ USP
else TRAP

Assembler Syntax: MOVE USP,An
MOVE An,USP

Attributes: Size = (Long)

Description: Moves the contents of the user stack pointer to or from the specified address register

Condition Codes:
Not affected

Instruction Format:

15	14	13	12	11	10	9	8	7	6	5	4	3	2	1	0
0	1	0	0	1	1	1	0	0	1	1	0	dr	REGISTER		

Instruction Fields:
dr field — Specifies the direction of transfer:
0 — Transfer the address register to the USP
1 — Transfer the USP to the address register
Register field — Specifies the address register for the operation

MOVEC

Move Control Register
(Privileged Instruction)

Operation: If supervisor state
then Rc ⬦ Rn or Rn ⬦ Rc
else TRAP

Assembler
Syntax: MOVEC Rc,Rn
MOVEC Rn,Rc

Attributes: Size = (Long)

Description: Moves the contents of the specified control register (Rc) to the specified general register (Rn) or copies the contents of the specified general register to the specified control register. This is always a 32-bit transfer even though the control register may be implemented with fewer bits. Unimplemented bits are read as zeros.

Condition Codes:
Not affected

Instruction Format:

15	14	13	12	11	10	9	8	7	6	5	4	3	2	1	0
0	1	0	0	1	1	1	0	0	1	1	1	1	1	1	dr
A/D	REGISTER			CONTROL REGISTER											

Instruction Fields:

dr field — Specifies the direction of the transfer:
0 — Control register to general register
1 — General register to control register

A/D field — Specifies the type of general register:
0 — Data register
1 — Address register

Register field — Specifies the register number
Control Register field — Specifies the control register

Hex	Control Register
000	Source Function Code (SFC) register
001	Destination Function Code (DFC) register
800	User Stack Pointer (USP)
801	Vector Base Register (VBR)

Any other code causes an illegal instruction exception.

MOVEM

Move Multiple Registers

Operation: Registers ⬦ Destination
Source ⬦ Registers

Assembler
Syntax: MOVEM register list,(ea)
MOVEM (ea),register list

Attributes: Size = (Word, Long)

Description: Moves the contents of selected registers to or from consecutive memory locations starting at the location specified by the effective address. A register is selected if the bit in the mask field corresponding to that register is set. The instruction size determines whether 16 or 32 bits of each register are transferred. In the case of a word transfer to either address or data registers, each word is sign-extended to 32 bits, and the resulting long word is loaded into the associated register.

Selecting the addressing mode also selects the mode of operation of the MOVEM instruction, and only the control modes, the predecrement mode, and the postincrement mode are valid. If the effective address is specified by one of the control modes, the registers are transferred starting at the specified address, and the address is incremented by the operand length (2 or 4) following each transfer. The order of the registers is from data register 0 to data register 7, then from address register 0 to address register 7.

If the effective address is specified by the predecrement mode, only a register to memory operation is allowed. The registers are stored starting at the specified address minus the operand length (2 or 4), and the address is decremented by the operand length following each transfer. The order of storing is from address register 7 to address register 0, then from data register 7 to data register 0. When the instruction has completed, the decremented address register contains the address of the last operand stored.

If the effective address is specified by the postincrement mode, only a memory to register operation is allowed. The registers are loaded starting at the specified address; the address is incremented by the operand length (2 or 4) following each transfer. The order of loading is the same as that of control mode addressing. When the instruction has completed, the incremented address register contains the address of the last operand loaded plus the operand length.

Condition Codes:
Not affected

Instruction Format:

15	14	13	12	11	10	9	8	7	6	5	4	3	2	1	0
0	1	0	0	1	dr	0	0	1	SIZE		MODE		REGISTER		
											EFFECTIVE ADDRESS				

15	14	13	12	11	10	9	8	7	6	5	4	3	2	1	0
						REGISTER LIST MASK									

MOVEM   Move Multiple Registers   MOVEM

Instruction Field:

dr field — Specifies the direction of the transfer:
0 — Register to memory
1 — Memory to register

Size field — Specifies the size of the registers being transferred:
0 — Word transfer
1 — Long transfer

Effective Address field — Specifies the memory address for the operation. For register to memory transfers, only control alterable addressing modes or the predecrement addressing mode are allowed as shown:

Addressing Mode	Mode	Register
Dn	—	—
An	—	—
(An)	010	reg. number:An
(An)+	—	—
-(An)	100	reg. number:An
(d16,An)	101	reg. number:An
(d8,An,Xn)	110	reg. number:An

Addressing Mode	Mode	Register
(xxx).W	111	000
(xxx).L	111	001
#<data>	—	—
(d16,PC)	—	—
(d8,PC,Xn)	—	—

For memory to register transfers, only control addressing modes or the postincrement addressing mode are allowed as shown:

Addressing Mode	Mode	Register
Dn	—	—
An	—	—
(An)	010	reg. number:An
(An)+	011	reg. number:An
-(An)	—	—
(d16,An)	101	reg. number:An
(d8,An,Xn)	110	reg. number:An

Addressing Mode	Mode	Register
(xxx).W	111	000
(xxx).L	111	001
#<data>	—	—
(d16,PC)	111	010
(d8,PC,Xn)	111	011

Register List Mask field — Specifies the registers to be transferred. The low order bit corresponds to the first register to be transferred; the high-order bit corresponds to the last register to be transferred. Thus, both for control modes and for the postincrement mode addresses, the mask correspondence is:

15	14	13	12	11	10	9	8	7	6	5	4	3	2	1	0
A7	A6	A5	A4	A3	A2	A1	A0	D7	D6	D5	D4	D3	D2	D1	D0

MOVEM   Move Multiple Registers   MOVEM

For the predecrement mode addresses, the mask correspondence is reversed:

15	14	13	12	11	10	9	8	7	6	5	4	3	2	1	0
D0	D1	D2	D3	D4	D5	D6	D7	A0	A1	A2	A3	A4	A5	A6	A7

NOTE

An extra read bus cycle occurs for memory operands. This accesses an operand at one address higher than the last register image required.

Operation: Source → Destination

Assembler
Syntax: MOVEP Dx,(d,Ay)
MOVEP (d,Ay),Dx

Attributes: Size = (Word, Long)

Description: Moves data between a data register and alternate bytes within the address space (typically assigned to a peripheral), starting at the location specified and incrementing by two. This instruction is designed for 8-bit peripherals on a 16-bit data bus. The high-order byte of the data register is transferred first and the low order byte is transferred last. The memory address is specified in the address register indirect plus 16-bit displacement addressing mode. If the address is even, all the transfers are to or from the high order half of the data bus; if the address is odd, all the transfers are to or from the low order half of the data bus. The instruction also accesses alternate bytes on an 8-bit bus.

Example: Long transfer to/from an even address

Byte Organization in Register

31	24	23	16	15	8	7	0
HI-ORDER		MID-UPPER		MID-LOWER		LOW-ORDER	

Byte Organization in Memory (Low Address at Top)

15	8	7	0
HI-ORDER			
MID-UPPER			
MID-LOWER			
LOW-ORDER			

Example: Word transfer to/from an odd address

Byte Organization in Register

31	24	23	16	15	8	7	0
				HI-ORDER		LOW-ORDER	

Byte Organization in Memory (Low Address at Top)

15	8	7	0
HI-ORDER			
LOW-UPPER			

Condition Codes:
Not affected

Instruction Format:

15	14	13	12	11	10	9	8	7	6	5	4	3	2	1	0
0	0	0	0	DATA REGISTER			OP-MODE			0	0	1	ADDRESS REGISTER		
DISPLACEMENT (16 BITS)															

Instruction Fields:
Data Register field — Specifies the data register for the instruction
Op-Mode field — Specifies the direction and size of the operation:
100 — Transfer word from memory to register
101 — Transfer long from memory to register
110 — Transfer word from register to memory
111 — Transfer long from register to memory
Address Register field — Specifies the address register which is used in the address register indirect plus displacement addressing mode
Displacement field — Specifies the displacement used in the operand address

MOVEQ   Move Quick   **MOVEQ**

Operation: Immediate Data ♦ Destination

Assembler Syntax: MOVEQ #⟨data⟩,Dn

Attributes: Size = (Long)

Description: Moves a byte of immediate data to a 32-bit data register. The data in an 8-bit field within the operation word is sign extended to a long operand in the data register as it is transferred.

Condition Codes:

X	N	Z	V	C
—	*	*	0	0

X Not affected
N Set if the result is negative. Cleared otherwise.
Z Set if the result is zero. Cleared otherwise.
V Always cleared
C Always cleared

Instruction Format:

15	14	13	12	11	10	9	8	7	6	5	4	3	2	1	0
0	1	1	1	REGISTER			0	DATA							

Instruction Fields:
Register field — Specifies the data register to be loaded
Data field — 8 bits of data, which are sign extended to a long operand

MOVES   Move Address Space   **MOVES**
(Privileged Instruction)

Operation: If supervisor state
then Rn ♦ Destination [DFC] or Source [SFC] ♦ Rn
else TRAP

Assembler Syntax: MOVES Rn,⟨ea⟩
MOVES ⟨ea⟩,Rn

Attributes: Size = (Byte, Word, Long)

Description: Moves the byte, word, or long operand from the specified general register to a location within the address space specified by the destination function code (DFC) register; or, moves the byte, word, or long operand from a location within the address space specified by the source function code (SFC) register to the specified general register.

If the destination is a data register, the source operand replaces the corresponding low-order bits of that data register, depending on the size of the operation. If the destination is an address register, the source operand is sign extended to 32 bits and then loaded into that address register.

Condition Codes:
Not affected

Instruction Format:

15	14	13	12	11	10	9	8	7	6	5	4	3	2	1	0
0	0	0	0	1	1	1	0	SIZE		EFFECTIVE ADDRESS					
										MODE			REGISTER		
A/D	REGISTER			dr	0	0	0	0	0	0	0	0	0	0	0

Instruction Fields:
Size field — Specifies the size of the operation:
00 — Byte operation
01 — Word operation
10 — Long operation

MULS Signed Multiply MULS

Operation: Source * Destination ♦ Destination

Assembler Syntax: MULS.W (ea),Dn 16×16 ♦ 32

Attributes: Size = (Word)

Description: Multiplies two signed operands yielding a signed result. The multiplier and multiplicand are both word operands, and the result is a long word operand. A register operand is the low order word; the upper word of the register is ignored. All 32 bits of the product are saved in the destination data register.

Condition Codes:

X	N	Z	V	C
—	*	*	0	0

X Not affected
N Set if the result is negative. Cleared otherwise.
Z Set if the result is zero. Cleared otherwise.
V Always cleared
C Always cleared

Instruction Format (word form):

15	14	13	12	11	10	9	8	7	6	5	4	3	2	1	0
1	1	0	0		REGISTER		1	1	1			EFFECTIVE ADDRESS MODE REGISTER			

Instruction Fields:
Register field — Specifies a data register as the destination
Effective Address field — Specifies the source operand. Only data addressing modes are allowed as shown:

Addressing Mode	Mode	Register
Dn	000	reg. number:Dn
An	—	—
(An)	010	reg. number:An
(An)+	011	reg. number:An
−(An)	100	reg. number:An
(d16,An)	101	reg. number:An
(d8,An,Xn)	110	reg. number:An

Addressing Mode	Mode	Register
(xxx).W	111	000
(xxx).L	111	001
#(data)	111	100
(d16,PC)	111	010
(d8,PC,Xn)	111	011

MOVES Move Address Space MOVES
(Privileged Instruction)

Effective Address Field — Specifies the source or destination location within the alternate address space. Only memory alterable addressing modes are allowed as shown:

Addressing Mode	Mode	Register
Dn	—	—
An	—	—
(An)	010	reg. number:An
(An)+	011	reg. number:An
−(An)	100	reg. number:An
(d16,An)	101	reg. number:An
(d8,An,Xn)	110	reg. number:An

Addressing Mode	Mode	Register
(xxx).W	111	000
(xxx).L	111	001
#(data)	—	—
(d16,PC)	—	—
(d8,PC,Xn)	—	—

A/D field — Specifies the type of general register:
0 — Data register
1 — Address register
Register field — Specifies the register number
dr field — Specifies the direction of the transfer:
0 — From (ea) to general register
1 — From general register to (ea)

NOTE

For either of the two following examples with the same address register as both source and destination
MOVES.x An,(An)+
MOVES.x An,−(An)
the value stored is undefined. The current implementation of the MC68010 stores the incremented or decremented value of An.

MULU

Unsigned Multiply

NBCD

Negate Decimal with Extend

NBCD

Operation: Source * Destination ♦ Destination $16 \times 16 \blacklozenge 32$

Operation: $0 - (\text{Destination}_{10}) - X \blacklozenge$ Destination

Assembler Syntax: MULU.W (ea),Dn

Assembler Syntax: NBCD (ea)

Attributes: Size = (Word)

Attributes: Size = (Byte)

Description: Multiplies two unsigned operands yielding an unsigned result. The multiplier and multiplicand are both word operands, and the result is a long word operand. A register operand is the low-order word; the upper word of the register is ignored. All 32 bits of the product are saved in the destination data register.

Description: Subtracts the destination operand and the extend bit from zero. The operation is performed using binary coded decimal arithmetic. The packed BCD result is saved in the destination location. This instruction produces the tens complement of the destination if the extend bit is zero, or the nines complement if the extend bit is one. This is a byte operation only.

Condition Codes:

X	N	Z	V	C
–	*	*	0	0

X Not affected
N Set if the result is negative. Cleared otherwise.
Z Set if the result is zero. Cleared otherwise.
V Always cleared
C Always cleared

Condition Codes:

X	N	Z	V	C
*	U	*	U	*

X Set the same as the carry bit
N Undefined
Z Cleared if the result is non-zero. Unchanged otherwise.
V Undefined
C Set if a decimal borrow occurs. Cleared otherwise.

NOTE

Normally the Z condition code bit is set via programming before the start of the operation. This allows successful tests for zero results upon completion of multiple precision operations.

Instruction Format (word form):

15	14	13	12	11	10	9	8	7	6	5	4	3	2	1	0
1	1	0	0	REGISTER			0	1	1	MODE			REGISTER		
										EFFECTIVE ADDRESS					

Instruction Format:

15	14	13	12	11	10	9	8	7	6	5	4	3	2	1	0
0	1	0	0	1	0	0	0	0	0	MODE			REGISTER		
										EFFECTIVE ADDRESS					

Instruction Fields:

Register field — Specifies a data register as the destination
Effective Address field — Specifies the source operand. Only data addressing modes are allowed as shown:

Addressing Mode	Mode	Register		Addressing Mode	Mode	Register
Dn	000	reg. number:Dn		(xxx).W	111	000
An	–	–		(xxx).L	111	001
(An)	010	reg. number:An		#(data)	111	100
(An) +	011	reg. number:An				
– (An)	100	reg. number:An				
(d16,An)	101	reg. number:An		(d16,PC)	111	010
(d8,An,Xn)	110	reg. number:An		(d8,PC,Xn)	111	011

Instruction Fields:

Effective Address field — Specifies the destination operand. Only data alterable addressing modes are allowed as shown:

Addressing Mode	Mode	Register		Addressing Mode	Mode	Register
Dn	000	reg. number:Dn		(xxx).W	111	000
An	–	–		(xxx).L	111	001
(An)	010	reg. number:An		#(data)	–	–
(An) +	011	reg. number:An				
– (An)	100	reg. number:An				
(d16,An)	101	reg. number:An		(d16,PC)	–	–
(d8,An,Xn)	110	reg. number:An		(d8,PC,Xn)	–	–

NEG

Negate

NEG

Operation: 0 − (Destination) ⟶ Destination

Assembler Syntax: NEG ⟨ea⟩

Attributes: Size = (Byte, Word, Long)

Description: Subtracts the destination operand from zero and stores the result in the destination location. The size of the operation is specified as byte, word, or long.

Condition Codes:

X	N	Z	V	C
*	*	*	*	*

X Set the same as the carry bit
N Set if the result is negative. Cleared otherwise.
Z Set if the result is zero. Cleared otherwise.
V Set if an overflow occurs. Cleared otherwise.
C Cleared if the result is zero. Set otherwise.

Instruction Format:

15	14	13	12	11	10	9	8	7	6	5	4	3	2	1	0
0	1	0	0	0	1	0	0	SIZE		EFFECTIVE ADDRESS					
										MODE			REGISTER		

Instruction Fields:

Size field — Specifies the size of the operation:
00 — Byte operation
01 — Word operation
10 — Long operation

Effective Address field — Specifies the destination operand. Only data alterable addressing modes are allowed as shown:

Addressing Mode	Mode	Register
Dn	000	reg. number: Dn
An	—	—
(An)	010	reg. number: An
(An)+	011	reg. number: An
−(An)	100	reg. number: An
(d16,An)	101	reg. number: An
(d8,An,Xn)	110	reg. number: An

NEGX

Negate with Extend

NEGX

Operation: 0 − (Destination) − X ⟶ Destination

Assembler Syntax: NEGX ⟨ea⟩

Attributes: Size = (Byte, Word, Long)

Description: Subtracts the destination operand and the extend bit from zero. Stores the result in the destination location. The size of the operation is specified as byte, word, or long.

Condition Codes:

X	N	Z	V	C
*	*	*	*	*

X Set the same as the carry bit
N Set if the result is negative. Cleared otherwise.
Z Cleared if the result is non-zero. Unchanged otherwise.
V Set if an overflow occurs. Cleared otherwise.
C Set if a borrow occurs. Cleared otherwise.

NOTE

Normally the Z condition code bit is set via programming before the start of the operation. This allows successful tests for zero results upon completion of multiple precision operations.

Instruction Format:

15	14	13	12	11	10	9	8	7	6	5	4	3	2	1	0
0	1	0	0	0	0	0	0	SIZE		EFFECTIVE ADDRESS					
										MODE			REGISTER		

Instruction Fields:

Size field — Specifies the size of the operation:
00 — Byte operation
01 — Word operation
10 — Long operation

Addressing Mode	Mode	Register
(xxx).W	111	000
(xxx).L	111	001
#⟨data⟩	—	—
(d16,PC)	—	—
(d8,PC,Xn)	—	—

NEGX

Negate with Extend

Effective Address field — Specifies the destination operand. Only data alterable addressing modes are allowed as shown:

Addressing Mode	Mode	Register
Dn	000	reg. number:Dn
An	—	—
(An)	010	reg. number:An
(An) +	011	reg. number:An
– (An)	100	reg. number:An
(d₁₆,An)	101	reg. number:An
(d₈,An,Xn)	110	reg. number:An

Addressing Mode	Mode	Register
(xxx).W	111	000
(xxx).L	111	001
#(data)	—	—
(d₁₆,PC)	—	—
(d₈,PC,Xn)	—	—

NOP

No Operation

Operation: None

Assembler Syntax: NOP

Attributes: Unsized

Description: Performs no operation. The processor state, other than the program counter, is unaffected. Execution continues with the instruction following the NOP instruction.

Condition Codes:
Not affected

Instruction Format:

15	14	13	12	11	10	9	8	7	6	5	4	3	2	1	0
0	1	0	0	1	1	1	0	0	1	1	1	0	0	0	1

504

NOT

Logical Complement

NOT

Operation: ~ Destination ♦ Destination

Assembler Syntax: NOT ⟨ea⟩

Attributes: Size = (Byte, Word, Long)

Description: Calculates the ones complement of the destination operand and stores the result in the destination location. The size of the operation is specified as byte, word, or long.

Condition Codes:

X	N	Z	V	C
–	*	*	0	0

X Not affected
N Set if the result is negative. Cleared otherwise.
Z Set if the result is zero. Cleared otherwise.
V Always cleared
C Always cleared

Instruction Format:

15	14	13	12	11	10	9	8	7	6	5	4	3	2	1	0
0	1	0	0	0	1	1	0	SIZE		EFFECTIVE ADDRESS					
										MODE			REGISTER		

Instruction Fields:

Size field — Specifies the size of the operation:
00 — Byte operation
01 — Word operation
10 — Long operation

Effective Address field — Specifies the destination operand. Only data alterable addressing modes are allowed as shown:

Addressing Mode	Mode	Register
Dn	000	reg. number:Dn
An	–	–
(An)	010	reg. number:An
(An)+	011	reg. number:An
–(An)	100	reg. number:An
(d₁₆,An)	101	reg. number:An
(d₈,An,Xn)	110	reg. number:An

Addressing Mode	Mode	Register
(xxx).W	111	000
(xxx).L	111	001
#⟨data⟩	–	–
(d₁₆,PC)	–	–
(d₈,PC,Xn)	–	–

OR

Inclusive OR Logical

OR

Operation: Source V Destination ♦ Destination

Assembler Syntax: OR ⟨ea⟩,Dn
OR Dn,⟨ea⟩

Attributes: Size = (Byte, Word, Long)

Description: Performs an inclusive OR operation on the source operand and the destination operand and stores the result in the destination location. The size of the operation is specified as byte, word, or long. The contents of an address register may not be used as an operand.

Condition Codes:

X	N	Z	V	C
–	*	*	0	0

X Not affected
N Set if the most significant bit of the result is set. Cleared otherwise.
Z Set if the result is zero. Cleared otherwise.
V Always cleared
C Always cleared

Instruction Format:

15	14	13	12	11	10	9	8	7	6	5	4	3	2	1	0
1	0	0	0	REGISTER			OP-MODE			EFFECTIVE ADDRESS					
										MODE			REGISTER		

Instruction Fields:

Register field — Specifies any of the eight data registers

Op-Mode field —

Byte	Word	Long	Operation
000	001	010	(⟨ea⟩) V (⟨Dn⟩) ♦ ⟨Dn⟩
100	101	110	(⟨Dn⟩) V (⟨ea⟩) ♦ ⟨ea⟩

OR

Inclusive OR Logical

OR

Effective Address field — If the location specified is a source operand, only data addressing modes are allowed as shown:

Addressing Mode	Mode	Register
Dn	000	reg. number:Dn
An	—	—
(An)	010	reg. number:An
(An)+	011	reg. number:An
−(An)	100	reg. number:An
(d₁₆,An)	101	reg. number:An
(d₈,An,Xn)	110	reg. number:An

Addressing Mode	Mode	Register
(xxx).W	111	000
(xxx).L	111	001
#⟨data⟩	111	100
(d₁₆,PC)	111	010
(d₈,PC,Xn)	111	011

If the location specified is a destination operand, only memory alterable addressing modes are allowed as shown:

Addressing Mode	Mode	Register
Dn	—	—
An	—	—
(An)	010	reg. number:An
(An)+	—	—
−(An)	100	reg. number:An
(d₁₆,An)	101	reg. number:An
(d₈,An,Xn)	110	reg. number:An

Addressing Mode	Mode	Register
(xxx).W	111	000
(xxx).L	111	001
#⟨data⟩	—	—
(d₁₆,PC)	—	—
(d₈,PC,Xn)	—	—

Notes:
1. If the destination is a data register, it must be specified using the destination Dn mode, not the destination ⟨ea⟩ mode.
2. Most assemblers use ORI when the source is immediate data.

ORI

Inclusive OR

ORI

Operation: Immediate Data V Destination → Destination

Assembler Syntax: ORI #⟨data⟩,⟨ea⟩

Attributes: Size = (Byte, Word, Long)

Description: Performs an inclusive OR operation on the immediate data and the destination operand and stores the result in the destination location. The size of the operation is specified as byte, word, or long. The size of the immediate data matches the operation size.

Condition Codes:

X	N	Z	V	C
—	*	*	0	0

X Not affected
N Set if the most significant bit of the result is set. Cleared otherwise.
Z Set if the result is zero. Cleared otherwise.
V Always cleared
C Always cleared

Instruction Format:

15	14	13	12	11	10	9	8	7	6	5	4	3	2	1	0
0	0	0	0	0	0	0	0		SIZE			EFFECTIVE ADDRESS			
												MODE			REGISTER
WORD DATA (16 BITS)												BYTE DATA (8 BITS)			
LONG DATA (32 BITS)															

Instruction Fields:
Size field — Specifies the size of the operation.
00 — Byte operation.
01 — Word operation.
10 — Long operation.

Inclusive OR

Effective Address field — Specifies the destination operand. Only data alterable addressing modes are allowed as shown:

Addressing Mode	Mode	Register
Dn	000	reg. number:Dn
An	—	—
(An)	010	reg. number:An
(An)+	011	reg. number:An
-(An)	100	reg. number:An
(d16,An)	101	reg. number:An
(d8,An,Xn)	110	reg. number:An

Addressing Mode	Mode	Register
(xxx).W	111	000
(xxx).L	111	001
#<data>	—	—
(d16,PC)	—	—
(d8,PC,Xn)	—	—

Immediate field — (Data immediately following the instruction):
If size = 00, the data is the low-order byte of the immediate word
If size = 01, the data is the entire immediate word
If size = 10, the data is the next two immediate words

Inclusive OR Immediate
to Condition Codes

Operation: Source V CCR ▸ CCR

**Assembler
Syntax:** ORI #(data),CCR

Attributes: Size = (Byte)

Description: Performs an inclusive OR operation on the immediate operand and the condition codes and stores the result in the condition code register (low-order byte of the status register). All implemented bits of the condition code register are affected.

Condition Codes:

X	N	Z	V	C
*	*	*	*	*

X Set if bit 4 of immediate operand is one. Unchanged otherwise.
N Set if bit 3 of immediate operand is one. Unchanged otherwise.
Z Set if bit 2 of immediate operand is one. Unchanged otherwise.
V Set if bit 1 of immediate operand is one. Unchanged otherwise.
C Set if bit 0 of immediate operand is one. Unchanged otherwise.

Instruction Format:

15	14	13	12	11	10	9	8	7	6	5	4	3	2	1	0
0	0	0	0	0	0	0	0	0	0	1	1	1	1	0	0
0	0	0	0	0	0	0	0	BYTE DATA (8 BITS)						0	0

ORI to SR

Inclusive OR Immediate to the Status Register
(Privileged Instruction)

Operation: If supervisor state
then Source V SR → SR
else TRAP

Assembler Syntax: ORI #⟨data⟩,SR

Attributes: Size = (Word)

Description: Performs an inclusive OR operation of the immediate operand and the contents of the status register and stores the result in the status register. All implemented bits of the status register are affected.

Condition Codes:

X	N	Z	V	C
*	*	*	*	*

X Set if bit 4 of immediate operand is one. Unchanged otherwise.
N Set if bit 3 of immediate operand is one. Unchanged otherwise.
Z Set if bit 2 of immediate operand is one. Unchanged otherwise.
V Set if bit 1 of immediate operand is one. Unchanged otherwise.
C Set if bit 0 of immediate operand is one. Unchanged otherwise.

Instruction Format:

15	14	13	12	11	10	9	8	7	6	5	4	3	2	1	0
0	0	0	0	0	0	0	0	0	1	1	1	1	1	0	0

WORD DATA (16 BITS)

PEA

Push Effective Address

Operation: Sp − 4 → SP; ⟨ea⟩ → (SP)

Assembler Syntax: PEA ⟨ea⟩

Attributes: Size = (Long)

Description: Computes the effective address and pushes it onto the stack. The effective address is a long word address.

Condition Codes: Not affected

Instruction Format:

15	14	13	12	11	10	9	8	7	6	5	4	3	2	1	0
0	1	0	0	1	0	0	0	0	1		MODE			REGISTER	

(bits 5–0: EFFECTIVE ADDRESS — MODE, REGISTER)

Instruction Fields:

Effective Address field — Specifies the address to be pushed onto the stack. Only control addressing modes are allowed as shown:

Addressing Mode	Mode	Register	Addressing Mode	Mode	Register
Dn	—	—	(xxx).W	111	000
An	—	—	(xxx).L	111	001
(An)	010	reg. number:An	#⟨data⟩	—	—
(An)+	—	—			
-(An)	—	—			
(d16,An)	101	reg. number:An	(d16,PC)	111	010
(d8,An,Xn)	110	reg. number:An	(d8,PC,Xn)	111	011

RESET

RESET

Reset External Devices
(Privileged Instruction)

Operation: If supervisor state
then Assert RESET Line
else TRAP

Assembler
Syntax: RESET

Attributes: Unsized

Description: Asserts the RESET signal for 124 clock periods, resetting all external devices. The processor state, other than the program counter, is unaffected and execution continues with the next instruction.

Condition Codes:
Not affected

Instruction Format:

15	14	13	12	11	10	9	8	7	6	5	4	3	2	1	0
0	1	0	0	1	1	1	0	0	1	1	1	0	0	0	0

ROL / ROR

ROL
ROR

Rotate (Without Extend)

Operation: Destination Rotated by ⟨count⟩ → Destination

Assembler
Syntax: ROd Dx,Dy
ROd #⟨data⟩,Dy
ROd ⟨ea⟩
where d is direction, L or R

Attributes: Size = (Byte, Word, Long)

Description: Rotates the bits of the operand in the direction specified (L or R). The extend bit is not included in the rotation. The rotate count for the rotation of a register is specified in either of two ways:
1. Immediate — The rotate count (1-8) is specified in the instruction.
2. Register — The rotate count is the value in the data register specified in the instruction, modulo 64.

The size of the operation for register destinations is specified as byte, word, or long. The contents of memory, ⟨ea⟩, can be rotated one bit only, and operand size is restricted to a word.

The ROL instruction rotates the bits of the operand to the left; the rotate count determines the number of bit positions rotated. Bits rotated out of the high-order bit go to the carry bit and also back into the low-order bit.

The ROR instruction rotates the bits of the operand to the right; the rotate count determines the number of bit positions rotated. Bits rotated out of the low-order bit go to the carry bit and also back into the high-order bit.

ROL
ROR

Rotate (Without Extend)

Condition Codes:

X	N	Z	V	C
—	.	.	0	.

X Not affected
N Set if the most significant bit of the result is set. Cleared otherwise.
Z Set if the result is zero. Cleared otherwise.
V Always cleared
C Set according to the last bit rotated out of the operand. Cleared when the rotate count is zero.

Instruction Format (Register Rotate):

15	14	13	12	11	10	9	8	7	6	5	4	3	2	1	0
1	1	1	0	COUNT/REGISTER			dr	SIZE		i:r	1	1	REGISTER		

Instruction Fields (Register Rotate):

Count/Register field:
If i:r = 0, this field contains the rotate count. The values 1-7 represent counts of 1-7, and 0 specifies a count of 8.
If i:r = 1, this field specifies a data register that contains the rotate count (modulo 64).
dr field — Specifies the direction of the rotate:
0 — Rotate right
1 — Rotate left
Size field — Specifies the size of the operation:
00 — Byte operation
01 — Word operation
10 — Long operation
i:r field — Specifies the rotate count location:
If i:r = 0, immediate rotate count
If i:r = 1, register rotate count
Register field — Specifies a data register to be rotated

Instruction Format (Memory Rotate):

15	14	13	12	11	10	9	8	7	6	5	4	3	2	1	0
1	1	1	0	0	1	1	dr	1	1	EFFECTIVE ADDRESS					
										MODE			REGISTER		

ROL
ROR

Rotate (Without Extend)

Instruction Fields (Memory Rotate):

dr field — Specifies the direction of the rotate:
0 — Rotate right
1 — Rotate left
Effective Address field — Specifies the operand to be rotated. Only memory alterable addressing modes are allowed as shown:

Addressing Mode	Mode	Register
Dn	—	—
An	—	—
(An)	010	reg. number:An
(An)+	011	reg. number:An
-(An)	100	reg. number:An
(d16,An)	101	reg. number:An
(d8,An,Xn)	110	reg. number:An

Addressing Mode	Mode	Register
(xxx).W	111	000
(xxx).L	111	001
#(data)	—	—
(d16,PC)	—	—
(d8,PC,Xn)	—	—

ROXL
ROXR

Rotate with Extend

ROXL
ROXR

Operation: Destination Rotated with X by ⟨count⟩ ➧ Destination

Assembler Syntax:
ROXd Dx,Dy
ROXd #⟨data⟩,Dy
ROXd ⟨ea⟩
where d is direction, L or R

Attributes: Size = (Byte, Word, Long)

Description: Rotates the bits of the operand in the direction specified (L or R). The extend bit is included in the rotation. The rotate count for the rotation of a register is specified in either of two ways:

1. Immediate — The rotate count (1-8) is specified in the instruction.
2. Register — The rotate count is the value in the data register specified in the instruction, modulo 64.

The size of the operation for register destinations is specified as byte, word, or long. The contents of memory, ⟨ea⟩, can be rotated one bit only, and operand size is restricted to a word.

The ROXL instruction rotates the bits of the operand to the left; the rotate count determines the number of bit positions rotated. Bits rotated out of the high-order bit go to the carry bit and the extend bit; the previous value of the extend bit rotates into the low-order bit.

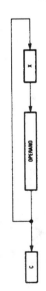

The ROXR instruction rotates the bits of the operand to the right; the rotate count determines the number of bit positions rotated. Bits rotated out of the low order bit go to the carry bit and the extend bit; the previous value of the extend bit rotates into the high order bit.

ROXL
ROXR

Rotate with Extend

ROXL
ROXR

Condition Codes:

X	N	Z	V	C
*	*	*	0	*

X — Set to the value of the last bit rotated out of the operand. Unaffected when the rotate count is zero.
N — Set if the most significant bit of the result is set. Cleared otherwise.
Z — Set if the result is zero. Cleared otherwise.
V — Always cleared
C — Set according to the last bit rotated out of operand. When the rotate count is zero, set to the value of the extend bit.

Instruction Format (Register Rotate):

15	14	13	12	11	10	9	8	7	6	5	4	3	2	1	0
1	1	1	0	COUNT/ REGISTER			dr	SIZE		i/r	1	0	REGISTER		

Instruction Fields (Register Rotate):

Count/Register field:
If i/r = 0, this field contains the rotate count. The values 1-7 represent counts of 1-7, and 0 specifies a count of 8.
If i/r = 1, this field specifies a data register that contains the rotate count (modulo 64).

dr field — Specifies the direction of the rotate:
0 — Rotate right
1 — Rotate left

Size field — Specifies the size of the operation:
00 — Byte operation
01 — Word operation
10 — Long operation

i/r field — Specifies the rotate count location:
If i/r = 0, immediate rotate count
If i/r = 1, register rotate count

Register field — Specifies a data register to be rotated

Instruction Format (Memory Rotate):

15	14	13	12	11	10	9	8	7	6	5	4	3	2	1	0
1	1	1	0	0	1	0	dr	1	1	MODE			REGISTER		
										EFFECTIVE ADDRESS					

Instruction Fields (Memory Rotate):

dr field — Specifies the direction of the rotate:
0 — Rotate right
1 — Rotate left

ROXL
ROXR

Rotate with Extend

Effective Address field — Specifies the operand to be rotated. Only memory alterable addressing modes are allowed as shown:

Addressing Mode	Mode	Register
Dn	—	—
An	—	—
(An)	010	reg. number:An
(An)+	011	reg. number:An
-(An)	100	reg. number:An
(d16,An)	101	reg. number:An
(d8,An,Xn)	110	reg. number:An

Addressing Mode	Mode	Register
(xxx).W	111	000
(xxx).L	111	001
#(data)	—	—
(d16,PC)	—	—
(d8,PC,Xn)	—	—

RTD

Return and Deallocate

Operation: (SP) ⬧ PC; SP + 4 + d ⬧ SP

Assembler Syntax: RTD #(displacement)

Attributes: Unsized

Description: Pulls the program counter value from the stack and adds the sign-extended 16-bit displacement value to the stack pointer. The previous program counter value is lost.

Condition Codes:
Not affected

Instruction Format:

15	14	13	12	11	10	9	8	7	6	5	4	3	2	1	0
0	1	0	0	1	1	1	0	0	1	1	1	0	1	0	0
						DISPLACEMENT (16 BITS)									

Instruction Field:
Displacement field — Specifies the twos complement integer to be sign extended and added to the stack pointer

RTE

**Return from Exception
(Privileged Instruction)**

Operation: If supervisor state
then (SP) ⬧ SR; SP + 2 ⬧ SP; (SP) ⬧ PC;
SP + 4 ⬧ SP;
restore state and deallocate stack according to (SP)
else TRAP

**Assembler
Syntax:** RTE

Attributes: Unsized

Description: Loads the processor state information stored in the exception stack frame located at the top of the stack into the processor. The instruction examines the stack format field in the format/offset word to determine how much information must be restored.

Condition Codes:
Set according to the condition code bits in the status register value restored from the stack

Instruction Format:

15	14	13	12	11	10	9	8	7	6	5	4	3	2	1	0
0	1	0	0	1	1	1	0	0	1	1	1	0	0	1	1

Format/Offset word (in stack frame):

15	14	13	12	11	10	9	8	7	6	5	4	3	2	1	0
FORMAT				0	0				VECTOR OFFSET						

Format Field of Format/Offset Word:
Contains the format code, which implies the stack frame size (including the format/offset word):

0000 — Short Format, removes four words. Loads the status register and the program counter from the stack frame.

1000 — MC68010 Long Format, removes 29 words

Any other value in this field causes the processor to take a format error exception.

RTR

Return and Restore Condition Codes

Operation: (SP) ⬧ CCR; SP + 2 ⬧ SP;
(SP) ⬧ PC; SP + 4 ⬧ SP

**Assembler
Syntax:** RTR

Attributes: Unsized

Description: Pulls the condition code and program counter values from the stack. The previous condition codes and program counter values are lost. The supervisor portion of the status register is unaffected.

Condition Codes:
Set to the condition codes from the stack

Instruction Format:

15	14	13	12	11	10	9	8	7	6	5	4	3	2	1	0
0	1	0	0	1	1	1	0	0	1	1	1	0	1	1	1

RTS

Return from Subroutine

Operation: (SP) ▶ PC; SP + 4 ▶ SP

Assembler
Syntax: RTS

Attributes: Unsized

Description: Pulls the program counter value from the stack. The previous program counter value is lost.

Condition Codes:
Not affected

Instruction Format:

15	14	13	12	11	10	9	8	7	6	5	4	3	2	1	0
0	1	0	0	1	1	1	0	0	1	1	1	0	1	0	1

SBCD

Subtract Decimal with Extend

Operation: Destination$_{10}$ − Source$_{10}$ − X ▶ Destination

Assembler SBCD Dx,Dy
Syntax: SBCD −(Ax), −(Ay)

Attributes: Size = (Byte)

Description: Subtracts the source operand and the extend bit from the destination operand and stores the result in the destination location. The subtraction is performed using binary coded decimal arithmetic; the operands are packed BCD numbers. The instruction has two modes:
1. Data register to data register: The data registers specified in the instruction contain the operands.
2. Memory to memory: The address registers specified in the instruction access the operands from memory using the predecrement addressing mode.

This operation is a byte operation only.

Condition Codes:

X	N	Z	V	C
*	U	*	U	*

X Set the same as the carry bit
N Undefined
Z Cleared if the result is non-zero. Unchanged otherwise.
V Undefined
C Set if a borrow (decimal) is generated. Cleared otherwise.

NOTE

Normally the Z condition code bit is set via programming before the start of an operation. This allows successful tests for zero results upon completion of multiple-precision operations.

Instruction Format:

15	14	13	12	11	10	9	8	7	6	5	4	3	2	1	0
1	0	0	0	REGISTER Ry			1	0	0	0	0	R/M	REGISTER Rx		

Instruction Fields:
Register Ry field — Specifies the destination register
If R/M = 0, specifies a data register
If R/M = 1, specifies an address register for the predecrement addressing mode
R/M field — Specifies the operand addressing mode:
0 — The operation is data register to data register
1 — The operation is memory to memory
Register Rx field — Specifies the source register:
If R/M = 0, specifies a data register
If R/M = 1, specifies an address register for the predecrement addressing mode

Scc

Set According to Condition

Scc

Operation: If Condition True
then 1s ⬧ Destination
else 0s ⬧ Destination

Assembler
Syntax: Scc ⟨ea⟩

Attributes: Size = (Byte)

Description: Tests the specified condition code; if the condition is true, sets the byte specified by the effective address to TRUE (all ones). Otherwise, sets that byte to FALSE (all zeros). Condition code cc specifies one of the following conditions:

CC	carry clear	0100	$\overline{C}$	LS	low or same	0011	$C+Z$
CS	carry set	0101	C	LT	less than	1101	$N\cdot\overline{V}+\overline{N}\cdot V$
EQ	equal	0111	Z	MI	minus	1011	N
F	never true	0001	0	NE	not equal	0110	$\overline{Z}$
GE	greater or equal	1100	$N\cdot V+\overline{N}\cdot\overline{V}$	PL	plus	1010	$\overline{N}$
GT	greater than	1110	$N\cdot V\cdot\overline{Z}+\overline{N}\cdot\overline{V}\cdot\overline{Z}$	T	always true	0000	1
HI	high	0010	$\overline{C}\cdot\overline{Z}$	VC	overflow clear	1000	$\overline{V}$
LE	less or equal	1111	$Z+N\cdot\overline{V}+\overline{N}\cdot V$	VS	overflow set	1001	V

Condition Codes:
Not affected

Instruction Format:

15	14	13	12	11	10	9	8	7	6	5	4	3	2	1	0
0	1	0	1		CONDITION			1	1		EFFECTIVE ADDRESS				
											MODE			REGISTER	

Instruction Fields:
Condition field — The binary code for one of the conditions listed in the table
Effective Address field — Specifies the location in which the true/false byte is to be stored. Only data alterable addressing modes are allowed as shown:

Addressing Mode	Mode	Register
Dn	000	reg. number:Dn
An	—	—
(An)	010	reg. number:An
(An)+	011	reg. number:An
-(An)	100	reg. number:An
(d₁₆,An)	101	reg. number:An
(d₈,An,Xn)	110	reg. number:An

Addressing Mode	Mode	Register
(xxx).W	111	000
(xxx).L	111	001
#⟨data⟩	—	—
(d₁₆,PC)	—	—
(d₈,PC,Xn)	—	—

Note: A subsequent NEG.B instruction with the same effective address can be used to change the Scc result from TRUE or FALSE to the equivalent arithmetic value (TRUE = 1, FALSE = 0). In the MC68000, MC68HC000, and MC68008 a memory destination is read before it is written to.

STOP

Load Status Register and Stop
(Privileged Instruction)

STOP

Operation: If supervisor state
then Immediate Data ⬧ SR; STOP
else TRAP

Assembler
Syntax: STOP #⟨data⟩

Attributes: Unsized

Description: Moves the immediate operand into the status register (both user and supervisor portions), advances the program counter to point to the next instruction, and stops the fetching and executing of instructions. A trace, interrupt, or reset exception causes the processor to resume instructions execution. A trace exception occurs if instruction tracing is enabled when the STOP instruction begins execution. If an interrupt request is asserted with a priority higher than the priority level set by the new status register value, an interrupt exception occurs; otherwise, the interrupt request is ignored. External reset always initiates reset exception processing.

Condition Codes:
Set according to the immediate operand

Instruction Format:

15	14	13	12	11	10	9	8	7	6	5	4	3	2	1	0
0	1	0	0	1	1	1	0	0	1	1	1	0	0	1	0
							IMMEDIATE DATA								

Instruction Fields:
Immediate field — Specifies the data to be loaded into the status register

SUB

Subtract

Operation: Destination − Source → Destination

Assembler Syntax:
SUB (ea),Dn
SUB Dn,(ea)

Attributes: Size = (Byte, Word, Long)

Description: Subtracts the source operand from the destination operand and stores the result in the destination. The size of the operation is specified as byte, word, or long. The mode of the instruction indicates which operand is the source, which is the destination, and which is the operand size.

Condition Codes:

X	N	Z	V	C
*	*	*	*	*

X Set to the value of the carry bit
N Set if the result is negative. Cleared otherwise.
Z Set if the result is zero. Cleared otherwise.
V Set if an overflow is generated. Cleared otherwise.
C Set if a borrow is generated. Cleared otherwise.

Instruction Format:

15	14	13	12	11	10	9	8	7	6	5	4	3	2	1	0
1	0	0	1	REGISTER			OP-MODE			MODE			REGISTER		
										EFFECTIVE ADDRESS					

Instruction Fields:

Register field — Specifies any of the eight data registers

Op-Mode field —

Byte	Word	Long	Operation
000	001	010	((Dn)) − ((ea)) → (Dn)
100	101	110	((ea)) − ((Dn)) → (ea)

SUB

Subtract

Effective Address field — Determines the addressing mode. If the location specified is a source operand, all addressing modes are allowed as shown:

Addressing Mode	Mode	Register	Addressing Mode	Mode	Register
Dn*	000	reg. number:Dn	(xxx).W	111	000
An*	001	reg. number:An	(xxx).L	111	001
(An)	010	reg. number:An	#(data)	111	100
(An)+	011	reg. number:An			
−(An)	100	reg. number:An			
(d16,An)	101	reg. number:An	(d16,PC)	111	010
(d8,An,Xn)	110	reg. number:An	(d8,PC,Xn)	111	011

*For byte size operation, address register direct is not allowed.

If the location specified is a destination operand, only memory alterable addressing modes are allowed as shown:

Addressing Mode	Mode	Register	Addressing Mode	Mode	Register
Dn	—	—	(xxx).W	111	000
An	—	—	(xxx).L	111	001
(An)	010	reg. number:An	#(data)	—	—
(An)+	011	reg. number:An			
−(An)	100	reg. number:An			
(d16,An)	101	reg. number:An	(d16,PC)	—	—
(d8,An,Xn)	110	reg. number:An	(d8,PC,Xn)	—	—

Notes:
1. If the destination is a data register, it must be specified as a destination Dn address, not as a destination (ea) address.
2. Most assemblers use SUBA when the destination is an address register, and SUBI or SUBQ when the source is immediate data.

SUBA

Subtract Address

Operation: Destination – Source ♦ Destination

**Assembler
Syntax:** SUBA (ea),An

Attributes: Size = (Word, Long)

Description: Subtracts the source operand from the destination address register and stores the result in the address register. The size of the operation is specified as word or long. Word size source operands are sign extended to 32-bit quantities prior to the subtraction.

Condition Codes:
Not affected

Instruction Format:

15	14	13	12	11	10	9	8	7	6	5	4	3	2	1	0
1	0	0	1	REGISTER			OP-MODE			EFFECTIVE ADDRESS					
										MODE			REGISTER		

Op-Mode Field:

Word	Long	Operation
011	111	(An) – (ea) ♦ (An)

Instruction Fields:

Register field — Specifies the destination, any of the eight address registers

Op-Mode field — Specifies the size of the operation:
011 — Word operation. The source operand is sign extended to a long operand and the operation is performed on the address register using all 32 bits.
111 — Long operation

Effective Address field — Specifies the source operand. All addressing modes are allowed as shown:

Addressing Mode	Mode	Register
Dn	000	reg. number:Dn
An	001	reg. number:An
(An)	010	reg. number:An
(An)+	011	reg. number:An
–(An)	100	reg. number:An
(d16,An)	101	reg. number:An
(d8,An,Xn)	110	reg. number:An

Addressing Mode	Mode	Register
(xxx).W	111	000
(xxx).L	111	001
#(data)	111	100
(d16,PC)	111	010
(d8,PC,Xn)	111	011

SUBI

Subtract Immediate SUBI

Operation: Destination – Immediate Data ♦ Destination

**Assembler
Syntax:** SUBI #(data),(ea)

Attributes: Size = (Byte, Word, Long)

Description: Subtracts the immediate data from the destination operand and stores the result in the destination location. The size of the operation is specified as byte, word, or long. The size of the immediate data matches the operation size.

Condition Codes:

X	N	Z	V	C
*	*	*	*	*

X Set to the value of the carry bit
N Set if the result is negative. Cleared otherwise.
Z Set if the result is zero. Cleared otherwise.
V Set if an overflow occurs. Cleared otherwise.
C Set if a borrow occurs. Cleared otherwise.

Instruction Format:

15	14	13	12	11	10	9	8	7	6	5	4	3	2	1	0
0	0	0	0	0	1	0	0	SIZE		EFFECTIVE ADDRESS					
										MODE			REGISTER		
WORD DATA (16 BITS)										BYTE DATA (8 BITS)					
LONG DATA (32 BITS)															

Instruction Fields:

Size field — Specifies the size of the operation:
00 — Byte operation
01 — Word operation
10 — Long operation

Effective Address field — Specifies the destination operand. Only data alterable addressing modes are allowed as shown:

Addressing Mode	Mode	Register
Dn	000	reg. number:Dn
An	—	—
(An)	010	reg. number:An
(An)+	011	reg. number:An
–(An)	100	reg. number:An
(d16,An)	101	reg. number:An
(d8,An,Xn)	110	reg. number:An

Addressing Mode	Mode	Register
(xxx).W	111	000
(xxx).L	111	001
#(data)	—	—
(d16,PC)	—	—
(d8,PC,Xn)	—	—

Immediate field — (Data immediately following the instruction)
If size = 00, the data is the low order byte of the immediate word
If size = 01, the data is the entire immediate word
If size = 10, the data is the next two immediate words

SUBQ

Subtract Quick

SUBQ

Operation: Destination – Immediate Data ♦ Destination

**Assembler
Syntax:** SUBQ #⟨data⟩,⟨ea⟩

Attributes: Size = (Byte, Word, Long)

Description: Subtracts the immediate data (1-8) from the destination operand. The size of the operation is specified as byte, word, or long. Only word and long operations are allowed with address registers, and the condition codes are not affected. When subtracting from address registers, the entire destination address register is used, regardless of the operation size.

Condition Codes:

X	N	Z	V	C
*	*	*	*	*

X Set to the value of the carry bit
N Set if the result is negative. Cleared otherwise.
Z Set if the result is zero. Cleared otherwise.
V Set if an overflow occurs. Cleared otherwise.
C Set if a borrow occurs. Cleared otherwise.

Instruction Format:

15	14	13	12	11	10	9	8	7	6	5	4	3	2	1	0
0	1	0	1		DATA		1		SIZE			EFFECTIVE ADDRESS			
												MODE		REGISTER	

Instruction Fields:

Data field — Three bits of immediate data; 1-7 represent immediate values of 1-7, and 0 represents 8
Size field — Specifies the size of the operation:
  00 — Byte operation
  01 — Word operation
  10 — Long operation
Effective Address field — Specifies the destination location. Only alterable addressing modes are allowed as shown:

Addressing Mode	Mode	Register
Dn	000	reg. number:Dn
An*	001	reg. number:An
(An)	010	reg. number:An
(An) +	011	reg. number:An
–(An)	100	reg. number:An
(d16,An)	101	reg. number:An
(d8,An,Xn)	110	reg. number:An

*Word and Long only.

SUBX

Subtract with Extend

SUBX

Operation: Destination – Source – X ♦ Destination

**Assembler
Syntax:** SUBX Dx,Dy
SUBX –(Ax),–(Ay)

Attributes: Size = (Byte, Word, Long)

Description: Subtracts the source operand and the extend bit from the destination operand and stores the result in the destination location. The instruction has two modes:
  1. Data register to data register: The data registers specified in the instruction contain the operands.
  2. Memory to memory: The address registers specified in the instruction access the operands from memory using the predecrement addressing mode.
The size of the operand is specified as byte, word, or long.

Condition Codes:

X	N	Z	V	C
*	*	*	*	*

X Set to the value of the carry bit
N Set if the result is negative. Cleared otherwise.
Z Cleared if the result is non-zero. Unchanged otherwise.
V Set if an overflow occurs. Cleared otherwise.
C Set if a carry occurs. Cleared otherwise.

NOTE

Normally the Z condition code bit is set via programming before the start of an operation. This allows successful tests for zero results upon completion of multiple-precision operations.

Instruction Format:

15	14	13	12	11	10	9	8	7	6	5	4	3	2	1	0
1	0	0	1		REGISTER Ry		1		SIZE	0	0	R/M		REGISTER Rx	

Instruction Fields:

Register Ry field — Specifies the destination register:
If R/M = 0, specifies a data register
If R/M = 1, specifies an address register for the predecrement addressing mode
Size field — Specifies the size of the operation:
  00 — Byte operation
  01 — Word operation
  10 — Long operation
R/M field — Specifies the operand addressing mode:
  0 — The operation is data register to data register
  1 — The operation is memory to memory
Register Rx field — Specifies the source register:
If R/M = 0, specifies a data register
If R/M = 1, specifies an address register for the predecrement addressing mode

Addressing Mode	Mode	Register
(xxx).W	111	000
(xxx).L	111	001
#⟨data⟩	—	—
(d16,PC)	—	—
(d8,PC,Xn)	—	—

518

SWAP

Swap Register Halves

SWAP

Operation: Register [31:16] ⬌ Register [15:0]

Assembler Syntax: SWAP Dn

Attributes: Size = (Word)

Description: Exchange the 16-bit words (halves) of a data register

Condition Codes:

X	N	Z	V	C
—	*	*	0	0

X Not affected
N Set if the most-significant bit of the 32-bit result is set. Cleared otherwise.
Z Set if the 32-bit result is zero. Cleared otherwise.
V Always cleared
C Always cleared

Instruction Format:

15	14	13	12	11	10	9	8	7	6	5	4	3	2	1	0
0	1	0	0	1	0	0	0	0	1	0	0	0	REGISTER		

Instruction Fields:
Register field — Specifies the data register to swap

TAS

Test and Set an Operand

TAS

Operation: Destination Tested ⬦ Condition Codes; 1 ⬦ bit 7 of Destination

Assembler Syntax: TAS (ea)

Attributes: Size = (Byte)

Description: Tests and sets the byte operand addressed by the effective address field. The instruction tests the current value of the operand and sets the N and Z condition bits appropriately. TAS also sets the high order bit of the operand. The operation uses a read-modify-write memory cycle that completes the operation without interruption. This instruction supports use of a flag or semaphore to coordinate several processors.

Condition Codes:

X	N	Z	V	C
—	*	*	0	0

X Not affected
N Set if the most significant bit of the operand is currently set. Cleared otherwise.
Z Set if the operand was zero. Cleared otherwise.
V Always cleared
C Always cleared

Instruction Format:

15	14	13	12	11	10	9	8	7	6	5	4	3	2	1	0
0	1	0	0	1	0	1	0	1	1	EFFECTIVE ADDRESS		MODE			REGISTER

Instruction Fields:
Effective Address field — Specifies the location of the tested operand. Only data alterable addressing modes are allowed as shown:

Addressing Mode	Mode	Register	Addressing Mode	Mode	Register
Dn	000	reg. number:Dn	(xxx).W	111	000
An	—	—	(xxx).L	111	001
(An)	010	reg. number:An	#(data)	—	—
(An)+	011	reg. number:An			
-(An)	100	reg. number:An			
(d16,An)	101	reg. number:An	(d16,PC)	—	—
(d8,An,Xn)	110	reg. number:An	(d8,PC,Xn)	—	—

TRAP

Trap

TRAP

Operation: 1 → S bit of SR;
SSP − 2 → SSP; Format/Offset → (SSP); — MC68010 only
SSP − 4 → SSP; PC → (SSP); SSP − 2 → SSP;
SR → (SSP); Vector Address → PC

Assembler Syntax: TRAP #⟨vector⟩

Attributes: Unsized

Description: Causes a TRAP #⟨vector⟩ exception. The instruction adds the immediate operand ⟨vector⟩ of the instruction to 32 to obtain the vector number. The range of vector values is 0-15, which provides 16 vectors.

Condition Codes: Not affected

Instruction Format:

15	14	13	12	11	10	9	8	7	6	5	4	3	2	1	0
0	1	0	0	1	1	1	0	0	1	0	0		VECTOR		

Instruction Fields:
Vector field — Specifies the trap vector to be taken

TRAPV

Trap on Overflow

TRAPV

Operation: If V then TRAP

Assembler Syntax: TRAPV

Attributes: Unsized

Description: If the overflow condition is set, causes a TRAPV exception (vector number 7). If the overflow condition is not set, the processor performs no operation and execution continues with the next instruction.

Condition Codes: Not affected

Instruction Format:

15	14	13	12	11	10	9	8	7	6	5	4	3	2	1	0
0	1	0	0	1	1	1	0	0	1	1	1	0	1	1	0

TST   Test an Operand   TST

Operation: Destination Tested ♦ Condition Codes

**Assembler
Syntax:** TST (ea)

Attributes: Size = (Byte, Word, Long)

Description: Compares the operand with zero and sets the condition codes according to the results of the test. The size of the operation is specified as byte, word, or long.

Condition Codes:

X	N	Z	V	C
–	*	*	0	0

X Not affected
N Set if the operand is negative. Cleared otherwise.
Z Set if the operand is zero. Cleared otherwise.
V Always cleared
C Always cleared

Instruction Format:

15	14	13	12	11	10	9	8	7	6	5	4	3	2	1	0
0	1	0	0	1	0	1	0	SIZE		EFFECTIVE ADDRESS					
										MODE			REGISTER		

Instruction Fields:

Size field — Specifies the size of the operation:
00 — Byte operation
01 — Word operation
10 — Long operation

Effective Address field — Specifies the destination operand. If the operation size is word or long, all addressing modes are allowed. If the operation size is byte, only data addressing modes are allowed as shown:

Addressing Mode	Mode	Register
Dn	000	reg. number:Dn
An	–	–
(An)	010	reg. number:An
(An) +	011	reg. number:An
– (An)	100	reg. number:An
(d16,An)	101	reg. number:An
(d8,An,Xn)	110	reg. number:An

Addressing Mode	Mode	Register
(xxx).W	111	000
(xxx).L	111	001
#:data⟩	–	–
(d16,PC)	111	010
(d8,PC,Xn)	111	011

UNLK   Unlink   UNLK

Operation: An ♦ SP; (SP) ♦ An; SP + 4 ♦ SP

**Assembler
Syntax:** UNLK An

Attributes: Unsized

Description: Loads the stack pointer from the specified address register then loads the address register with the long word pulled from the top of the stack.

Condition Codes:
Not affected

Instruction Format:

15	14	13	12	11	10	9	8	7	6	5	4	3	2	1	0
0	1	0	0	1	1	1	0	0	1	0	1	1	REGISTER		

Instruction Fields:
Register field — Specifies the address register for the instruction

APPENDIX C

68000 Instruction Execution Times

MC68000 AND MC68HC000 INSTRUCTION EXECUTION TIMES

This section contains listings of the instruction execution times in terms of external clock (CLK) periods. In this data, it is assumed that both memory read and write cycles consist of four clock periods. A longer memory cycle causes the generation of wait states that must be added to the total instruction times.

The number of bus read and write cycles for each instruction is also included with the timing data. This data is shown as

$$n(r/w)$$

where
    n is the total number of clock periods
    r is the number of read cycles
    w is the number of write cycles

For example, a timing number shown as 18(3/1) means that the total number of clock periods is 18. Of the 18 clock periods, 12 are used for the three read cycles (four periods per cycle). Four additional clock periods are used for the single write cycle, for a total of 16 clock periods. The bus is idle for two clock periods during which the processor completes the internal operations required for the instruction.

NOTE
The total number of clock periods (n) includes instruction fetch and all applicable operand fetches and stores.

8.1 OPERAND EFFECTIVE ADDRESS CALCULATION TIMES

Table 8-1 lists the numbers of clock periods required to compute the effective addresses for instructions. The total includes fetching any extension words, computing the address, and fetching the memory operand. The total number of clock periods, the number of read cycles, and the number of write cycles (zero for all effective address calculations) are shown in the previously described format.

8.2 MOVE INSTRUCTION EXECUTION TIMES

Tables 8-2 and 8-3 list the numbers of clock periods for the move instructions. The totals include instruction fetch, operand reads, and operand writes. The total number of clock periods, the number of read cycles, and the number of write cycles are shown in the previously described format.

Table 8-1. Effective Address Calculation Times

	Addressing Mode	Byte, Word	Long
	Register		
Dn	Data Register Direct	**0**(0/0)	**0**(0/0)
An	Address Register Direct	**0**(0/0)	**0**(0/0)
	Memory		
(An)	Address Register Indirect	**4**(1/0)	**8**(2/0)
(An) +	Address Register Indirect with Postincrement	**4**(1/0)	**8**(2/0)
− (An)	Address Register Indirect with Predecrement	**6**(1/0)	**10**(2/0)
(d$_{16}$, An)	Address Register Indirect with Displacement	**8**(2/0)	**12**(3/0)
(d$_8$, An, Xn)*	Address Register Indirect with Index	**10**(2/0)	**14**(3/0)
(xxx).W	Absolute Short	**8**(2/0)	**12**(3/0)
(xxx).L	Absolute Long	**12**(3/0)	**16**(4/0)
(d$_8$, PC)	Program Counter Indirect with Displacement	**8**(2/0)	**12**(3/0)
(d$_{16}$, PC, Xn)*	Program Counter Indirect with Index	**10**(2/0)	**14**(3/0)
#(data)	Immediate	**4**(1/0)	**8**(2/0)

*The size of the index register (Xn) does not affect execution time.

Table 8-2. Move Byte and Word Instruction Execution Times

Source	Destination								
	Dn	An	(An)	(An) +	− (An)	(d$_{16}$, An)	(d$_8$, An, Xn)*	(xxx).W	(xxx).L
Dn	**4**(1/0)	**4**(1/0)	**8**(1/1)	**8**(1/1)	**8**(1/1)	**12**(2/1)	**14**(2/1)	**12**(2/1)	**16**(3/1)
An	**4**(1/0)	**4**(1/0)	**8**(1/1)	**8**(1/1)	**8**(1/1)	**12**(2/1)	**14**(2/1)	**12**(2/1)	**16**(3/1)
(An)	**8**(2/0)	**8**(2/0)	**12**(2/1)	**12**(2/1)	**12**(2/1)	**16**(3/1)	**18**(3/1)	**16**(3/1)	**20**(4/1)
(An) +	**8**(2/0)	**8**(2/0)	**12**(2/1)	**12**(2/1)	**12**(2/1)	**16**(3/1)	**18**(3/1)	**16**(3/1)	**20**(4/1)
− (An)	**10**(2/0)	**10**(2/0)	**14**(2/1)	**14**(2/1)	**14**(2/1)	**18**(3/1)	**20**(3/1)	**18**(3/1)	**22**(4/1)
(d$_{16}$, An)	**12**(3/0)	**12**(3/0)	**16**(3/1)	**16**(3/1)	**16**(3/1)	**20**(4/1)	**22**(4/1)	**20**(4/1)	**24**(5/1)
(d$_8$, An, Xn)*	**14**(3/0)	**14**(3/0)	**18**(3/1)	**18**(3/1)	**18**(3/1)	**22**(4/1)	**24**(4/1)	**22**(4/1)	**26**(5/1)
(xxx).W	**12**(3/0)	**12**(3/0)	**16**(3/1)	**16**(3/1)	**16**(3/1)	**20**(4/1)	**22**(4/1)	**20**(4/1)	**24**(5/1)
(xxx).L	**16**(4/0)	**16**(4/0)	**20**(4/1)	**20**(4/1)	**20**(4/1)	**24**(5/1)	**26**(5/1)	**24**(5/1)	**28**(6/1)
(d$_{16}$, PC)	**12**(3/0)	**12**(3/0)	**16**(3/1)	**16**(3/1)	**16**(3/1)	**20**(4/1)	**22**(4/1)	**20**(4/1)	**24**(5/1)
(d$_8$, PC, Xn)*	**14**(3/0)	**14**(3/0)	**18**(3/1)	**18**(3/1)	**18**(3/1)	**22**(4/1)	**24**(4/1)	**22**(4/1)	**26**(5/1)
#(data)	**8**(2/0)	**8**(2/0)	**12**(2/1)	**12**(2/1)	**12**(2/1)	**16**(3/1)	**18**(3/1)	**16**(3/1)	**20**(4/1)

*The size of the index register (Xn) does not affect execution time.

Table 8-3. Move Long Instruction Execution Times

Source	Destination								
	Dn	An	(An)	(An) +	− (An)	(d$_{16}$, An)	(d$_8$, An, Xn)*	(xxx).W	(xxx).L
Dn	**4**(1/0)	**4**(1/0)	**12**(1/2)	**12**(1/2)	**12**(1/2)	**16**(2/2)	**18**(2/2)	**16**(2/2)	**20**(3/2)
An	**4**(1/0)	**4**(1/0)	**12**(1/2)	**12**(1/2)	**12**(1/2)	**16**(2/2)	**18**(2/2)	**16**(2/2)	**20**(3/2)
(An)	**12**(3/0)	**12**(3/0)	**20**(3/2)	**20**(3/2)	**20**(3/2)	**24**(4/2)	**26**(4/2)	**24**(4/2)	**28**(5/2)
(An) +	**12**(3/0)	**12**(3/0)	**20**(3/2)	**20**(3/2)	**20**(3/2)	**24**(4/2)	**26**(4/2)	**24**(4/2)	**28**(5/2)
− (An)	**14**(3/0)	**14**(3/0)	**22**(3/2)	**22**(3/2)	**22**(3/2)	**26**(4/2)	**28**(4/2)	**26**(4/2)	**30**(5/2)
(d$_{16}$, An)	**16**(4/0)	**16**(4/0)	**24**(4/2)	**24**(4/2)	**24**(4/2)	**28**(5/2)	**30**(5/2)	**28**(5/2)	**32**(6/2)
(d$_8$, An, Xn)*	**18**(4/0)	**18**(4/0)	**26**(4/2)	**26**(4/2)	**26**(4/2)	**30**(5/2)	**32**(5/2)	**30**(5/2)	**34**(6/2)
(xxx).W	**16**(4/0)	**16**(4/0)	**24**(4/2)	**24**(4/2)	**24**(4/2)	**28**(5/2)	**30**(5/2)	**28**(5/2)	**32**(6/2)
(xxx).L	**20**(5/0)	**20**(5/0)	**28**(5/2)	**28**(5/2)	**28**(5/2)	**32**(6/2)	**34**(6/2)	**32**(6/2)	**36**(7/2)
(d, PC)	**16**(4/0)	**16**(4/0)	**24**(4/2)	**24**(4/2)	**24**(4/2)	**28**(5/2)	**30**(5/2)	**28**(5/2)	**32**(5/2)
(d, PC, Xn)*	**18**(4/0)	**18**(4/0)	**26**(4/2)	**26**(4/2)	**26**(4/2)	**30**(5/2)	**32**(5/2)	**30**(5/2)	**34**(6/2)
#(data)	**12**(3/0)	**12**(3/0)	**20**(3/2)	**20**(3/2)	**20**(3/2)	**24**(4/2)	**26**(4/2)	**24**(4/2)	**28**(5/2)

*The size of the index register (Xn) does not affect execution time.

Reprinted with permission of Motorola Inc.

8.3 STANDARD INSTRUCTION EXECUTION TIMES

The numbers of clock periods shown in Table 8-4 indicate the times required to perform the operations, store the results, and read the next instruction. The total number of clock periods, the number of read cycles, and the number of write cycles are shown in the previously described format. The number of clock periods, the number of read cycles, and the number of write cycles, respectively, must be added to those of the effective address calculation where indicated by a plus sign (+).

In Table 8-4, the following notation applies:
- An - Address register operand
- Dn - Data register operand
- ea - An operand specified by an effective address
- M - Memory effective address operand

Table 8-4. Standard Instruction Execution Times

Instruction	Size	op<ea>, An†	op<ea>, Dn	op Dn, <M>
ADD/ADDA	Byte, Word	8(1/0) +	4(1/0) +	8(1/1) +
	Long	6(1/0) + * *	6(1/0) + * *	12(1/2) +
AND	Byte, Word	—	4(1/0) +	8(1/1) +
	Long	—	6(1/0) + * *	12(1/2) +
CMP/CMPA	Byte, Word	6(1/0) +	4(1/0) +	—
	Long	6(1/0) +	6(1/0) +	—
DIVS	—	—	158(1/0) + *	—
DIVU	—	—	140(1/0) + *	—
EOR	Byte, Word	—	4(1/0) * * *	8(1/1) +
	Long	—	8(1/0) * * *	12(1/2) +
MULS	—	—	70(1/0) + *	—
MULU	—	—	70(1/0) + *	—
OR	Byte, Word	—	4(1/0) +	8(1/1) +
	Long	—	6(1/0) + * *	12(1/2) +
SUB	Byte, Word	8(1/0) +	4(1/0) +	8(1/1) +
	Long	6(1/0) + * *	6(1/0) + * *	12(1/2) +

NOTES:
+ add effective address calculation time
† word or long only
* indicates maximum basic value added to word effective address time.
* * The base time of six clock periods is increased to eight if the effective address mode is register direct or immediate (effective address time should also be added).
* * * Only available effective address mode is data register direct.
DIVS, DIVU — The divide algorithm used by the MC68000 provides less than 10% difference between the best and worst case timings.
MULS, MULU — The multiply algorithm requires 38 + 2n clocks where n is defined as:
    MULU: n = the number of ones in the <ea>
    MULS: n = concatanate the <ea> with a zero as the LSB; n is the resultant number of 10 or 01 patterns in the 17-bit source; i.e., worst case happens when the source is $5555.

8.4 IMMEDIATE INSTRUCTION EXECUTION TIMES

The numbers of clock periods shown in Table 8-5 include the times to fetch immediate operands, perform the operations, store the results, and read the next operation. The total number of clock periods, the number of read cycles, and the number of write cycles are shown in the previously described format. The number of clock periods, the number of read cycles, and the number of write cycles, respectively, must be added to those of the effective address calculation where indicated by a plus sign (+).

In Table 8-5, the following notation applies:
- \# - Immediate operand
- Dn - Data register operand
- An - Address register operand
- M - Memory operand

Table 8-5. Immediate Instruction Execution Times

Instruction	Size	op #,Dn	op #,An	op #,M
ADDI	Byte, Word	8(2 0)	—	12(2 1) +
	Long	16(3 0)	—	20(3 2) +
ADDQ	Byte, Word	4(1 0)	4(1 0)*	8(1 1) +
	Long	8(1 0)	8(1 0)	12(1 2) +
ANDI	Byte, Word	8(2 0)	—	12(2 1) +
	Long	14(3 0)	—	20(3 2) +
CMPI	Byte, Word	8(2 0)	—	8(2 0) +
	Long	14(3 0)	—	12(3 0) +
EORI	Byte, Word	8(2 0)	—	12(2 1) +
	Long	16(3 0)	—	20(3 2) +
MOVEQ	Long	4(1 0)	—	—
ORI	Byte, Word	8(2 0)	—	12(2 1) +
	Long	16(3 0)	—	20(3 2) +
SUBI	Byte, Word	8(2 0)	—	12(2 1) +
	Long	16(3 0)	—	20(3 2) +
SUBQ	Byte, Word	4(1 0)	8(1 0)*	8(1 1) +
	Long	8(1 0)	8(1 0)	12(1 2) +

8.5 SINGLE OPERAND INSTRUCTION EXECUTION TIMES

Table 8-6 lists the timing data for the single operand instructions. The total number of clock periods, the number of read cycles, and the number of write cycles are shown in the previously described format. The number of clock periods, the number of read cycles, and the number of write cycles, respectively, must be added to those of the effective address calculation where indicated by a plus sign (+).

Table 8-6. Single Operand Instruction Execution Times

Instruction	Size	Register	Memory
CLR	Byte, Word	4(1/0)	8(1/1) +
	Long	6(1/0)	12(1/2) +
NBCD	Byte	6(1/0)	8(1/1) +
NEG	Byte, Word	4(1/0)	8(1/1) +
	Long	6(1/0)	12(1/2) +
NEGX	Byte, Word	4(1/0)	8(1/1) +
	Long	6(1/0)	12(1/2) +
NOT	Byte, Word	4(1/0)	8(1/1) +
	Long	6(1/0)	12(1/2) +
Scc	Byte, False	4(1/0)	8(1/1) +
	Byte, True	6(1/0)	8(1/1) +
TAS	Byte	4(1/0)	14(2/1) +
TST	Byte, Word	4(1/0)	4(1/0) +
	Long	4(1/0)	4(1/0) +

+ add effective address calculation time

8.6 SHIFT/ROTATE INSTRUCTION EXECUTION TIMES

Table 8-7 lists the timing data for the shift and rotate instructions. The total number of clock periods, the number of read cycles, and the number of write cycles are shown in the previously described format. The number of clock periods, the number of read cycles, and the number of write cycles, respectively, must be added to those of the effective address calculation where indicated by a plus sign (+).

Table 8-7. Shift/Rotate Instruction Execution Times

Instruction	Size	Register	Memory
ASR, ASL	Byte, Word	6 + 2n(1/0)	8(1/1) +
	Long	8 + 2n(1/0)	—
LSR, LSL	Byte, Word	6 + 2n(1/0)	8(1/1) +
	Long	8 + 2n(1/0)	—
ROR, ROL	Byte, Word	6 + 2n(1/0)	8(1/1) +
	Long	8 + 2n(1/0)	—
ROXR, ROXL	Byte, Word	6 + 2n(1/0)	8(1/1) +
	Long	8 + 2n(1/0)	—

+ add effective address calculation time for word operands
n is the shift count

8.7 BIT MANIPULATION INSTRUCTION EXECUTION TIMES

Table 8-8 lists the timing data for the bit manipulation instructions. The total number of clock periods, the number of read cycles, and the number of write cycles are shown in the previously described format. The number of clock periods, the number of read cycles, and the number of write cycles, respectively, must be added to those of the effective address calculation where indicated by a plus sign (+).

Table 8-8. Bit Manipulation Instruction Execution Times

Instruction	Size	Dynamic		Static	
		Register	Memory	Register	Memory
BCHG	Byte	–	8(1/1) +	–	12(2/1) +
	Long	8(1/0) *	–	12(2/0) *	–
BCLR	Byte	–	8(1/1) +	–	12(2/1) +
	Long	10(1/0) *	–	14(2/0) *	–
BSET	Byte	–	8(1/1) +	–	12(2/1) +
	Long	8(1/0) *	–	12(2/0) *	–
BTST	Byte	–	4(1/0) +	–	8(2/0) +
	Long	6(1/0)	–	10(2/0)	–

+ add effective address calculation time
* indicates maximum value; data addressing mode only

8.8 CONDITIONAL INSTRUCTION EXECUTION TIMES

Table 8-9 lists the timing data for the conditional instructions. The total number of clock periods, the number of read cycles, and the number of write cycles are shown in the previously described format.

Table 8-9. Conditional Instruction Execution Times

Instruction	Displacement	Branch Taken	Branch Not Taken
Bcc	Byte	10(2/0)	8(1/0)
	Word	10(2/0)	12(2/0)
BRA	Byte	10(2/0)	–
	Word	10(2/0)	–
BSR	Byte	18(2/2)	–
	Word	18(2/2)	–
DBcc	cc true	–	12(2/0)
	cc false, Count Not Expired	10(2/0)	–
	cc false, Counter Expired	–	14(3/0)

8.9 JMP, JSR, LEA, PEA, AND MOVEM INSTRUCTION EXECUTION TIMES

Table 8-10 lists the timing data for the jump (JMP), jump to subroutine (JSR), load effective address (LEA), push effective address (PEA), and move multiple registers (MOVEM) instructions. The total number of clock periods, the number of read cycles, and the number of write cycles are shown in the previously described format.

Table 8-10. JMP, JSR, LEA, PEA, and MOVEM Instruction Execution Times

Instruction	Size	(An)	(An)+	−(An)	(d₁₆, An)	(d₈, An, Xn)+	(xxx).W	(xxx).L	(d₁₆ PC)	(d₈, PC, Xn)*
JMP	—	8(2/0)	—	—	10(2/0)	14(3/0)	10(2/0)	12(3/0)	10(2/0)	14(3/0)
JSR	—	16(2/2)	—	—	18(2/2)	22(2/2)	18(2/2)	20(3/2)	18(2/2)	22(2/2)
LEA	—	4(1/0)	—	—	8(2/0)	12(2/0)	8(2/0)	12(3/0)	8(2/0)	12(2/0)
PEA	—	12(1/2)	—	—	16(2/2)	20(2/2)	16(2/2)	20(3/2)	16(2/2)	20(2/2)
MOVEM M♦R	Word	12+4n (3+n/0)	12+4n (3+n/0)	—	16+4n (4+n/0)	18+4n (4+n/0)	16+4n (4+n/0)	20+4n (5+n/0)	16+4n (4n/0)	18+4n (4+n/0)
	Long	12+8n (3+2n/0)	12+8n (3+2n/0)	—	16+8n (4+2n/0)	18+8n (4+2n/0)	16+8n (4+2n/0)	20+8n (5+2n/0)	16+8n (4+2n/0)	18+8n (4+2n/0)·
MOVEM R♦M	Word	8+4n (2/n)	—	8+4n (2/n)	12+4n (3/n)	14+4n (3/n)	12+4n (3/n)	16+4n (4/n)	—	—
	Long	8+8n (2/2n)	—	8+8n (2/2n)	12+8n (3/2n)	14+8n (3/2n)	12+8n (3/2n)	16+8n (4/2n)	—	—

n is the number of registers to move.
* The size of the index register (Xn) does not affect the instruction's execution time.

8.10 MULTI-PRECISION INSTRUCTION EXECUTION TIMES

Table 8-11 lists the timing data for multi-precision instructions. The number of clock periods includes the time to fetch both operands, perform the operations, store the results, and read the next instructions. The total number of clock periods, the number of read cycles, and the number of write cycles are shown in the previously described format.

The following notation applies in Table 8-11:
Dn - Data register operand
M - Memory operand

Table 8-11. Multi-Precision Instruction Execution Times

Instruction	Size	op Dn, Dn	op M, M
ADDX	Byte, Word	4(1/0)	18(3/1)
	Long	8(1/0)	30(5/2)
CMPM	Byte, Word		12(3/0) ·
	Long		20(5/0)
SUBX	Byte, Word	4(1/0)	18(3/1)
	Long	8(1/0)	30(5/2)
ABCD	Byte	6(1/0)	18(3/1)
SBCD	Byte	6(1/0)	18(3/1)

8.11 MISCELLANEOUS INSTRUCTION EXECUTION TIMES

Tables 8-12 and 8-13 list the timing data for miscellaneous instructions. The total number of clock periods, the number of read cycles, and the number of write cycles are shown in the previously described format. The number of clock periods, the number of read cycles, and the number of write cycles, respectively, must be added to those of the effective address calculation where indicated by a plus sign (+).

Table 8-12. Miscellaneous Instruction Execution Times

Instruction	Size	Register	Memory
ANDI to CCR	Byte	20(3/0)	–
ANDI to SR	Word	20(3/0)	–
CHK (No Trap)	–	10(1/0) +	
EORI to CCR	Byte	20(3/0)	–
EORI to SR	Word	20(3/0)	–
ORI to CCR	Byte	20(3/0)	–
ORI to SR	Word	20(3/0)	–
MOVE from SR	–	6(1/0)	8(1/1) +
MOVE to CCR	–	12(1/0)	12(1/0) +
MOVE to SR	–	12(2/0)	12(2/0) +
EXG	–	6(1/0)	–
EXT	Word	4(1/0)	–
	Long	4(1/0)	–
LINK	–	16(2/2)	–
MOVE from USP	–	4(1/0)	–
MOVE to USP	–	4(1/0)	–
NOP	–	4(1/0)	–
RESET	–	132(1/0)	–
RTE	–	20(5/0)	–
RTR	–	20(5/0)	–
RTS	–	16(4/0)	–
STOP	–	4(0/0)	–
SWAP	–	4(1/0)	–
TRAPV	–	4(1/0)	–
UNLK	–	12(3/0)	–

+ add effective address calculation time

Table 8-13. Move Peripheral Instruction Execution Times

Instruction	Size	Register → Memory	Memory → Register
MOVEP	Word	16(2/2)	16(4/0)
	Long	24(2/4)	24(6/0)

8.12 EXCEPTION PROCESSING EXECUTION TIMES

Table 8-14 lists the timing data for exception processing. The numbers of clock periods include the times for all stacking, the vector fetch, and the fetch of the first instruction of the handler routine. The total number of clock periods, the number of read cycles, and the number of write cycles are shown in the previously described format. The number of clock periods, the number of read cycles, and the number of write cycles, respectively, must be added to those of the effective address calculation where indicated by a plus sign (+).

Table 8-14. Exception Processing Execution Times

Exception	Periods
Address Error	50(4/7)
Bus Error	50(4/7)
CHK Instruction	40(4/3) +
Divide by Zero	38(4/3) +
Illegal Instruction	34(4/3)
Interrupt	44(5/3) *
Privilege Violation	34(4/3)
RESET **	40(6/0)
Trace	34(4/3)
TRAP Instruction	34(4/3)
TRAPV Instruction	34(5/3)

+ add effective address calculation time

* The interrupt acknowledge cycle is assumed to take four clock periods

** Indicates the time from when RESET and HALT are first sampled as negated to when instruction execution starts

APPENDIX D

68000 Peripheral Data Sheets

Technical Summary

Parallel Interface/Timer (PI/T)

The MC68230 parallel interface/timer (PI/T) provides versatile double buffered parallel interfaces and a system oriented timer for MC68000 systems. The parallel interfaces operate in unidirectional or bidirectional modes, either 8 or 16 bits wide. In the unidirectional modes, an associated data direction register determines whether each port pin is an input or output. In the bidirectional modes the data direction registers are ignored and the direction is determined dynamically by the state of four handshake pins. These programmable handshake pins provide an interface flexible enough for connection to a wide variety of low, medium, or high speed peripherals or other computer systems. The PI/T ports allow use of vectored or autovectored interrupts, and also provide a DMA request pin for connection to the MC68450 direct memory access controller (DMAC) or a similar circuit. The PI/T timer contains a 24-bit wide counter and a 5-bit prescaler. The timer may be clocked by the system clock (PI/T CLK pin) or by an external clock (TIN pin), and a 5-bit prescaler can be used. It can generate periodic interrupts, a square wave, or a single interrupt after a programmed time period. It can also be used for elapsed time measurement or as a device watchdog.

Features of the PI/T include:

- M68000 Bus Compatible
- Port Modes Include:
  Bit I/O
  Unidirectional 8 Bit and 16 Bit
  Bidirectional 8 Bit and 16 Bit
- Programmable Handshaking Options
- 24-Bit Programmable Timer Modes
- Five Separate Interrupt Vectors, Four of Which May be Dedicated to External Interrupt Service Requests
- Separate Port and Timer Interrupt Service Requests
- Registers are Read/Write and Directly Addressable
- Registers are Addressed for MOVEP (Move Peripheral) and DMAC Compatibility

MC68230

INTRODUCTION

The PI/T consists of two logically independent sections: the ports and the timer. The port section consists of port A (PAO-PA7), port B (PBO-PB7), four handshake pins (H1, H2, H3, and H4), two general input/output (I/O) pins, and six dual-function pins. The dual-function pins can individually operate as a third port (port C) or an alternate function related to either port A, port B, or the timer. The four programmable handshake pins, depending on the mode, can control data transfer to and from the ports, can be used as general-purpose I/O pins, or can be used as interrupt-generating edge-sensitive inputs with corresponding interrupt vector numbers. Refer to Figure 1.

The timer consists of a 24-bit counter, optionally clocked by a 5-bit prescaler. Three pins provide complete timer I/O: PC2/TIN, PC3/TOUT, and PC7/TIACK. Only the ones needed for the given configuration perform the timer function, while the others remain port C I/O.

The system bus interface provides for asynchronous transfer of data from the PI/T to a bus master over the data bus (DO-D7). Data transfer acknowledge ($\overline{DTACK}$), register selects (RS1-RS5), timer interrupt acknowledge ($\overline{TIACK}$), read/write line ($R/\overline{W}$), chip select ($\overline{CS}$), or port interrupt acknowledge ($\overline{PIACK}$) control data transfers between the PI/T and an M68000 processor.

PORT MODE DESCRIPTION

The primary focus of most applications will be on port A, port B, the handshake pins, the port interrupt pins, and the DMA request pin. They are controlled in the following way: the port general control register contains a 2-bit field that specifies one of four operation modes. These govern the overall operation of the ports and determine

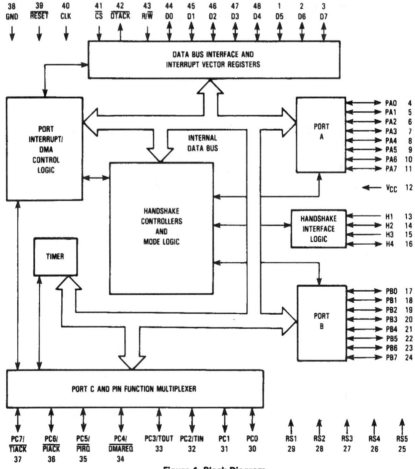

Figure 1. Block Diagram

their interrelationships. Some modes require additional information from each port's control register to further define its operation. In each port control register, there is a 2-bit submode field that serves this purpose. Each port mode/submode combination specifies a set of programmable characteristics that fully define the behavior of that port and two of the handshake pins. This structure is summarized in Table 1 and Figure 2.

Table 1. Port Mode Control Summary

Mode 0 (Unidirectional 8-Bit Mode)
Port A
Submode 00 — Pin-Definable Double-Buffered Input or Single-Buffered Output
H1 — Latches input data
H2 — Status/interrupt generating input, general-purpose output, or operation with H1 in the interlocked or pulsed handshake protocols
Submode 01 — Pin-Definable Double-Buffered Output or Non-Latched Input
H1 — Indicates data received by peripheral
H2 — Status/interrupt generating input, general-purpose output, or operation with H1 in the interlocked or pulsed handshake protocols
Submode 1X — Pin-Definable Single-Buffered Output or Non-Latched Input
H1 — Status/interrupt generating input
H2 — Status/interrupt generating input or general-purpose output
Port B
H3 and H4 — Identical to port A, H1 and H2
Mode 1 (Unidirectional 16-Bit Mode)
Port A — Most Significant Data Byte or Non-Latched Input or Single Buffered Output
Submode XX — (Not Used)
H1 — Status/interrupt generating input
H2 — Status/interrupt generating input or general-purpose output
Port B — Least-Significant Data Byte
Submode X0 — Pin-Definable Double-Buffered Input or Single- Buffered Output
H3 — Latches input data
H4 — Status/interrupt generating input, general-purpose output, or opereation with H3 in the interlocked or pulsed handshake protocols
Submode X1 — Pin-Definable Double-Buffered Output or Non-Latched Input
H3 — Indicates data received by peripheral
H4 — Status/interrupt generating input, general-purpose output, or operation with H3 in the interlocked or pulsed handshake protocols
Mode 2 (Bidirectional 8-Bit Mode)
Port A — Bit I/O
Submode XX — (Not Used)
Port B — Double-Buffered Bidirectional Data
Submode XX — (Not Used)
H1 — Indicates output data received by the peripheral and controls output drivers
H2 — Operating with H1 in the interlocked or pulsed output handshake protocols
H3 — Latches input data
H4 — Operation with H3 in the interlocked or pulsed input protocols
Mode 3 (Bidirectional 16-Bit Mode)
Port A — Double-Buffered Bidirectional Data (Most-Significant Data Byte)
Submode XX — (Not Used)
Port B — Double-Buffered Bidirectional Data (Least-Significant Data Byte)
Submode XX — (Not Used)
H1 — Indicates output data received by peripheral and controls output drivers
H2 — Operation with H1 in the interlocked or pulsed output handshake protocols
H3 — Latches input data
H4 — Operation with H3 in the interlocked or pulsed input handshake protocols

MC68230

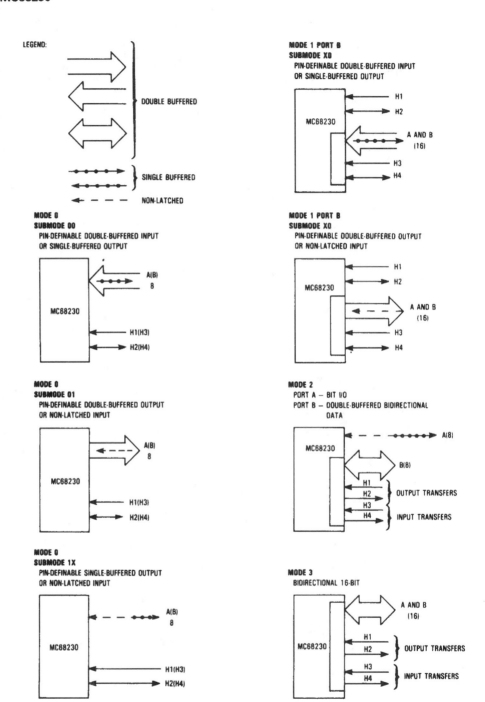

Figure 2. Port Mode Layout

SIGNAL DESCRIPTION

The input and output signals are illustrated functionally in Figure 3 and described in the following paragraphs.

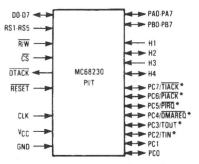

* Individually Programmable Dual-Function Pin

Figure 3. Logical Pin Assignment

BIDIRECTIONAL DATA BUS (D0-D7)

The data bus pins, D0-D7, form an 8-bit bidirectional data bus to/from an M68000 bus master. These pins are active high.

REGISTER SELECTS (RS1-RS5)

The register select pins, RS1-RS5, are active high high-impedance inputs that determine which of the 23 internal registers is being selected. They are provided by the M68000 bus master or other bus master.

READ/WRITE (R/W̄)

R/W̄ is a high-impedance read/write input signal from the M68000 bus master, indicating whether the current bus cycle is a read (high) or write (low) cycle.

CHIP SELECT (C̄S̄)

C̄S̄ is a high-impedance input that selects the PI/T registers for the current bus cycle. The data strobe (upper or lower) of the bus master, along with the appropriate address bits, must be included in the chip-select equation. A low level corresponds to an asserted chip select.

DATA TRANSFER ACKNOWLEDGE (D̄T̄ĀC̄K̄)

D̄T̄ĀC̄K̄ is an active low output that signals the completion of the bus cycle. During read or interrupt acknowledge cycles, D̄T̄ĀC̄K̄ is asserted after data has been provided on the data bus; during write cycles it is asserted after data has been accepted at the data bus. Data transfer acknowledge is compatible with the MC68000 and with other M68000 bus mastsers such as the MC68450 direct memory access controller (DMAC). A pullup resistor is required to maintain D̄T̄ĀC̄K̄ high between bus cycles.

RESET (R̄ĒS̄ĒT̄)

R̄ĒS̄ĒT̄ is a high-impedance input used to initialize all PI/T functions. All control and data direction registers are cleared and most internal operations are disabled by the assertion of R̄ĒS̄ĒT̄ (low).

CLOCK (CLK)

The clock pin is a high-impedance TTL-compatible signal with the same specifications as the MC68000. The PI/T contains dynamic logic throughout, and hence this clock must not be gated off at any time. It is not necessary that this clock maintain any particular phase relationship with the M68000 system clock. It may be connected to an independent frequency source (faster or slower) as long as all bus specifications are met.

PORT A AND PORT B (PA0-PA7 and PB0-PB7)

Ports A and B are 8-bit ports that may be concatenated to form a 16-bit port in certain modes. The ports may be controlled in conjunction with the handshake pins H1-H4. For stabilization during system power up, ports A and B have internal pullup resistors to V_{CC}. All port pins are active high.

HANDSHAKE PINS (H1-H4)

Handshake pins H1-H4 are multi-purpose pins that (depending on the operational mode) may provide an interlocked handshake, a pulsed handshake, interrupt-generating edge-sensitive inputs (independent of data transfers), or simple I/O pins. For stabilization during system power up, H2 and H4 have internal pullup resistors to V_{CC}. The sense of H1-H4 (active high or low) may be programmed in bits 3-0 of the port general control register. Independent of the mode, the instantaneous level of the handshake pins can be read from the port status register.

PORT C (PC0-PC7/ALTERNATE FUNCTION)

This port can be used as eight general purpose I/O pins (PC0-PC7) or any combination of six special function pins and two general purpose I/O pins (PC0-PC1). Each dual-function pin can be a standard I/O or a special function independent of the other port C pins. When used as a port C pin, these pins are active high. They may be individually programmed as inputs or outputs by the port C data direction register. The dual function pins are defined in the following paragraphs.

The alternate functions TIN, TOUT, and T̄ĪĀC̄K̄ are timer I/O pins. TIN may be used as a rising-edge triggered external clock input or an external run/halt control pin (the timer is in the run state if run/halt is high and in the halt state if run/halt is low). TOUT may provide an active low timer interrupt request output or a general-purpose square-wave output, initially high. T̄ĪĀC̄K̄ is an active low high-impedance input used for timer interrupt acknowledge.

The port functions of the PI/T (ports A and B) have an independent pair of active low interrupt request (P̄ĪR̄Q̄) and interrupt acknowledge (P̄ĪĀC̄K̄) pins.

MC68230

The $\overline{DMAREQ}$ (direct memory access request) pin provides an active low direct memory access controller request pulse for three clock cycles, completely compatible with the MC68450 DMAC. Note that if these pins are used for an alternate function, the corresponding bit in the Port C Data Direction Register must be programmed as an input (0).

SIGNAL SUMMARY

Table 2 is a summary of all the signals discussed in the previous paragraphs.

BUS INTERFACE OPERATION

The PI/T has an asynchronous bus interface, primarily designed for use with an MC68000 microprocessor. With care, however, it can be connected to synchronous microprocessor buses.

In an asynchronous system the PI/T clock may operate at a significantly different frequency, either higher or lower, than the bus master and other system components, as long as all bus specifications are met. The MC68230 CLK pin has the same specifications as the MC68000 CLK pin, and must not be gated off at any time.

The following signals generate normal read and write cycles to the PI/T: $\overline{CS}$ (chip select), R/$\overline{W}$ (read/write), RS1-RS5 (five register select bits), DO-D7 (the 8-bit bidirectional data bus), and $\overline{DTACK}$ (data transfer acknowledge). To generate interrupt acknowledge cycles, PC6/$\overline{PIACK}$ or PC7/$\overline{TIACK}$ is used instead of $\overline{CS}$, and the register select pins are ignored. No combination of the following pin functions may be asserted simultaneously: $\overline{CS}$, $\overline{PIACK}$, or $\overline{TIACK}$.

TIMER OPERATION

The MC68230 timer can provide several facilities needed by MC68000 operating systems. It can generate periodic interrupts, a square wave, or a single interrupt after a programmed time period. Also, it can be used for elapsed time measurement or as a device watchdog.

The PI/timer contains a 24-bit synchronous down counter that is loaded from three 8-bit counter preload registers. The 24-bit counter may be clocked by the output of a 5-bit (divide-by-32) prescaler or by an external timer input (TIN). If the prescaler is used, it may be clocked by the system clock (CLK pin) or by the TIN external input. The counter signals the occurrence of an event primarily

Table 2. Signal Summary

Signal Name	Input/Output	Active State	Edge/Level Sensitive	Output States
CLK	Input		Falling and Rising Edge	
$\overline{CS}$	Input	Low	Level	
D0-D7	Input/Output	High = 1, low = 0	Level	High, Low, High Impedance
$\overline{DMAREQ}$	Output	Low		High, Low
$\overline{DTACK}$	Output	Low		High, Low, High Impedance*
H1(H3)***	Input	Low or High	Asserted Edge	
H2)H4**	Input or Output	Low or High	Asserted Edge	High, Low, High Impedance
PA0-PA7**, PB0-PB7** PC0-PC7	Input/Output Input or Output	High = 1, Low = 0	Level	High, Low, High Impedance
$\overline{PIACK}$	Input	Low	Level	
$\overline{PIRQ}$	Output	Low		Low, High Impedance*
RS1-RS5	Input	High = 1, Low = 0	Level	
R/$\overline{W}$	Input	High Read, Low Write	Level	
$\overline{RESET}$	Input	Low	Level	
$\overline{TIACK}$	Input	Low	Level	
TIN (External Clock)	Input		Rising Edge	
TIN (Run/Halt)	Input	High	Level	
TOUT (Square Wave)	Output	Low		High, Low
TOUT ($\overline{TIRQ}$)	Output	Low		Low, High Impedance*

*Pullup resistors required.
**Note these pins have internal pullup resistors.
***H1 is level sensitive for output buffer control in modes 2 and 3.

Reprinted with permission of Motorola Inc.

through zero detection. (A zero is when the counter of the 24-bit timer is equal to zero.) This sets the zero detect status (ZDS) bit in the timer status register. It may be checked by the processor or may be used to generate a timer interrupt. The ZDS bit can be reset by writing a one to the timer status register in that bit position independent of timer operation.

The general operation of the timer is flexible and easily programmable. The timer is fully configured and controlled by programming the 8-bit timer control register. It controls: 1) the choice between the port C operation of three timer pins, 2) whether the counter is loaded from the counter preload register or rolls over when zero detect is reached, 3) the clock input, 4) whether the prescaler is used, and 5) whether the timer is enabled.

REGISTER MODEL

A register model that includes the corresponding register selects is shown in Table 3.

Table 3. Register Model (Sheet 1 of 2)

Register Select Bits 5 4 3 2 1	7	6	5	4	3	2	1	0	Register Value After RESET (Hex Value)	
0 0 0 0 0	Port Mode Control		H34 Enable	H12 Enable	H4 Sense	H3 Sense	H2 Sense	H1 Sense	0	0	Port General Control Register
0 0 0 0 1	*	SVCRQ Select		IPF Select		Port Interrupt Priority Control			0	0	Port Service Request Register
0 0 0 1 0	Bit 7	Bit 6	Bit 5	Bit 4	Bit 3	Bit 2	Bit 1	Bit 0	0	0	Port A Data Direction Register
0 0 0 1 1	Bit 7	Bit 6	Bit 5	Bit 4	Bit 3	Bit 2	Bit 1	Bit 0	0	0	Port B Data Direction Register
0 0 1 0 0	Bit 7	Bit 6	Bit 5	Bit 4	Bit 3	Bit 2	Bit 1	Bit 0	0	0	Port C Data Direction Register
0 0 1 0 1	Interrupt Vector Number						*	**	0	F	Port Interrupt Vector Register
0 0 1 1 0	Port A Submode		H2 Control			H2 Int Enable	H1 SVCRQ Enable	H1 Stat Control	0	0	Port A Control Register
0 0 1 1 1	Port B Submode		H4 Control			H4 Int Enable	H3 SVCRQ Enable	H3 Stat Control	0	0	Port B Control Register
0 1 0 0 0	Bit 7	Bit 6	Bit 5	Bit 4	Bit 3	Bit 2	Bit 1	Bit 0	*	*	Port A Data Register
0 1 0 0 1	Bit 7	Bit 6	Bit 5	Bit 4	Bit 3	Bit 2	Bit 1	Bit 0	*	*	Port B Data Register
0 1 0 1 0	Bit 7	Bit 6	Bit 5	Bit 4	Bit 3	Bit 2	Bit 1	Bit 0	***		Port A Alternate Register
0 1 0 1 1	Bit 7	Bit 6	Bit 5	Bit 4	Bit 3	Bit 2	Bit 1	Bit 0	***		Port B Alternate Register
0 1 1 0 0	Bit 7	Bit 6	Bit 5	Bit 4	Bit 3	Bit 2	Bit 1	Bit 0	****		Port C Data Register
0 1 1 0 1	H4 Level	H3 Level	H2 Level	H1 Level	H4S	H3S	H2S	H1S	****		Port Status Register
0 1 1 1 0	*	*	*	*	*	*	*	*	0	0	(Null)
0 1 1 1 1	*	*	*	*	*	*	*	*	0	0	(Null)

*Unused, read as zero
**Value before RESET
***Current value on pins
****Undetermined value

MC68230

Table 3. Register Model (Sheet 2 of 2)

Register Select Bits 5 4 3 2 1	7	6	5	4	3	2	1	0	Register Value After $\overline{RESET}$ (Hex Value)		Register
1 0 0 0 0	TOUT/$\overline{TIACK}$ Control			ZD Control	Clock Control			Timer Enable	0	0	Timer Control Register
1 0 0 0 1	Bit 7	Bit 6	Bit 5	Bit 4	Bit 3	Bit 2	Bit 1	Bit 0	0	F	Timer Interrupt Vector Register
1 0 0 1 0	*	*	*	*	*	*	*	*	0	0	(Null)
1 0 0 1 1	Bit 23	Bit 22	Bit 21	Bit 20	Bit 19	Bit 18	Bit 17	Bit 16	*	*	Counter Preload Register (High)
1 0 1 0 0	Bit 15	Bit 14	Bit 13	Bit 12	Bit 11	Bit 10	Bit 9	Bit 8	*	*	Counter Preload Register (Mid)
1 0 1 0 1	Bit 7	Bit 6	Bit 5	Bit 4	Bit 3	Bit 2	Bit 1	Bit 0	*	*	Counter Preload Register (Low)
1 0 1 1 0	*	*	*	*	*	*	*	*	0	0	(Null)
1 0 1 1 1	Bit 23	Bit 22	Bit 21	Bit 20	Bit 19	Bit 18	Bit 17	Bit 16	*	*	Count Register (High)
1 1 0 0 0	Bit 15	Bit 14	Bit 13	Bit 12	Bit 11	Bit 10	Bit 9	Bit 8	*	*	Count Register (Mid)
1 1 0 0 1	Bit 7	Bit 6	Bit 5	Bit 4	Bit 3	Bit 2	Bit 1	Bit 0	*	*	Count Register (Low)
1 1 0 1 0	*	*	*	*	*	*	*	ZDS	0	0	Timer Status
1 1 0 1 1	*	*	*	*	*	*	*	*	0	0	(Null)
1 1 1 0 0	*	*	*	*	*	*	*	*	0	0	(Null)
1 1 1 0 1	*	*	*	*	*	*	*	*	0	0	(Null)
1 1 1 1 0	*	*	*	*	*	*	*	*	0	0	(Null)
1 1 1 1 1	*	*	*	*	*	*	*	*	0	0	(Null)

*Unused, read as zero
**Value before $\overline{RESET}$

ELECTRICAL SPECIFICATIONS

MAXIMUM RATINGS

Characteristic	Symbol	Value	Unit
Supply Voltage	V_{CC}	− 0.3 to +7.0	V
Input Voltage	V_{in}	− 0.3 to +7.0	V
Operating Temperature Range	T_A	0 to 70	°C
Storage Temperature	T_{stg}	− 55 to + 150	°C

This device contains circuitry to protect the inputs against damage due to high static voltages or electric fields; however, it is advised that normal precautions be taken to avoid application of any voltage higher than maximum-rated voltages to this high-impedance circuit. Reliability of operation is enhanced if unused inputs are tied to an appropriate logic voltage level (e.g., either GND or V_{CC}).

THERMAL CHARACTERISTICS

Characteristic	Symbol	Value (Max)	Symbol	Value (Max)	Rating
Thermal Resistance Ceramic (L/LC) Plastic (P)	θ_{JA}	40 40	θ_{JC}	15* 20*	°C/W

TechnicalSummary

Dual Asynchronous
Receiver/Transmitter (DUART)

The MC68681 dual universal asynchronous receiver/transmitter (DUART) is part of the M68000 Family of peripherals and directly interfaces to the MC68000 processor via an asynchronous bus structure. The MC68681 consists of eight major sections: internal control logic, timing logic, interrupt control logic, a bidirectional 8-bit data bus buffer, two independent communication channels (A and B), a 6-bit parallel input port, and an 8-bit parallel output port.

Figure 1 illustrates the basic block diagram of the MC68681 and should be referred to during the discussion of its features which include the following:

- M68000 Bus Compatible
- Two Independent Full-Duplex Asynchronous Receiver/Transmitter Channels
- Maximum Data Transfer
  - 1X—1 MB/second
  - 16X—125 kB/second
- Quadruple-Buffered Receiver Data Registers
- Double-Buffered Transmitter Data Registers
- Independently Programmable Baud Rate for Each Receiver and Transmitter Selectable From:
  - 18 Fixed Rates: 50 to 38.4k Baud
  - One User Defined Rate Derived from a Programmable Timer/Counter
  - External 1X Clock or 16X Clock
- Programmable Data Format
  - Five to Eight Data Bits plus Parity
  - Odd, Even, No Parity, or Force Parity
  - One, One and One-Half, or Two Stop Bits Programmable in One-Sixteenth Bit Increments
- Programmable Channel Modes
  - Normal (Full Duplex)
  - Automatic Echo
  - Local Loopback
  - Remote Loopback
- Automatic Wake-up Mode for Multidrop Applications
- Multi-Function 6-Bit Input Port
  - Can Serve as Clock or Control Inputs
  - Change-of-State Detection on Four Inputs
- Multi-Function 8-Bit Output Port
  - Individual Bit Set/Reset Capability
  - Outputs Can be Programmed to be Status/Interrupt Signals
- Multi-Function 16-Bit Programmable Counter/Timer
- Versatile Interrupt System
  - Single Interrupt Output with Eight Maskable Interrupting Conditions
  - Interrupt Vector Output on Interrupt Acknowledge
  - Output Port Can be Configured to Provide a Total of Up to Six Separate Wire-ORable Interrupt Outputs

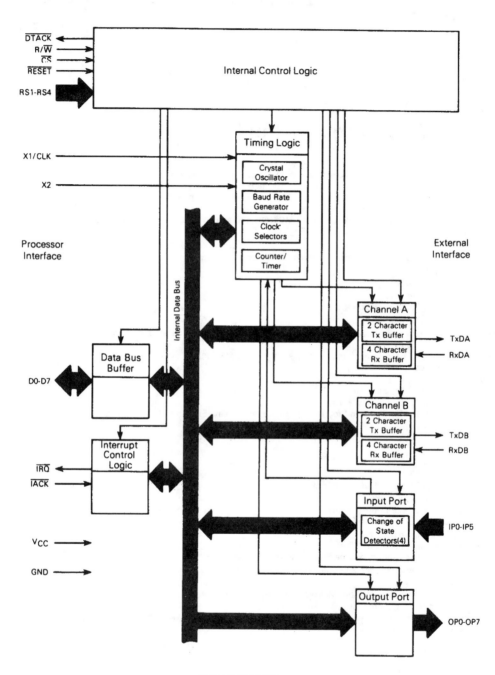

Figure 1. Block Diagram

MC68681

FEATURES (Continued)

- Parity, Framing, and Overrun Error Detection
- False-Start Bit Detection
- Line-Break Detection and Generation
- Detects Break Which Originates in the Middle of a Character
- Start-End Break Interrupt/Status
- On-Chip Crystal Oscillator
- TTL Comaptible
- Single +5 V Power Supply

INTERNAL CONTROL LOGIC

The internal control logic receives operation commands from the central processing unit (CPU) and generates appropriate signals to the internal sections to control device operation. It allows the registers within the DUART to be accessed and various commands to be performed by decoding the four register-select lines (RS1 through RS4). Besides the four register-select lines, there are three other inputs to the internal control logic from the CPU: read/write (R/W̄), which allows read and write transfers between the CPU and DUART via the data bus buffer; chip select (C̄S̄), which is the DUART chip select; and reset (R̄ĒS̄ĒT̄),which is used to initialize or reset the DUART. Output from the internal control logic is the data transfer acknowledge (D̄T̄ĀC̄K̄) signal which is asserted during read, write, or interrupt acknowledge cycles. D̄T̄ĀC̄K̄ indicates to the CPU that data has been latched on a CPU write cycle or that valid data is present on the data bus during a CPU read cycle or interrupt acknowledge (ĪĀC̄K̄) cycle.

TIMING LOGIC

The timing logic consists of a crystal oscillator, a baud-rate generator (BRG), a programmable 16-bit counter/timer (C/T), and four clock selectors. The crystal oscillator operates directly from a 3.6864 MHz crystal connected across the X1/CLK and X2 inputs or from an external clock of the appropriate frequency connected to X1/CLK. The clock serves as the basic timing reference for the baud-rate generator, the counter/timer, and other internal circuits. A clock signal, within the limits given in **ELECTRICAL SPECIFICATIONS**, must always be supplied to the DUART.

The baud-rate generator operates from the oscillator or external clock input and is capable of generating 18 commonly used data communication baud rates ranging from 50 to 38.4k by producing internal clock outputs at 16 times the actual baud rate. The counter/timer can be used in the timer mode to produce a 16X clock for any other baud rate by counting down the crystal clock or external clock. Other baud rates may also be derived by connecting 16X or 1X clocks to certain input port pins which have alternate functions as receiver or transmitter

clock inputs. The four clock selectors allow the independent selection, for each receiver and transmitter, of any of these baud rates.

The 16-bit counter/timer (C/T) included within the DUART and timing logic can be programmed to use one of several timing sources as its input. The output of the counter/timer is available to the internal clock selectors and can also be programmed to be a parallel output at OP3. In the timer mode, the counter/timer acts as a programmable divider and can be used to generate a square-wave output at OP3. In the counter mode, the contents of the counter/timer can be read by the CPU and it can be stopped and started under program control. The counter counts down the number of pulses stored in the concatenation of the counter/timer upper register and counter/timer lower register and produces an interrupt. This is a system oriented feature which may be used to keep track of timeouts when implementing various application protocols.

INTERRUPT CONTROL LOGIC

The following registers are associated with the interrupt control logic: interrupt mask register (IMR), interrupt status register (ISR), auxiliary control register (ACR), and interrupt vector register (IVR).

A single active-low interrupt output (ĪR̄Q̄) is provided which can be used to notify the processor that any of eight internal events has occurred. The interrupt mask register (IMR) can be programmed to select only certain conditions which cause ĪR̄Q̄ to be asserted while the interrupt status register (ISR) can be read by the CPU to determine all currently active interrupting conditions. When an active-low interrupt acknowledge signal (ĪĀC̄K̄) from the processor is assserted while the DUART has an interrupt pending, the DUART will place the contents of the interrupt vector register (IVR) (i.e., the interrupt vector) on the data bus and assert the data transfer acknowledge signal (D̄T̄ĀC̄K̄).

In addition, the DUART offers the ability to program the parallel outputs OP3 through OP7 to provide discrete interrupt outputs for the transmitters, the receivers, and the counter/timer.

DATA BUS BUFFER

The data bus buffer provides the interface between the external and internal data buses. It is controlled by the internal control logic to allow read and write data transfer operations to take place between the controlling CPU and DUART by way of the eight parallel data lines (D0 through D7).

COMMUNICATION CHANNELS A AND B

Each communication channel comprises a full-duplex asynchronous receiver/transmitter (UART). The operating frequency for each receiver and each transmitter can be selected independently from the baud-rate generator, the counter/timer, or from an external clock.

The transmitter accepts parallel data from the CPU, converts it to a serial bit stream, inserts the appropriate start, stop, and optional parity bits, and outputs a composite serial stream of data on the TxD output pin. The receiver accepts serial data on the RxD pin, converts this serial input to parallel format, checks for a start bit, stop bit, parity bit (if any), or break condition, and transfers an assembled character to the CPU during read operations.

INPUT PORT

The inputs to this unlatched 6-bit port (IP0 through IP5) can be read by the CPU by performing a read operation. High or low inputs to the input port result in the CPU reading a logic one or logic zero, respectively; that is there is no inversion of the logic level. Since the input port is a 6-bit port, performing a read operation will result in D7 being read as a logic one and D6 reflecting the logic level of IACK. Besides general-purpose inputs, the inputs to this port can be individually assigned specific auxiliary functions serving the communication channels.

Four change-of-state detectors, also provided within the input port, are associated with inputs IP0, IP1, IP2, and IP3. A high-to-low or low-to-high transition of these inputs lasting longer than 25 to 30 microseconds (best-to-worst case times) will set the corresponding bit in the input port change register (IPCR). The bits are cleared when the register is read by the CPU. Also, the DUART can be programmed so any particular change of state can generate an interrupt to the CPU. The DUART recognizes a level change on an input pin internally only after it has sampled the new level on the pin for two successive pulses of the sampling clock. The sampling clock is 38.4 kHz and is derived from one of the baud-rate generator taps. The resulting sampling period is slightly more than 25 microseconds (this assumes that the clock input is 3.6864 MHz). Subsequently, if the level change occurs on or just before a sampling pulse, it will be recognized internally after 25 microseconds. However, if the level change occurs just after a sampling pulse, it will be sampled the first time after 25 microseconds. Thus, in this case the level change will not be recognized internally until 50 microseconds after the level change took place on the pin.

OUTPUT PORT

This 8-bit multi-purpose output port can be used as a general-purpose output port. Associated with the output port is an output port register (OPR).

All bits of the output port register can be individually set and reset. A bit is set by performing a write operation at the appropriate address with the accompanying data specifying the bits to be set (one equals set and zero equals no change). Similarly, a bit is reset by performing a write operation at another address with the accompanying data specifying the bits to be reset (one equals reset and zero equals no change).

The output port register stores data that is to be output at the output port pins. Unlike the input port, if a particular bit of the output port register is set to a logic one or logic zero the output pin will be at a low or high level, respectively. Thus, a **logic inversion** takes place internal to the DUART with respect to this register. The outputs are complements of the data contained in the output port register.

Besides general-purpose outputs, the outputs can be individually assigned specific auxiliary functions serving the communication channels. The assignment is accomplished by appropriately programming the channel A and B mode registers (MR1A, MR1B, MR2A, and MR2B) and the output port configuration register (OPCR).

SIGNAL DESCRIPTION

The following paragraphs contain a brief description of the input and output signals.

NOTE

The terms **assertion** and **negation** will be used extensively. This is done to avoid confusion when dealing with a mixture of "active low" and "active high" signals. The term assert or assertion is used to indicate that a signal is active or true, independent of whether that level is represented by a high or low voltage. The term negate or negation is used to indicate that a signal is inactive or false.

V_{CC} AND GND

Power is supplied to the DUART using these two signals. V_{CC} is power (+5 volts) and GND is the ground connection.

CRYSTAL INPUT OR EXTERNAL CLOCK (X1/CLK)

This input is one of two connections to a crystal or a connection to an external clock. A crystal or a clock, within the specified limits, must be supplied at all times. If a crystal is used, a capacitor of approximately 10 to 15 picofarads should be connected from this pin to ground.

CRYSTAL INPUT (X2)

This input is an additional connection to a crystal. If an external TTL-level clock is used, this pin should be tied to ground. If a crystal is used, a capacitor of approximately 0 to 5 picofarads should be connected from this pin to ground.

RESET (RESET)

The DUART can be reset by asserting the RESET signal or by programming the appropriate command register. A hardware reset, assertion of RESET, clears status registers A and B (SRA and SRB), the interrupt mask register (IMR), the interrupt status register (ISR), the output port register (OPR), and the output port configuration register (OPCR). RESET initializes the interrupt vector register (IVR) to $0F_{16}$, places parallel outputs OP0 through OP3 in the high state, places the counter/timer in timer mode, and places channels A and B in the inactive state with the channel A transmitter serial-data output (TxDA) and channel B transmitter serial-data output (TxDB) in the mark (high) state.

Software resets are not as encompassing and are achieved by appropriately programming the channel A and/or B command register. Reset commands can be programmed through the command register to reset the receiver, transmitter, error status, or break-change interrupts for each channel.

CHIP SELECT (CS)

This active low input signal, when low, enables data transfers between the CPU and DUART on the data lines (D0 through D7). These data transfers are controlled by read/write (R/W) and the register-select inputs (RS1 through RS4). When chip select is high the D0 through D7 data lines are placed in the high-impedance state.

READ/WRITE (R/W)

When high, this input indicates a read cycle, and when low, it indicates a write cycle. A cycle is initiated by assertion of the chip-select input.

DATA TRANSFER ACKNOWLEDGE (DTACK)

This three-state active low open-drain output is asserted in read, write, or interrupt acknowledge (IACK) cycles to indicate the proper transfer of data between the CPU and DUART.

REGISTER-SELECT BUS (RS1 THROUGH RS4)

The register-select bus lines during read write operations select the DUART internal registers, ports, or commands.

DATA BUS (D0 THROUGH D7)

These bidirectional three-state data lines are used to transfer commands, data, and status between the CPU and DUART. D0 is the least-significant bit.

INTERRUPT REQUEST (IRQ)

This active low, open-drain output signals the CPU that one or more of the eight maskable interrupting conditions are true.

INTERRUPT ACKNOWLEDGE (IACK)

This active low input indicates an interrupt acknowledge cycle. If there is an interrupt pending (IRQ asserted) and this pin asserted, the DUART responds by placing the interrupt vector on the data bus and then asserting DTACK. If there is not an interrupt pending (IRQ negated), the DUART ignores the status of this pin.

CHANNEL A TRANSMITTER SERIAL-DATA OUTPUT (TxDA)

This signal is the transmitter serial-data output for channel A. The least-significant bit is transmitted first. This output is held high (mark condition) when the transmitter is disabled, idle, or operating in the local loopback

mode. (Mark is high and space is low.) Data is shifted out this pin on the falling edge of the programmed clock source.

CHANNEL A RECEIVER SERIAL-DATA INPUT (RxDA)

This signal is the receiver serial-data input for channel A. The least-significant bit is received first. Data on this pin is sampled on the rising edge of the programmed clock source.

CHANNEL B TRANSMITTER SERIAL-DATA OUTPUT (TxDB)

This signal is the transmitter serial-data output for channel B. The least-significant bit is transmitted first. The output is held high (mark condition) when the transmitter is disabled, idle, or operating in the local loopback mode. Data is shifted out this pin on the falling edge of the programmed clock source.

CHANNEL B RECEIVER SERIAL-DATA INPUT (RxDB)

This signal is the receiver serial-data input for channel B. The least-significant bit is received first. Data on this pin is sampled on the rising edge of the programmed clock source.

PARALLEL INPUTS (IP0 THROUGH IP5)

Each of the parallel inputs (IP0 through IP5) can be used as general-purpose inputs. However, each one has an alternate function(s) which is described in the following paragraphs.

IP0 This input can be used as the channel A clear-to-send active low input (CTSA). A change-of-state detector is also sassociated with this input.

IP1 This input can be used as the channel B clear-to-send active low input (CTSB). A change-of-state detector is also associated with this input.

IP2 This input can be used as the channel B receiver external clock input (RxCB), or the counter timer external clock input. When this input is used as the external clock by the receiver, the received data is sampled on the rising edge of the clock. A change-of-state detector is also associated with this input.

IP3 This input can be used as the channel A transmitter external clock input (TxCA). When this input is used as the external clock by the transmitter, the transmitted data is clocked on the falling edge of the clock. A change-of-state detector is also associated with this input.

IP4 This input can be used as the channel A receiver external clock input (RxCA). When this input is used as the external clock by the receiver, the received data is sampled on the rising edge of the clock.

IP5 This input can be used as the channel B transmitter external clock (TxCB). When this input is used as the external clock by the transmitter, the transmitted data is clocked on the falling edge of the clock.

PARALLEL OUTPUTS (OP0 THROUGH OP7)

Each of the parallel outputs can be used as general-purpose outputs. However, each one has an alternate function(s) which is described in the following paragraphs.

OP0 This output can be used as the channel A active low request-to-send ($\overline{\text{RTSA}}$) output. When used for this function, it is automatically negated and reasserted by either the receiver or transmitter.

OP1 This output can be used as the channel B active low request-to-send ($\overline{\text{RTSB}}$) output. When used for this function, it is negated and reasserted automatically by either the receiver or transmitter.

OP2 This output can be used as the channel A transmitter 1X-clock or 16X-clock output, or the channel A receiver 1X-clock output.

OP3 This output can be used as the open-drain active low counter-ready output, the open-drain timer output, the channel B transmitter 1X-clock output, or the channel B receiver 1X-clock output.

OP4 This output can be used as the channel A open-drain active-low receiver-ready or buffer-full interrupt outputs ($\overline{\text{RxRDYA}}/\overline{\text{FFULLA}}$) by appropriately programming bit 6 of mode register 1A.

OP5 This output can be used as the channel B open-drain active-low receiver-ready or buffer-full interrupt outputs ($\overline{\text{RxRDYB}}/\overline{\text{FFULLB}}$) by appropriately programming bit 6 of mode register 1B.

OP6 This output can be used as the channel A open-drain active-low transmitter-ready interrupt output ($\overline{\text{TxRDYA}}$) by appropriately programming bit 6 of the output port configuration regiser.

OP7 This output can be used as the channel B open-drain active-low transmitter-ready interrupt output ($\overline{\text{TxRDYB}}$) by appropriately programming bit 7 of the output port configuration register.

SIGNAL SUMMARY

Table 1 provides a summary of all the MC68681 signals described above.

Table 1. Signal Summary (Sheet 1 of 2)

Signal Name	Mnemonic	Pin No.	Input/Output	Active State
Power Supply (+5 V)	V_{CC}	40	Input	High
Ground	GND	20	Input	Low
Crystal Input or External Clock	X1/CLK	32	Input	—
Crystal Input	X2	33	Input	—
Reset	$\overline{\text{RESET}}$	34	Input	Low
Chip Select	$\overline{\text{CS}}$	35	Input	Low
Read/Write	R/$\overline{\text{W}}$	8	Input	High/Low
Data Transfer Acknowledge	$\overline{\text{DTACK}}$	9	Output*	Low
Register-Select Bus Bit 4	RS4	6	Input	High
Register-Select Bus Bit 3	RS3	5	Input	High
Register-Select Bus Bit 2	RS2	3	Input	High
Register-Select Bus Bit 1	RS1	1	Input	High
Bidirectional-Data Bus Bit 7	D7	19	Input/Output	High
Bidirectional-Data Bus Bit 6	D6	22	Input/Output	High
Bidirectional-Data Bus Bit 5	D5	18	Input/Output	High
Bidirectional-Data Bus Bit 4	D4	23	Input/Output	High
Bidirectional-Data Bus Bit 3	D3	17	Input/Output	High
Bidirectional-Data Bus Bit 2	D2	24	Input/Output	High
Bidirectional-Data Bus Bit 1	D1	16	Input/Output	High
Bidirectional-Data Bus Bit 0 (Least-Significant Bit)	D0	25	Input/Output	High
Interrupt Request	$\overline{\text{IRQ}}$	21	Output*	Low
Interrupt Acknowledge	$\overline{\text{IACK}}$	37	Input	Low
Channel A Transmitter Serial Data	TxDA	30	Output	—
Channel A Receiver Serial Data	RxDA	31	Input	—

Table 1. Signal Summary (Sheet 2 of 2)

Signal Name	Mnemonic	Pin No.	Input/Output	Active State
Channel B Transmitter Serial Data	TxDB	11	Output	—
Channel B Receiver Serial Data	RxDB	10	Input	—
Parallel Input 5	IP5	38	Input	—
Parallel Input 4	IP4	39	Input	—
Parallel Input 3	IP3	2	Input	—
Parallel Input 2	IP2	36	Input	—
Parallel Input 1	IP1	4	Input	—
Parallel Input 0	IP0	7	Input	—
Parallel Output 7	OP7	15	Output**	—
Parallel Output 6	OP6	26	Output**	—
Parallel Output 5	OP5	14	Output**	—
Parallel Output 4	OP4	27	Output**	—
Parallel Output 3	OP3	13	Output**	—
Parallel Output 2	OP2	28	Output	—
Parallel Output 1	OP1	12	Output	—
Parallel Output 0	OP0	29	Output	—

*Requires a pullup resistor.
**May require a pullup resistor, depending upon its programmed function.

PROGRAMMING AND REGISTER DESCRIPTION

The operation of the DUART is programmed by writing control words into the appropriate registers. Operational feedback is provided by way of the status registers which can be read by the CPU. The DUART register address and address-triggered commands are described in Table 2.

Figure 2 illustrates a block diagram of the DUART from a programming standpoint and details the register configuration for each block. The locations marked "do not access" should never be read during normal operation. They are used by the factory for testing purposes.

Tables 3 and 4 are provided to illustrate the various input port pin functions and output port pin functions respectively.

Table 5 is provided to illustrate the various clock sources which may be selected for the counter and timer. More detailed information can be obtained from Table 6.

Care should be exercised if the contents of a register is changed during receiver/transmitter operation since certain changes may cause undesired results. For example, changing the number of bits-per-character while the transmitter is active may cause the transmission of an incorrect character. The contents of the mode registers (MR), the clock-select register (CSR), the output port configuration register (OPCR), and bit 7 of the auxiliary control register (ACR[7]) should only be changed after the receiver(s) and transmitter(s) have been issued software Rx and Tx reset commands. Similarly, certain changes to the auxiliary control register (ACR bits six through four) should only be made while the counter/timer (C/T) is not used (i.e., stopped if in counter mode, output and/or interrupt masked in timer mode).

Mode registers one and two of each channel are accessed via independent auxiliary pointers. The pointer is set to channel A mode register one (MR1A) and channel B mode register one (MR1B) by RESET or by issuing a "reset pointer" command via the corresponding command register. Any read of write of the mode register while the pointer is at MR1A or MR1B switches the pointer to channel A mode register two (MR2A) or channel B mode register 2 (MR2B). The pointer then remains at MR2A or MR2B. So, subsequent accesses will address MR2A or MR2B, unless the pointer is reset to MR1A or MR1B as described above.

Mode, command, clock-select, and status register are duplicated for each channel to provide total independent operation and control. Refer to Table 6 for descriptions of the register and input and output port bits.

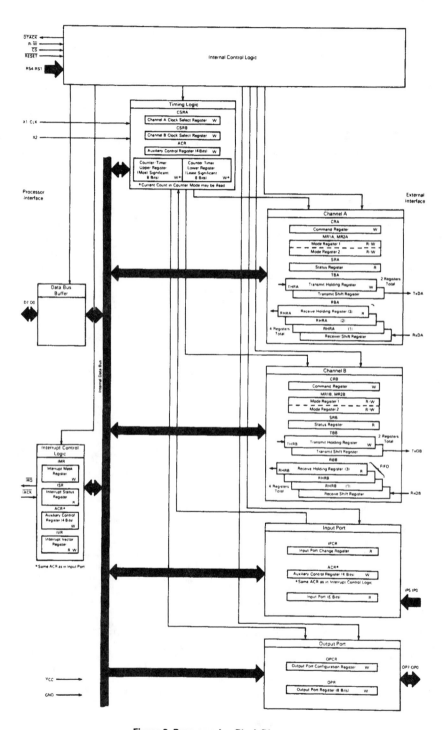

Figure 2. Programming Block Diagram

Table 2. Register Addressing and Address-Triggered Commands

RS4	RS3	RS2	RS1	Read (R/$\overline{W}$ = 1)		Write (R/$\overline{W}$ = 0)	
0	0	0	0	Mode Register A	(MR1A, MR2A)	Mode Register A	(MR1A, MR2A)
0	0	0	1	Status Register A	(SRA)	Clock-Select Register A	(CSRA)
0	0	1	0	Do Not Access*		Command Register A	CRA)
0	0	1	1	Receiver Buffer A	(RBA)	Transmitter Buffer A	(TBA)
0	1	0	0	Input Port Change Register	(IPCR)	Auxiliary Control Register	(ACR)
0	1	0	1	Interrupt Status Register	(ISR)	Interrupt Mask Register	(IMR)
0	1	1	0	Counter Mode: Current MSB of Counter	(CUR)	Counter/Timer Upper Register	(CTUR)
0	1	1	1	Counter Mode/ Current LSB of Counter	(CLR)	Counter/Timer Lower Register	(CTLR)
1	0	0	0	Mode Register B	(MR1B, MR2B)	Mode Register B	(MR1B, MR2B)
1	0	0	1	Status Register B	(SRB)	Clock-Select Register B	(CSRB)
1	0	1	0	Do Not Access*		Comand Register B	(CRB)
1	0	1	1	Receiver Buffer B	(RBB)	Transmitter Buffer B	(TBB)
1	1	0	0	Interrupt-Vector Register	(IVR)	Interrupt-Vector Register	(IVR)
1	1	0	1	Input Port (Unlatched)		Output Port Configuration Register	(OPCR)
1	1	1	0	Start-Counter Command**		Output Port Register (OPR)	Bit Set Command**
1	1	1	1	Stop-Counter Command**		Output Port Register (OPR)	Bit Reset Command**

*This address location is used for factory testing of the DUART and should not be read. Reading this location will result in undesired effects and possible incorrect transmission or reception of characters. Register contents may also be changed.
**Address triggered commands.

Table 3. Programming of Input Port Functions

Function	Input Port Pin					
	IP5	IP4	IP3	IP2	IP1	IP0
General Purpose	Default	Default	Default	Default	Default	Default
Change-of-State Detector			Default	Default	Default	Default
External Counter 1X Clock Input				ACR[6:4]* = 000		
External Timer 16X Clock Input				ACR[6:4]* = 100		
External Timer 1X Clock Input				ACR[6:4]* = 101		
RxCA 16X		CSRA[7:4] = 1110				
RxCA 1X		CSRA[7:4] = 1111				
TxCA 16X			CSRA[3:0] = 1110			
TxCA 1X			CSRA[3:0] = 1111			
RxCB 16X				CSRB[7:4] = 1110		
RxCB 1X				CSRB[7:4] = 1111		
TxCB 16X	CSRB[3:0] = 1110					
TxCB 1X	CSRB[3:0] = 1111					
TxCTSA						MR2A[4] = 1
TxCTSB				MR2B[4] = 1		

NOTE: Default refers to the function the input port pins perform when not used in one of the other modes. Only those functions which show the register programming are available for use.

*In these modes, because IP2 is used for the counter/timer-clock input, it is not available for use as the channel B receiver-clock input.

Reprinted with permission of Motorola Inc.

Table 4. Programming of Output Port Functions

Function	Output Port Pin							
	OP7	OP6	OP5	OP4	OP3	OP2	OP1	OP0
General Purpose	OPCR[7] = 0	OPCR[6] = 0	OPCR[5] = 0	OPCR[4] = 0	OPCR[3:2] = 00	OPCR[1:0] = 00	MR1B[7] = 0 MR2B[5] = 0	MR1A[7] = 0 MR2A[5] = 0
C̅T̅R̅D̅Y̅					OPCR[3:2] = 01, ACR[6] = 0*			
Timer Output					OPCR[3:2] = 01, ACR[6] = 1*			
TxCB 1X					OPCR[3:2] = 10			
RxCB 1X					OPCR[3:2] = 11			
TxCA 16X						OPCR[1:0] = 01		
TxCA 1X						OPCR[1:0] = 10		
RxCA 1X						OPCR[1:0] = 11		
T̅x̅R̅D̅Y̅A̅		OPCR[6] = 1*						
T̅x̅R̅D̅Y̅B̅	OPCR[7] = 1*							
R̅x̅R̅D̅Y̅A̅				OPCR[4] = 1, MR1A[6] = 0*				
R̅x̅R̅D̅Y̅B̅			OPCR[5] = 1, MR1B[6] = 0*					
F̅F̅U̅L̅L̅A̅				OPCR[4] = 1, MR1A[6] = *				
F̅F̅U̅L̅L̅B̅			OPCR[5] = 1, MR1B[6] = 1*					
R̅x̅R̅T̅S̅A̅								MR1A[7] = 1
T̅x̅R̅T̅S̅A̅								MR2A[5] = 1
R̅x̅R̅T̅S̅B̅							MR1B[7] = 1	
T̅x̅R̅T̅S̅B̅							MR2B[5] = 1	

Note: Only those functions which show the register programming are available for use.
* Pin requires a pullup resistor if used for this function.

Table 5. Selection of Clock Sources for the Counter and Timer Modes

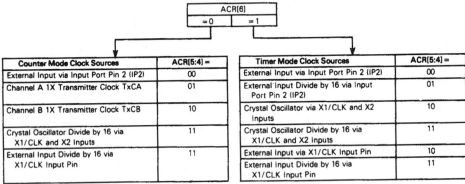

Counter Mode Clock Sources	ACR[5:4] =
External Input via Input Port Pin 2 (IP2)	00
Channel A 1X Transmitter Clock TxCA	01
Channel B 1X Transmitter Clock TxCB	10
Crystal Oscillator Divide by 16 via X1/CLK and X2 Inputs	11
External Input Divide by 16 via X1/CLK Input Pin	11

Timer Mode Clock Sources	ACR[5:4] =
External Input via Input Port Pin 2 (IP2)	00
External Input Divide by 16 via Input Port Pin 2 (IP2)	01
Crystal Oscillator via X1/CLK and X2 Inputs	10
Crystal Oscillator Divide by 16 via X1/CLK and X2 Inputs	11
External Input via X1/CLK Input Pin	10
External Input Divide by 16 via X1/CLK Input Pin	11

NOTE: Only those functions which show the register programming are available for use.

Table 6. Register Bit Formats (Sheet 1 of 5)

CHANNEL A MODE REGISTER 1 (MR1A) AND CHANNEL B MODE REGISTER 1 (MR1B)

Rx RTS Control	Rx IRQ Select	Error Mode	Parity Mode		Parity Type	Bits-per-Character	
Bit 7	Bit 6	Bit 5	Bit 4	Bit 3	Bit 2 With Parity 0 = Even 1 = Odd	Bit 1	Bit 0
0 = Disabled 1 = Enabled	0 = RxRDY 1 = FFULL	0 = Char 1 = Block	0 0 = With Parity 0 1 = Force Parity 1 0 = No Parity 1 1 = Multidrop Mode*		Force Parity 0 = Low 1 = High Multidrop Mode 0 = Data 1 = Address	0 0 = 5 0 1 = 6 1 0 = 7 1 1 = 8	

*The parity bit is used as the address/data bit in multidrop mode.

CHANNEL A MODE REGISTER 2 (MR2A) AND CHANNEL B MODE REGISTER 2 (MR2B)

Channel Mode		Tx RTS Control	CTS Enable Transmitter	Stop Bit Length			
Bit 7	Bit 6	Bit 5	Bit 4	Bit 3	Bit 2	Bit 1	Bit 0
0 0 = Normal 0 1 = Automatic Echo 1 0 = Local Loopback 1 1 = Remote Loopback		0 = Disabled 1 = Enabled	0 = Disabled 1 = Enabled			6-8 Bits/ Character	5-Bits/ Character
NOTE: If an external 1X clock is used for the transmitter, MR2 bit 3 = 0 selects one stop bit and MR2 bit 3 = 1 selects two stop bits to be transmitted.				(0) 0 0 0 0 = (1) 0 0 0 1 = (2) 0 0 1 0 = (3) 0 0 1 1 = (4) 0 1 0 0 = (5) 0 1 0 1 = (6) 0 1 1 0 = (7) 0 1 1 1 = (8) 1 0 0 0 = (9) 1 0 0 1 = (A) 1 0 1 0 = (B) 1 0 1 1 = (C) 1 1 0 0 = (D) 1 1 0 1 = (E) 1 1 1 0 = (F) 1 1 1 1 =		0.563 0.625 0.688 0.750 0.813 0.875 0.938 1.000 1.563 1.625 1.688 1.750 1.813 1.875 1.938 2.000	1.063 1.125 1.188 1.250 1.313 1.375 1.438 1.500 1.563 1.625 1.688 1.750 1.813 1.875 1.938 2.000

Table 6. Register Bit Formats (Sheet 2 of 5)

CLOCK-SELECT REGISTER A (CSRA)

Receiver-Clock Select	Baud Rate Set 1 ACR Bit 7=0	Baud Rate Set 2 ACR Bit 7=1
Bit 7 Bit 6 Bit 5 Bit 4		
0 0 0 0	50	75
0 0 0 1	110	110
0 0 1 0	134.5	134.5
0 0 1 1	200	150
0 1 0 0	300	300
0 1 0 1	600	600
0 1 1 0	1200	1200
0 1 1 1	1050	2000
1 0 0 0	2400	2400
1 0 0 1	4800	4800
1 0 1 0	7200	1800
1 0 1 1	9600	9600
1 1 0 0	38.4k	19.2k
1 1 0 1	Timer	Timer
1 1 1 0	IP4-16X	IP4-16X
1 1 1 1	IP4-1X	IP4-1X

NOTE: Receiver clock is always a 16X clock except when CSRA bits seven through four equal 1111.

Transmitter-Clock Select	Baud Rate Set 1 ACR Bit 7=0	Baud Rate Set 2 ACR Bit 7=1
Bit 3 Bit 2 Bit 1 Bit 0		
0 0 0 0	50	75
0 0 0 1	110	110
0 0 1 0	134.5	134.5
0 0 1 1	200	150
0 1 0 0	300	300
0 1 0 1	600	600
0 1 1 0	1200	1200
0 1 1 1	1050	2000
1 0 0 0	2400	2400
1 0 0 1	4800	4800
1 0 1 0	7200	1800
1 0 1 1	9600	9600
1 1 0 0	38.4k	19.2k
1 1 0 1	Timer	Timer
1 1 1 0	IP3-16X	IP3-16X
1 1 1 1	IP3-1X	IP3-1X

NOTE: Transmitter clock is always a 16X clock except when CSRA bits three through zero equal 1111.

CLOCK-SELECT REGISTER B (CSRB)

Receiver-Clock Select	Baud Rate Set 1 ACR Bit 7=0	Baud Rate Set 2 ACR Bit 7=1
Bit 7 Bit 6 Bit 5 Bit 4		
0 0 0 0	50	75
0 0 0 1	110	110
0 0 1 0	134.5	134.5
0 0 1 1	200	150
0 1 0 0	300	300
0 1 0 1	600	600
0 1 1 0	1200	1200
0 1 1 1	1050	2000
1 0 0 0	2400	2400
1 0 0 1	4800	4800
1 0 1 0	7200	1800
1 0 1 1	9600	9600
1 1 0 0	38.4k	19.2k
1 1 0 1	Timer	Timer
1 1 1 0	IP2-16X	IP2-16X
1 1 1 1	IP2-1X	IP2-1X

NOTE: Receiver clock is always a 16X clock except when CSRB bits seven through four equal 1111.

Transmitter-Clock Select	Baud Rate Set 1 ACR Bit 7=0	Baud Rate Set 2 ACR Bit 7=1
Bit 3 Bit 2 Bit 1 Bit 0		
0 0 0 0	50	75
0 0 0 1	110	110
0 0 1 0	134.5	134.5
0 0 1 1	200	150
0 1 0 0	300	300
0 1 0 1	600	600
0 1 1 0	1200	1200
0 1 1 1	1050	2000
1 0 0 0	2400	2400
1 0 0 1	4800	4800
1 0 1 0	7200	1800
1 0 1 1	9600	9600
1 1 0 0	38.4k	19.2k
1 1 0 1	Timer	Timer
1 1 1 0	IP5-16X	IP5-16X
1 1 1 1	IP5-1X	IP5-1X

NOTE: Transmitter clock is always a 16X clock except when CSRB bits three through zero equal 1111.

Table 6. Register Bit Formats (Sheet 3 of 5)

CHANNEL A COMMAND REGISTER (CRA) AND CHANNEL B COMMAND REGISTER (CRB)

Not Used*				Transmitter Commands		Receiver Commands	
	Miscellaneous Commands						
Bit 7	Bit 6	Bit 5	Bit 4	Bit 3	Bit 2	Bit 1	Bit 0
X	0 0 0 No Command			0 0 No Action, Stays in Present Mode		0 0 No Action, Stays in Present Mode	
	0 0 1 Reset MR Pointer to MR1			0 1 Transmitter Enabled		0 1 Receiver Enabled	
	0 1 0 Reset Receiver			1 0 Transmitter Disabled		1 0 Receiver Disabled	
	0 1 1 Reset Transmitter			1 1 Don't Use, Indeterminate		1 1 Don't Use, Indeterminate	
	1 0 0 Reset Error Status						
	1 0 1 Reset Channel's Break-Change Interrupt						
	1 1 0 Start Break						
	1 1 1 Stop Break						

*Bit seven is not used and may be set to either zero or one.

CHANNEL A STATUS REGISTER (SRA) AND CHANNEL B STATUS REGISTER (SRB)

Received Break	Framing Error	Parity Error	Overrun Error	TxEMT	TxRDY	FFULL	RxRDY
Bit 7*	Bit 6*	Bit 5*	Bit 4	Bit 3	Bit 2	Bit 1	Bit 0
0 = No 1 = Yes	0 = No 1 = Yes	0 = No 1 = Yes	0 = No 1 = Yes	0 = No 1 = Yes	0 = No 1 = Yes	0 = No 1 = Yes	0 = No 1 = Yes

*These status bits are appended to the corresponding data character in the receive FIFO and are valid only when the RxRDY bit is set. A read of the status register provides these bits (seven through five) from the top of the FIFO together with bits four through zero. These bits are cleared by a reset error status command. In character mode, they are discarded when the corresponding data character is read from the FIFO.

OUTPUT PORT CONFIGURATION REGISTER (OPCR)

OP7	OP6	OP5	OP4	OP3		OP2	
Bit 7	Bit 6	Bit 5	Bit 4	Bit 3	Bit 2	Bit 1	Bit 0
0 = OPR Bit 7 1 = TxRDYB	0 = OPR Bit 6 1 = TxRDYA	0 = OPR Bit 5 1 = RxRDYB / FFULLB	0 = OPR Bit 4 1 = RxRDYA / FFULLA	0 0 = OPR Bit 3 0 1 = C/T Output * 1 0 = TxCB (1X) 1 1 = RxCB (1X)		0 0 = OPR Bit 2 0 1 = TxCA (16X) 1 0 = TXCA (1X) 1 1 = RxCA (1X)	

*If OP3 is to be used for the timer output, the counter/timer should be programmed for timer mode (ACR[6] = 1), the counter/timer preload registers (CTUR and CTLR) initialized, and the start counter command issued before setting OPCR[3:2] = 01.

NOTE: OP1 and OP0 can be used as transmitter and receiver RTS control lines by appropriately programming the mode registers (MR1[7] for the receiver RxRTS, and MR2[5] for the transmitter TxRTS). OP1 is used for channel B's RTS control line and OP0 for channel A's RTS control line. When OP1 and OP0 are not used for RTS control, they may be used as general-purpose outputs. (See Table 4-3.)

OUTPUT PORT REGISTER (OPR)

OPR7	OPR6	OPR5	OPR4	OPR3	OPR2	OPR1	OPR0
Bit 7	Bit 6	Bit 5	Bit 4	Bit 3	Bit 2	Bit 1	Bit 0

Table 6. Register Bit Formats (Sheet 4 of 5)

AUXILIARY CONTROL REGISTER (ACR)

BRG SET Select*	Counter/Timer Mode and Source**			Delta*** IP3 IRQ	Delta*** IP2 IRQ	Delta*** IP1 IRQ	Delta*** IP0 IRQ
Bit 7	Bit 6	Bit 5	Bit 4	Bit 3	Bit 2	Bit 1	Bit 0
0 = Set 1	Mode	Clock Source		0 = Disabled	0 = Disabled	0 = Disabled	0 = Disabled
1 = Set 2	0 0 0 Counter	External (IP2)****		1 = Enabled	1 = Enabled	1 = Enabled	1 = Enabled
	0 0 1 Counter	TxCA – 1X Clock of Channel A Transmitter					
	0 1 0 Counter	TxCB – 1X Clock of Channel B Transmitter					
	0 1 1 Counter	Crystal or External Clock (X1/CLK) Divided by 16					
	1 0 0 Timer	External (IP2)****					
	1 0 1 Timer	External (IP2) Divided by 16****					
	1 1 0 Timer	Crystal or External Clock (X1/CLK)					
	1 1 1 Timer	Crystal or External Clock (X1/CLK) Divided by 16					

   * Should only be changed after both channels have been reset and are disabled.
   ** Should only be altered while the counter/timer is not in use (i.e., stopped if in counter mode, output and/or interrupt masked if in timer mode).
   *** Delta is equivalent to change-of-state.
   **** In these modes, because IP2 is used for the counter/timer clock input, it is not available for use as the channel B receiver-clock input.

INPUT PORT CHANGE REGISTER (IPCR)

Delta* Detected IP3	Delta* Detected IP2	Delta* Detected IP1	Delta* Detected IP0	Level IP3	Level IP2	Level IP1	Level IP0
Bit 7	Bit 6	Bit 5	Bit 4	Bit 3	Bit 2	Bit 1	Bit 0
0 = No	0 = No	0 = No	0 = No	0 = Low	0 = Low	0 = Low	0 = Low
1 = Yes	1 = Yes	1 = Yes	1 = Yes	1 = High	1 = High	1 = High	1 = High

* Delta is equivalent to change-of-state.

INTERRUPT STATUS REGISTER (ISR)

Input Port Change	Delta Break B	RxRDYB/ FFULLB	TxRDYB	Counter/ Timer Ready	Delta Break A	RxRDYA/ FFULLA	TxRDYA
Bit 7	Bit 6	Bit 5	Bit 4	Bit 3	Bit 2	Bit 1	Bit 0
0 = No	0 = No	0 = No	0 = No	0 = No	0 = No	0 = No	0 = No
1 = Yes	1 = Yes	1 = Yes	1 = Yes	1 = Yes	1 = Yes	1 = Yes	1 = Yes

INTERRUPT MASK REGISTER (IMR)

Input Port Change IRQ	Delta Break B IRQ	RxRDYB/ FFULLB IRQ	TxRDYB IRQ	Counter/ Timer Ready IRQ	Delta Break A IRQ	RxRDYA/ FFULLA IRQ	TxRDYA IRQ
Bit 7	Bit 6	Bit 5	Bit 4	Bit 3	Bit 2	Bit 1	Bit 0
0 = Masked	0 = Masked	0 = Masked	0 = Masked	0 = Masked	0 = Masked	0 = Masked	0 = Masked
1 = Pass	1 = Pass	1 = Pass	1 = Pass	1 = Pass	1 = Pass	1 = Pass	1 = Pass

Table 6. Register Bit Formats (Sheet 5 of 5)

COUNTER/TIMER UPPER REGISTER (CTUR)

C/T[15]	C/T[14]	C/T[13]	C/T[12]	C/T[11]	C/T[10]	C/T[9]	C/T[8]
Bit 7	Bit 6	Bit 5	Bit 4	Bit 3	Bit 2	Bit 1	Bit 0

COUNTER/TIMER LOWER REGISTER (CTLR)

C/T[7]	C/T[6]	C/T[5]	C/T[4]	C/T[3]	C/T[2]	C/T[1]	C/T[0]
Bit 7	Bit 6	Bit 5	Bit 4	Bit 3	Bit 2	Bit 1	Bit 0

INTERRUPT VECTOR REGISTER (IVR)

IVR[7]	IVR[6]	IVR[5]	IVR[4]	IVR[3]	IVR[2]	IVR[1]	IVR[0]
Bit 7	Bit 6	Bit 5	Bit 4	Bit 3	Bit 2	Bit 1	Bit 0

INPUT PORT

*	**	IP5	IP4	IP3	IP2	IP1	IP0
Bit 7	Bit 6	Bit 5	Bit 4	Bit 3	Bit 2	Bit 1	Bit 0

*Bit seven has no external pin. Upon reading the input port, bit seven will always be read as a one.
**Bit six has no external pin. Upon reading the input port, bit six will reflect the current logic level of $\overline{IACK}$.

OUTPUT PORT

OP7	OP6	OP5	OP4	OP3	OP2	OP1	OP0
Bit 7	Bit 6	Bit 5	Bit 4	Bit 3	Bit 2	Bit 1	Bit 0
$\overline{OPR[7]}$	$\overline{OPR[6]}$	$\overline{OPR[5]}$	$\overline{OPR[4]}$	$\overline{OPR[3]}$	$\overline{OPR[2]}$	$\overline{OPR[1]}$	$\overline{OPR[0]}$

Technical Summary
HCMOS Floating-Point Coprocessor

The MC68881 floating-point coprocessor is a full implementation of the *IEEE Standard for Binary Floating-Point Arithmetic* (754) for use with the Motorola M68000 Family of microprocessors. It is implemented using VLSI technology to give systems designers the highest possible functionality in a physically small device.

Intended primarily for use as a coprocessor to the MC68020 32-bit microprocessor unit (MPU), the MC68881 provides a logical extension to the main MPU integer data processing capabilities. It does this by providing a very high performance floating-point arithmetic unit and a set of floating-point data registers that are utilized in a manner that is analogous to the use of the integer data registers. The MC68881 instruction set is a natural extension of all earlier members of the M68000 Family, and supports all of the addressing modes of the host MPU. Due to the flexible bus interface of the M68000 Family, the MC68881 can be used with any of the MPU devices of the M68000 Family, and it may also be used as a peripheral to non-M68000 processors.

The major features of the MC68881 are:

- Eight general purpose floating-point data registers, each supporting a full 80-bit extended precision real data format (a 64-bit mantissa plus a sign bit, and a 15-bit signed exponent).

- A 67-bit arithmetic unit to allow very fast calculations, with intermediate precision greater than the extended precision format.

- A 67-bit barrel shifter for high-speed shifting operations (for normalizing etc.).

- Forty-six instructions, including 35 arithmetic operations.

- Full conformation to the IEEE 754 standard, including all requirements and suggestions.

- Support of functions not defined by the IEEE standard, including a full set of trigonometric and transcendental functions.

- Seven data types: byte, word and long integers; single, double, and extended precision real numbers; and packed binary coded decimal string real numbers.

- Twenty-two constants available in the on-chip ROM, including π, e, and powers of 10.

- Virtual memory machine operations.

- Efficient mechanisms for procedure calls, context switches, and interrupt handling.

- Fully concurrent instruction execution with the main processor.

- Use with any host processor, on an 8-, 16-, or 32-bit data bus.

MC68881

THE COPROCESSOR CONCEPT

The MC68881 functions as a coprocessor in systems where the MC68020 or MC68030 is the main processor via the M68000 coprocessor interface.* It functions as a peripheral processor in systems where the main processor is the MC68000, MC68008, or MC68010.

The MC68881 utilizes the M68000 Family coprocessor interface to provide a logical extension of the MC68020 registers and instruction set in a manner which is transparent to the programmer. The programmer perceives the MC68020/MC68881 execution model as if both devices are implemented on one chip.

A fundamental goal of the M68000 Family coprocessor interface is to provide the programmer with an execution model based upon sequential instruction execution by the MC68020 and the MC68881. For optimum performance, however, the coprocessor interface allows concurrent operations in the MC68881 with respect to the MC68020 whenever possible. In order to simplify the programmer's model, the coprocessor interface is designed to emulate, as closely as possible, non-concurrent operation between the MC68020 and the MC68881.

The MC68881 is a non-DMA type coprocessor which uses a subset of the general purpose coprocessor interface supported by the MC68020. Features of the interface implemented in the MC68881 are as follows:

- The main processor(s) and MC68881 communicate via standard M68000 bus cycles.

- The main processor(s) and MC68881 communications are not dependent upon the instruction sets or internal details of the individual devices (e.g., instruction pipes or caches, addressing modes).

- The main processor(s) and MC68881 may operate at different clock speeds.

- MC68881 instructions utilize all addressing modes provided by the main processor; all effective addresses are calculated by the main processor at the request of the coprocessor.

- All data transfers are performed by the main processor at the request of the MC68881; thus memory management, bus errors, address errors, and bus arbitration function as if the MC68881 instructions are executed by the main processor.

- Overlapped (concurrent) instruction execution enhances throughput while maintaining the programmer's model of sequential instruction execution.

- Coprocessor detection of exceptions which require a trap to be taken are serviced by the main processor at the request of the MC68881; thus exception processing functions as if the MC68881 instructions were executed by the main processor.

- Support of virtual memory/virtual machine systems is provided via the FSAVE and FRESTORE instructions.

- Up to eight coprocessors may reside in a system simultaneously; multiple coprocessors of the same type are also allowed.

- Systems may use software emulation of the MC68881 without reassembling or relinking user software.

HARDWARE OVERVIEW

The MC68881 is a high performance floating-point device designed to interface with the MC68020 as a coprocessor. This device fully supports the MC68020 virtual machine architecture, and is implemented in HCMOS, Motorola's low power, small geometry process. This process allows CMOS and HMOS (high-density NMOS) gates to be combined on the same device. CMOS structures are used where speed and low power is required, and HMOS structures are used where minimum silicon area is desired. The HCMOS technology enables the MC68881 to be very fast while consuming less power than comparable HMOS, and still have a reasonably small die size.

With some performance degradation, the MC68881 can also be used as a peripheral processor in systems where the MC68020 is not the main processor (e.g., MC68000, MC68008, MC68010). The configuration of the MC68881 as a peripheral processor or coprocessor may be completely transparent to user software (i.e., the same object code may be executed in either configuration).

The architecture of the MC68881 appears to the user as a logical extension of the M68000 Family architecture. Coupling of the coprocessor interface, allows the MC68020 programmer to view the MC68881 registers as though the registers are resident in the MC68020. Thus, a MC68020/MC68881 device pair appears to be one processor that supports seven floating-point and integer data types, and has eight integer data registers, eight address registers, and eight floating-point data registers.

The MC68881 programming model is shown in Figures 1 through 6, and consists of the following:

- Eight 80-bit floating-point data registers (FP0-FP7). These registers are analogous to the integer data registers (D0-D7) and are completely general purpose (i.e., any instruction may use any register).

- A 32-bit control register that contains enable bits for each class of exception trap, and mode bits to set the user-selectable rounding and precision modes.

- A 32-bit status register that contains floating-point condition codes, quotient bits, and exception status information.

- A 32-bit instruction address register that contains the main processor memory address of the last floating-point instruction that was executed. This address is used in exception handling to locate the instruction that caused the exception.

The connection between the MC68020 and the MC68881 is a simple extension of the M68000 bus interface. The MC68881 is connected as a coprocessor to the MC68020, and the selection of the MC68881 is based upon a chip select ($\overline{\text{CS}}$), which is decoded from the MC68020 function codes and address bus. Figure 7 illustrates the MC68881/MC68020 configuration.

As shown in Figure 8, the MC68881 is internally divided into three processing elements; the bus interface unit (BIU), the execution control unit (ECU), and the microcode control unit (MCU). The BIU communicates with the MC68020, and the ECU and MCU execute all MC68881 instructions.

---

*All references to the MC68020, throughout this technical summary, also apply to the MC68030.

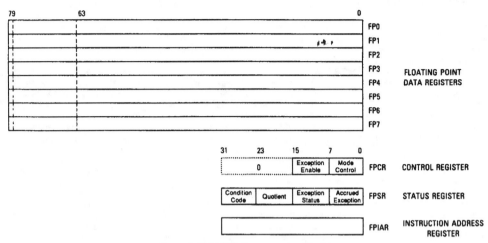

Figure 1. MC68881 Programming Model

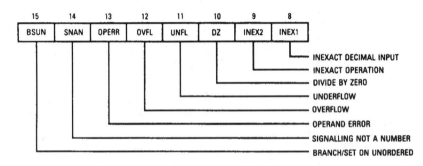

Figure 2. Exception Status/Enable Byte

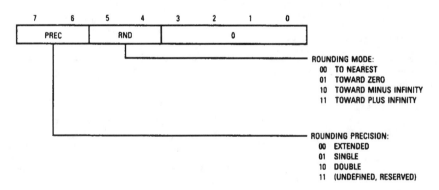

Figure 3. Mode Control Byte

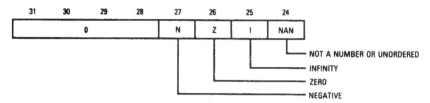

Figure 4. Condition Code Byte

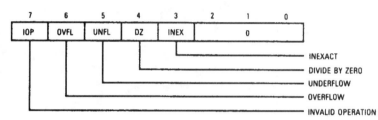

Figure 5. Quotient Byte

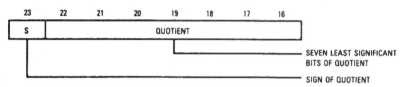

Figure 6. Accrued Exception Byte

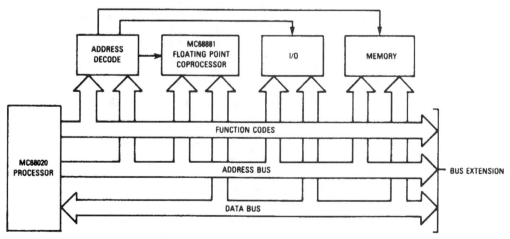

Figure 7. Typical Coprocessor Configuration

559

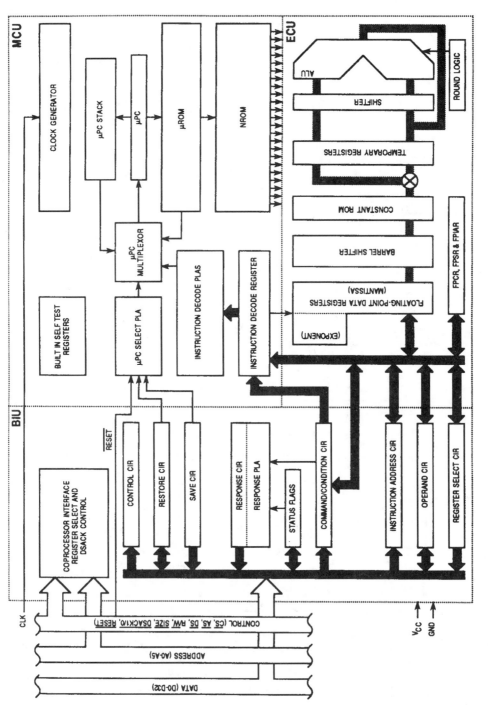

Figure 8. MC68881 Simplified Block Diagram

MC68881

The BIU contains the coprocessor interface registers, and the 32-bit control, status, and instruction address registers. In addition to these registers, the register select and DSACK timing control logic is contained in the BIU. Finally, the status flags used to monitor the status of communications with the main processor are contained in the BIU.

The eight 80-bit floating-point data registers (FP0-FP7) are located in the ECU. In addition to these registers, the ECU contains a high-speed 67-bit arithmetic unit used for both mantissa and exponent calculations, a barrel shifter that can shift from 1 bit to 67 bits in one machine cycle, and ROM constants (for use by the internal algorithms or user programs).

The MCU contains the clock generator, a two-level microcoded sequencer that controls the ECU, the microcode ROM, and self-test circuitry. The built-in self-test capabilities of the MC68881 enhance reliability and ease manufacturing requirements; however, these diagnostic functions are not available to the user.

BUS INTERFACE UNIT

All communications between the MC68020 and the MC68881 occur via standard M68000 Family bus transfers. The MC68881 is designed to operate on 8-, 16-, or 32-bit data buses.

The MC68881 contains a number of coprocessor interface registers (CIRs) which are addressed in the same manner as memory by the main processor. The M68000 Family coprocessor interface is implemented via a protocol of reading and writing to these registers by the main processor. The MC68020 implements this general purpose coprocessor interface protocol in hardware and microcode.

When the MC68020 detects a typical MC68881 instruction, the MC68020 writes the instruction to the memory-mapped command CIR, and reads the response CIR. In this response, the BIU encodes requests for any additional action required of the MC68020 on behalf of the MC68881. For example, the response may request that the MC68020 fetch an operand from the evaluated effective address and transfer the operand to the operand CIR. Once the MC68020 fulfills the coprocessor request(s), the MC68020 is free to fetch and execute subsequent insructions.

A key concern in a coprocessor interface that allows concurrent instrucion execution is synchronization during main processor and coprocessor communication. If a subsequent instruction is written to the MC68881 before the ECU has completed execution of the previous instruction, the response instructs the MC68020 to wait. Thus, the choice of concurrent or nonconcurrent instruction execution is determined on an instruction-by-instruction basis by the coprocessor.

The only difference between a coprocessor bus transfer and any other bus transfer is that the MC68020 issues a function code to indicate the CPU address space during the cycle (the function codes are generated by the M68000 Family processors to identify eight separate address spaces). Thus, the memory-mapped coprocessor interface registers do not infringe upon instruction or data address spaces. The MC68020 places a coprocessor ID field from the coprocessor instruction onto three of the upper address lines during coprocessor accesses. This ID, along with the CPU address space function code, is decoded to select one of eight coprocessors in the system.

Since the coprocessor interface protocol is based solely on bus transfers, the protocol is easily emulated by software when the MC68881 is used as a peripheral with any processor capable of memory-mapped I/O over an M68000 style bus. When used as a peripheral processor with the 8-bit MC68008, or the 16-bit MC68000, or MC68010, all MC68881 instructions are trapped by the main processor to an exception handler at execution time. Thus, the software emulation of the coprocessor interface protocol can be totally transparent to the user. The system can be quickly upgraded by replacing the main processor with an MC68020 without changes to the user software.

Since the bus is asynchronous, the MC68881 need not run at the same clock speed as the main processor. Total system performance may therefore be customized. For example, a system requiring very fast floating-point arithmetic with relatively slow integer arithmetic can be designed with an inexpensive main processor and a fast MC68881.

COPROCESSOR INTERFACE

The M68000 Family coprocessor interface is an integral part of the MC68881 and MC68020 design, with the interface tasks shared between the two. The interface is fully compatible with all present and future M68000 Family products. Tasks are partitioned such that the MC68020 does not have to decode coprocessor instructions, and the MC68881 does not have to duplicate main processor functions such as effective address evaluation.

This partitioning provides an orthogonal extension of the instruction set by permitting MC68881 instructions to utilize all MC68020 addressing modes and to generate execution time exception traps. Thus, from the programmer's view, the CPU and coprocessor appear to be integrated onto a single chip. While the execution of the majority of MC68881 instructions may be overlapped with the execution of MC68020 instructions, concurrency is completely transparent to the programmer. The MC68020 single-step and program flow (trace) modes are fully supported by the MC68881 and the M68000 Family coprocessor interface.

While the M68000 Family coprocessor interface permits coprocessors to be bus masters, the MC68881 is never a bus master. The MC68881 requests that the MC68020 fetch all operands and store all results. In this manner, the MC68020 32-bit data bus provides high speed transfer of floating-point operands and results while simplifying the design of the MC68881.

Since the coprocessor interface is based solely upon bus cycles and the MC68881 is never a bus master, the MC68881 can be placed on either the logical or physical side of the system memory management unit. This provides a great deal of flexibility in the system design.

The virtual machine architecture of the MC68020 is supported by the coprocessor interface and the MC68881 through the FSAVE and FRESTORE instructions. If the MC68020 detects a page fault and/or task time out, the MC68020 can force the MC68881 to stop whatever operation is in process at any time (even in the middle of the execution of an instruction) and save the MC68881 internal state in memory.

The size of the saved internal state of the MC68881 is dependent upon what the ECU is doing at the time that the FSAVE is executed. If the MC68881 is in the reset state when the FSAVE instruction is received, only one word of state is transferred to memory, which may be examined by the operating system to determine that the coprocessor programmer's model is empty. If the coprocessor is idle when the save instruction is received, only a few words of internal state are transferred to memory. If the MC68881 is in the middle of executing an instruction, it may be necessary to save the entire internal state of the machine. Instructions that can complete execution in less time than it would take to save the larger state in mid-instruction are allowed to complete execution and then save the idle state. Thus the size of the saved internal state is kept to a minimum. The ability to utilize several internal state sizes greatly reduces the average context switching time.

The FRESTORE instruction permits reloading of an internal state that was saved earlier, and continues any operation that was previously suspended. Restoring of the reset internal state functions just like a hardware reset to the MC68881 in that defaults are re-established.

OPERAND DATA FORMATS

The MC68881 supports the following data formats:
    Byte Integer (B)
    Word Integer (W)
    Long Word Integer(L)
    Single Precision Real (S)
    Double Precision Real(D)
    Extended Precision Real (X)
    Packed Decimal String Real(P)
The capital letters contained in parenthesis denote suffixes added to instructions in the assembly language source to specify the data format to be used.

INTEGER DATA FORMATS

The three integer data formats (byte, word, and long word) are the standard data formats supported in the M68000 Family architecture. Whenever an integer is used in a floating-point operation, the integer is automatically converted by the MC68881 to an extended precision floating-point operation, the integer is automatically converted by the MC68881 to an extended precision floating-point number before being used. For example, to add an integer constant of five to the number contained in floating-point data register 3 (FP3), the following instruction can be used:

FADD.W #5,FP3
(The Motorola assembler syntax "#" is used to is used to denote immediate addressing.)

The ability to effectively use integers in floating-point operaions saves user memory since an integer representation of a number, if respresentable, is usually smaller than the equivalent floating-point representation.

FLOATING-POINT DATA FORAMTS

The floating-point data formats single precision (32-bits) and double precision (64-bits) are as defined by the IEEE standard. These are the main floating-point formats and should be used for most calculations involving real numbers. Table 1 lists the exponent and mantissa size for single, double, and extended precision. The exponent is biased, and the mantissa is in sign and magnitude form. Since single and double precision require normalized numbers, the most significant bit of the mantissa is implied as one and is not included, thus giving one extra bit of precision.

Table 1. Exponent and Mantissa Sizes

Data Format	Exponent Bits	Mantissa Bits	Bias
Single	8	23(+ 1)	127
Double	11	52(+ 1)	1023
Extended	15	64	16383

The extended precision data format is also in conformance with the IEEE standard, but the standard does not specify this format to the bit level as it does for single and double precision. The memory format on the MC68881 consists of 96 bits (three long words). Only 80 bits are actually used, the other 16 bits are for future expandability and for long-word alignment of floating-point data structures. Extended format has a 15-bit exponent, a 64-bit mantissa, and a 1-bit mantissa sign.

Extended precision numbers are intended for use as temporary variables, intermediate values, or in places where extra precision is needed. For example, a compiler might select extended precision arithmetic for evaluation of the right side of an equation with mixed sized data and then convert the answer to the data type on the left side of the equation. It is anticipated that extended precision data will not be stored in large arrays, due to the amount of memory required by each number.

PACKED DECIMAL STRING REAL DATA FORMAT

The packed decimal data format allows packed BCD strings to be input to and output from the MC68881. The strings consist of a 3-digit base 10 exponent and a 17-digit base 10 mantissa. Both the exponent and mantissa have a separate sign bit. All digits are packed BCD, such that an entire string fits in 96 bits (three long words). As is the case with all data formats, when packed BCD strings are input to the MC68881, the strings are automatically converted to extended precision real values. This allows packed BCD numbers to be used as inputs to any operation. For example:

FADD.P #-6.023E + 24,FP5

BCD numbers can be output from the MC68881 in a format readily used for printing by a program generated by a high-level language compiler. For example:

FMOVE.P FP3,BUFFER{#-5}

instructs the MC68881 to convert the floating-point data register 3 (FP3) contents into a packed BCD string with five digits to the right of the decimal point (FORTRAN F format).

MC68881

DATA FORMAT SUMMARY

All data formats described above are supported orthogonally by all arithmetic and transcendental operations, and by all appropriate MC68020 addressing modes. For example, all of the following are legal instructions:

```
FADD.B    #3,FP0
FADD.W    D2,FP3
FADD.L    BIGINT,FP7
FADD.S    #3.14159,FP5
FADD.D    (SP) + ,FP6
```

```
FADD.X    [(TEMP__PTR,A7)],FP3
FADD.P    #1.23E25,FP0
```

On-chip calculations are performed to extended precision format, and the eight floating point data registers always contain exetended precision values. All data used in an operation is converted to extended precision by the MC68881 before the specific operation is performed, and all results are in extended precision. This ensures maximum accuracy without sacrificing performance.

Refer to Figure 9 for a summary of the memory formats for the seven data formats supported by the MC68881.

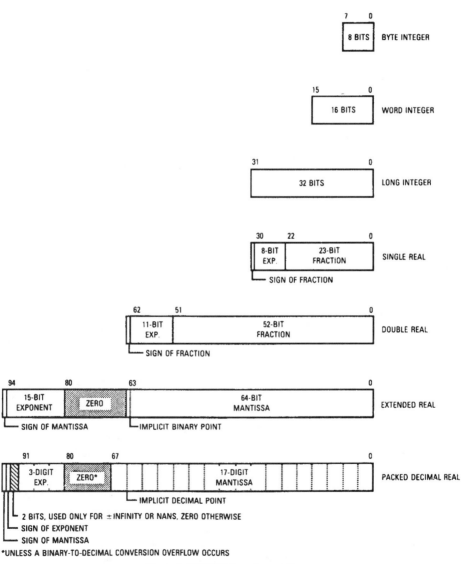

Figure 9. MC68881 Data Format Summary

INSTRUCTION SET

The MC68881 instruction set is organized into six major classes:

1. Moves between the MC68881 and memory or the MC68020 (in and out),
2. Move multiple registers (in and out),
3. Monadic operations,
4. Dyadic operations,
5. Branch, set, or trap conditionally, and
6. Miscellaneous.

MOVES

All moves from memory (or from an MC68020 data register) to the MC68881, cause data conversion from the source data format to the internal extended precision format.

All moves from the MC68881 to memory (or to an MC68020 data register), cause data conversion from the internal extended precison format to the destination data format.

Note that data movement instructions perform arithmeic operations, since the result is always rounded to the precision selected in the FPCR mode contol byte. The result is rounded using the selected rounding mode, and is checked for overflow and underflow.

The syntax for the move is:

FMOVE.(fmt)	(ea),FPn	Move to MC68881
FMOVE.(fmt)	FPm,(ea)	Move from MC68881
FMOVE.X	FPm,FPn	Move within MC68881

where:

(ea) is an MC68020 effective address operand and (fmt) is the data format size. FPm and FPn are floating-point data registers.

MOVE MULTIPLES

The floating-point move multiple instructions on the MC68881 are much like the integer counterparts on the M68000 Family processors. Any set of the floating-point registers FP0 through FP7 can be moved to or from memory with one instruction. These registers are always moved as 96-bit extended data with no conversion (hence no possibility of conversion errors). Some move multiple examples are as follows:

| FMOVEM | (ea),FP0-FP3/FP7 |
| FMOVEM | FP2/FP4/FP6,(ea) |

Move multiples are useful during context switches and interrupts to save or restore the state of a program. These moves are also useful at the start and end of a procedure to save and restore the calling routine's register set. In order to reduce procedure call overhead, the list of registers to be saved or restored can be contained in a data register. This allows run-time optimization by allowing a called routine to save as few registers as possible. Note that no rounding or overflow/underflow checking is performed by these operations.

MONADIC OPERATIONS

Monadic operations have one operand. This operand may be in a floating-point data register, memory, or in

an MC68020 data register. The result is always stored in a floating-point data register. For example, the syntax for square root is:

FSQRT.(fmt)	(ea),FPn or,
FSQRT.X	FPm,FPn or,
FSQRT.X	FPn

The MC68881 monadic operations available are as follows:

FABS	Absolute Value
FACOS	Arc Cosine
FASIN	Arc Sine
FATAN	Arc Tangent
FATANH	Hyperbolic Arc Tangent
FCOS	Cosine
FCOSH	Hyperbolic Cosine
FETOX	e to the x Power
FETOXM1	e to the x Power − 1
FGETEXP	Get Exponent
FGETMAN	Get Mantissa
FINT	Integer Part
FINTRZ	Integer Part (Truncated)
FLOG10	Log Base 10
FLOG2	Log Base 2
FLOGN	Log Base e
FLOGNP1	Log Base e of(x + 1)
FNEG	Negate
FSIN	Sine
FSINCOS	Simultaneous Sine and Cosine
FSINH	Hyperbolic Sine
FSQRT	Square Root
FTAN	Tangent
FTANH	Hyperbolic Tangent
FTENTOX	10 to the x Power
FTST	Test
FTWOTOX	2 to the x Power

DYADIC OPERATIONS

Dyadic operations have two input operands. The first input operand comes from a floating-point data register, memory, or an MC68020 data register. The second input operand comes from a floating-point data register. The destination is the same floating-point data register used for the second input. For example, the syntax for add is:

| FADD.(fmt) | (ea),FPn or, |
| FADD.X | FPm,FPn |

The MC68881 dyadic operations available are as follows:

FADD	Add
FCMP	Compare
FDIV	Divide
FMOD	Modulo Remainder
FMUL	Multiply
FREM	IEEE Remainder
FSCALE	Scale Exponent
FSGLDIV	Single Precision Divide
FSGLMUL	Single Precision Multiply
FSUB	Subtract

BRANCH, SET, AND TRAP-ON CONDITION

The floating-point branch, set, and trap-on condition instructions implemented by the MC68881 are similar to

MC68881

the equivalent integer instructions of the M68000 Family processors, except that more conditions exist due to the special values in IEEE floating-point arithmetic. When a conditional instruction is executed, the MC68881 performs the necessary condition checking and tells the MC68020 whether the condition is true or false; the MC68020 then takes the appropriate action. Since the MC68881 and MC68020 are closely coupled, the floating-point branch operations executed by the pair are very fast.

The MC68881 conditional operations are:

FBcc	Branch
FDBcc	Decrement and Branch
FScc	Set Byte According to Condition
FTRAPcc	Trap-on Condition (with an Optional Parameter)

where:

cc is one of the 32 floating-point conditional test specifiers as shown in Table 2.

Table 2. Floating-Point Conditional Test Specifiers

Mnemonic	Definition
NOTE	
The following conditional tests do not set the BSUN bit in the status register exception byte under any circumstances.	
F	False
EQ	Equal
OGT	Ordered Greater Than
OGE	Ordered Greater Than or Equal
OLT	Ordered Less Than
OLE	Ordered Less Than or Equal
OGL	Ordered Greater or Less Than
OR	Ordered
UN	Unordered
UEQ	Unordered or Equal
UGT	Unordered or Greater Than
UGE	Unordered or Greater or Equal
ULT	Unordered or Less Than
ULE	Unordered or Less or Equal
NE	Not Equal
T	True
NOTE	
The following conditional tests set the BSUN bit in the status register exception byte if the NAN condition code bit is set when a conditional instruction is executed.	
SF	Signaling False
SEQ	Signaling Equal
GT	Greater Than
GE	Greater Than or Equal
LT	Less Than
LE	Less Than or Equal
GL	Greater or Less Than
GLE	Greater Less or Equal
NGLE	Not (Greater, Less or Equal)
NGL	Not (Greater or Less)
NLE	Not (Less or Equal)
NLT	Not (Less Than)
NGE	Not (Greater or Equal)
NGT	Not (Greater Than)
SNE	Signaling Not Equal
ST	Signaling True

MISCELLANEOUS INSTRUCTIONS

Miscellaneous instructions include moves to and from the status, control, and instruction address registers. Also included are the virtual memory/machine FSAVE and FRESTORE instructions that save and restore the internal state of the MC68881.

FMOVE	(ea),FPcr	Move to Control Register(s)
FMOVE	FPcr,(ea)	Move from Control Register(s)
FSAVE	(ea)	Virtual Machine State Save
FRESTORE	(ea)	Virtual Machine State Restore

ADDRESSING MODES

The MC68881 does not perform address calculations. This satifies the criterion that an M68000 Family coprocessor must not depend on certain features or capabilities that may or may not be implemented by a given main processor. Thus, when the MC68881 instructs the MC68020 to transfer an operand via the coprocessor interface, the MC68020 performs the addressing mode calculations requested in the instruction. In this case, the instruction is encoded specifically for the MC68020, and the execution of the MC68881 is not dependent on that encoding, but only on the value of the command word written to the MC68881 by the main processor.

This interface is quite flexible and allows any addressing mode to be used with floating-point instructions. For the M68000 Family, these addressing modes include immediate, postincrement, predecrement, data or address register direct, and the indexed/indirect addressing modes of the MC68020. Some addressing modes are restricted for some instructions in keeping with the M68000 Family architectural definitions (e.g. PC relative addressing is not allowed for a destination operand).

The orthogonal instruction set of the MC68881, along with the flexible branches and addressing modes, allows a programmer writing assembly language code, or a compiler writer generating object or source code for the MC68020/MC68881 device pair, to think of the MC68881 as though the MC68881 is part of the MC68020. There are no special restrictions imposed by the coprocessor interface, and floating-point arithmetic is coded exactly like integer arithmetic.

TIMING TABLES FOR TYPICAL EXECUTION

This set of tables allows a quick determination of the typical execution time for any MC68881 instruction when the MC68020 is used as the main processor. The first table presented is for effective address caculations performed by the MC68020. Entries from this table are added to entries in the other tables, if necessary, to obtain the total number of clock cycles for an operation. The assumptions for the following tables are:

- The main processor is an MC68020 and operates on the same clock as the MC68881. Instruction prefetches do not hit in the MC68020 cache (or it is disabled) and the instruction is aligned such that a prefetch occurs before the command CIR is written by the MC68020.

- A 32-bit memory interface is used, and memory accesses occur with zero wait states. All memory operands, as well as the stack pointers, are long-word aligned.

- Accesses to the MC68881 require 3 clock cycles, with the exception of read accesses to the response and save CIRs, which require 5 clock cycles.

- No instruction overlap is utilized, so the coprocessor interface overhead is 11 clocks. This can be reduced to 2 clock cycles if optimized code sequences are used, or may be 11 clock cycles if overlap is attempted and a synchronization delay is required.

- Typical operand conversion and calculation times are used (i.e. input operands are assumed to be normalized numbers in the legal range for a given fuction).

- No exceptions are enabled or occur, and the default rounding mode and precision of round-to-nearest, extended precision, is used.

EFFECTIVE ADDRESS CALCULATIONS

For any instruction that requires an operand external to the MC68881, an evaluate effective address and transfer data response primitive is issued by the MC68881 during the dialog for that instruction. The amount of time that is required for the MC68020 to calculate the effective address while processing this primive for each addressing mode, excluding the transfer of the data to the MC68881, is shown in Table 3.

ARITHMETIC OPERATIONS

The Table 4 gives the typical instruction execution time for each arithmetic instruction. This group of instructions includes the majority of the MC68881 operations such as FADD, FSUB, etc. In addition to the instructions that perform arithmetic calculations as part of their function, the FCMP, FMOVE and FTST instructions are also included,

Table 3. Effective Address Calculations Execution Timing

Addressing Mode	Best Case	Cache Case	Worst Case
Dn or An	0	0	0
(An)	0	2	2
(An) +	3	6	6
– (An)	3	6	6
(d_{16},An) or (d_{16},PC)	0	2	3
(xxx.W)	0	2	3
(xxx).L	1	4	5
#(data)	0	0	0
(d_8,An,Xn) or (d_8,PC,Xn)	1	4	5
(d_{16},An,Xn) or (d_{16},PC,Xn)	3	6	7
(B)	3	6	7
(d_{16},B)	5	8	9
(d_{32},B)	11	14	16
([B],I)	8	11	12
$([B],I,d_{16})$	8	11	12
$([B],I,d_{32})$	10	13	15
$([d_{16},B],I)$	10	13	14
$([d_{16},B],I,d_{16})$	10	13	15
$([d_{16},B],I,d_{32})$	12	15	17
$([d_{32},B],I)$	16	19	21
$([d_{32},B],I,d_{16})$	16	19	21
$([d_{32},B],I,d_{32})$	18	21	24

B = Base address; 0, An, PC, Xn, An + Xn, PC + Xn. Form does not affect timing.
I = Index; 0 or Xn.
Note that Xn cannot be in B and I at the same time. Scaling and size of Sn does not affect timing.

Table 4. Arithmetic Operations Execution Timing

Instruction	FPm Source	Memory Source or Destination Operand Format*				
		Integer**	Single**	Double	Extended	Packed
FABS	35	62	54	60	58	872
FACOS	652	625	644	650	648	1462
FADD	51	80	72	78	76	888
FASIN	581	608	600	606	604	1418
FATAN	403	430	422	428	426	1240
FATANH	693	720	712	718	716	1530
FCMP	33	62	54	60	58	870
FCOS	391	418	410	416	414	1228
FCOSH	607	634	626	632	630	1444
FDIV	103	132	124	130	128	940
FETOX	497	524	516	522	520	1334
FETOXM1	545	572	564	570	568	1382
FGETEXP	45	72	64	70	68	882
FGETMAN	31	58	50	56	54	868
FINT	55	82	74	80	78	892
FINTRZ	55	82	78	80	74	892
FLOGN	525	552	544	550	548	1362
FLOGNP1	571	598	590	596	594	1408
FLOG10	581	608	600	606	604	1418
FLOG2	581	608	600	606	604	1418
FMOD	67	94	86	92	90	902
FMOVE to FPn	33	60	52	58	56	870
FMOVE to Memory***	—	100	80	86	72	1996
FMOVECR****	29	—	—	—	—	—
FMUL	71	100	92	98	96	908
FNEG	35	62	54	60	58	872
FREM	67	94	86	92	90	902
FSCALE	41	70	62	68	66	878
FSGLDIV	69	98	90	96	94	906
FSGLMUL	59	88	80	86	84	896
FSIN	391	418	410	416	414	1228
FSINCOS	451	478	470	476	474	1288
FSINH	687	714	706	712	710	1524
FSQRT	107	134	126	132	130	944
FSUB	51	80	72	78	76	888
FTAN	473	500	492	498	496	1310
FTANH	661	688	680	686	684	1498
FTENTOX	567	594	586	592	590	1404
FTST	33	60	52	58	56	870
FTWOTOX	567	594	586	592	590	1404

*Add the appropriate effective address calculation time.

**If the source or destination is an MC68020 data register, subtract 5 or 2 clock cycles, respectively.

***Assume a static K-factor is used if the destination data format is packed decimal. Add 14 clock cycles if a dynamic K-factor is used.

****The source operand is from the constant ROM rather than a floating-point data register.

since an implicit conversion is performed by those operations. For memory operands, the timing for the appropriate effective addressing mode must be added to the numbers in this table to determine the overall instruction execution times.

MOVE CONTROL REGISTER AND MOVE MULTIPLE OPERATIONS

The Table 5 gives the execution times for the FMOVE FPcr and FMOVEM instructions. The timing for the appropriate effective addressing mode must be added to the numbers in this table to determine the overall instruction execution times.

CONDITIONAL OPERATIONS

The Table 6 gives the execution times for the MC68881 conditional instructions. Each entry in this table, except those for the FScc instruction, is complete and does not require the addition of values from any other table. For the FScc instruction, the only additional factor that must

be included is the calculate effective address time for the operand to be modified.

FSAVE AND FRESTORE INSTRUCTIONS

The time requred for a context save or restore operation is given in Table 7. The appropriate calculate effective address times must be added to the values in this table to obtain the total execution time for these operations. For the FSAVE instruction, the MC68881 may use the not ready format code to force the MC68020 to wait while internal operations are completed in order to reduce the size of the saved state frame or reach a point where a save operation can be performed. The idle (minimum) time occurs if the MC68881 is in the idle phase when the save CIR is written. A time between the idle (minimum) and the idle (maximum) occurs if an instruction is in the end phase when the save CIR is read. The busy (minimum) time occurs if the MC68881 is in the initial phase, or at a save boundary in the middle phase, when the save CIR is read. Finally, the busy (maximum) time occurs if the MC68881 has just passed a save boundary in the middle phase when the save CIR is read.

Table 5. Move Control Register and FMOVEM Operations Execution Timing

Operation*		Best Case	Cache Case	Worst Case
FMOVE	FPcr,Rn	29	31	34
	FPcr,(ea)	31	33	36
	Rn,FPcr	26	28	31
	(ea),FPcr	31	33	36
	#(data),FPcr	30	30	31
FMOVEM	FPcr__list,(ea)	25 + 6n	27 + 6n	30 + 6n
	(ea),FPcr__list	25 + 6n	27 + 6n	30 + 6n
	#(data),FPcr__list	24 + 6n	25 + 6n	29 + 6n
FMOVEM	FPdr__list,(ea)	35 + 25n	37 + 25n	40 + 25n
	(ea),FPdr__list	33 + 23n	35 + 23n	38 + 23n
	Dn,(ea)	49 + 25n	51 + 25n	54 + 25n
	(ea),Dn	47 + 23n	49 + 23n	52 + 23n

*Add the appropriate effective address calculation time.
n is the number of registers transferred.

Table 6. Conditional Instructions Execution Timing

Operation	Comments	Best Case	Cache Case	Worst Case
FBcc.W	Branch Taken Branch Not Taken	18 16	20 18	23 19
FBcc.L	Branch Taken Branch Not Taken	18 16	20 18	23 21
FDBcc	True, Not Taken False, Not Taken False, Taken	18 22 18	20 24 20	24 32 26
FNOP	No Operation	16	18	19
FScc	Dn (An) + or − (An)* Memory**	16 18 16	18 22 20	21 25 23
FTRAPcc	Trap Taken Trap Not Taken	36 16	39 18	47 22
FTRAPcc.W	Trap Taken Trap Not Taken	38 18	41 20	45 23
FTRAPcc.L	Trap Taken Trap Not Taken	40 20	43 22	52 27

*For condition true; subtract one clock for condition false.
**Add the appropriate effective address calculation time.

Table 7. FSAVE and FRESTORE Instructions Execution Timing

Operation*	State Frame	Best Case	Cache Case	Worst Case
FRESTORE	Null Idle Busy	19 55 312	21 57 314	22 58 315
FSAVE	Null Idle (Minimum) Idle (Maximum) Busy (Minimum) Busy (Maximum)	14 50 286 316 552	16 52 218 318 554	18 54 290 320 556

*Add the appropriate effective address calculation time.

FUNCTIONAL SIGNAL DESCRIPTIONS

This section contains a brief description of the input and output signals for the MC68881 floating-point coprocessor. The signals are functionally organized into groups as shown in Figure 10.

NOTE

The terms **assertion** and **negation** are used extensively. This is done to avoid confusion when describing "active-low" and "active-high" signals. The term **assert** or **assertion** is used to indicate that a signal is active or true, independent of whether that level is represented by a high or low voltage. The term **negate** or **negation** is used to indicate that a signal is inactive or false.

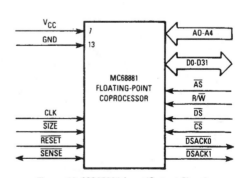

Figure 10. MC68881 Input/Output Signals

569

Table 8. Coprocessor Interface Register Selection

A4-A0	Offset	Width	Type	Register
0000x	$00	16	Read	Response
0001x	$02	16	Write	Control
0010x	$04	16	Read	Save
0011x	$06	16	R/W̄	Restore
0100x	$08	16	—	(Reserved)
0101x	$0A	16	Write	Command
0110x	$0C	16	—	(Reserved)
0111x	$0E	16	Write	Condition
100xx	$10	32	R/W̄	Operand
1010x	$14	16	Read	Register Select
1011x	$16	16	—	(Reserved)
110xx	$18	32	Read	Instruction Address
111xx	$1C	32	R/W̄	Operand Address

ADDRESS BUS (A0 through A4)

These active-high address line inputs are used by the main processor to select the coprocessor interface register locations located in the CPU address space. These lines control the register selection as listed in Table 8.

When the MC68881 is configured to operate over an 8-bit data bus, the A0 pin is used as an address signal for byte accesses of the coprocessor interface registers. When the MC68881 is configured to operate over a 16- or 32-bit system data bus, both the A0 and SIZE pins are strapped high and/or low as listed in Table 9.

Table 9. System Data Bus Size Configuration

A0	SIZE	Data Bus
—	Low	8-Bit
Low	High	16-Bit
High	High	32-Bit

DATA BUS (D0 through D31)

This 32-bit, bidirectional, three-state bus serves as the general purpose data path between the MC68020 and the MC68881. Regardless of whether the MC68881 is operated as a coprocessor or a peripheral processor, all interprocessor transfers of instruction information, operand data, status information, and requests for service occur as standard M68000 bus cycles.

The MC68881 will operate over an 8-, 16-, or 32-bit system data bus. Depending upon the system data bus configuration, both the A0 and SIZE pins are configured specifically for the applicable bus configuration. (Refer to **ADDRESS BUS (A0 through A4)** and **SIZE (SIZE)** for further details.)

SIZE (SIZE)

This active-low input signal is used in conjunction with the A0 pin to configure the MC68881 for operation over an 8-, 16-, or 32-bit system data bus. When the MC68881 is configured to operate over a 16- or 32-bit system data bus, both the SIZE and A0 pins are strapped high and/or low as listed in Table 9.

ADDRESS STROBE (AS)

This active-low input signal indicates that there is a valid address on the address bus, and both the chip select (CS) and read/write (R/W̄) signal lines are valid.

CHIP SELECT (CS)

This active-low input signal enables the main processor access to the MC68881 coprocessor interface registers. When operating the MC68881 as a peripheral processor, the chip select decode is system dependent (i.e., like the chip select on any peripheral). The CS signal must be valid (either asserted or negated) when AS is asserted. Refer to **CHIP SELECT TIMING** for further discussion of timing restrictions for this signal.

READ/WRITE (R/W̄)

This input signal indicates the direction of a bus transaction (read/write) by the main processor. A logic high (1) indicates a read from the MC68881, and a logic low (0) indicates a write to the MC68881. The R/W̄ signal must be valid when AS is asserted.

DATA STROBE (DS)

This active-low input signal indicates that there is valid data on the data bus during a write bus cycle.

MC68881

DATA TRANSFER AND SIZE ACKNOWLEDGE (DSACK0, DSACK1)

These active-low, three-state output signals indicate the completion of a bus cycle to the main processor. The MC68881 asserts both the DSACK0 and DSACK1 signals upon assertion of CS.

If the bus cycle is a main processor read, the MC68881 asserts DSACK0 and DSACK1 signals to indicate that the information on the data bus is valid. (Both DSACK signals may be asserted in advance of the valid data being placed on the bus.) If the bus cycle is a main processor write to the MC68881, DSACK0 and DSACK1 are used to acknowledge acceptance of the data by the MC68881.

The MC68881 also uses DSACK0 and DSACK1 signals to dynamically indicate to the MC68020 the "port" size (system data bus width) on a cycle-by-cycle basis. Depending upon which of the two DSACK pins are asserted in a given bus cycle, the MC68020 assumes data has been transferred to/from an 8-, 16-, or 32-bit wide data port. Table 10 lists the DSACK assertions that are used by the MC68881 for the various bus cycles over the various bus cycles over the various system data bus configurations.

Table 13 indicates that all accesses over a 32-bit bus where A4 equals zero are to 16-bit registers. The MC68881 implements all 16-bit coprocessor interface registers on data lines D16-D31 (to eliminate the need for on-chip multiplexers); however, the MC68020 expects 16-bit registers that are located in a 32-bit port at odd word addresses (A1 = 1) to be implemented on data lines D0-D15. For accesses to these registers when configured for 32-bit bus operation, the MC68881 generates DSACK signals as listed in Table 10 to inform the MC68020 of valid data on D16-D31 instead of D0-D15.

An external holding resistor is required to maintain both DSACK0 and DSACK1 high between bus cycles. In order to reduce the signal rise time, the DSACK0 and DSACK1 lines are actively pulled up (negated) by the MC68881 following the rising edge of AS or DS, and both DSACK lines are then three-stated (placed in the high-impedance state) to avoid interference with the next bus cycle.

RESET (RESET)

This active-low input signal causes the MC68881 to initialize the floating-point data registers to non-signaling not-a-numbers (NANs) and clears the floating-point control, status, and instruction address registers.

When performing a power-up reset, external circuitry should keep the RESET line asserted for a minimum of four clock cycles after V_{CC} is within tolerance. This assures correct initialization of the MC68881 when power is applied. For compatibility with all M68000 Family devices, 100 milliseconds should be used as the minimum.

When performing a reset of the MC68881 after V_{CC} has been within tolerance for more than the initial power-up time, the RESET line must have an asserted pulse width which is greater than two clock cycles. For compatability with all M68000 Family devices, 10 clock cycles should be used as the minimum.

CLOCK (CLK)

The MC68881 clock input is a TTL-compatable signal that is internally buffered for development of the internal clock signals. The clock input should be a constant frequency square wave with no stretching or shaping techniques required. The clock should not be gated off at any time and must conform to minimum and maximum period and pulse width times.

SENSE DEVICE (SENSE)

This pin may be used optionally as an additional GND pin, or as an indicator to external hardware that the MC68881 is present in the system. This signal is internally connected to the GND of the die, but it is not necessary to connect it to the external ground for correct device operation. If a pullup resistor (which should be larger than 10 kohm) is connected to this pin location, external hardware may sense the presence of the MC68881 in a system.

POWER (V_{CC} and GND)

These pins provide the supply voltage and system reference level for the internal circuitry of the MC68881. Care should be taken to reduce the noise level on these pins with appropriate capacitive decoupling.

NO CONNECT (NC)

One pin of the MC68881 package is designated as a no connect (NC). This pin position is reserved for future use by Motorola, and should not be used for signal routing or connected to V_{CC} or GND.

SIGNAL SUMMARY

Table 11 provides a summary of all the MC68881 signals described in this section.

Table 10. DSACK Assertions

Data Bus	A4	DSACK1	DSACK0	Comments
32-Bit	1	L	L	Valid Data on D31-D0
32-Bit	0	L	H	Valid Data on D31-D16
16-Bit	x	L	H	Valid Data on D31-D16 or D15-D0
8-Bit	x	H	L	Valid Data on D31-D24, D23-D16, D15-D8, or D7-D0
All	x	H	H	Insert Wait States in Current Bus Cycle

Table 11. Signal Summary

Signal Name	Mnemonic	Input/Output	Active State	Three State
Address Bus	A0-A4	Input	High	—
Data Bus	D0-D13	Input/Output	High	Yes
Size	$\overline{SIZE}$	Input	Low	—
Address Strobe	$\overline{AS}$	Input	Low	—
Chip Select	$\overline{CS}$	Input	Low	—
Read/Write	R/$\overline{W}$	Input	High/Low	—
Data Strobe	$\overline{DS}$	Input	Low	—
Data Transfer and Size Acknowledge	$\overline{DSACK0}$, $\overline{DSACK1}$	Output	Low	Yes
Reset	$\overline{RESET}$	Input	Low	—
Clock	CLK	Input	—	—
Sense Device	$\overline{SENSE}$	Input/Output	Low	No
Power Input	V_{CC}	Input	—	—
Ground	GND	Input	—	—

INTERFACING METHODS

MC68881/MC68020 INTERFACING

The following paragraphs describe how to connect the MC68881 to an MC68020 for coprocessor operation via an 8-, 16-, or 32-bit data bus.

32-Bit Data Bus Coprocessor Connection

Figure 11 illustrates the coprocessor interface connection of an MC68881 to an MC68020 via a 32-bit data bus. The MC68881 is configured to operate over a 32-bit data bus when both the A0 and $\overline{SIZE}$ pins are connected to V_{CC}.

16-Bit Data Bus Coprocessor Connection

Figure 12 illustrates the coprocessor interface connection of an MC68881 to an MC68020 via a 16-bit data bus. The MC68881 is configured to operate over a 16-bit data bus when the $\overline{SIZE}$ pin is connected to V_{CC}, and the A0 pin is connected to GND. The sixteen least significant data pins (D0-D15) must be connected to the sixteen most significant data pins (D16-D31) when the MC68881 is configured to operate over a 16-bit data bus (i.e., connect D0 to D16, D1 to D17, ... and D15 to D31). The DSACK pins of the two devices are directly connected, although it is not necessary to connect the $\overline{DSACK0}$ pin since the MC68881 never asserts it in this configuration.

8-Bit Data Bus Coprocessor Connection

Figure 13 illustrates the connection of an MC68881 to an MC68020 as a coprocessor over an 8-bit data bus. The MC68881 is configured to operate over an 8-bit data bus when the $\overline{SIZE}$ pin is connected to GND. The twenty four least significant data pins (D0-D23) must be connected to

the eight most significant data pins (D24-D31) when the MC68881 is configured to operate over an 8-bit data bus (i.e., connect D0 to D8, D16 and D24; D1 to D9, D17, and D25; ... and D7 to D15, D23 and D31). The DSACK pins of the two devices are directly connected, although it is not necessary to connect the $\overline{DSACK1}$ pin since the MC68881 never asserts it in this configuration.

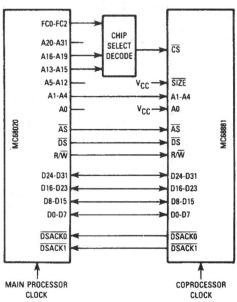

Figure 11. 32-Bit Data Bus Coprocessor Connection

MC68881

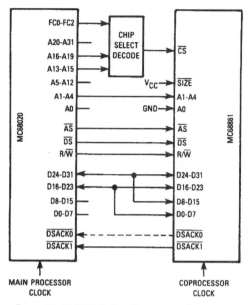

Figure 12. 16-Bit Data Bus Coprocessor Connection

MC68881-MC68000/MC68008/MC68010 INTERFACING

The following paragraphs describe how to connect the MC68881 to an MC68000, MC68008, or MC68010 processor for operation as a peripheral via an 8- or 16-bit data bus.

16-Bit Data Bus Peripheral Processor Connection

Figure 14 illustrates the connection of an MC68881 to an MC68000 or MC68010 as a peripheral processor over a 16-bit data bus. The MC68881 is configured to operate over a 16-bit data bus when the $\overline{\text{SIZE}}$ pin is connected to V_{CC}, and the A0 pin is connected to GND. The sixteen least significant data pins (D0-D15) must be connected to the sixteen most significant data pins (D16-D31) when the MC68881 is configured to operate over a 16-bit data bus (i.e., connect D0 to D16, D1 to D17, ... and D15 to D31). The $\overline{\text{DSACK1}}$ pin of the MC68881 is connected to the $\overline{\text{DTACK}}$ pin of the main processor, and the DSACK0 pin is not used.

When connected as a peripheral processor, the MC68881 chip select ($\overline{\text{CS}}$) decode is system dependent. If the MC68000 is used as the main processor, the MC68881 $\overline{\text{CS}}$ must be decoded in the supervisor or user data spaces. However, if the MC68010 is used for the main processor, the MOVES instruction may be used to emulate any CPU space access that the MC68020 generates for coprocessor communications. Thus, the $\overline{\text{CS}}$ decode logic for such systems may be the same as in an MC68020 system, such that the MC68881 will not use any part of the data address spaces.

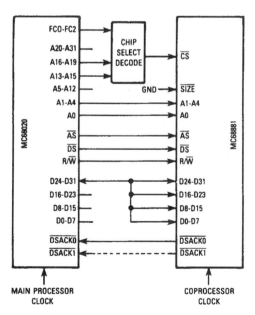

Figure 13. 8-Bit Data Bus Coprocessor Connection

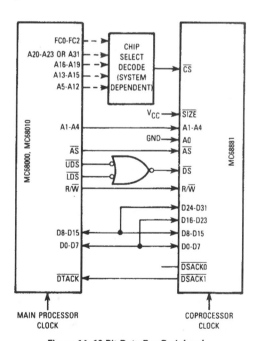

Figure 14. 16-Bit Data Bus Peripheral Processor Connection

8-Bit Data Bus Peripheral Processor Connection

Figure 15 illustrates the connection of an MC68881 to an MC68008 as a peripheral processor over an 8-bit data bus. The MC68881 is configured to operate over an 8-bit data bus when the $\overline{\text{SIZE}}$ pin is connected to GND. The eight least significant data pins (D0-D7) must be connected to the twenty four most significant data pins (D8-D31) when the MC68881 is configured to operate over an

8-bit data bus (i.e. connect D0 to D8, D16, and D24; D1 to D9, D17, and D25; ... and D7 to D15, D23, and D31). The $\overline{\text{DSACK0}}$ pin of the MC68881 is connected to the $\overline{\text{DTACK}}$ pin of the MC68008, and the $\overline{\text{DSACK1}}$ pin is not used.

When connected as a peripheral processor, the MC68881 chip select ($\overline{\text{CS}}$) decode is system dependent, and the $\overline{\text{CS}}$ must be decoded in the supervisor or user data spaces.

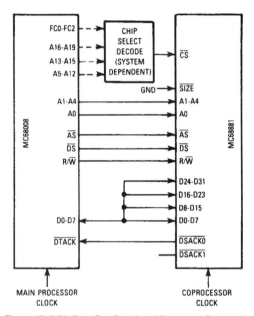

Figure 15. 8-Bit Data Bus Peripheral Processor Connection

APPENDIX E

A Note About ASM68K and EMU68K

The ASM68K assembler and EMU68K emulator programs included on the companion disk are constantly going through changes. Many improvements and fixes have been made to each since the third edition of this book was written.

On the companion disk there are two directories: ASM68K and EMU68K. Inside each you will find a .DOC file containing documentation about each program. The documentation contains fixes and changes since the original versions were released with the third edition. The most recent versions are always available from the author's Web site at:

http://www.sunybroome.edu/~antonakos_j

Navigate to the 68000 page and look for the download area.

Some of the additions to EMU68K are:

- New commands (breakpoint, unassemble, hex math, fill)
- A larger emulator memory (up to 64KB)
- New DOS interrupt interface via TRAP #10

Some of the changes to ASM68K were cosmetic (tab expansions in list file), others were corrective (BGT/BLT are *not* swapped anymore).

Additional comments and suggestions regarding ASM68K and EMU68K are welcome. Send e-mail from the Web page or e-mail the author at:

antonakos_j@sunybroome.edu

APPENDIX F

A Review of Number Systems and Binary Arithmetic

When working with microprocessors, you will find it necessary to have a good grasp of the binary and hexadecimal number systems, and the arithmetic associated with them. This appendix is intended as a review (or a brief introduction) to this material.

DECIMAL VERSUS BINARY

A decimal number is composed of one or more digits chosen from a set of 10 digits {0, 1, 2, 3, 4, 5, 6, 7, 8, 9}. Each digit in a decimal number has an associated *weight* that is used to give the digit meaning. For example, the decimal number 357 contains three 100s, five 10s, and seven 1s. The weight of the digit 3 is 100, the weight of the digit 5 is 10, and the weight of the digit 7 is 1. The weights of each digit in a decimal number are related to its *base*. Decimal numbers are base-10 numbers. Thus, the weights are all multiples of 10. Look at our example decimal number again:

3	5	7 —digits
10^2	10^1	10^0—weights as powers of 10
100	10	1 —actual weight value
300	50	7 —components of number

Notice that the weights are all powers of 10 beginning with 0. The components of the number are found by multiplying each digit value by its respective weight. The number itself is found by adding the individual components together. This technique applies to numbers in *any* base.

Now, a binary number is a number composed of digits (called **bits**) chosen from a set of only **two** digits {0, 1}. Base-2 is used for binary numbers because there are only two legal digits in a binary number. This means that the weights of the bits in a binary number will all be multiples of 2! Consider the binary number 10110. The associated weights are as follows:

1	0	1	1	0 —bits
2^4	2^3	2^2	2^1	2^0—weights as powers of 2
16	8	4	2	1 —actual weight value
16		4	2	—components of number

The components are again found by multiplying each bit in the number by its respective power of 2. The individual components add up to 22. Thus, 10110 binary equals 22 decimal. We now have a technique for determining the value of any binary number.

Going from one base to another requires a *conversion*. As we just saw, going from 10110 to 22 required us to perform a **binary-to-decimal** conversion. How do we go the other way? For example, what binary number represents the decimal number 37? This requires a **decimal-to-binary** conversion. One way to do this conversion is as follows:

$$37 / 2 = 18 \text{ with 1 left over}$$
$$18 / 2 = \ 9 \text{ with 0 left over}$$
$$9 / 2 = \ 4 \text{ with 1 left over}$$
$$4 / 2 = \ 2 \text{ with 0 left over}$$
$$2 / 2 = \ 1 \text{ with 0 left over}$$
$$1 / 2 = \ 0 \text{ with 1 left over}$$

The number is repeatedly divided by 2 and the remainder recorded. When we get to *1 / 2 = 0 with 1 left over* we are done dividing. The binary result is found by reading the remainder bits from the bottom up. So, 37 decimal equals 100101 binary. To check, we use the binary-to-decimal conversion technique:

1	0	0	1	0	1
32	16	8	4	2	1
32			4		1

Adding 32, 4, and 1 gives 37, the original number!

Here are some common binary and decimal numbers:

Binary	Decimal
1010	10
1111	15
1100100	100
10000000	128
11111111	255
1111101000	1000

Clearly, it helps to have a good understanding of the various powers of 2 to perform conversions. The first 20 powers of 2 are as follows:

$$2^0 = 1 \qquad 2^5 = 32$$
$$2^1 = 2 \qquad 2^6 = 64$$
$$2^2 = 4 \qquad 2^7 = 128$$
$$2^3 = 8 \qquad 2^8 = 256$$
$$2^4 = 16 \qquad 2^9 = 512$$
$$2^{10} = 1024 \qquad 2^{15} = 32768$$
$$2^{11} = 2048 \qquad 2^{16} = 65536$$
$$2^{12} = 4096 \qquad 2^{17} = 131072$$
$$2^{13} = 8192 \qquad 2^{18} = 262144$$
$$2^{14} = 16384 \qquad 2^{19} = 524288$$

THE HEXADECIMAL NUMBER SYSTEM

It is not easy to remember large binary numbers. For instance, examine the 20-bit binary number shown here for 5 seconds. Then close your eyes and try to repeat it:

1 0 1 0 1 1 1 1 0 1 1 0 1 0 1 1 0 0 1 1

Were you able to do it? Most people cannot, because the short-term memory is not capable of storing so many bits of information. It is possible, however, to remember shorter sequences of characters. Try this character sequence:

A F 6 B 3

Were you able to remember it? Those who have difficulty with the 20-bit example can usually do the five-character example easily. Here is where the trick comes in: The 20-bit binary number *is the same* as the five-character sequence. This is because AF6B3 represents a **hexadecimal** number (base 16). Here is how the examples relate to each other:

- Separate the number into groups of 4 bits each

    1 0 1 0   1 1 1 1   0 1 1 0   1 0 1 1   0 0 1 1

- Find the individual decimal equivalents of each group

    1 0 1 0   1 1 1 1   0 1 1 0   1 0 1 1   0 0 1 1
      10   15   6   11   3

- Replace each 10...15 value with its A...F equivalent

    1 0 1 0   1 1 1 1   0 1 1 0   1 0 1 1   0 0 1 1
      10   15   6   11   3
      A   F   6   B   3

This technique makes it possible to work with large binary numbers through their hexadecimal equivalents.

The word **hexadecimal** refers to "6" and "10," or "16." This number system contains numbers composed of digits and letters chosen from the set {0, 1, 2, 3, 4, 5, 6, 7, 8, 9, A, B, C, D, E, F}. The decimal number 10 is represented in hexadecimal as **A.** Letters B, C, D, E, and F are equivalent to 11 through 15, respectively. Each digit or letter in a hexadecimal number represents 4 binary bits (as we have seen). The binary patterns associated with the 16 hexadecimal symbols are as follows:

0 = 0 0 0 0	4 = 0 1 0 0	8 = 1 0 0 0	C = 1 1 0 0
1 = 0 0 0 1	5 = 0 1 0 1	9 = 1 0 0 1	D = 1 1 0 1
2 = 0 0 1 0	6 = 0 1 1 0	A = 1 0 1 0	E = 1 1 1 0
3 = 0 0 1 1	7 = 0 1 1 1	B = 1 0 1 1	F = 1 1 1 1

It is much more convenient (and easier to remember) to use two hexadecimal symbols than 8 binary bits. For instance, 3EH (the H stands for Hex) means 00111110B (B for Binary). Larger binary numbers prove this point even better (as our AF6B3H example showed). Because microprocessor address and data lines commonly utilize 8, 16, or even 32 bits of data, the 2-, 4-, or 8-symbol hexadecimal equivalents are easier to deal with.

BINARY ADDITION

Once a decimal number has been translated into binary, what can we do with it? Usually, a number is loaded into a computer so that it can be manipulated. When two or more numbers are input, we often need to find their sum. So, it is necessary for the microprocessor to know how to add in binary. The rules for binary addition of 2 bits are short and straightforward:

$$
\begin{array}{ccccc}
 & & & & 1 \\
0 & 0 & 1 & 1 & +1 \\
+0 & +1 & +0 & +1 & +1 \\
\hline
0 & 1 & 1 & 10 & 11
\end{array}
$$

Because the binary number system allows only the two symbols 0 and 1, the sum of 1 + 1 cannot be 2! So, it is 0 with *1 to carry* into the next column of bits.

Likewise, the sum of 1 + 1 + 1 is not 3, but 1 with 1 to carry. These are all the rules we have to remember. Let us use them to add two 8-bit numbers:

$$
\begin{array}{ll}
1\ 1\ 1\ 1\ 1 & \text{—carry bits} \\
1\ 0\ 0\ 1\ 0\ 1\ 1\ 0 & \text{—first number } = 150 \text{ (or 96H)} \\
+\ 0\ 0\ 1\ 1\ 1\ 0\ 1\ 1 & \text{—second number } = \ 59 \text{ (or 3BH)} \\
\hline
1\ 1\ 0\ 1\ 0\ 0\ 0\ 1 & \text{—result } = 209 \text{ (or D1H)}
\end{array}
$$

This is the type of addition performed by microprocessors.

BINARY SUBTRACTION

Oddly enough, we use binary *addition* to perform subtraction in binary. For example, subtracting these two numbers:

$$
\begin{array}{r}
50 \\
- 18 \\
\hline
\end{array}
$$

is the same as **adding** these two numbers:

$$
\begin{array}{r}
50 \\
+ -18 \\
\hline
\end{array}
$$

So, we need a method for converting positive 18 into negative 18. We use a technique called *2's complement* to perform this conversion. To find the 2's complement of 18, we do the following:

$$
\begin{array}{l}
\quad\; 0\,0\,0\,1\,0\,0\,1\,0 \text{—positive 18 in 8-bit binary} \\
\quad\; 1\,1\,1\,0\,1\,1\,0\,1 \text{—complement of each bit} \\
+ \qquad\qquad\quad 1 \text{—add 1} \\
\hline
\quad\; 1\,1\,1\,0\,1\,1\,1\,0 \text{—the 2's complement of 18}
\end{array}
$$

It is not possible to distinguish a 2's complement binary number from an ordinary binary number. Two's complement refers to the way in which we *interpret* a binary number. The −18 representation 11101110 must be interpreted as a **signed** binary number. A signed binary number uses its MSB as a *sign* bit, where 0 means positive and 1 means negative. So, we have:

$$
\begin{array}{ll}
\mathbf{0}\,0\,0\,1\,0\,0\,1\,0 & +18, \text{ sign bit is } 0 \\
\mathbf{1}\,1\,1\,0\,1\,1\,1\,0 & -18, \text{ sign bit is } 1
\end{array}
$$

The value of the sign bit in these two numbers supports our interpretation.

Getting back to the original problem, we can now subtract 18 from 50 by adding the 2's complement of 18 (which is −18) to 50:

$$
\begin{array}{r}
1\,1\,1\,1\,1\,1 \quad \text{—carry bits} \\
0\,0\,1\,1\,0\,0\,1\,0\text{—} \quad 50 \\
+\; 1\,1\,1\,0\,1\,1\,1\,0\text{—} -18 \\
\hline
0\,0\,1\,0\,0\,0\,0\,0\text{—} \quad 32
\end{array}
$$

The carry out of the MSB is ignored. Note that the MSB of the result is 0, indicating a positive result.

This brief discussion should have familiarized you with the types of numbers encountered when dealing with microprocessors. Now, when you encounter instructions like:

```
CODES     DC.W      $14, $3EA9, 2200

          MOVE.B    $4C,D4
          MOVEA.L   #$8000,A3
```

you will have an idea what the numbers mean.

APPENDIX G

ASCII Character Set

Character*	Code	Character	Code	Character	Code	Character	Code
NUL	0 ($00)	blank	32 ($20)	@	64 ($40)	`	96 ($60)
SOH	1	!	33	A	65	a	97
STX	2	"	34	B	66	b	98
ETX	3	#	35	C	67	c	99
EOT	4	$	36	D	68	d	100
ENQ	5	%	37	E	69	e	101
ACK	6	&	38	F	70	f	102
BEL	7	'	39	G	71	g	103
BS	8	(	40	H	72	h	104
HT	9	)	41	I	73	i	105
LF	10	*	42	J	74	j	106
VT	11	+	43	K	75	k	107
FF	12	,	44	L	76	l	108
CR	13	–	45	M	77	m	109
SO	14	.	46	N	78	n	110
SI	15	/	47	O	79	o	111
DLE	16 ($10)	0	48 ($30)	P	80 ($50)	p	112 ($70)
DC1	17	1	49	Q	81	q	113
DC2	18	2	50	R	82	r	114
DC3	19	3	51	S	83	s	115
DC4	20	4	52	T	84	t	116
NAK	21	5	53	U	85	u	117
SYN	22	6	54	V	86	v	118
ETB	23	7	55	W	87	w	119
CAN	24	8	56	X	88	x	120
EM	25	9	57	Y	89	y	121
SUB	26	:	58	Z	90	z	122
ESC	27	;	59	[	91	{	123
FS	28	<	60	\	92			124
GS	29	=	61	]	93	}	125
RS	30	>	62	↑	94	-	126
US	31	?	63	_	95	DEL	127 ($7F)

*These 32 characters (code numbers 0 through 31) are known as **control characters.**

APPENDIX H

DOS/BIOS Function Calls

These interrupt references are designed to remind you what registers are used for a particular interrupt function. These functions are for use with EMU68K.

VIDEO SERVICES: SELECTED FUNCTIONS

INT 10H, Function 00H: Set video mode
 Input: AH = 00H
   AL = Video mode (0 to 7)
 Output: None

INT 10H, Function 01H: Set cursor type
 Input: AH = 01H
   CH = Top scan line
    Bits 0 to 4: Start line of cursor
    Bits 5 to 6: 00, Normal
    01, Cursor not displayed
    Bit 7: Reserved
   CL = Bottom scan line
    Bits 0 to 4: End line of cursor
    Bits 5 to 7: Reserved
 Output: None

INT 10H, Function 02H: Set cursor position
 Input: AH = 02H
   BH = Display page number (0 to 7)
   DH = Row (first is 0)
   DL = Column (first is 0)
 Output: None

INT 10H, Function 03H: Read cursor position

Input: AH = 03H
            BH = Display page number (0 to 7)

Output: AX = 00H
            CH = Starting cursor scan line
            CL = Ending cursor scan line
            DH = Row
            DL = Column

INT 10H, Function 05H: Select new video page

Input: AH = 05H
            AL = Page number (0 to 7)

Output: None

INT 10H, Function 06H: Scroll page up

Input: AH = 06H
            AL = Scroll distance in rows (0 for all)
            BH = Attribute to use on blanked lines
            CH = Top row
            CL = Leftmost column
            DH = Bottom row
            DL = Rightmost column

Output: None

INT 10H, Function 07H: Scroll page down

Input: AH = 07H
            AL = Scroll distance in rows (0 for all)
            BH = Attribute to use on blanked lines
            CH = Top row
            CL = Leftmost column
            DH = Bottom row
            DL = Rightmost column

Output: None

INT 10H, Function 08H: Read character/attribute from screen

Input: AH = 08H
            BH = Display page (0 to 7)

Output: AH = Attribute
            AL = Character

INT 10H, Function 09H: Write character/attribute to screen

Input: AH = 09H
            AL = Character to write
            BH = Display page (0 to 7)
            BL = Attribute

                              CX = Number of characters
            Output: None
INT 10H, Function 0AH: Write character only to screen
            Input: AH = 0AH
                      AL = Character to write
                      BH = Display page (0 to 7)
                      CX = Number of characters
            Output: None

INT 10H, Function 0FH: Read video status
            Input: AH = 0FH
            Output: AH = Number of columns
                      AL = Video mode
                      BH = Display page

KEYBOARD FUNCTIONS

INT 16H, Function 00H: Read keyboard input
            Input: AH = 00H
            Output: AH = Scan code or character ID for special character
                      AL = ASCII code or other translation of character

INT 16H, Function 02H: Return shift flag status
            Input: AH = 02H
            Output: AL = Current shift status
                              Bit 0: Right Shift
                              Bit 1: Left Shift
                              Bit 2: Ctrl
                              Bit 3: Alt
                              Bit 4: Scroll Lock
                              Bit 5: Num Lock
                              Bit 6: Caps Lock
                              Bit 7: Insert

INT 16H, Function 03H: Set typematic rate and delay
            Input: AH = 03H
                      AL = 05H
                      BH = Delay in milliseconds
                          00H: 250, 01H: 500, 02H: 750, 03H: 1000
                      BL = Typematic rate in characters per second

00H: 30.0,	08H: 15.0,	10H: 7.5,	18H: 3.7
01H: 26.7,	09H: 13.3,	11H: 6.7,	19H: 3.3
02H: 24.0,	0AH: 12.0,	12H: 6.0,	1AH: 3.0
03H: 21.8,	0BH: 10.9,	13H: 5.5,	1BH: 2.7

04H: 20.0,	0CH: 10.0,	14H: 5.0,	1CH: 2.5
05H: 18.5,	0DH: 9.2,	15H: 4.6,	1DH: 2.3
06H:	17.1,0EH:8.6,	16H:	4.3, 1EH: 2.1
07H:	16.0,0FH:8.0,	17H:	4.0, 1FH: 2.0

PARALLEL PRINTER FUNCTIONS

INT 17H, Function 00H: Print character
    Input: AH = 00H
             AL = Character to print
             DX = Printer number (0, 1, or 2)
    Output: AH = Printer status
             Bit 0: Time out
             Bit 1: Reserved
             Bit 2: Reserved
             Bit 3: I/O error
             Bit 4: Printer selected
             Bit 5: Out of paper
             Bit 6: Acknowledgment from printer
             Bit 7: Printer not busy

INT 17H, Function 01H: Initialize printer
    Input: AH = 01H
             DX = Printer number (0, 1, or 2)
    Output: AH = Printer status
             (same bit assignments as Function 00H)

INT 17H, Function 02H: Read printer status
    Input: AH = 02H
             DX = Printer number (0, 1, or 2)
    Output: AH = Printer status
             (same bit assignments as Function 00H)

SELECTED DOS FUNCTIONS

INT 21H, Function 01H: Keyboard input
    Input: AH = 01H
    Output: AL = Keyboard character

INT 21H, Function 02H: Display output
    Input: AH = 02H
             DL = Character to output
    Output: None

INT 21H, Function 05H: Printer output
     Input: AH = 05H
                DL = Character to output
     Output: None

INT 21H, Function 08H: Console input without echo
     Input: AH = 08H
     Output: AL = Character from console

INT 21H, Function 0BH: Check standard input status
     Input: AH = 0BH
     Output: AL = 0FFH: Character available
                      00H: No character available

INT 21H, Function 19H: Get current drive
     Input: AH = 19H
     Output: AL = Drive ID

INT 21H, Function 2AH: Get date
     Input: AH = 2AH
     Output: AL = Day of the week
                CX = Year
                DH = Month
                DL = Day

INT 21H, Function 2BH: Set date
     Input: AH = 2BH
                CX = Year
                DH = Month
                DL = Day
     Output: AL = 00H: Date is valid
                      0FFH: Date is invalid

INT 21H, Function 2CH: Get time
     Input: AH = 2CH
     Output: CH = Hour
                CL = Minutes
                DH = Seconds
                DL = 100ths of seconds

INT 21H, Function 2DH: Set time
     Input: AH = 2DH
                CH = Hour
                CL = Minutes
                DH = Seconds
                DL = 100ths of seconds
     Output: AL = 00H: Time is valid
                      0FFH: Time is invalid

INT 21H, Function 30H: Get DOS version number
            Input: AH = 30H
            Output: AL = Major version number
                          AH = Minor version number
                          BX = 0000H
                          CX = 0000H

Solutions and Answers to Selected Odd-Numbered Study Questions

1.1: 1. Microwave oven
2. Digital television
3. Portable heart monitor
4. FAX machines
5. UPC label scanners
6. Cash registers
7. Automobiles
8. Cameras
9. Sporting equipment
10. Answering machines

These are just a few examples. There are many correct answers not listed here, and the list is growing every day.

1.3: In all three cases, timing is required to enable transfer of data at the proper time. Specifically, the serial I/O section uses timing signals to generate its baud rate clock. Memory section access time is also controlled by system timing (and set by the devices). Interrupt timing is vital to proper program operation. Interrupts may be lost if not latched at the proper time.

1.7: A cash register without recordkeeping would require less RAM than one with recordkeeping, since all transactions do not have to be stored.

1.9: Doors: any opening/closing
Windows: any opening/closing/breaking
Elevators: all switches (up/down/open/close/floor number) and a timing interrupt to prevent the elevator from getting stuck on any floor

1.11: Downloading of main program; status information; serial number of part being assembled; commands from host computer

1.13: The data from each device will collide on the data bus, giving unpredictable results.

1.15: CPU-1: Graphics control
CPU-2: Sounds and player I/O
CPU-3: Game control

1.17: Image processing, animation, signal processing, simulation, and scientific calculations (weather prediction, astronomy).

1.19: The new CPU must execute instructions available on the earlier model. Any unused bit patterns on the older CPU could be decoded by the new CPU as additional (new) instructions. Hardware is added to implement the new instructions.

1.21: The MOVEA.L instruction takes the longest to execute due to its length of three words.

1.23: HELLO.LST, HELLO.OBJ, and HELLO.HEX.

1.25: Microcontroller advantages:
- On-board EPROM/RAM/timers/A-D/I-O (serial/parallel)
- All functions on one chip; therefore, less cost to lay out, build, and troubleshoot (repair)

Disadvantages:
- Limited instruction sets
- Special hardware needed to program on-board EPROM

2.1: Users should not be allowed to STOP/RESET the computer due to danger to the system, and themselves. In a multiuser system, all users stop if one user executes STOP. All users may lose their data if RESET is executed.

2.3: Each user gets 32KB of RAM. The operating system does indeed require some RAM for its own use. One way to do this is to allocate a small portion of each user's RAM for operating system functions.

2.5: The .B, .W, and .L extensions specify the operand size for an instruction.

2.7: Longword at address 30 is 9AFC3007. Word at address 30 is 9AFC.

2.9: The $ signifies that $100 is a hexadecimal number. 100 is a decimal number and is smaller than $100.

2.11: a) longword
b) longword
c) longword
d) word or longword
e) byte, word, or longword

2.13: The condition codes are contained in the user byte of the status register.

2.15: XNZVC are, respectively, 01100.

2.17: All 68000 registers are 32 bits wide. They may be used to store a 24-bit address. The 8086 cannot do this because its address bus is wider than its registers.

3.1: ORG: sets starting address for assembly
DC.W: reserve a word of storage for constant
DS.B: reserve block of bytes in memory
END: finish pass (or finish assembly)

3.3: The assembler creates an object file and a list file.

3.5: Relocatable instructions must be modified when a program is moved to a new location in memory.

3.7: A linker connects many separate object files together to make a single file.

3.9: a) register direct
b) register direct
c) register indirect
d) register indirect with postincrement
e) register indirect with predecrement
f) program counter relative
g) address register indirect with index

3.11: They enable the processor to execute different sections of program code depending on recognized conditions.

3.13: The data at location A3 − 1 is copied into itself and A3 remains unchanged upon completion.

3.15: a) 00000492
b) 00000491
c) 00000492
d) 00000494
e) 00000490

3.17: a) 004000
b) 007C00
c) FF8400
d) FFCB00
e) 140000
f) EF0000

3.19: a) D0 = 11223344, D1 = 99AA5566
b) D0 = 00001122, D1 = 00003344, D2 = FFFF99AA, D3 = 00005566

3.21: Less space is needed to store the operand (only 3 bits are needed to specify the immediate data).

3.23: The lower word of D0 is copied into the upper word of D1.
The upper word of D0 is copied into the lower word of D0.
The lower word of D1 remains the same and is also copied into the upper word of D0.

3.25: CLR.B D0 is a logical clear and does not affect the flags the way SUB.B D0,D0 does.

3.27: Computes D2/(D0*D1). Result in D2.

3.29: 00005678, FFFFDEF0

3.31: 65535, 32767

3.33: ANDI.L #$23F9,D2

3.35: 1st inst: D5 = 3B25AC08
2nd inst: D5 = 3B25BC4D
3rd inst: D5 = 43BF0093
4th inst: D5 = 43BFFF6C

3.37: 1st inst: D4 = F935C474
2nd inst: D4 = F93523A6

3.39: To keep the sign of its binary number the same.

3.41: a) Z flag = 1
b) illegal instruction
c) Z flag = 0
d) illegal instruction

3.43: 00000048

3.45: SCC (A6)

3.47: Only RTS correctly reads data back from the stack and resumes execution from previous PC.

3.49: $8000. Use the OR operation.

4.1: The processor's context is made up of the CPU registers, condition codes, and PC. It is the exact state of the machine at interrupt time. The context must be saved at the beginning of an interrupt service in order to resume execution at a later time.

4.3: A privilege violation exception is generated.

4.5: FC_2 is low for user operations, high for supervisor.

4.7: ANDI.W   #$5FFF,SR

4.11: $118 through $11B.

4.13: It is not possible for any of the group-2 exceptions to occur simultaneously.

4.15: All data registers require 16 words of stack memory. All address registers (except for A7) require an additional 14 words. The program counter and condition codes require 6 more words, totaling 36 words for each exception. This gives a nesting level of 28 (just divide 1024 by 36).

4.17: Not returning a vector number during an interrupt acknowledge cycle is one way to generate a bus error.

4.19: 1F0 through 1F3.

4.23: $20

4.25: No. The overflow bit is clear.

4.27: MOVEM.L   D1/D4/D5,SAVEAREA

4.29: The level-2 interrupt is not acknowledged.

4.31: They contain the interrupt level utilized during exception processing.

4.37:
```
ADDEX   ADD.L   D0,D7
        ADD.L   D1,D7
        ADD.L   D2,D7
        ADD.L   D3,D7
        ADD.L   D4,D7
        ADD.L   D5,D7
        ADD.L   D6,D7
        RTE
```
Note: The address of ADDEX must be stored in the memory locations reserved for TRAPV.

5.9: Y1 is at 8C00, Y2 is at 8C06, and Y3 is at 8C1A.

5.13:
```
DC.B   1, 16
DC.B   2, 32
DC.B   4, 64
DC.B   8, 128
```

5.15: a) 8E1C

b) 8E60

c) 8E80

d) 8EBC

5.19: The length of the string will be incorrect.

5.23: Advantage: Does not waste memory. Disadvantage: Requires a lot of searching to find data.

5.29: No. The only destination operand allowed with MOVEA is A*n*.

6.1: A program specification describes exactly what the program must do, and what its inputs and outputs are.

6.3: Processor registers are better suited for passing small amounts of data or the address of a block of data items.

6.5: MOVEQ requires only one word of machine code, which allows it to execute very quickly. However, only byte values may be specified.

6.7: The location of the end-of-string marker determines the length of a text string.

6.9: At least two ways: use an additional byte to store the sign (i.e., 00 for positive, 01 for negative), or use a bit in the exponent byte. Real numbers can use the same BCD format, but special attention must be paid to alignment (of the imaginary decimal point).

6.11: Because the X flag is used as one of the inputs to ABCD.

6.13: The initial status is needed so that future changes can be detected with the exclusive OR operation.

6.15: The length of the instruction, the addressing mode used, and the operation being performed all affect the execution time of an instruction.

6.17: So that normal program execution may resume when the exception completes.

7.13: $\overline{AS}$ is low. $\overline{UDS}$ is low. $\overline{LDS}$ is high. $R/\overline{W}$ is low.

7.15: $\overline{AS}$ is low. $\overline{UDS}$ and $\overline{LDS}$ are low. $R/\overline{W}$ is low.

7.17: A_1 through A_3 provide the interrupt level during interrupt acknowledge.

7.25: Existing bus circuitry can be used by other masters.

7.27: FC_2 controls (indicates) the user/supervisor status. It could be used as an input to a memory address decoder to enable/disable certain user/supervisor memory accesses.

7.29: The internal A_0 bit is low when accessing even locations and high when accessing odd locations.

8.1: a) A_0 is used to activate $\overline{UDS}$ or $\overline{LDS}$.
  b) A_0 is ignored, $\overline{UDS}$ and $\overline{LDS}$ are both low.

8.3: a) $\overline{UDS} = 0, \overline{LDS} = 1$
  b) $\overline{UDS} = 1, \overline{LDS} = 0$
  c) $\overline{UDS} = 0, \overline{LDS} = 0$

8.5: $8(250 \text{ ns}) = 2 \text{ μs}$.

8.7: a) 2 levels of delay: $15 + 20 = 35 \text{ ns}$
  b) 3 levels of delay: $15 + 15 + 20 = 50 \text{ ns}$
  c) 3 levels of delay: $15 + 20 + 15 = 50 \text{ ns}$
  d) 2 levels of delay: $15 + 20 = 35 \text{ ns}$

8.9: $(250 \text{ ns})(8 + 2) = 2.5 \text{ ms}$.

8.11: E04FFF

8.15: a) 2A0000 through 2AFFFF
  b) 540000 through 57FFFF
  c) 400000 through 5FFFFF
  d) E00000 through EFFFFF

8.17: $\overline{A}$: 600000 through 63FFFF
$\overline{B}$: 640000 through 67FFFF
$\overline{C}$: 680000 through 6BFFFF
$\overline{D}$: 6C0000 through 6FFFFF
$\overline{E}$: 700000 through 73FFFF
$\overline{F}$: 740000 through 77FFFF
$\overline{G}$: 780000 through 7BFFFF
$\overline{H}$: 7C0000 through 7FFFFF

8.19: RAM allows the interrupt vectors to be loaded/modified at will.

8.23: The synchronous cycles take much longer. The delay is built-in.

8.27: Each new address pin adds 2 address bits to the DRAM, one for RAS, the other for CAS.

8.29: The interrupt service routine could maintain a counter used to address the rows during refresh. This counter could be written to the refresh circuitry each time an interrupt occurs.

8.31: Waiting.

9.1: A memory location exists within a RAM or EPROM, and is used for storing program instructions and data. A memory-mapped I/O location is a hardware circuit designed to input or output data from the real world.

9.3: 360000 to 37FFFF

9.5: Because the output data is only present on the data bus for a few hundred nanoseconds.

9.19: 76.8 KHz divided by 16 gives 4,800. The baud rate is 4,800 bits/sec.

9.21: At 9,600 baud, the bit time is 104 µs. This gives a character time of 1.04 ms.

9.25: When 80 characters have been entered on a single line, or when a CR code is received.

9.31: (342*512*24) / 3 equals 1,368KB.

10.1: Add $40 to all register addresses except for the GCR.

10.3: $09

10.7: 64KB (for byte move operations). Use a chain operation or multiple block moves to accommodate larger block sizes.

10.9: $FC00

10.11: Maps to Exxxxx.

10.13:
```
LBA: 0010 1100 0000 0000
LAM: 1111 1111 0000 0000
PBA: 1100 1001 0000 0000
```
Address maps from 2C0000 to C90000.

10.15: $53

10.17: Write B_5 into the clock select register.

10.23: A sample sine wave data group is:
```
80  98  B0  C6  D9  E9  F5  FC
FF  FC  F5  E9  D9  C6  B0  98
7F  67  4F  39  26  16  0A  03
01  03  0A  16  26  39  4F  67
```
Note: Use something like V = 127*SIN(ANGLE) + 128 and convert V into hex.

10.25: Scan all keys, making a table of those held down. Ignore any keys that were previously held down.

10.27: To detect any key, output a zero on port A. When any key is pressed, $\overline{INT}$ will go low.

10.33: Internal circuitry allows pins to be configured for either direction by programming a bit in the data direction register.

10.35: The receiver ready status bit does not have to be sampled repeatedly. B_3 is the receiver interrupt vector.

10.37: Four daisy-chained 68901s will provide:
28 parallel I/O lines
64 interrupts
4 serial channels

10.39: Only the third control register (B_7 pattern).

10.41: The last two (patterns B_7 and $_{22}$).

10.47: Use the other half of data bus and $\overline{UDS}$ in the $\overline{RD}$ logic.

10.49: Program the interrupt system, initialize both serial devices (baud rate, parity, etc.), reset the timer, and program direction of I/O.

10.51: The keypad scanner can be expanded into the required number of keys required by the calculator (25 keys require 5 bits on ports A and B, for example). The multiplexed display can also be easily expanded to the required number of digits (with eight digits allowed before a hardware change is needed). The software routines that control the keypad and display could be integrated, so that scanning a row (or column) of keys is performed as each display is updated.

11.1: The two NAND gates compose an RS latch. When the switch is in the upper position, the lower NAND gate outputs a 0. This causes $\overline{RESET}$ and $\overline{HALT}$ to go low. When the switch is thrown, the lower gate outputs a 1, negating $\overline{RESET}$ and $\overline{HALT}$.

11.3: ENABLE can be used to force INT high.

11.13: Implementation of byte-wide data tables, such as ASCII string messages.

11.17: A RAM write to address 8FFx places a copy of D_0 into the flip-flop.

12.1: The 68008 only reads eight bits at a time from memory. All instruction accesses will take twice as long.

12.3: Virtual memory is a technique that allows a small physical memory to operate as a large logical memory.

12.5: A logical address can reside anywhere in the entire addressing space of the processor, whereas a physical address is limited to the range of physical memory implemented in hardware.

12.7: Information contained within the exception stack frame.

12.9: The 68020 has full 32-bit address and data buses, an instruction cache, and many new instructions and addressing modes.

12.11: Lower address lines are used to directly access a specific cache entry in a direct-mapped cache.

12.13: If all of the instructions within the loop fit inside the instruction cache, the processor is able to fetch the loop instructions very quickly.

12.15: The 68030 contains a memory management unit, a 256-byte data cache, and an enhanced bus controller.

12.17: Demand-paging is a memory allocation technique that only allocates frames of physical memory to a program when the program demands it. A table of page frames must be maintained to support demand paging.

12.19: The 68040 contains an internal coprocessor, larger instruction and data caches, and is capable of bus snooping in multiprocessor systems.

12.21: Bus snooping allows the 68040 to eavesdrop on the data cache of other 68040s it shares common buses with. It is used to help maintain cache coherency in a multiprocessor system.

12.23: A set-associative cache acts like multiple direct-mapped caches running in parallel. Multiple tags are checked, one for each set, during a cache access.

12.25: The 500-byte program will completely fit inside the 68040's instruction cache. The caches on the 68020 and 68030 are not large enough to hold the entire program, and will require numerous updates with each pass through the loop as the instructions get cycled into and out of the cache.

INDEX